VOLUME ONE HUNDRED AND EIGHTY TWO

PROGRESS IN
**MOLECULAR BIOLOGY
AND TRANSLATIONAL
SCIENCE**
Curing Genetic Diseases Through
Genome Reprogramming

VOLUME ONE HUNDRED AND EIGHTY TWO

PROGRESS IN
MOLECULAR BIOLOGY AND TRANSLATIONAL SCIENCE

Curing Genetic Diseases Through Genome Reprogramming

Edited by

GIANLUCA PETRIS
Medical Research Council Laboratory of Molecular Biology (MRC LMB), Cambridge, United Kingdom

Academic Press is an imprint of Elsevier
50 Hampshire Street, 5th Floor, Cambridge, MA 02139, United States
525 B Street, Suite 1650, San Diego, CA 92101, United States
The Boulevard, Langford Lane, Kidlington, Oxford OX5 1GB, United Kingdom
125 London Wall, London EC2Y 5AS, United Kingdom

First edition 2021

Copyright © 2021 Elsevier Inc. All rights reserved.

No part of this publication may be reproduced or transmitted in any form or by any means, electronic or mechanical, including photocopying, recording, or any information storage and retrieval system, without permission in writing from the publisher. Details on how to seek permission, further information about the Publisher's permissions policies and our arrangements with organizations such as the Copyright Clearance Center and the Copyright Licensing Agency, can be found at our website: www.elsevier.com/permissions.

This book and the individual contributions contained in it are protected under copyright by the Publisher (other than as may be noted herein).

Notices
Knowledge and best practice in this field are constantly changing. As new research and experience broaden our understanding, changes in research methods, professional practices, or medical treatment may become necessary.

Practitioners and researchers must always rely on their own experience and knowledge in evaluating and using any information, methods, compounds, or experiments described herein. In using such information or methods they should be mindful of their own safety and the safety of others, including parties for whom they have a professional responsibility.

To the fullest extent of the law, neither the Publisher nor the authors, contributors, or editors, assume any liability for any injury and/or damage to persons or property as a matter of products liability, negligence or otherwise, or from any use or operation of any methods, products, instructions, or ideas contained in the material herein.

ISBN: 978-0-323-85301-9
ISSN: 1877-1173

For information on all Academic Press publications
visit our website at https://www.elsevier.com/books-and-journals

Publisher: Zoe Kruze
Acquisitions Editor: Ashlie M. Jackman
Developmental Editor: Jhon Michael Peñano
Production Project Manager: James Selvam
Cover Designer: Matthew Limbert

Typeset by SPi Global, India

Contents

Contributors xi
Preface xvii

1. **Making sense of heritable human genome editing: Scientific and ethical considerations**   1
   Andy Greenfield

   1. Introduction   2
   2. Heritable human genome editing (HHGE): Initial considerations   4
   3. HHGE as an assisted reproductive technology   8
   4. Ethics of HHGE: Beyond safety and harm-benefit analyses   13
   5. HHGE: Impacts on societal norms   19
   6. Governance: Who decides?   22
   7. Conclusion   24
   Acknowledgments   25
   References   25

2. **CRISPR genome engineering for retinal diseases**   29
   Ariel Kantor, Michelle E. McClements, Caroline F. Peddle, Lewis E. Fry, Ahmed Salman, Jasmina Cehajic-Kapetanovic, Kanmin Xue, and Robert E. MacLaren

   1. Introduction   31
   2. CRISPR genome editing   38
   3. DNA base editing   40
   4. RNA editing   45
   5. Epigenetic editing   52
   6. Toward clinical use of CRISPR genome editing for the treatment of retinal disease   58
   7. Challenges and future perspectives   65
   References   70

3. **Advances in gene editing strategies for epidermolysis bullosa**   81
   Thomas Kocher and Ulrich Koller

   1. The blistering skin disease epidermolysis bullosa   82
   2. Gene therapeutic applications for EB   86

3. Gene editing development for the treatment of genodermatoses — 87
4. Conclusion and considerations for future gene editing applications in EB — 101
Acknowledgment — 103
References — 104

4. Targeted genome editing for the correction or alleviation of primary Immunodeficiencies — 111
Christopher J. Sipe, Patricia N. Claudio Vázquez, Joseph G. Skeate, R. Scott McIvor, and Branden S. Moriarity

1. Introduction — 114
2. Classifications of PIDs — 117
3. Current PID therapies and clinical trials — 123
4. Next generation of gene-editing tools — 132
References — 145

5. Genome editing approaches to β-hemoglobinopathies — 153
Mégane Brusson and Annarita Miccio

1. Introduction — 155
2. Genome editing tools — 157
3. Correcting β-globin gene mutations to treat β-hemoglobinopathies — 162
4. De-repression of the endogenous fetal γ-globin genes — 169
Author contributions — 177
Acknowledgments — 177
Conflicts of interest — 177
References — 177

6. Rewriting *CFTR* to cure cystic fibrosis — 185
Giulia Maule, Marjolein Ensinck, Mattijs Bulcaen, and Marianne S. Carlon

1. Cystic fibrosis—Where do we stand today — 187
2. Gene editing—Towards rewriting *CFTR* — 192
3. Modulation of *CFTR* expression levels as an alternative next-generation approach — 203
4. From Petri-dish to patient — 206
5. Conclusion — 211
Acknowledgments — 212
References — 212

7. Gene editing and modulation for Duchenne muscular dystrophy — 225
Anthony A. Stephenson and Kevin M. Flanigan

1. Introduction to Duchenne muscular dystrophy — 226
2. First-generation gene therapies for DMD — 232
3. Second-generation approaches to DMD — 237
4. Challenges for current and future DMD therapies — 244
5. Conclusions — 247
References — 248

8. Genome editing in the human liver: Progress and translational considerations — 257
Samantha L. Ginn, Sharntie Christina, and Ian E. Alexander

1. Introduction — 258
2. The liver: A high value therapeutic target — 259
3. Liver biology and implications for genome editing — 263
4. Genome editing technologies — 266
5. Disease specific challenges of genome editing in the liver — 270
6. Strategies to achieve genome editing outcomes in the liver — 275
7. Translational considerations — 279
8. Concluding remarks — 281
References — 282

9. Genome editing in lysosomal disorders — 289
Luisa Natalia Pimentel-Vera, Edina Poletto, Esteban Alberto Gonzalez, Fabiano de Oliveira Poswar, Roberto Giugliani, and Guilherme Baldo

1. Introduction: Lysosomal disorders — 291
2. Physiopathology, clinical manifestations and natural history — 292
3. Diagnosis and molecular genetics — 293
4. Current treatments: Impact and limitations — 297
5. Genome editing tools — 301
6. Genome-edited cell models — 303
7. Genome-edited animal models — 306
8. Therapies based on genome editing — 307
9. Clinical trials for lysosomal disorders — 317
10. Conclusions — 319
References — 319

10. Genome editing in mucopolysaccharidoses and mucolipidoses · 327
Hallana Souza Santos, Edina Poletto, Roselena Schuh, Ursula Matte, and Guilherme Baldo

   1. Clinical and molecular characteristics of mucopolysaccharidoses and mucolipidoses · 328
   2. Genome editing: General concepts · 332
   3. Genome editing for MPS and ML: In vitro studies · 338
   4. In vivo genome editing · 341
   5. Clinical trials · 344
   6. Conclusions · 347
   References · 347

11. Gene and epigenetic editing in the treatment of primary ciliopathies · 353
Elisa Molinari and John A. Sayer

   1. Introduction · 355
   2. Primary ciliopathies · 359
   3. CRISPR/Cas systems and their therapeutic applications · 365
   4. CRISPR/Cas gene editing for the treatment of primary ciliopathies · 379
   5. CRISPR/Cas epigenetic editing for the treatment of primary ciliopathies · 383
   6. Further considerations and concluding remarks · 387
   Acknowledgments · 389
   References · 389

12. Genome editing in stem cells for genetic neurodisorders · 403
Claudia Dell' Amico, Alice Tata, Enrica Pellegrino, Marco Onorati, and Luciano Conti

   1. Introduction: The basics of genome editing technologies · 405
   2. Pluripotent stem cells and neural stem cells: An overview · 408
   3. Genome editing for neurodevelopmental disorders · 416
   4. Genome editing of neurodegenerative disorders · 423
   5. Future directions and concluding remarks · 429
   Acknowledgments · 430
   References · 431

13. Reprogramming translation for gene therapy — 439
Chiara Ambrosini, Francesca Garilli, and Alessandro Quattrone

1. Introduction — 440
2. Translational regulation in eukaryotes — 442
3. Kozak consensus sequence — 445
4. uORFs — 455
5. Internal ribosomal entry sites (IRESs) — 463
6. Concluding remarks — 466
Acknowledgments — 466
References — 467

14. Synthetic genomics for curing genetic diseases — 477
Simona Grazioli and Gianluca Petris

1. Introduction to synthetic genomics — 478
2. Synthetic genomics techniques — 487
3. Synthetic genomics applications for the treatment of genetic diseases — 500
4. Perspectives — 507
Acknowledgments — 508
References — 508

Index — *521*

Contributors

Ian E. Alexander
Gene Therapy Research Unit, Children's Medical Research Institute, Faculty of Medicine and Health, The University of Sydney and Sydney Children's Hospitals Network; Discipline of Child and Adolescent Health, Sydney Medical School, Faculty of Medicine and Health, The University of Sydney, Westmead, NSW, Australia

Chiara Ambrosini
Laboratory of Translational Genomics, CIBIO—Department of Cellular, Computational and Integrative Biology, University of Trento, Trento, Italy

Guilherme Baldo
Postgraduate Program in Genetics and Molecular Biology, UFRGS; Gene Therapy Center; Biodiscovery Research Group, Experimental Research Center, HCPA; Department of Physiology, UFRGS, Porto Alegre; Laboratório Células, Tecidos e Genes do Hospital de Clínicas de Porto Alegre, Porto Alegre, RS, Brazil

Mégane Brusson
Université de Paris, Imagine Institute, Laboratory of Chromatin and Gene Regulation During Development, INSERM UMR 1163, Paris, France

Mattijs Bulcaen
Molecular Virology and Gene Therapy, Department of Pharmaceutical and Pharmacological Sciences, KU Leuven, Flanders, Belgium

Marianne S. Carlon
Molecular Virology and Gene Therapy, Department of Pharmaceutical and Pharmacological Sciences, KU Leuven, Flanders, Belgium

Jasmina Cehajic-Kapetanovic
Nuffield Laboratory of Ophthalmology, Nuffield Department of Clinical Neurosciences & NIHR Oxford Biomedical Research Centre, University of Oxford; Oxford Eye Hospital, Oxford University Hospitals NHS Foundation Trust, Oxford, United Kingdom

Sharntie Christina
Gene Therapy Research Unit, Children's Medical Research Institute, Faculty of Medicine and Health, The University of Sydney and Sydney Children's Hospitals Network, Westmead, NSW, Australia

Patricia N. Claudio Vázquez
Department of Pediatrics; Masonic Cancer Center; Center for Genome Engineering; Stem Cell Institute; Department of Genetics, Cell Biology and Development, University of Minnesota, Minneapolis, MN, United States

Luciano Conti
Department of Cellular, Computational and Integrative Biology—CIBIO, University of Trento, Trento, Italy

Claudia Dell' Amico
Unit of Cell and Developmental Biology, Department of Biology, University of Pisa, Pisa, Italy

Fabiano de Oliveira Poswar
Medical Genetics Service, HCPA, Porto Alegre, Brazil

Marjolein Ensinck
Molecular Virology and Gene Therapy, Department of Pharmaceutical and Pharmacological Sciences, KU Leuven, Flanders, Belgium

Kevin M. Flanigan
Center for Gene Therapy, Abigail Wexner Research Institute, Nationwide Children's Hospital; Department of Pediatrics; Department of Neurology, College of Medicine, The Ohio State University, Columbus, OH, United States

Lewis E. Fry
Nuffield Laboratory of Ophthalmology, Nuffield Department of Clinical Neurosciences & NIHR Oxford Biomedical Research Centre, University of Oxford; Oxford Eye Hospital, Oxford University Hospitals NHS Foundation Trust, Oxford, United Kingdom

Francesca Garilli
Laboratory of Translational Genomics, CIBIO—Department of Cellular, Computational and Integrative Biology, University of Trento, Trento, Italy

Samantha L. Ginn
Gene Therapy Research Unit, Children's Medical Research Institute, Faculty of Medicine and Health, The University of Sydney and Sydney Children's Hospitals Network, Westmead, NSW, Australia

Roberto Giugliani
Postgraduate Program in Genetics and Molecular Biology, UFRGS; Gene Therapy Center; Medical Genetics Service; INAGEMP; Biodiscovery Research Group, Experimental Research Center, HCPA; Department of Genetics, UFRGS, Porto Alegre, Brazil

Esteban Alberto Gonzalez
Postgraduate Program in Genetics and Molecular Biology, UFRGS; Gene Therapy Center, HCPA, Porto Alegre, Brazil

Simona Grazioli
Scuola Superiore Sant'Anna, Pisa, Italy

Andy Greenfield
MRC Mammalian Genetics Unit, Harwell Institute, Oxfordshire, United Kingdom

Ariel Kantor
Nuffield Laboratory of Ophthalmology, Nuffield Department of Clinical Neurosciences & NIHR Oxford Biomedical Research Centre, University of Oxford, Oxford, United Kingdom

Thomas Kocher
EB House Austria, Research Program for Molecular Therapy of Genodermatoses, Department of Dermatology and Allergology, University Hospital of the Paracelsus Medical University Salzburg, Salzburg, Austria

Ulrich Koller
EB House Austria, Research Program for Molecular Therapy of Genodermatoses, Department of Dermatology and Allergology, University Hospital of the Paracelsus Medical University Salzburg, Salzburg, Austria

Robert E. MacLaren
Nuffield Laboratory of Ophthalmology, Nuffield Department of Clinical Neurosciences & NIHR Oxford Biomedical Research Centre, University of Oxford; Oxford Eye Hospital, Oxford University Hospitals NHS Foundation Trust, Oxford, United Kingdom

Ursula Matte
Laboratório Células, Tecidos e Genes do Hospital de Clínicas de Porto Alegre, Porto Alegre, RS, Brazil

Giulia Maule
Department CIBIO, University of Trento; Institute of Biophysics, National Research Council, Trento, Italy

Michelle E. McClements
Nuffield Laboratory of Ophthalmology, Nuffield Department of Clinical Neurosciences & NIHR Oxford Biomedical Research Centre, University of Oxford, Oxford, United Kingdom

R. Scott McIvor
Department of Pediatrics; Masonic Cancer Center; Center for Genome Engineering; Stem Cell Institute; Department of Genetics, Cell Biology and Development, University of Minnesota, Minneapolis, MN, United States

Annarita Miccio
Université de Paris, Imagine Institute, Laboratory of Chromatin and Gene Regulation During Development, INSERM UMR 1163, Paris, France

Elisa Molinari
Translational and Clinical Research Institute, Faculty of Medical Sciences, Newcastle University, International Centre for Life, Central Parkway, Newcastle upon Tyne, United Kingdom

Branden S. Moriarity
Department of Pediatrics; Masonic Cancer Center; Center for Genome Engineering; Stem Cell Institute; Department of Genetics, Cell Biology and Development, University of Minnesota, Minneapolis, MN, United States

Marco Onorati
Unit of Cell and Developmental Biology, Department of Biology, University of Pisa, Pisa, Italy

Caroline F. Peddle
Nuffield Laboratory of Ophthalmology, Nuffield Department of Clinical Neurosciences & NIHR Oxford Biomedical Research Centre, University of Oxford, Oxford, United Kingdom

Enrica Pellegrino
Unit of Cell and Developmental Biology, Department of Biology, University of Pisa, Pisa, Italy; Host-Pathogen Interactions in Tuberculosis Laboratory, The Francis Crick Institute, London, United Kingdom

Gianluca Petris
Medical Research Council Laboratory of Molecular Biology (MRC LMB), Cambridge, United Kingdom

Luisa Natalia Pimentel-Vera
Postgraduate Program in Genetics and Molecular Biology, UFRGS; Gene Therapy Center, HCPA, Porto Alegre, Brazil

Edina Poletto
Postgraduate Program in Genetics and Molecular Biology, UFRGS; Gene Therapy Center, HCPA, Porto Alegre; Laboratório Células, Tecidos e Genes do Hospital de Clínicas de Porto Alegre, Porto Alegre, RS, Brazil

Alessandro Quattrone
Laboratory of Translational Genomics, CIBIO—Department of Cellular, Computational and Integrative Biology, University of Trento, Trento, Italy

Ahmed Salman
Nuffield Laboratory of Ophthalmology, Nuffield Department of Clinical Neurosciences & NIHR Oxford Biomedical Research Centre, University of Oxford, Oxford, United Kingdom

Hallana Souza Santos
Laboratório Células, Tecidos e Genes do Hospital de Clínicas de Porto Alegre, Porto Alegre, RS, Brazil

John A. Sayer
Translational and Clinical Research Institute, Faculty of Medical Sciences, Newcastle University, International Centre for Life, Central Parkway; Renal Services, The Newcastle Hospitals NHS Foundation Trust; NIHR Newcastle Biomedical Research Centre, Newcastle upon Tyne, United Kingdom

Roselena Schuh
Laboratório Células, Tecidos e Genes do Hospital de Clínicas de Porto Alegre, Porto Alegre, RS, Brazil

Christopher J. Sipe
Department of Pediatrics; Masonic Cancer Center; Center for Genome Engineering; Stem Cell Institute; Department of Genetics, Cell Biology and Development, University of Minnesota, Minneapolis, MN, United States

Joseph G. Skeate
Department of Pediatrics; Masonic Cancer Center; Center for Genome Engineering; Stem Cell Institute, University of Minnesota, Minneapolis, MN, United States

Anthony A. Stephenson
Center for Gene Therapy, Abigail Wexner Research Institute, Nationwide Children's Hospital, Columbus, OH, United States

Alice Tata
Department of Cellular, Computational and Integrative Biology—CIBIO, University of Trento, Trento, Italy

Kanmin Xue
Nuffield Laboratory of Ophthalmology, Nuffield Department of Clinical Neurosciences & NIHR Oxford Biomedical Research Centre, University of Oxford; Oxford Eye Hospital, Oxford University Hospitals NHS Foundation Trust, Oxford, United Kingdom

Preface

This book, entitled *Curing Genetic Diseases Through Genome Reprogramming*, captures an historic moment in the field of gene therapy: the dawn of a new age in which the dream of curing genetic diseases has become realizable.[1] This age has begun only because generations of scientists over the past two centuries dedicated their lives to understand how cells and their genomes function. Over the years they discovered which genes are involved in which diseases and they invented genetic surgery methods to correct these defective genes. This work required the harnessing of human pathogens and their transformation into safe vectors and a source of tools.[2,3] It also required the unrelated study of the billions of years old fight between bacteria and their viruses.[4]

For this book I have invited interdisciplinary contributions from a great many established and younger scientists working on a number of different aspects of gene and cell therapies, all passionate about finally curing genetic diseases. It is only thanks to these multidisciplinary efforts, investments, and knowledge that we are entering the era of genome reprogramming, when medicine is shifting from treating genetic diseases with palliative therapies to effectively curing the source of the disease by directly editing the cellular genome.

For this reason, it was not a surprise that a few months before the publication of this book, the 2020 Nobel Prize in Chemistry was awarded to Emmanuelle Charpentier and Jennifer A. Doudna for their development, in 2012, of the most revolutionary tool for genome editing: CRISPR-Cas. Moreover, the efforts of scientists, investors, and patients worldwide have produced a number of gene and genome reprogrammed cell therapy trials that have skyrocketed to several hundred in the past few years. This enthusiasm has been sustained by the success of some of the most recent gene therapy products.

2020 is also the year for the first systemic gene therapy approval by both the FDA and the EMA for spinal muscular atrophy (Zolgensma), a very common and life-threatening genetic disease, while at the end of 2017 the FDA, and a year later the EMA, approved the Luxturna gene therapy treatment for the retinal dystrophy RPE65-associated Leber congenital amaurosis.[1] Earlier in 2016 the first ex vivo gene addition therapy using retroviral vectors (Strimvelis) was EMA-approved for adenosine deaminase-deficient severe combined immunodeficiency, a very rare genetic disorder. Beyond these viral vector-based gene addition therapies, genome editing tools are also entering clinical use for ex vivo and in vivo applications[5,6]; in 2020 the first in vivo CRISPR genome engineering clinical trial started for an inherited retinal disease and the first CRISPR therapy was injected to patients for the CEP290-associated Leber congenital amaurosis.[6]

Thanks to these cutting-edge gene therapeutic products and other examples, as you will read throughout the many chapters of this book, it will be evident that the horizon of genome reprogramming really is drawing close as a therapy for several important genetic diseases, even though for a general and broad application of this revolution in medicine many technical, legislative and ethical challenges need to be addressed.

Indeed, the era of genome reprogramming comes not without concerns; the first rationally gene edited people were born in 2018, raising tremendous ethical concerns about the real possibility of genetically designing the human being.[7] Such fears have accompanied discoveries and technological developments about DNA since at least 1967[8,9] but, of course, it is only nowadays that this can be technically done. Now mankind may wonder whether the ability to genetically eradicate certain diseases, even before their onset, is acceptable using existing, not future, technologies.

Furthermore, genetic therapies are also controversial for their actual accessibility and price that, on the current market, is far in excess of certain already expensive biological drugs; the price for a treatment with Luxturna or Zolgensma is $425,000 and $2.1 million, respectively.[10] These costs are, obviously, out of the reach for the very vast majority of patients and their families, and may drain tremendous resources from national health systems. Besides, as you will read in this book, many genetic disorders are not truly single genetic diseases, but result from myriad genetic defects leading to a kind of similar phenotype; each of these mutations may affect a very limited number of patients and may require a dedicated gene therapy approach, posing serious practical and economical hurdles for current drug development regulatory pipelines. The risk could be that drugs will be developed that will cease to be produced, as recently happened for the Glybera treatment for lipoprotein lipase deficiency,[11] or have potentially curable diseases involving only few patients which will not be treated because economic reasons prevent viable advanced therapy medicinal products from being developed.

For all these concerns, before diving into the different chapters describing specific diseases where genome reprogramming is primarily deployed to cure somatic cells, Chapter 1 is an accurate perspective of the field literally for *making sense of human genome editing* from a scientific, regulatory, and ethical perspective. Indeed, the author Andy Greenfield entitled his chapter "Making sense of heritable human genome editing: scientific and ethical considerations." His contribution is fundamental to understanding and thinking about all aspects of modifying the human genome including the potentially most controversial one of genetically designing humans.

The following chapters will help the readers to move from moral and controversial aspects to more concrete practical examples of why we need to fulfill our ability to manipulate the genome for good. Chapters 2–5 will discuss genetic diseases whose characteristics and the features of the affected organs have made them more amenable to be at the vanguard of the entire field. The eye is one of these privileged examples and Chapter 2 "CRISPR genome engineering for retinal diseases" by Ariel Kantor, Michelle E. McClements, Caroline F. Peddle, Lewis E. Fry, Ahmed Salman, Jasmina Cehajic-Kapetanovic, Kanmin Xue, and Robert E. MacLaren will guide you through the successes, hurdles, and new hopes created by the CRISPR toolkit in fighting blindness.

In Chapter 3, Thomas Kocher and Ulrich Koller touch upon current and future therapies for treating a family of very painful and demanding conditions of the skin, by describing the "Advances in gene editing strategies for epidermolysis bullosa." The accessibility of this organ, as in the case of the eye, make it an extremely interesting target.

Chapter 4 "Targeted genome editing for the correction or alleviation of primary immunodeficiencies" by Christopher J. Sipe, Patricia N. Claudio Vázquez, Joseph G. Skeate, R. Scott McIvor, and Branden S. Moriarity describes the long-term medical battle against a range of syndromes preventing the development of effective immune defenses. Despite remarkable achievements, pitfalls in treating these life-threatening immune diseases had halted the entire field for almost a decade before of the current technological recovery.

Next, the development of therapeutic solutions for another product of the bone marrow will be addressed in Chapter 5 by Mégane Brusson and Annarita Miccio "Genome editing approaches to β-hemoglobinopathies." Many therapies, currently in trials, are soon expected to be approved for the treatment of these blood disorders. Their approval will breathe new life for many patients, given the high frequency of β-hemoglobinopathies.

Chapter 6 "Rewriting CFTR to cure cystic fibrosis" provides a very thorough analysis of gene therapy progress for a very common monogenic disease by Giulia Maule, Marjolein Ensinck, Mattijs Bulcaen, and

Marianne S. Carlon. Despite cystic fibrosis patients have seen their life quality and expectancy improved by the available symptomatic and pharmacological modulator treatments, a genetic correction may cure the disease once and for all, though successful delivery of current genetic surgery tools is the major obstacle preventing the run from bench to bedside.

The ability to run and move freely is a goal to reach for all patients affected by muscular dystrophies, in Chapter 7 Anthony A. Stephenson and Kevin M. Flanigan in "Gene editing and modulation for Duchenne muscular dystrophy" will describe the step taken in this direction and which are still missing.

Chapter 8, "Genome editing in the human liver: progress and translational considerations" by Samantha L. Ginn, Sharntie Christina, and Ian E. Alexander illustrates the prospects of how in the near future the unique regenerative and biological properties of the liver make it a suitable target to cure by genome reprogramming genetic diseases affecting this organ and beyond.

The next two chapters will deal with the extremely complex challenges related to lysosomal diseases. In Chapter 9 "Genome editing in lysosomal disorders," the authors Luisa Natalia Pimentel-Vera, Edina Poletto, Esteban Alberto Gonzalez, Fabiano de Oliveira Poswar, Roberto Giugliani, and Guilherme Baldo will describe how, for many of these diseases, attempts to perform genetic surgeries are still hindered by a lack of knowledge of disease pathogenesis and how genome reprogramming is still required to confirm genotype–phenotype relationship and to create all the essential cell and animal models necessary to understand the different patients' therapeutic needs. Among lysosomal storage diseases, a focused picture on mucopolysaccharidoses and mucolipidoses will be given in Chapter 10, "Genome editing in mucopolysaccharidoses and mucolipidoses" by Hallana Souza Santos, Edina Poletto, Roselena Schuh, Ursula Matte, and Guilherme Baldo.

In Chapter 11 "Gene and epigenetic editing in the treatment of primary ciliopathies," Elisa Molinari and John A. Sayer describe the different challenges posed by these diseases, which are extremely arduous when the kidney is the organ affected by the ciliopathies.

Chapter 12, "Genome editing in stem cells for genetic neurodisorders" by Claudia Dell' Amico, Alice Tata, Enrica Pellegrino, Marco Onorati, and Luciano Conti, will present the frontiers of current gene editing and stem cells methods that may be exploited for the understanding and future treatment of neurological disorders, which, with the outstanding exception of spinal muscular atrophy, still require thorough unraveling of their molecular mechanisms.

The continuously expanding genome reprogramming toolbox offers new opportunities to finely tune the expression of disease causing and disease associated genes, in particular when they require precise translation regulation. These approaches are reviewed by Chiara Ambrosini, Francesca Garilli, and Alessandro Quattrone in Chapter 13 "Reprogramming translation for gene therapy."

The book ends with a provocative chapter (Chapter 14), "Synthetic genomics for curing genetic diseases" that I wrote with the help of Simona Grazioli, a very brilliant young scientist, on synthetic genomics and the use of such technologies to move genome editing therapies beyond their current limits.

I would like to thank all the authors and reviewers for their invaluable contribution in the making of this book.

Serving as an Editor for this book has been a pleasant opportunity to expand my knowledge of the state of the art, future approaches and applications of therapeutic genome manipulation. Genetic diseases exist affecting all organs, but much exciting technological progresses have been made toward curing many of them.

I am sure that the readers will enjoy all the information and the different authors' perspectives collected in this book; may them contribute to fostering new ideas and new therapies.

I am deeply and passionately convinced that, progressively, many more future medicines will be based on curing disease through genome reprogramming, offering a permanent and unprecedented solution for a vast range of human disorders with unmet medical needs.

Gianluca Petris
Cambridge

References

1. Anguela XM, High KA. Entering the modern era of gene therapy. *Annu Rev Med.* 2019;70:273–288. https://doi.org/10.1146/annurev-med-012017-043332.
2. Thomas CE, Ehrhardt A, Kay MA. Progress and problems with the use of viral vectors for gene therapy. *Nat Rev Genet.* 2003;4(5):346–358. https://doi.org/10.1038/nrg1066.
3. Terns RM, Terns MP. CRISPR-based technologies: prokaryotic defense weapons repurposed. *Trends Genet.* 2014;30(3):111–118. https://doi.org/10.1016/j.tig.2014.01.003.
4. Klompe SE, Sternberg SH. Harnessing "a billion years of experimentation": the ongoing exploration and exploitation of CRISPR–Cas immune systems. *CRISPR J.* 2018; 1(2):141–158. https://doi.org/10.1089/crispr.2018.0012.

5. Ashmore-Harris C, Fruhwirth GO. The clinical potential of gene editing as a tool to engineer cell-based therapeutics. *Clin Transl Med*. 2020;9(1). https://doi.org/10.1186/s40169-020-0268-z.
6. Harrison C. First CRISPR therapy dosed. *Nat Biotechnol*. 2020;38(4):382. https://doi.org/10.1038/d41587-020-00003-1.
7. Greely HT. CRISPR'd babies: human germline genome editing in the "He Jiankui affair". *J Law Biosci*. 2019;6(1):111–183. https://doi.org/10.1093/jlb/lsz010.
8. Nirenberg MW. Will society be prepared? *Science*. 1967;157(3789):633. https://doi.org/10.1126/science.157.3789.633.
9. Berg P. Asilomar 1975: DNA modification secured. *Nature*. 2008;455(7211):290–291. https://doi.org/10.1038/455290a.
10. DeJoy SA, Bohl MG, Mahoney K, Blake C. Estimating the financial impact of gene therapy. *medRxiv*. 2020. https://doi.org/10.1101/2020.10.27.20220871.
11. Dunbar CE, High KA, Joung JK, Kohn DB, Ozawa K, Sadelain M. Gene therapy comes of age. *Science*. 2018;359(6372). https://doi.org/10.1126/science.aan4672.

CHAPTER ONE

Making sense of heritable human genome editing: Scientific and ethical considerations

Andy Greenfield*
MRC Mammalian Genetics Unit, Harwell Institute, Oxfordshire, United Kingdom
*Corresponding author: e-mail address: a.greenfield@har.mrc.ac.uk

Contents

1. Introduction — 2
   1.1 The use of novel therapies — 2
   1.2 The promise of genome editing: Somatic interventions — 3
2. Heritable human genome editing (HHGE): Initial considerations — 4
   2.1 HHGE: Lessons from model organisms — 5
   2.2 Genome editing in human research embryos — 6
   2.3 Assessing the acceptability and governance of HHGE — 7
3. HHGE as an assisted reproductive technology — 8
   3.1 Preimplantation genetic testing (PGT): A germline technology in all but name — 8
   3.2 "GM" in ART: What's in a name? — 10
   3.3 "The therapeutic fallacy": Take your pick of fallacies — 12
4. Ethics of HHGE: Beyond safety and harm-benefit analyses — 13
   4.1 HHGE: Beyond the monogenic disease-prevention paradigm — 14
   4.2 Human dignity and the meaning of "human" — 16
   4.3 Science and ethics—An unhappy marriage? — 18
5. HHGE: Impacts on societal norms — 19
   5.1 Disease and disability — 19
   5.2 The impact of HHGE on our understanding of parental responsibility — 21
6. Governance: Who decides? — 22
7. Conclusion — 24
Acknowledgments — 25
References — 25

Abstract

Genome editing, particularly the use of CRISPR-Cas9-based methodologies, is revolutionizing biology through its impacts on research and the translation of these into applications in biomedicine. Somatic genome editing aimed at treating individuals with disease raises some significant ethical issues, but proposed heritable interventions, through the use of genome editing in gametes or embryos, raise a number of distinct social, ethical and political issues. This review will consider some proposed uses of heritable human genome editing (HHGE) and several of the objections to these that have been raised. Making sense of such proposed uses requires viewing HHGE as an assisted reproductive technology (ART) that, like preimplantation genetic testing (PGT) and mitochondrial replacement techniques (MRT), aims to prevent disease transmission during sexual reproduction, rather than acting as a therapy for an existing individual. Applications beyond the paradigm of disease prevention raise even more difficult scientific and ethical questions. Here, I will discuss various themes that are prominent in discussions of the science and ethics of HHGE, including impacts on human dignity and society, the language of HHGE used for public dialogue and the governance of HHGE.

Abbreviation

ART	assisted reproductive technology
HFEA	Human Fertilization & Embryology Authority
HHGE	heritable human genome editing
MRT	mitochondrial replacement techniques
PGT	preimplantation genetic testing
PNT	prenatal testing

1. Introduction
1.1 The use of novel therapies

The use of innovative interventions in medicine attracts attention from scientists, clinicians, bioethicists, and citizens in general, for numerous reasons. On the one hand, this use reflects the noble aim of using the human intellect, in the form of the scientific method, to develop new treatments that aim to reduce the burden of human existence, mostly in the form of combatting diseases that can shatter the lives of people living with them. On the other hand, "innovation" implies "new" and this is likely to entail a good deal of uncertainty: will the intervention be safe and effective? Will it be made accessible to all those who need it, or will it be only for the few who can afford to purchase it? More broadly, does the nature of the intervention align with broader societal values, or is it disruptive to these, and who decides?

One area that continues to attract attention is regenerative medicine—generally a term applied to the use of gene or cell therapies, including genetically-altered cells, as experimental interventions in the clinic. The outcomes of such therapies have been mixed, raising a series of questions concerning regenerative medicine"s social license to operate.[1] It has been argued that "a combination of poor quality science, unclear funding models, unrealistic hopes, and unscrupulous private clinics" have threatened this social license and must be addressed.[1] I will expand on this in just one area: informed consent. An individual must give permission, verbally or in writing, *before* they receive any medical treatment; and for this consent to be *valid* it must be voluntary, informed, and the consenting individual must have the capacity to make the decision. Securing the *informed consent* of participants in a trial of an experimental therapy is absolutely vital for its ethical acceptability, but it raises a number of difficulties. Firstly, the science of regenerative medicine is fiendishly complex, and the pre-clinical data concerning safety and efficacy may be scant or difficult to interpret, such as in the case of the use of animal models. Thus, whether a prospective participant is "truly informed" is always open to question. Secondly, the individuals who want such therapies are often desperate; this may be a "last chance saloon" scenario and their emotional commitment to finding a "miracle cure", perhaps fuelled by media hype, may cloud their judgment. Finally, those endorsing or promoting such therapies may have any number of conflicting interests, including financial, which result in conscious or unconscious "over-selling" of a product. All of these factors threaten the possibility of legitimately securing informed consent in the context of new therapies and make the development of effective oversight and regulation of regenerative medicine a high priority for any jurisdiction in which it is permitted.

1.2 The promise of genome editing: Somatic interventions

Genome editing is a disruptive technology that has revolutionized our ability to make precise genetic alterations, by modifying DNA sequences in a controlled fashion, both in research and in the context of clinical translation. Genome editing can be used in combination with organoid approaches to probe human gene function in vitro, and more sophisticated animal models of human disease can now be produced with comparative ease by the use, in particular, of CRISPR-Cas9 genome editing tools. In terms of clinical application, one spectacular example of cellular therapy employing genome editing concerns successful immunotherapy in children with acute lymphoblastic leukemia using T cells that had undergone genome editing to produce tumor-targeting chimeric antigen receptor (CAR T-cells).[2,3] There are now a growing number of therapies involving genome editing ex vivo and in vivo that show great promise,[4] although it will take time to establish safety. The degree of efficacy is also likely to vary, with some treatments having more modest benefits. Moreover, costs of treatment are likely to be very high. As an example, *Zolgensma*, the one-time gene therapy treatment for spinal muscular atrophy in children under 2 years of age has been priced at up to US$2.1M.[a] Such pricing partly reflects the requirement to recoup investment in development of such a drug, and this will also be a factor in the pricing of future genome editing interventions. Such pricing will continue to be a challenge for healthcare providers, and insurance companies, moving forward. Given such ongoing uncertainty about somatic genome editing treatments for disease, in terms of safety, efficacy and accessibility, the use of genome editing to prevent disease transmission remains a topic that continues to attract attention.

2. Heritable human genome editing (HHGE): Initial considerations

The use of genome editing to influence the inheritance of specific human traits would involve editing in gametes (or their precursors), zygotes or early embryos in order to convert a pathogenic mutation into an allele that is present in unaffected family members or common in the relevant population and known not to cause disease. Establishing a pregnancy with an embryo that had been edited directly, or derived from gametes containing edited DNA, may result in the creation of an individual free of the relevant monogenic disease and with an edited genome. Since such an intervention would result in individuals that *might* themselves transmit an edited sequence to offspring, I will call it heritable human genome editing (HHGE), although it is sometimes called "human germline genome editing." "HHGE" makes it clear that the intention here is to establish pregnancies that result in the births of individuals with edited genomes, as opposed to the use of genome editing in human oocytes or zygotes for research purposes. This nomenclature follows the 2020 Academies of Sciences International Commission report on heritable genome editing in humans.[5] It is widely accepted that HHGE could only be considered if two conditions are met: (i) pre-clinical data would need to establish that a methodology exists that can reliably introduce the desired edit at the target site with no additional changes, avoid any off-target edits and meet these two requirements in every cell; (ii) HHGE would need to be lawful in the nation in

---

[a] https://www.forbes.com/sites/joshuacohen/2019/06/05/at-over-2-million-zolgensma-is-the-worlds-most-expensive-therapy-yet-relatively-cost-effective/?sh=2526002045f5.

which it was performed, following extensive public engagement and consultation, and be performed with regulatory oversight to prevent malpractice.

2.1 HHGE: Lessons from model organisms

Scientific consensus at the time of writing is that the first condition cannot yet be met: no methodology exists that is sufficiently precise and efficient for use in human embryos intended for assisted reproduction in the clinic.[5] The laboratory mouse is the most commonly used model organism for human biology, and genome editing tools, allied to increasingly sophisticated phenotyping,[6] promise to yield better models of a range of human genetic diseases. Such models are vital for understanding disease mechanisms, and also permit pre-clinical testing of novel therapies. This means that the mouse is probably the most frequently genome-edited organism on the planet. These data have revealed that CRISPR-Cas9 approaches, in particular those involving the introduction of DNA nicks or double-strand breaks, are highly efficient at introducing changes at the target locus, but control of the editing is not perfect and insertions or deletions (indels) or other complex rearrangements at the target site are not uncommon, presumably due to the frequency with which non-homologous end-joining (NHEJ) results in repair to the DNA break, rather than homology-directed repair (HDR) from a template. Aberrant repair at the target locus may be missed by standard genotyping protocols and therefore careful quality control of genome-edited mouse lines is key to valid experimentation and subsequent phenotyping.[7–9] Whilst off-target events, including the introduction of indels and single nucleotide variants, were once thought to be likely, it is now clear that these can be effectively eliminated by careful design and selection of guide RNAs.[10] Mosaicism, in which genome editing occurs after the first zygotic cell division to yield cells with distinct genotypes, is still commonly observed in mouse experiments. Various approaches to minimize this have been devised,[11] but ultimately, mosaicism does not present a major problem to the mouse geneticist, since F_0 mice generated by genome editing of embryos can be bred and F_1 animals carrying the desired edit, and only the desired edit, can simply be selected for further breeding and experimentation. In the context of human embryonic editing, however, mosaicism may well be problematic, since if cells of the target organ are not appropriately edited, the disease-prevention strategy may be undermined. In conclusion, genome editing in this closely related mammalian model has allowed the development of a framework for the consideration of safe and efficient genome editing in human tissues, including embryos.[12]

2.2 Genome editing in human research embryos

Early studies using genome editing in human research embryos utilized non-viable tri-pronuclear (3PN) embryos as a response to ethical concerns about the use of human embryos, since 3PN embryos are not suitable for clinical use.[13,14] (I will not discuss whether human embryo research *should* be permitted: it is lawful in the UK, where it is regulated by the Human Fertilization & Embryology Authority (HFEA), and also permitted in the US, Japan and some European countries, including Sweden and Belgium.[15] Moreover, it is a topic that has received much attention in the context of discussions of the convention of the 14-day culture limit[16,17]). However, the response of such embryos to DNA damage introduced by editing reagents may be different to that of diploid (2PN) embryos. Subsequent studies using viable embryos fell into two broad categories: gene function studies and proof-of-principle mutation correction studies. Fogarty et al[18] performed a loss-of-function study of the *POU5F1* gene using CRISPR-Cas9 to introduce indels at the target gene. They described efficient targeting of the *POU5F1* locus and reported that off-target mutations were undetectable above background PCR error rates. Some embryos had unedited cells, so mosaic outcomes were reported. Another study on viable embryos attempted to correct a pathogenic mutation in the *MYBPC3* gene.[19] Introduction of editing reagents into the oocyte was performed at the same time as sperm in a modified form of intracytoplasmic sperm injection (ICSI). High levels of correction of the mutated gene were reported by HDR, although this was attributed not to the exogenous repair template provided, but to the maternal allele. Several manuscripts appearing in 2020 reported data from CRISPR-Cas9 genome editing in human embryos and revealed unexpectedly high levels of structural rearrangements at the target site, some encompassing large chromosomal segments.[20-22] A satisfactory explanation of these events requires precise understanding of the way in which human embryos repair damage to DNA, which may result in large deletions, although there have also been disputed claims about gene conversion accounting for extended regions exhibiting loss of heterozygosity.[19] Clearly, much more research is required in human embryos, including the use of genome editing,[23] in order to inform our understanding of the earliest events in human development and shed light on a range of phenomena, such as totipotency, cell fate restriction and aneuploidy. Research in this area will hopefully impact efficiencies of IVF, lead to a better understanding of the causes of miscarriage, and pave the way for improved techniques in stem cell and regenerative medicine. Such research is also a prerequisite of exerting fine control over DNA repair, if this is required for control of genome editing in embryos. Base editing, a variant of CRISPR-Cas that involves the direct enzymatic conversion of DNA bases to introduce point mutations, is a highly promising alternative to conventional cut-and-paste approaches,[24] one which has been reported to be efficient in human embryos.[25] The CRISPR-Cas toolkit is likely to expand further and be optimized in coming years, but it still remains unclear whether it can attain the levels of precision and efficiency required for clinical use of HHGE. Of course, novel genome editing methodologies may at some point supersede CRISPR-Cas approaches.

2.3 Assessing the acceptability and governance of HHGE

Meeting the second condition above, i.e. ensuring that HHGE is lawful in the nation in which it is performed, following extensive public engagement and consultation, and performed with strict regulatory oversight, is currently a work in progress with no guaranteed outcome. Two international bodies were established in 2019 to address some of these matters. The International Commission convened by the US National Academies of Medicine and Sciences and the UK Royal Society published its report in 2020, detailing the nature of a responsible clinical pathway toward initial uses of HHGE.[5] A World Health Organization (WHO) expert panel will make recommendations on the governance of human genome editing, both somatic and heritable, in 2021. The international nature of these bodies indicates what is considered to be at stake in making decisions about HHGE. Humanity as a whole has an interest in the controlled alteration of human genomes and the trajectory of this technology, not least because "reproductive tourism" can mean access to such technologies can be had across borders. It should not be assumed that even if a safe HHGE methodology becomes available, the decision to use it is inevitable. Of course, there are many possibilities between an outright ban on HHGE and embracing routine and wholesale alteration of the human genome for *any* desired purpose. Fine-grained decision-making here, enabling some interventions and drawing some boundaries, is as likely an outcome as the two limiting cases. Neither of the aforementioned bodies had ethics as their primary concern, but a number of ethical assessments of prospective uses of HHGE have already been produced in recent years, and some of these will be discussed below.

3. HHGE as an assisted reproductive technology

There is a tradition (the ethical, legal and social implications (ELSI) model) in which a new technology is considered by asking (roughly) the question: what can I do with this new technology and would it be ethically acceptable? This question takes a perspective from the technology looking out into the world. Another way of asking ethical questions about new technologies is to ask a more "critical" question: what challenges exist in our society/world and would use of this technology be an ethical response to that challenge? This requires a perspective from the world toward the technology, and is an approach more likely to consider potential technological solutions *alongside* other possible interventions that promise to assist. It might also allow for a more considered assessment of what the challenges are and whether all of them are worth rising to.

3.1 Preimplantation genetic testing (PGT): A germline technology in all but name

In the context of potential uses of HHGE, the critical approach would ask what societal challenges exist to which HHGE is a possible solution. One obvious challenge is the prevalence of major birth defects among live born infants, generally regarded to be in the range of 3–5%.[26] But if prevention of serious heritable disease transmission from parents to offspring is the challenge, there are existing solutions: *i)* avoid having children; *ii)* adopt; *iii)* use unaffected gamete donors; *iv)* use either prenatal or preimplantation genetic testing (PNT or PGT, respectively). How one responds to these potential solutions will be affected by all manner of personal attitudes and cultural norms. (Even the label "serious" requires a degree of judgment and may be contested). For somebody who wishes to "have children" in the traditional sense of becoming pregnant and giving birth to a related child, the first two would not be seen as solutions. For somebody who wishes to be genetically related to the children they have, in the conventional manner, only the last two can offer the prospect of the birth of a genetically-related child free of the particular disease they risk transmitting. PGT would do so without the possible need to abort a pregnancy, but will result in familiar financial and emotional costs associated with in vitro fertilization (IVF), and success is not guaranteed.[27]

Factors affecting the likely success of PGT include: *i)* the circumstances of inheritance i.e. the proportion of embryos that are predicted to be unaffected; and *ii)* the number and quality of embryos produced, which will likely be lower in older women or couples with reduced fertility. A combination of these means that some couples may have no, or very few, transferrable embryos after PGT treatment.[27] Finally, there are some circumstances in which PGT is simply not an option, because all embryos produced by a couple will be affected due to the genotypes of the parents (see Section 4 for more on this). The recent Academies of Sciences Commission report on HHGE identifies such circumstances and suggests that these are cases that, whilst very rare, offer the most favorable balance of potential harms and benefits when considering *initial* uses of HHGE; however, it also indicates that this conclusion applies only to the prevention of transmission of *serious* diseases in the absence of good alternatives, where "serious disease" is defined as "one that causes severe morbidity or premature death."[5] Of course, for HHGE or PGT to be suitable interventions, these diseases need to be compatible with individuals surviving to reproductive age and being able to reproduce, such as Huntington's disease, cystic fibrosis, sickle-cell anemia, and beta thalassemia.

Now we can see that HHGE, if used to prevent disease transmission, is an assisted reproductive technology (ART) that, like PGT, aims to prevent heritable disease transmission to a genetically-related child. Two points are immediately worth making in this context: (*i*) Whilst HHGE has understandably attracted much attention because it is a "germline genetic technology", PGT is not usually discussed as a technology in this category. This is despite the fact that it is an ART, one requiring IVF, and biopsy and genetic testing of embryos, prior to selection of unaffected embryos to establish a pregnancy. Indeed, a number of commentaries on the acceptability of HHGE make the point that genome editing in embryos is not required precisely *because* PGT is a safe and lawful procedure that *already* allows the prevention of disease transmission. Why we do not routinely describe PGT as a germline technology, therefore, is unclear, but it does mean that it often escapes some of the ethical criticism that HHGE attracts. (*ii*) The extent to which HHGE (and PGT) can be seen as useful solutions to a societal challenge will depend partly on whether one sees the desire for a genetically-related child free of disease as sufficiently important to warrant the inevitable investment in time and money to optimize, regulate and legislate for the technology. It is relatively commonplace to see the "desire" for genetically-related children described, not as a potential right worthy of protection, but as a mere "want", whose importance in this context can be dismissed, especially when there are so many other demands made on limited science and healthcare budgets.[28,29] But at the same time, having genetically-related children has been a way of life for humans for millennia. It is not compulsory to see this practice as a dispensable cultural artifact, or an expression of some dubious related desire, such as respect for kinship[30]; rather, we can value an ancient human form of life and acknowledge both the sheer delight and "meaning" that family resemblances offer to people every day, and the misery that their absence can entail. We may even, in time, come to recognize this desire as a "need", often unmet, and/or acknowledge the existence of a "right" to a genetically-related child, in the same way that other human rights have slowly emerged, historically. Of course, this possibility inevitably draws our attention to the fact that assessing the ethical acceptability of HHGE requires the ethical (re)assessment of many other practices and norms, and that there is no "neutral" place from which to begin such assessment—we are immersed in ethics at all times.

3.2 "GM" in ART: What's in a name?

As we have seen, our attitudes to technologies that influence human inheritance are guided by complex ethical, social and emotional responses, both to the technologies themselves, but also to networks of associated practices, desires and norms. PGT appears not to be viewed so commonly as disruptive to certain cherished ethical norms, despite the similarities between the intentions with which PGT and a prospective HHGE would be used. Unlike HHGE, PGT would not produce genetically modified (GM) humans, since it merely selects from the available cohort of embryos. It may be that the very term "GM" evokes certain negative responses—partly by association with contentious debates over GM crops or GM farmed animals—or by association with any number of fictional portrayals of dystopian human futures in which GM humans feature. One response to this might be to query the use of the term "GM"; to suggest that genome editing simply results in genomic changes that can occur *anyway*, since such edits are simply the response of the cell to induced DNA damage and the requirement to repair it. This response might be called the "there is nothing to see here" response, because DNA damage and repair occur countless times in countless cells every day, due to many internal and external natural genotoxic insults, including but not limited to: *i)* cosmic and natural ionizing radiations; *ii)* the production of endogenous cellular genotoxic molecules like aldehydes; *iii)* the spontaneous DNA mutations arising from deamination of cytidine and adenine; *iv)* the intrinsic error rate of DNA replication; *v)* DNA damage due to transcriptional stress; *vi)* expansion and contraction of satellite DNA; *vii)* natural genome editing due to the activity of endogenous retroviruses and transposons. Mutation rates vary strikingly between different groups of organisms; in humans it has been estimated that each zygote will have 64 new (de novo) mutations not found in the parents, primarily due to mutations occurring in cells that generate male gametes.[31] A debate about the distinctions between agricultural uses of genome editing, based on indel production by DNA repair, and traditional GM—in which transgenes from other species were introduced to create some GM crops—is currently a high profile one in a number of jurisdictions.

The description of a technology that intervenes in the inheritance of human traits as a "GM" technology was also an aspect of the recent debates over the acceptability of mitochondrial replacement techniques (MRT). These techniques involve the manipulation of the mature oocyte either before or after fertilization in order to prevent the inheritance of mitochondria from the prospective mother by replacing them with mitochondria from a donated egg.[32] In a sense, ooplasm replacement might have been a more accurate term, since the ooplasm of the prospective mother is replaced, though not entirely, by the donor ooplasm through transfer of nuclear material into the donor ooplasm. MRT raises many ethical and social issues, some of which will be explored later, but here I am making the point that how a technology is described, the metaphors and tropes that are drawn on when discussing it, can influence responses to it. Some claim that MRT is not a GM technology, because it does not involve the modification of mitochondrial DNA (mtDNA) sequences as GM is conventionally understood to involve. MRT, by contrast, generates new combinations of mtDNA, reducing the ratio of pathogenic mtDNA to non-pathogenic mtDNA. Whilst this seems reasonable, it is also understandable why some will see the refusal to use the term "GM" as a piece of political "spin", an attempt to avoid the criticisms that cluster around GM in a number of contexts. These comments relate to the issue of "framing" and how framing can influence responses. Some think that the nuclear DNA transfer at the heart of MRT makes it a "cloning technology", given the similarities between it and somatic cell nuclear transfer (SCNT)[28]; and given that human cloning has been almost universally rejected as ethically unacceptable, a re-branding exercise was required, they suggest, to distance MRT from reproductive cloning. We will discuss matters of framing later, in Section 6.

3.3 "The therapeutic fallacy": Take your pick of fallacies

We have seen that use of HGGE, in at-risk couples, would involve conversion of a pathogenic mutation(s) into a more common, functional allele, resulting in an unaffected embryo. If safe and effective, it would ensure that any birth was of a genetically-related child free of a specified serious disease. No *existing* individual with the disease is treated by this intervention and so in this case HHGE, like PGT and MRT, is not therapeutic. The belief that HHGE is a therapy is sometimes called the *therapeutic fallacy*.[33] However, whilst it is unclear how many people actually hold that belief, it is important that in avoiding this particular mistake we do not over-correct and conclude that HHGE is not an intervention in the broader space of medicine, because it is not a therapy—a fallacy deserving of its own name, perhaps. Apart from the incidental fact that HHGE would involve all of the usual clinical aspects of IVF, a number of features of its scientific rationale and ethical acceptability are related directly to the inheritance and pathology of the diseases whose transmission it aims to prevent. For example, the Academies Commission on HHGE categorized prospective uses of HHGE on the basis of such features they might possess, relating to the seriousness of diseases they aimed to prevent and their modes of inheritance.[5] These categories were themselves relevant to the commission's consideration of ethical acceptability, since they concluded that *initial* uses of HHGE should be restricted to those categories in which serious monogenic diseases were under consideration, and then *only* when alternatives to HHGE were non-existent or very poor. So, we see that traditional medical considerations concerning minimizing harm and limiting first-use to circumstances with the most favorable harm-to-benefit analyses are central to how many professionals think about and evaluate possible applications of HHGE. This response to the therapeutic fallacy draws our attention to a much broader use of the term "medical", one in which there is clearly room for vaccination, preventative dietary measures, subsidized yoga classes and cycle-to-work schemes, anti-malarial prophylaxis for vacations, contraception etc. None of these are traditional therapies, but all are considered by most to be important for individual and public health, physical and mental. HHGE is not a therapy; neither is PGT, and a regulated form of PGT is relatively common in Europe and elsewhere in the world. Choosing whether to have children and choosing to have children to which they are related and that are free of specified diseases, whilst often contested, turn out to be choices that matter a great deal to many people in different walks of life.[34]

4. Ethics of HHGE: Beyond safety and harm-benefit analyses

As we saw earlier, the National Academies Commission on HHGE relied on a harm–benefit analysis framework when considering the possibility of delineating a responsible clinical pathway to first-in-human uses of HHGE; and it concluded that such analyses were only favorable in very limited circumstances. This was primarily due to considerations of safety, where uncertainty is still too great to be offset by the potential benefits of HHGE in most cases. The outcome was that a responsible path could only be envisaged in certain circumstances, dominated by couples where all embryos they produce would be affected by a serious monogenic disease and for whom, therefore, PGT would not be an option. Examples include couples in which both individuals are homozygous for a recessive deleterious mutation causing such a disease, or in which one individual is homozygous for a dominant mutation. These circumstances are very rare, but their frequency in geographic terms will vary substantially depending on a number of factors: the frequency of the disease-causing mutation in a given population; the rates of consanguineous unions in such a population; whether heterozygous carriers of recessive mutations enjoy a benefit in certain environments. An example of the latter is when individuals with a single sickle cell allele (sickle cell trait) who contract malaria are less likely to die. Because of this, the prevalence of sickle cell trait is high in many populations in sub-Saharan Africa, and there could be hundreds to potentially thousands of homozygous couples of reproductive age (with sickle cell disease) across those areas where sickle cell trait is most prevalent.[5]

As the preceding discussion indicates, there are genetic and environmental factors that complicate the analysis of potential harms and benefits arising from the use of HHGE to prevent serious disease transmission. This suggests that decisions about the appropriateness of HHGE to prevent transmission of the many *thousands* of monogenic diseases, for most of which no effective treatment exists, will need to be taken on a case-by-case basis, with acknowledgment of the scientific, clinical and ethical complexities that arise in particular circumstances. But ethical evaluations go beyond harm-benefit considerations. The following sections will explore some of these in more detail: the nature of human dignity; societal impacts on the vulnerable and impacts on parents; the governance of HHGE and democracy.

4.1 HHGE: Beyond the monogenic disease-prevention paradigm

We have focused on the comparatively straightforward cases in which HHGE would be used to prevent transmission of a monogenic disease, such as cystic fibrosis or Huntington's disease. These cases may be acceptable in certain circumstances: where pre-clinical data indicate the availability of a safe and effective methodology for editing; where prospective parents have no, or particularly poor, alternatives to avoiding disease transmission to a genetically-related child; where that disease is highly penetrant and serious, i.e. severe in its impact on the health of the child and the lives of the family. This framework for initial uses is founded on secure scientific knowledge—of the effects of the methodology and the consequences of the editing on the individual created—and on favorable harm-benefit analyses.

More difficult questions arise when considering other scenarios. If initial clinical uses prove to be safe and effective, should HHGE be gradually "rolled out" as an alternative to PGT? This would amount to prospective parents having a *choice* about how best to avoid disease transmission. Some might prefer PGT, perhaps because it simply selects from embryos that are already there, without the requirement to intervene in the embryonic genome. They may see it as less invasive, or even more natural, although viewing naturalness as a placeholder for "good", whilst common, is problematic.[35] On the other hand, some may view HHGE as preferable, on the basis that, in the case of a highly efficient methodology, no embryos would need to be discarded due to their genotype. The only selection required would be for those general features most likely to support an ongoing pregnancy and live birth (although these are notoriously difficult to elucidate). Does such parental choice raise the specter of a certain form of consumerism, one that already blights some aspects of assisted reproduction? I will discuss this further in Section 5 below.

More challenging still are questions about different types of application of HHGE. Many of the most common diseases that cause suffering and mortality are not inherited in a simple fashion. Rather, hundreds or thousands of gene variants contribute to the risk of getting the disease. This risk is also influenced by the environment and life history. Our understanding of such multi-factorial, polygenic inheritance is not currently secure enough to justify the editing of hundreds or thousands of gene variants with a view to altering disease risk in some predictable fashion, even if a methodology for so doing was available. But if our understanding of genomics were to result in the knowledge of which combination of variants in a given genome could be engineered to significantly reduce the risk of diseases such as Type I diabetes, dementia and coronary heart disease, without raising the risk of other diseases due to pleiotropic effects of the edited genes, would such HHGE be justified? Whilst the "disease prevention" paradigm that justified the simpler uses of HHGE would seem to play the same justificatory role here, there are complications. One of these is that most of us will inherit some combination of genetic variants that predisposes us to some sort of ill-health at some point in our lives, which is why the polygenic diseases mentioned above, and numerous others, are so common. Should this lead us to conclude that HHGE would be justified on a routine basis, as a public health intervention, if its use could lead to an overall population reduction in the instances of these diseases? Whilst it is still unclear whether polygenic human genome editing could play this imagined role, due to uncertainties in the evolution of editing methodologies and our understanding of genome biology, the problem envisaged here is one of the *transformation* of human reproduction, indeed, the transformation of human societies, by HHGE, as often imagined by dystopian literature. And we have not included in this discussion of possible future uses of HHGE those that might *enhance* the function of certain genes for different purposes, those which introduce novel biological functions from other species, perhaps even for cosmetic purposes, or those that attempt to radically alter human biology with a view to creating a new sub-species. These might appear to be topics more at home in a science fiction novel, but if we find them repellent it would still be useful to know why.

I will discuss other potential societal impacts of HHGE, and how to consider their acceptability, below. But we appear to have come a long way from the use of HHGE to prevent a monogenic disease. Two points here: firstly, one might be legitimately concerned about the prospects for rational discussion and evaluation of these "simpler" uses of HHGE because they may well inherit the concerns associated with more outlandish applications, like a form of ethical bystander effect. Secondly, we should remind ourselves that transformation of society might also be the upshot of *opposition* to HHGE if the premises of arguments made against it are applied across the board. Could IVF, PGT and MRT survive as ARTs if HHGE is opposed (and made unlawful or not made lawful) on the basis of its unjustifiably privileging the interests of parents who merely *desire* genetically-related offspring? And could contraception, or at least state-funded contraception, survive widespread political opposition to "medical" interventions that are not therapeutic?

4.2 Human dignity and the meaning of "human"

A reliance on harm–benefit analyses as the primary means of assessing ethical acceptability has been criticized by many commentators, due to the possibility that almost any potential harm might be justifiably worth risking if the potential benefits are sufficiently probable and significant. This utilitarian model is opposed by human rights approaches to HHGE, which emphasize that some "moral goods are resistant to trade-offs."[36] Human dignity is a value concept characterized by a similar immunity to ethical "trade-offs." It has been valued and honored historically, both in a secular and religious sense. The Universal Declaration on the Human Genome and Human Rights of UNESCO states that interventions in the human germline "could be contrary to human dignity".[37] If this is read as "*might* be contrary," then it would be difficult to disagree. We can easily imagine interventions using HHGE that would be contrary to human dignity, such as genomic alterations that predictably and intentionally lead to high levels of suffering, or that limit the freedom of an individual by imposing severe physical or psychological constraints on their life course. Such constraints have been explored in literary fiction[38] and also in the philosophical literature that discusses the impact of potential loss of autonomy of genetically engineered humans on their membership of moral and political communities[39] and the risk of commodification of our children in the name of human mastery of nature, with its Promethean resonance.[40] But we can also imagine interventions that do *not* have such negative implications for autonomy or dignity. Given that serious diseases themselves constitute an all too familiar assault on human dignity, one often caused by a random role of the dice of inheritance, HHGE to prevent disease transmission would be an example.

These considerations suggest that the concept of human dignity, if such a unitary concept does indeed exist, is unlikely to be decisive in terms of shaping our attitudes to HHGE. Unless, that is, one considers the *very idea* of altering the DNA of a gamete or embryo to be an affront to human dignity. The reasons for maintaining such a position are not easy to elaborate, at least independently of notions of the intrinsic value of an open life, free from the influence of parental plans and desires; or the idea that altering DNA in such a planned way amounts to another instance of scientific hubris. In its 2018 report on genome editing and human reproduction, the Nuffield Council on Bioethics working group concluded that HHGE "could be ethically acceptable in some circumstances."[41] The circumstances in question require that HHGE is "intended to secure, and is consistent with, the welfare of a person" born following use of HHGE, and that this use "upholds principles of social justice and solidarity." These conclusions suggest that, in ways which are now familiar given earlier discussions, *circumstances* are all-important when considering the ethical acceptability of HHGE. Of course, how one meets the guiding principles that the circumstances demand, and who decides if they are met, are matters of continued discussion and ethical evaluation themselves.

Finally, how are we to understand the concept of the human being as an ethically relevant category in and of itself? Many have argued for its importance, not least in explaining certain divergent attitudes we have to human beings and fellow creatures.[42] But it does not seem realistic to make a definition of "human being", complete with necessary and sufficient conditions. The human genome, sometimes viewed as *the* common heritage of humanity, does not exist: rather, there are countless trillions of genomes in the cells of the billions of humans on the planet, constantly changing due to the impacts of DNA replication error and environmental mutagens, as discussed earlier—this is the stuff of evolution, of course. Is it possible to unite around such a mutable collection of biological "objects"? To show solidarity with dynamic human genomes? Probably not in any meaningful way. But to counter this biologically reductive view of the human, we can reflect on the idea that we all know a human when we see one. In the words of the philosopher Bernard Williams: "Told that there are human beings trapped in a burning building, *on the strength of that fact alone* we mobilize as many resources as we can to rescue them."[43] Williams is surely right that our primitive reactions to other humans, our implicit understanding of our commitments and obligations to them, is a fundamental element of human life that positively guides or constrains our actions, and one that ethics should acknowledge. This is not to deny that knowledge that animals were in a burning building would also elicit a response, but there may be more cogitation before acting and the risks that most humans would be prepared to take to rescue them would probably be more limited. Finally, if told that humans with edited genomes were trapped in a burning building, I take it that our response would, and should, be the same as for any other humans. To me, it is clear that, in those lawful circumstances that are most likely to represent the initial uses of any HHGE, understanding what it means to be human does *not* exclude creating humans with edited genomes.

4.3 Science and ethics—An unhappy marriage?

We have seen that assessments of safety and efficacy are important in determining when a new technology is ready for clinical use. But can a scientific assessment of safety and efficacy be a matter of scientific judgment *alone*? Can scientific data unambiguously support a particular policy decision, or will such decision always incorporate non-scientific evaluations? These are important questions, because scientific assessment often benefits from a presumption of objectivity, both in terms of the data—they are what they are irrespective of whether they are convenient or not—and in terms of the scientific experts that such assessments require. Questions concerning the role of science in the development of policy and how data are interpreted during such a process were prominent during the discussion of how best to manage the Covid-19 pandemic in 2020-2021.

An example of such a scientific assessment in the ART sector was the expert panel convened by the UK HFEA on four occasions between 2011 and 2016 to examine the safety and efficacy of MRT. How MRT was introduced into the UK clinic is a highly complex matter. But one important component, that influenced the general public, law-makers and policy developers, was the conclusion of the 2016 expert panel that MRT should be offered in the UK clinic in a cautious fashion, in particular parental circumstances only, with strict oversight by the HFEA.[44] Indeed, some elements of these conclusions are familiar in the context of the recent recommendations by the International Commission on Human Genome Editing: MRT should be used only for the prevention of transmission of serious mitochondrial disease, and only when prospective parents have no or poor alternatives to have a genetically-related child. But whilst these conclusions were supported by pre-clinical data from model organisms and human research embryos, it seems clear that scientific considerations were not the only ones in play. To offer examples given by the philosopher Tim Lewens,[45] the panel's conclusion that preclinical safety data supported clinical use of MRT, given the potential benefits, implicitly endorses the idea that desiring a genetically-related child is an ethically significant "unmet need" and that satisfying this desire is therefore suitably beneficial. But this conclusion on the ethical significance of desiring genetic relatedness is contested, as we have seen. A similar point can be made about any consideration of whether a new technique is "safe enough" i.e. evidential thresholds and the role played by uncertainty: whilst any judgment is informed by the relevant scientific data, the conclusion "safe enough" will likely be based on a harm-benefit analysis that requires the identification of benefits, one that is an evaluative process. The combination, in such safety assessments, of scientific-empirical conclusions, such as "technique X has been safely used in mice and macaques", with implicit evaluative presumptions, such as "we have an obligation to assist someone in having a genetically-related child", yields a certain hybridity. Lewens cautions against such hybridity being obscured by a "division of advisory labor.", characterized by ethical assessments that do not concern themselves with scientific-technical considerations, and "purely" scientific assessments that do not make explicit their ethical commitments. Both parties in such independent assessments might argue that they are sticking to what they know. The risk here is that essentially hybrid questions of safety and efficacy may not receive sufficient attention. The claim that hybrid questions are key to using science in public policy is not, for the avoidance of doubt, the same as stating that science and ethics essentially do the same thing. It is important to state that there is still room for the identification of true empirical sentences—statements about the way things are as revealed by experimental data, such as "mutations in *CFTR* exons can cause cystic fibrosis," for example. I do not think that everyday scientific realism is a casualty of these remarks. Rather, they caution against glib statements about science being "value-free", especially when scientific data are employed in the service of policy advice, in contested areas such as climate science and genetic modification in agriculture. This caution is warranted because the data in question, that form evidence, are not used *on their own* to draw conclusions, as we have seen. Their *interpretation* is politically sensitive and they are generated in social contexts in which only certain types of scientific research proposals receive funding and support, an outcome which is itself based partly on evaluative or political decision-making, along the lines of "which science *matters* most?"

5. HHGE: Impacts on societal norms
5.1 Disease and disability

One of the most frequently discussed concerns about the impacts of technologies such as HHGE and PGT relates to potential negative impacts on those now living with disease or disability. If prospective parents choose to use genetic testing, or in future, HHGE, to avoid the birth of a child with a disease or disability, what negative attitudes toward individuals living with that disease/disability does that choice *express*? If HHGE were to result in a gradual reduction in the numbers of individuals living with certain diseases and disabilities, what would the impacts be on individuals living with these diseases/disabilities? Would they feel more social isolation? Would there be increased discrimination toward them? Would less support be available, or would there be reduced focus on treatment, perhaps due to reduced research in a rare disease area with fewer instances? These are important questions and answering them is difficult. Firstly, it should be remembered that HHGE, like PGT, will not entirely eliminate disease because a significant proportion of genetic diseases arise de novo, attributable to either mutations specific to the parental germline(s) or mutations arising in the first mitotic divisions of embryonic development (as discussed earlier and[46]). We should also examine whether the advent of PGT, which was first reported in 1990 as a means of preventing transmission of adrenoleukodystrophy and X-linked mental retardation,[47] and other prenatal screening technologies, has coincided with increased or decreased discrimination toward those living with disease and disability. This is a complicated empirical question, the answer to which is likely to vary from country to country.

I can see no contradiction between support for parental choice in this matter and respect for the rights of individuals living with disease and disability. If there is any lack of respect, it is toward the disease or disability, not the person living with it.[34] But this sentiment, whilst appealing, is complicated by the fact that many people with a disability see it as part of their identity: they cannot conceive of *being themselves* in the absence of the disease. Matters are also complicated by the contribution that social context plays in the severity of a disease or disability. The suffering associated with requiring a wheelchair for mobility can be greatly exacerbated if one lives in a society that makes no effort to create accessible spaces for wheelchairs. A life-saving bone marrow transplant is only going to save a life if it is actually available or affordable. This social dimension is important, but it should not be misunderstood. Some heritable diseases seem to be intrinsically awful, with no good therapies, requiring multiple hospital admissions in early childhood and resulting in chronic suffering and premature death.[b] I think it is wrong to characterize the desire to avoid such an outcome for one's own child as an expression of disrespect, or part of a project of

---

[b] https://www.theguardian.com/science/2014/aug/01/genetic-disorder-funding-david-cameron.

perfecting our children. Such condemnation is surely a failure of moral imagination, a failure to even try to understand what it would be like to be a parent in such circumstances. The fact that difficult cases exist, cases in which the suffering could be greatly ameliorated by changes in society, very much resulting in a life worth living, should not prevent us from seeing the compelling case to avoid those diseases that have the most appalling impacts on the quality of life of a person. What is required here is a sufficiently fine-grained response to disease and disability, one that does justice to its complexity. Whilst gene therapies for some conditions are promising, they may not result in a "cure." As we have seen, they are also likely to be incredibly expensive due to the rarity of some of the conditions and the business model for production of the intervention, and so the impulse to prevent is likely to persist. "Honoring fragility" is something that people tend to do when they are thinking in the abstract, rather than thinking about their own children.

5.2 The impact of HHGE on our understanding of parental responsibility

The safety of HHGE is paramount, i.e. its potential impact on individuals created by its use. We have also considered possible impacts of widespread use of HHGE on others in society i.e. those living with diseases and disability. Another potential impact is on parents who might seek HHGE. If the use of HHGE becomes more commonplace, will all parents come to feel a responsibility to use it? Will this sense of responsibility extend to performing genetic testing prior to attempting reproduction, and the use of technologies like HHGE to prevent disease transmission, or even to a feeling of guilt if such technologies are not used to improve certain traits in their children? This creeping sense of responsibility, known as responsibilization, might account for the gradual extension of cases of PGT beyond the most serious disorders to encompass late onset and low penetrance diseases. It might also account for a sense of pressure being felt to use technologies such as HHGE. We should also consider whether HHGE could disrupt inter-generational relationships, by imagining future scenarios in which children resent their parents for *not* having intervened in their embryonic genomes. Such a scenario appears to concede that a degree of commercialization and commodification of our children is almost inevitable. But we should remember that these will be the consequence of choices made that were not compulsory. Ongoing discussions about such topics will be important when considering, as a society, whether and how to proceed with HHGE.

6. Governance: Who decides?

In a country like the UK, HHGE will only be permitted if Parliament amends the Human Fertilization & Embryology Act and thereby makes it lawful. Other nations will require similar processes; other will be very different. Oversight of such a technology is likely to be key to public and political support for HHGE. But whilst governance is associated with institutions that make laws, and bodies that scrutinize activities by checking lists and ticking boxes, whilst referring to the relevant laws or regulatory guidelines, there is a broader conception of governance available. As historian and philosopher Benjamin Hurlbut argues: "Modes of deliberation and governance shape what questions are asked and what questions go unasked, what perspectives are privileged or silenced in public debate, and what priorities and practices come to be embedded in our institutions."[48] Hurlbut is especially motivated to make explicit and analyze certain framings that characterize discussions of HHGE and scientific innovation more broadly: scientific progress, the inevitability of technological innovation, society and ethics lagging behind such innovation. He questions the right of science (scientists) to self-govern, to act as gate-keepers to technological change i.e. to make decisions that are in their own interests, including their own financial interests, rather than in the public interest. His alternative vision is that "we the people" should claim the authority to decide on technological trajectories through "democratic governance." There is much here to admire, but also reasons for concern. In societies characterized by diverse ethical commitments and values, a consensus on what "the people" want may be elusive. If, in some imagined future, "the people" choose to embed HHGE in their institutions and lives following a democratic process, is it even possible to say that something has gone wrong? Does the "authority of democracy" model have the resources to articulate the possibility of a "wrong answer" here, or does democracy always deliver the "right" answer? This seems implausible, given a casual glance at history. Perhaps democracy is best viewed as a mechanism for coping with diversity, for avoiding very unpleasant alternatives? Of course, "deliberative democracy" is more of an ideal in this context, a placeholder for a genuinely inclusive political process, characterized by broad discussion, that addresses a very wide range of questions and not just a privileged few. But there are signs here that those promoting this ideal in this context, which is in and of itself unobjectionable, sense that such a process will deliver their preferred outcome.

Hurlbut's narrative is one in which power is a central theme. Who gets to decide whether we, as a society, legitimately use technologies such as HHGE? And if power is central to these considerations, then forms of politics are too. Such politics requires engaging with interest groups, listening to marginalized voices and engaging themes that are often submerged, in new and interesting ways.[49] It is clear now why "framing" and language are central concerns here for Hurlbut and other bioethicists too. So much ordinary politics consists of the production of narrow framings, empowered by the use of social media (and a limited number of characters); of one side contesting the framings of another: "what they didn't tell you is that…"; of newspapers jostling to influence public opinion with the narrowest of all framings, the headline. But I hope that any intellectually substantive process of deliberation in the space of "the ethics of HHGE" will be expansive and broad and yield conclusions that are effectively immune to the standard charge of being framed too narrowly. The conversations we need to have are big and important. They will need to engage all those with an interest in HHGE and related technologies, which means effectively *everyone*. The technical language of academic journals, through which scientists, sociologists and bioethicists communicate with each other, will not be fit for this purpose. We will need ways of communicating about complex issues that allow anybody to make sense of us.

Such big conversations should allow for diversity in all its senses. Hurlbut's characterization of the scientific community as complicit in the illegitimate and unlawful use of HHGE in China in 2018, one which implicitly or explicitly encouraged He Jankui to violate societal norms, "break the glass" and be a genome editing "pioneer"[48], is not the only characterization of scientists that is available. Can scientists be driven by a lust for money and/or the rewards of a prestigious career? Can such drives distort the quest for scientific truth and corrupt? The answer to both of these questions is: yes, of course. Can sociologists and philosophers be similarly driven? Of course. But scientists can also be driven by a genuine desire to discover facts (true statements) and act in the public interest, as can anyone. The scientific community, even the community of the scientific elite, is not monolithic. And the framing of elite scientists as a "powerful cabal" trades on certain ugly historical tropes about the illegitimate concentration of power in the hands of the few, and should be used sparingly.

Of course, in those societies in which overt regulation of technologies such as HHGE is non-existent or piecemeal, concerns about the improper influence or action of unconstrained individuals will be more common.

Using an expression attributed to the bioethicist John Harris, if we are on a slippery slope, the question is whether we should be using "skis or crampons?"[c] It seems likely that jurisdictions with strict regulation of HHGE and associated ARTs will see fewer concerns expressed about slippery slopes leading to wholesale consumerism in the ART clinic, especially if such regulation has public trust because it is widely considered to be in the public interest. However, one final comment on governance that reflects societal values and "the will of the people": we should be wary of this finding expression in excessive centralization and control. I shudder at the thought of a future "Ministry for Acceptable Research and Innovation" that controls all stages of the innovation process, in the name of "the people." This image has some unfortunate historical connotations and reminds us that human ingenuity and thirst for knowledge are sometimes best left to run free, because serendipity will yield novel insights that cannot be anticipated by even the most thorough planning.

7. Conclusion

Although we have surveyed a number of ethical concerns about the implications of HHGE, many of which merit further discussion, we should avoid automatic "genome editing exceptionalism", a variant of "genetic exceptionalism"[50], which considers genome editing to be special or unique in comparison to other interventions that influence human biology or culture, such that it should be singled out for special attention and criticism. Many of the arguments against HHGE that we have considered could, with only minor modification, be easily applied to numerous other clinical and societal practices that are commonplace and apparently uncontentious in the main. The impact of education, for example, on the autonomy of children or young adults fails to attract similar high-profile scrutiny. There are other examples of areas that appear as problematic as HHGE in terms of implications, such as the ways in which parental cultures influence the next generation and subsequent generations to reinforce cultural norms. The whole machinery of non-genetic human inheritance, of the environment, wealth and social prospects, of life horizons and cultural expectations, all of which affect our biology, perhaps through epigenetic impacts or in ways that we may not yet understand, surely warrants the same level of ethical scrutiny, for the same reasons.[51] This should not imply that novel

---

[c] https://blogs.bmj.com/bmj/2015/02/04/kate-adlington-mitochondrial-donation/.

technologies such as HHGE should be immune from criticism because of the existence of objectionable ethical or cultural norms; rather, it is to insist on argumentative consistency, through a thorough examination of the implications of arguments against HHGE, investigating whether, far from failing to support the purported conclusions, they succeed in achieving more than desired, calling into question the validity or acceptability of numerous cultural norms in ways that threaten societal cohesion more than HHGE itself.

Some may think that genome editing of the human genome to control or influence reproductive outcomes, whilst apparently innocent in the case of the prevention of transmission of monogenic diseases, threatens a more dystopian trajectory, one in which the latest version of humanity's ongoing quest to seize control over its fate becomes used excessively, despite opposed voices. The prospect of extensive and widespread editing of embryonic genomes, fuelled by hype and consumerism, is so grim that we should refuse to let the genie out of the bottle and ban *any* applications, or so the argument goes. Is this prudence? Or simply a familiar example of knee-jerk opposition to technology? Perhaps it is hopeless idealism in the face of one of history's lessons about humanity: humans will explore, innovate and seek improvement in their lives. And they will make mistakes in so doing. I conclude, as one "voice" among many, that we should permit a safe and effective HHGE, in specified circumstances and with strict oversight, to be another tool in the battle against disease transmission, but that we should take small steps only, and we should not assume or fear any inexorable slide toward HHGE applications that threaten to harm social cohesion. Instead, we should simply prevent them through the exercise of legal provisions. If we do not do that, we will only have *ourselves* to blame.

Acknowledgments

I thank the UK Medical Research Council for supporting my research programme through core funding to the Mammalian Genetics Unit at MRC Harwell. I thank numerous colleagues at the HFEA, the Nuffield Council on Bioethics, the Academies of Sciences International Commission on HHGE and elsewhere for discussions on the topics of ART, MRT and HHGE over the last 10 years. Opinions expressed in this review are mine and not necessarily those of any organizations with which I am associated.

References

1. Cossu G, Birchall M, Brown T, et al. Lancet commission: stem cells and regenerative medicine. *Lancet.* 2018;391(10123):883–910.

2. Qasim W, Zhan H, Samarasinghe S, et al. Molecular remission of infant B-ALL after infusion of universal TALEN gene-edited CAR T cells. *Sci Transl Med.* 2017;9(374): eaaj2013.
3. Delhove J, Qasim W. Genome-edited T cell therapies. *Curr Stem Cell Rep.* 2017;3 (2):124–136.
4. Ernst MPT, Broeders M, Herrero-Hernandez P, Oussoren E, van der Ploeg AT, Pijnappel W. Ready for repair? Gene editing enters the Clinic for the treatment of human disease. *Mol Ther Methods Clin Dev.* 2020;18:532–557.
5. National Academy of Sciences. *Heritable Human Genome Editing.* Washington, DC: The National Academies Press; 2020. https://doi.org/10.17226/25665.
6. Cacheiro P, Haendel MA, Smedley D, The International Mouse Phenotyping Consortium and The Monarch Initiative. New models for human disease from the International Mouse Phenotyping Consortium. *Mamm Genome.* 2019;30(5–6):143–150.
7. Kosicki M, Tomberg K, Bradley A. Repair of double-strand breaks induced by CRISPR-Cas9 leads to large deletions and complex rearrangements. *Nat Biotechnol.* 2018;36(8):765–771.
8. Teboul L, Greenfield A. CRISPR/Cas9-mediated mutagenesis: mind the gap? *CRISPR J.* 2018;1(4):263–264.
9. Codner GF, Mianne J, Caulder A, et al. Application of long single-stranded DNA donors in genome editing: generation and validation of mouse mutants. *BMC Biol.* 2018;16 (1):70.
10. Nutter LMJ, Heaney JD, Lloyd KCK, et al. Response to "unexpected mutations after CRISPR-Cas9 editing in vivo". *Nat Methods.* 2018;15(4):235–236.
11. Li Y, Weng Y, Bai D, et al. Precise allele-specific genome editing by spatiotemporal control of CRISPR-Cas9 via pronuclear transplantation. *Nat Commun.* 2020;11 (1):4593.
12. Teboul L, Herault Y, Wells S, Qasim W, Pavlovic G. Variability in genome editing outcomes: challenges for research reproducibility and clinical safety. *Mol Ther.* 2020;28 (6):1422–1431.
13. Liang P, Xu Y, Zhang X, et al. CRISPR/Cas9-mediated gene editing in human tripronuclear zygotes. *Protein Cell.* 2015;6(5):363–372.
14. Kang X, He W, Huang Y, et al. Introducing precise genetic modifications into human 3PN embryos by CRISPR/Cas-mediated genome editing. *J Assist Reprod Genet.* 2016;33(5):581–588.
15. Matthews KR, Morali D. National human embryo and embryoid research policies: a survey of 22 top research-intensive countries. *Regen Med.* 2020;15(7):1905–1917.
16. Hurlbut JB, Hyun I, Levine AD, et al. Revisiting the Warnock rule. *Nat Biotechnol.* 2017;35(11):1029–1042.
17. Appleby JB, Bredenoord AL. Should the 14-day rule for embryo research become the 28-day rule? *EMBO Mol Med.* 2018;10(9):e9437.
18. Fogarty NME, McCarthy A, Snijders KE, et al. Genome editing reveals a role for OCT4 in human embryogenesis. *Nature.* 2017;550(7674):67–73.
19. Ma H, Marti-Gutierrez N, Park SW, et al. Correction of a pathogenic gene mutation in human embryos. *Nature.* 2017;548(7668):413–419.
20. Alanis-Lobato G, Zohren J, McCarthy A, et al. Frequent loss-of-heteozygosity in CRISPR-Cas9-edited early human embryos. *Proc Natl Acad Sci U S A.* 2020. in the press.
21. Zuccaro MV, Xu J, Mitchell C, et al. Allele-specific chromosome removal after Cas9 cleavage in human embryos. *Cell.* 2020;183(6):P1650–1664.E15. https://doi.org/10.1016/j.cell.2020.10.025.
22. Ledford H. CRISPR gene editing in human embryos wreaks chromosomal mayhem. *Nature.* 2020;583(7814):17–18.

23. Lea R, Niakan K. Human germline genome editing. *Nat Cell Biol.* 2019;21(12):1479–1489.
24. Rees HA, Liu DR. Base editing: Precision chemistry on the genome and transcriptome of living cells. *Nat Rev Genet.* 2018;19(12):770–788.
25. Zhang M, Zhou C, Wei Y, et al. Human cleaving embryos enable robust homozygotic nucleotide substitutions by base editing. *Genome Biol.* 2019;20(1):101.
26. Kirby RS. The prevalence of selected major birth defects in the United States. *Semin Perinatol.* 2017;41(6):338–344.
27. Steffann J, Jouannet P, Bonnefont JP, Chneiweiss H, Frydman N. Could failure in preimplantation genetic diagnosis justify editing the human embryo genome? *Cell Stem Cell.* 2018;22(4):P481–P482. https://doi.org/10.1016/j.stem.2018.01.004.
28. Baylis F. Human nuclear genome transfer (so-called mitochondrial replacement): clearing the underbrush. *Bioethics.* 2017;31(1):7–19.
29. Baylis F, Ikemoto L. The Council of Europe and the prohibition on human germline genome editing. *EMBO Rep.* 2017;18(12):2084–2085.
30. Greenfield A. Carry on editing. *Br Med Bull.* 2018;127(1):23–31.
31. Drake JW, Charlesworth B, Charlesworth D, Crow JF. Rates of spontaneous mutation. *Genetics.* 1998;148(4):1667–1686.
32. Greenfield A, Braude P, Flinter F, Lovell-Badge R, Ogilvie C, Perry ACF. Assisted reproductive technologies to prevent human mitochondrial disease transmission. *Nat Biotechnol.* 2017;35(11):1059–1068.
33. Mills PF. Genome editing and human reproduction: the therapeutic fallacy and the "most unusual case". *Perspect Biol Med.* 2020;63:126–140.
34. Glover J. *Choosing Children: Genes, Disability, And Design.* Oxford: Clarendon Press; 2006.
35. Nuffield Council on Bioethics. Naturalness. In: *London*; 2015.
36. German Ethics Council. Intervening in the human germline. In: *English Translation of Executive Summary and Recommendations.* Berlin: German Ethics Council; 2019. https://www.ethikrat.org/fileadmin/Publikationen/Stellungnahmen/englisch/opinion-intervening-in-the-human-germline-summary.pdf.
37. UNESCO. *Universal Declaration on the Human Genome and Human Rights*; 1997. Article 24.
38. Ishiguro K. *Never Let Me Go.* London: Faber and Faber; 2005.
39. Habermas J. *The Future of Human Nature.* Cambridge: Polity Press; 2003.
40. Sandel MJ. *The Case against Perfection: Ethics in the Age of Genetic Engineering.* Cambridge: Harvard University Press; 2007.
41. Nuffield Council on Bioethics. *Genome editing and human reproduction: Social and ethical issues. London*; 2018.
42. Diamond C. Eating meat and eating people. In: *Realism and the Realistic Spirit.* MIT Press; 1991.
43. Williams B. The human prejudice. In: *Philosophy as a Humanistic Discipline.* Princeton University Press; 2006.
44. Human Fertilisation & Embryology Authority (HFEA). *Scientific review of the safety and efficacy of methods to avoid mitochondrial disease through assisted conception: 2016 update*; 2016. https://www.hfea.gov.uk/media/2611/fourth_scientific_review_mitochondria_2016.pdf.
45. Lewens T. The division of advisory labour: the case of 'mitochondrial donation'. *Eur J Philos Sci.* 2019;9:10.
46. Jonsson H, Sulem P, Arnadottir GA, et al. Multiple transmissions of *de novo* mutations in families. *Nat Genet.* 2018;50(12):1674–1680.
47. Handyside AH, Kontogianni EH, Hardy K, Winston RM. Pregnancies from biopsied human preimplantation embryos sexed by Y-specific DNA amplification. *Nature.* 1990;344(6268):768–770.

48. Hurlbut JB. Imperatives of governance: Human genome editing and the problem of progress. *Perspect Biol Med.* 2020;63(1):177–194.
49. Dryzek JS, Nicol D, Niemeyer S, et al. Global citizen deliberation on genome editing. *Science.* 2020;369(6510):1435–1437.
50. Murray TH. Is genetic exceptionalism past its sell-by date? On genomic diaries, context, and content. *Am J Bioeth.* 2019;19(1):13–15.
51. Lewens T. Blurring the germline: Genome editing and transgenerational epigenetic inheritance. *Bioethics.* 2020;34(1):7–15.

CHAPTER TWO

CRISPR genome engineering for retinal diseases

Ariel Kantor[a,*], Michelle E. McClements[a], Caroline F. Peddle[a], Lewis E. Fry[a,b], Ahmed Salman[a], Jasmina Cehajic-Kapetanovic[a,b], Kanmin Xue[a,b], and Robert E. MacLaren[a,b]

[a]Nuffield Laboratory of Ophthalmology, Nuffield Department of Clinical Neurosciences & NIHR Oxford Biomedical Research Centre, University of Oxford, Oxford, United Kingdom
[b]Oxford Eye Hospital, Oxford University Hospitals NHS Foundation Trust, Oxford, United Kingdom
*Corresponding author: e-mail address: enquiries@eye.ox.ac.uk

Contents

1. Introduction	31
2. CRISPR genome editing	38
3. DNA base editing	40
3.1 Base editor systems	42
4. RNA editing	45
4.1 ADAR physiology	46
4.2 RNA editing tools	47
4.3 Exogenous ADAR approaches	47
4.4 Endogenous ADAR approaches	50
4.5 Clinical applications and challenges for RNA editing treatment of inherited retinal disease	51
5. Epigenetic editing	52
5.1 Application to retinal disease: Targeting pathogenic genes in dominant diseases	54
5.2 Targeting disease pathways	55
5.3 Comparisons to other methods	56
5.4 Challenges	57
6. Toward clinical use of CRISPR genome editing for the treatment of retinal disease	58
6.1 Immunogenic concerns	59
6.2 Undesired genome editing	60
6.3 Improving safety of CRISPR-Cas systems	61
6.4 Delivery of CRISPR reagents to the retina	63
7. Challenges and future perspectives	65
References	70

Abstract

Novel gene therapy treatments for inherited retinal diseases have been at the forefront of translational medicine over the past couple of decades. Since the discovery of CRISPR mechanisms and their potential application for the treatment of inherited human conditions, it seemed inevitable that advances would soon be made using retinal models of disease. The development of CRISPR technology for gene therapy and its increasing potential to selectively target disease-causing nucleotide changes has been rapid. In this chapter, we discuss the currently available CRISPR toolkit and how it has been and can be applied in the future for the treatment of inherited retinal diseases. These blinding conditions have until now had limited opportunity for successful therapeutic intervention, but the discovery of CRISPR has created new hope of achieving such, as we discuss within this chapter.

Abbreviations

AAV	adeno-associated viral vectors
ABCA4	ATP-binding cassette transporter protein family members 4
AD	autosomal dominant
ADAR	adenosine deaminase acting on RNA
AMD	age-related macular degeneration
APOBEC1	rat-derived cytosine deaminase apolipoprotein B MRNA editing enzyme catalytic subunit 1
AR	autosomal recessive
BEST1	bestrophin 1
BG	benzyl-guanine
CNGA3	cyclic nucleotide gated channel subunit alpha 3
CNGB3	cyclic nucleotide gated channel subunit beta 3
CPP	cell-penetrating peptides
CRISPR	clustered regularly interspaced short palindromic repeats
CRISPRi	CRISPR interference
crRNA	CRISPR RNA
CRX	cone-rod homeobox
dCas13	catalytically deactivated CRISPR associated RNA binding protein 13
dCas9	catalytically deficient CRISPR associated DNA binding protein 9
DSB	double strand break
dsRBD	double-stranded RNA binding domain
dsRNA	double-stranded RNA
ELOV4	elongation of very long chain fatty acids
gRNA	guide RNA
HDR	homology-directed repair
HITI	homology-independent targeted integration
INDELs	insertion or deletions
iPSCs	induced pluripotent stem cells
KRAB	Krüppel-associated box
LCA	leber congenital amaurosis
MCP	MS2 bacteriophage coat protein
MERTK	MER proto-oncogene tyrosine kinase

M-MLV RT	Moloney murine leukemia virus reverse transcriptase
NHEJ	non-homologous end joining
NHP	non-human primate
NPs	nanoparticles
NR2E3	nuclear receptor subfamily 2 group E member 3
OTC	ornithine transcarbamylase
PAM	protospacer adjacent motif
PE gRNA	prime editing guide RNA
PRPF31	pre-mRNA processing factor 31
PRPH2	peripherin 2
RecA	recombinase A
REP1	rab escort protein 1
RHO	rhodopsin
RNAi	RNA interference
RNPs	ribonucleoproteins
RP	retinitis pigmentosa
RP1	retinitis pigmentosa 1 axonemal microtubule associated protein
RPE	retinal pigment epithelium
RPGR	retinitis pigmentosa GTPase regulator
scFv	single-chain fragment variable
sgRNA	small guide RNA
SpCas9	*Streptococcus pyogenes* Cas9
TALEN	transcription activator-like effector nucleases
tracrRNA	trans-activating RNA
UDG	Uracil-DNA glycosylase
UGI	DNA glycosylase inhibitor
USH2A	Usherin
X	X-linked
ZFN	zinc finger nucleases

1. Introduction

Inherited retinal diseases are prime targets for intervention by genome engineering techniques. The retina is a specialized multi-layered structure that lines the inside surface of the eye (Fig. 1). It is composed of numerous cell types that develop in a specific sequence, creating a neural tissue that is able to respond to visual stimuli. Inherited retinal diseases are both genetically and clinically heterogeneous and arise from mutations in genes important to the functioning of the cells that form the retina.[1] Such diseases are currently known to be caused by mutations in over 300 genes/loci (https://sph.uth.edu/retnet/sum-dis.htm#A-genes). These mutations can lead to autosomal recessive, dominant and X-linked inherited forms that cause either dysfunction and/or loss of cells within the

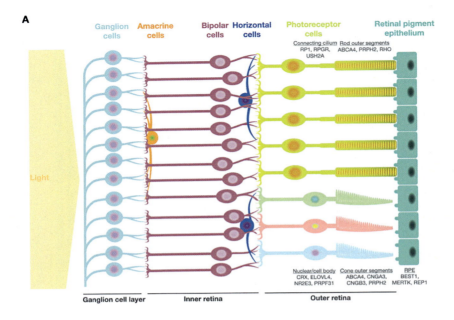

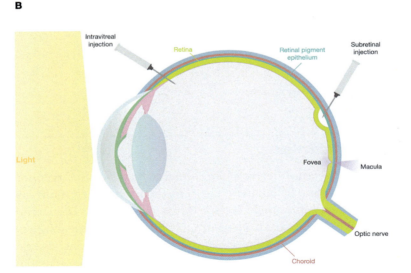

Fig. 1 Anatomy of the mammalian retina. (A) The retina is a highly structured neural tissue composed of multiple cell types. Light must pass through the retinal ganglion cell layer and inner retina to reach the outer segments of the rod and cone photoreceptors. These cells respond to the light and send the signal back through the bipolar cells and the ganglion cells, which converge into the optic nerve. Selected gene expression

retina, which ultimately prevent the retina detecting or conveying responses to visual stimuli. It is important to have an understanding of the composition and function of the retina when considering potential treatments for inherited retinal diseases. The light-sensitive rod and cone photoreceptor cells lie toward the back of the eye and form the outer retina (Fig. 1). These cells have a highly specialized morphology that include an outer segment of stacked discs that increase the surface area of the cell. The membrane-bound discs contain the light-sensitive pigments that absorb different wavelengths of light, the response to which is to initiate a transduction cascade that causes hyperpolarization of the photoreceptor cell. This in turn signals the cells of the inner retina, the bipolar cells, which then transfer the signal on to the retinal ganglion cells. Cross-communication between these cell types occurs by way of horizontal and amacrine cells with the resulting combined responses passing through the optic nerve to the brain for visual image processing. Given this complex retinal structure and interaction of multiple cell types, it is clear how dysfunction in one or many of these cell populations might disrupt the important sequence of events that enables vision. The functioning retina is also dependent on the health of the retinal pigment epithelium (RPE), a layer of cells lining the outer retina critical in supporting photoreceptor cell survival. Genetic mutations can influence the function and survival of all these cell types, creating a wide variety of retinal phenotypes that vary in severity. However, the majority of known mutations directly impact the photoreceptor and RPE cells, making these the primary targets of therapeutic interventions.

In humans, the composition of photoreceptor cells is variable across the outer retina with a concentrated area of cone photoreceptor cells responsible for visual acuity and color vision found in the central region, known as the

---

products in which mutations cause retinal disease are included: ATP-binding cassette transporter protein family members 4 (ABCA4), bestrophin 1 (BEST1), cyclic nucleotide gated channel subunit alpha 3 (CNGA3) and beta 3 (CNGB3), cone-rod homeobox (CRX), elongation of very long chain fatty acids (ELOVL4), MER proto-oncogene tyrosine kinase (MERTK), nuclear receptor subfamily 2 group E member 3 (NR2E3), peripherin 2 (PRPH2), pre-mRNA processing factor 31 (PRPF31), choroideremia Rab escort protein 1 (REP1), rhodopsin (RHO), retinitis pigmentosa 1 axonemal microtubule associated protein (RP1), retinitis pigmentosa GTPase regulator (RPGR), usherin (USH2A). (B) The retina forms the inner layer of the eye. The foveal depression occurs in the central region of the eye known as the macula, in which there is a high density of cone photoreceptor cells that enable color vision and visual acuity. The peripheral region beyond the macula is composed predominantly of rod photoreceptor cells. Gene therapy interventions can be provided by intravitreal or subretinal delivery routes.

macula. The density of cone photoreceptors then decreases outwardly from this zone leaving the peripheral outer retina composed largely of rod photoreceptor cells. The genetic origin of disease may lead to the dysfunction or death of one cell type of the retina, such as with cone dystrophies in which a predominant loss of central vision occurs,[2] or it may affect multiple cell types within the retina, such as with retinitis pigmentosa in which the peripheral retina is lost first, often followed by the central retina.[3] In the severe condition Leber congenital amaurosis (LCA), patients suffer a rapid loss of both the macular and peripheral retina at an early age.[4] Understanding the natural physiological progression that results from specific mutations enables us to measure the success of a therapeutic intervention using methods that are non-invasive.[5]

As a gene therapy target, the retina exhibits many benefits compared to other tissue types given its surgical access and relative immune privilege thanks to the blood-retina barrier.[6] Indeed, in the past decade the biggest strides in gene therapy development have been achieved in the eye largely due to these key features. While lentiviral vectors have been used in clinical trials and have indicated safety, to date there is an absence of reports of efficacy (for example from the recently terminated (ClinicalTrials.gov, NCT01367444; Stargardt's Macular Degeneration). The majority of human trials for the treatment of retinal disease use adeno-associated viral vectors (AAV) and dozens of such trials have now been completed, are ongoing or are currently recruiting (https://clinicaltrials.gov/). With growing evidence of both safety and efficacy[7–10] it is not surprising the first Food and Drug Administration-approved AAV gene therapy is for the treatment of an inherited retinal disease (LCA). Application of AAV gene therapy vectors in clinical trials have focussed on gene supplementation strategies for autosomal recessive and X-linked conditions although such approaches are also being applied for multifactorial diseases such as age-related macular degeneration (ClinicalTrials.gov, NCT03846193; Geographic Atrophy). The vectors share common characteristics and carry transgenes that contain the genetic elements that enable expression of a correct copy of the gene of interest in the desired cell type of the retina. Data published over the past decade from clinical trials have been critical for developing appropriate surgical techniques and understanding the safe doses of AAV that can be delivered to the retina.[11,12] Trials are ongoing using AAV gene supplementation strategies for the treatment of X-linked retinitis pigmentosa (*RPGR*), choroideremia (*REP1*) and retinoschisis (*RS1*) plus autosomal recessive retinitis pigmentosa (*MERTK, PDE6B*) and achromatopsia (*CNGA3, CNGB3*).

However, the application of gene supplementation strategies is relatively limited as a large proportion of inherited retinal diseases will not benefit from such treatments. There are scenarios where gene supplementation might be desired but cannot be implemented due to the packaging limitation of AAV vectors. For example, the most common form of childhood-onset macular degeneration arises from mutations in *ABCA4*, but the coding sequence for this gene exceeds the packaging capacity of AAV and therefore requires a less efficient dual vector[13,14] or lentiviral vector[15,16] approach. While the dual AAV vector system appears feasible, developing new strategies that specifically target and correct the genome may be beneficial for this particular disorder and indeed other forms of inherited retinal disease caused by mutations in large genes (*e.g.*, >4.5 kb; *USH2A, EYS, ABCA4, CDH23, MYO7A*). Design of appropriate treatment strategies for given retinal diseases is largely guided by knowledge and understanding of the genetic causation (Fig. 2).

With gene supplementation approaches limited to the treatment of loss of function mutations, new treatment options are required for the large number of retinal diseases that result from gain of function and/or dominant negative effects, which cannot be overcome by provision of a correct form of the gene. In such a situation, the temptation might be to attempt silencing of the mutated allele, but the issue of haploinsufficiency needs to be considered as one correct copy of a gene is not always enough to enable normal function, for example in the case of mutations in *PRPF31*.[17] In situations where mutations lead to dominant negative or toxic gain of function effects, for example with the *RHO* variant P23H,[18,19] gene augmentation has proved ineffective.[20] More encouraging therapeutic strategies have employed a combined mutation-independent RNA interference knockdown and supplementation approach.[21] However, achieving the appropriate balance of knockdown and supplementation has proved difficult in pre-clinical testing with further optimisations still required. For such diseases, mutation correction would be a highly desirable therapeutic strategy.

Despite concerns of haploinsufficiency, some forms of inherited retinal disease may benefit from an allele-specific gene silencing approach (as discussed in more detail in a later section), but others will rely on the development of direct genome editing methods. Clustered regularly interspaced short palindromic repeats (CRISPR) genome engineering therefore holds great promise in being the solution to the current absence of treatment options for many retinal diseases. However, expectations need to be tempered by consideration of translational feasibility. Given the large number

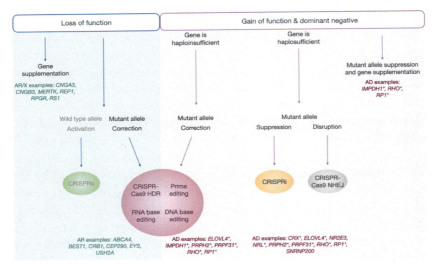

Fig. 2 Gene therapy approaches for retinal disease. Appropriate treatment options for particular forms of inherited disease need to be determined from their genetic origin and guided by knowledge of the molecular consequences and subsequent pathophysiological outcomes. Gene supplementation strategies show success for autosomal recessive (AR) and X-linked disorders (X), but not all genes are deliverable by AAV vectors. Autosomal dominant (AD) mutations are good candidates for genome correction and editing with some also appropriate for epigenetic targeting. ATP-binding cassette transporter protein family members 4 (ABCA4), bestrophin 1 (BEST1), centrosomal protein 290 (CEP290), crumbs cell polarity complex component 1 (CRB1), cyclic nucleotide gated channel subunit alpha 3 (CNGA3) and beta 3 (CNGB3), cone-rod homeobox (CRX), elongation of very long chain fatty acids (ELOVL4), eyes shut homolog (EYS), inosine monophosphate dehydrogenase 1 (IMPDH1), MER proto-oncogene tyrosine kinase (MERTK), nuclear receptor subfamily 2 group E member 3 (NR2E3), neural retina leucine zipper (NRL), peripherin 2 (PRPH2), pre-mRNA processing factor 31 (PRPF31), choroideremia Rab escort protein 1 (REP1), retinoschisin 1 (RS1), rhodopsin (RHO), retinitis pigmentosa 1 axonemal microtubule associated protein (RP1), retinitis pigmentosa GTPase regulator (RPGR), small nuclear ribonucleotprotein U5 subunit 200 (SNRNP200), usherin (USH2A). *Mutation dependent.

of genes in which mutations arise and cause inherited retinal disease and the allelic heterogeneity within those genes, the potential number of therapeutic vectors to be designed and optimized are vast. Currently more than 1200 disease-causing variants in the gene *ABCA4* alone have been identified (www.lovd.nl/ABCA4) with patients carrying different combinations of disease-causing alleles.[22] We therefore discuss in this chapter not only the expanding toolkit available for CRISPR genome engineering, but also the translational implications and considerations of using such molecular mechanisms to treat inherited retinal disease.

Despite only being presented to the scientific community in 2012,[23] the potential of CRISPR-based strategies for the treatment of inherited retinal diseases was very quickly realized. Editing in mouse models has already been achieved in proof-of-principle studies as well as targeting of common mutations that cause inherited retinal disease.[24-27] Successful pre-clinical genome editing of a deep-intronic mutation in the gene *CEP290*, which causes LCA10,[28] has led to the first approved CRISPR genome engineering clinical trial for inherited retinal disease and the first delivery into a patient was achieved in early 2020.[29] Further discussion on this trial and pre-clinical developments relevant to CRISPR genome engineering in retinal disease will be discussed throughout this chapter, but it is evident that the CRISPR toolkit is rapidly expanding and therefore so too is the scope of treatment pipelines based around these molecular mechanisms. In this chapter we discuss the current CRISPR catalogue and strategies being developed for the treatment of inherited retinal diseases, including allele-specific silencing by epigenetic repression or by direct DNA editing and mutation correction by single-base editing, prime editing and RNA base editing.

To advance the treatment options for inherited retinal diseases, current ambitions aim to achieve safety and efficacy of the rapidly developing CRISPR strategies in proof-of-principle pre-clinical studies. It is clear that AAV delivery to different retinal cell targets is readily achievable, but the expression profile and impact of CRISPR elements in the retina needs to be better understood. As a bacterial-derived molecular system, the expression of such components in mammalian tissue needs to be carefully assessed, not only for immune responses to such products but also the desired and undesired effects of their genome editing activity. While AAV delivery is currently well tolerated, it may be that other delivery options of the CRISPR components may be required. As genome editors, the safety of CRISPR strategies will be paramount to their ultimate success. On-target editing will provide the genome engineering required for therapeutic intervention but assessment and understanding of off-target editing rates will be just as important. These aspects will be discussed throughout this chapter as we review the recent developments and current options available for translational researchers developing treatments for inherited retinal disease through CRISPR genome engineering strategies.

2. CRISPR genome editing

Genome engineering using zinc finger nucleases (ZFNs), or transcription activator-like effector nucleases (TALENs) have existed for some time, but the clinical application of these techniques has been limited by the generally low *in vivo* editing efficiencies.[30] The discovery of the mechanisms of CRISPR-Cas9 in the bacterial immune system, and subsequent adaptation into a powerful gene editing tool has revolutionized the field of molecular biology and generated excitement for the potential of novel therapeutic approaches to treat human conditions.[23] The CRISPR-Cas system encompasses a variety of components which differ widely in mechanisms of action and offer therapeutic potential by direct genome interaction and/or editing. This developing genome engineering toolkit offers great potential for the treatment of inherited retinal diseases.

Broadly, there are two classes of CRISPR systems, each containing multiple CRISPR types. Class 1 contains type I, type III and type IV systems, while Class 2 contains types II, V, and VI CRISPR systems. The most widely used platform for genome targeting applications derives from the Class 2 type II CRISPR-associated enzyme, Cas9, which acts as a single effector protein; in contrast, the Class I Cas enzymes operate as multi-subunit protein complexes. Despite the complexity of the Cas family, all systems share a requirement for CRISPR RNA (crRNA) for defined target specificity while type II variants have an additional requirement for trans-activating RNA (tracrRNA), which forms a scaffold structure.[31] For gene editing applications, the two CRISPR RNAs are joined as one small guide RNA (sgRNA), which greatly simplifies delivery. The Cas9: sgRNA complex randomly interrogates DNA in the cell, searching first for the appropriate protospacer adjacent motif (PAM), a short motif adjacent to the target sequence. Upon recognition of the PAM sequence, the Cas9 protein unwinds the DNA, allowing the Cas9-associated sgRNA to hybridize with the exposed DNA strand (the protospacer). If the DNA sequence matches the sgRNA target sequence, the HNH and RuvC catalytic domains of the endonuclease cleave both strands of the target DNA, generating a double strand break (DSB).[23]

In eukaryotic organisms, Cas9-induced DSBs are repaired either by error prone non-homologous end joining (NHEJ) or homology-directed repair (HDR).[32] Typically, CRISPR-induced DSBs are repaired by NHEJ, an efficient and prevalent mechanism in human cells that results in random insertions or deletions (indels) and gene disruption in the target region. HDR is a well-established mechanism that can be harnessed to insert a specific DNA template for precise restoration of the DNA sequence, with limited off-target activity or on-target alterations.[33] HDR-mediated insertion, however, requires the presence of the correct template and is typically characterized by lower efficiency than NHEJ repair. In addition, the correction of a point-mutation by HDR has been shown to be highly ineffective particularly in non-dividing cells, such as those present in the retina, and the dsDNA break formation by Cas9 nuclease generates undesired indel mutations at a substantial frequency that annuls the potential benefit from corrected mutation.[34] Thus, correction of patient derived iPSCs, followed up autologous transplantation of the corrected retinal cells, has been readily employed for CRISPR-mediated HDR in the retina. Bassuk and colleagues utilized CRISPR-Cas9-triggered HDR to repair an *RPGR* mutation that causes X-linked RP in primary fibroblasts that were transduced to produce patient-specific iPSCs.[35] The authors showed 13% correction of the c.3070 G > T mutation and conversion of the wild-type allele, providing proof-of-concept for a targeted iPSC-based transplantation approach for retinal disease. Similar studies have demonstrated correction of mutations in patient-derived iPSCs in *RPGR, Pde6β, MAK, CEP290, N2RE3,* and *RHO* genes.[27,36–39] Recently, Cai and colleagues reported HDR-mediated correction of a mutation causing autosomal dominant retinitis pigmentosa (adRP) *in vivo*.[24] The authors developed a Cas9/*Escherichia coli* recombinase A (RecA) protein technique, which incorporated *Streptococcus pyogenes* Cas9 (SpCas9) and sgRNA-targeted RecA, to precisely correct the naturally occurring *Pde6b* missense mutation in postnatal *rd1* mice. The authors demonstrated that Cas9/RecA-mediated correction promoted the survival of rod and cone photoreceptors, restored Pde6b in rod photoreceptors, and enhanced visual functions of *rd1* mice. Similarly, Suzuki and colleagues previously devised a homology-independent targeted integration (HITI) strategy, which boosts DNA knock-in efficiency in non-dividing cells.[40] The group demonstrated *in vivo* proof-of-concept in the Royal College of Surgeons rat, a model for retinitis pigmentosa resulting from a 1.9 kb deletion in the *Mertk* gene. Following dual delivery of a HITI-AAV, there was a detectable increase in *Mertk* mRNA and protein expression levels as well as thickening of the degenerating photoreceptor layer, although in contrast to Cai and colleagues the rescue was only partial and not enough to completely restore vision.

The NHEJ DNA repair pathway is currently the more commonly adopted approach in developing treatments for retinal diseases, especially for mutations arising from autosomal dominant diseases. Through careful design of the guide RNA, the mutant allele can be preferentially disrupted while the wild-type allele remains preserved. In the absence of haploinsufficiency mechanisms, this could ameliorate disease phenotypes and has been successfully demonstrated across several genetic targets and retinal disease models. Bakondi and colleagues provided the first *in vivo* functional correction of an inherited dominant mutation using CRISPR,[41] providing proof-of-concept that CRISPR/Cas9 can be utilized to treat retinal degenerations. The authors showed that a single subretinal injection of gRNA/Cas9 plasmid in combination with electroporation generated allele-specific disruption of the S334ter mutation (Rho^{S334}) in a transgenic mouse model, which prevented retinal degeneration and improved visual function. Subsequently, the Broccoli, Liu, and Li groups described a spacer-mediated allele-specific knockout approach to specifically inactivate the P23H *RHO* mutant, which represents approximately half of all *RHO*-associated adRP cases.[25,26,41] To broaden the breadth of targetable *Rho* mutations, Tsai and colleagues developed a two-pronged ablate-and-replace strategy that destroyed expression of endogenous genes in a mutation-independent manner and enabled expression of functional protein through exogenous cDNA.[42] The authors validated the approach in two kinds of human *RHO* mutation knock-in mouse models: Rho^{P23H} and Rho^{D190N}, leading to significantly greater survival of functioning photoreceptors in comparison to the scrambled gRNA control vector. More recently, a similar approach was utilized to target autosomal dominant cone-rod dystrophy (CORD6) in mice and primates.[43]

3. DNA base editing

Although HDR-mediated gene editing can be harnessed to insert a specific DNA template for precise restoration of the DNA sequence, this pathway is characterized by low efficiency and high rates of undesired indel mutations that nullify the potential benefit from repairing the mutation.[33] Recently, CRISPR-Cas-mediated single base pair editing systems (or 'base editing' systems) have been devised to bypass these limitations.[44–46] DNA base editing and the more recent prime editing technologies have remarkable potential as therapeutic tools to correct disease-causing mutations in the human genome. CRISPR-mediated HDR are confined to editing within dividing-cells, since these pathways are restricted to the S and G2 phases of the cell cycle. Conversely, base editing employs cellular mismatch repair machinery and can be applied to reverse these defects in both dividing and terminally differentiated cell types.[47] Gene therapy is a major area where DNA base editing and prime editing toolkits can be applied because they have already been adapted to characterize, model, and correct the underlying causes of human genetic conditions.

DNA base-editing and prime-editing may prove to be particularly well adapted for correction of large genes (>4 kb). For many retinal diseases, vector-mediated gene delivery is not feasible due to the cargo limit of viral vectors as well as the expression of multiple heterogeneous variants of the target gene.[48] The *ABCA4* coding sequence (~6.8 kb) and *USH2A* coding sequence (~15.6 kb) genes together account for almost 25% of all inherited retinal diseases.[49] The therapeutic promise of prime editors may be especially useful in addressing the heterogeneity of inherited retinal disease (IRD)-mutations, such as the clustered mutation-spectrum of *ABCA4*. Despite the therapeutic potential of prime editors, however, such applications would also be constrained by a number of clinical and commercial factors such as prevalence, clinical trial development and enrolment feasibility, and industry interest in targeting treatments for ultra-rare diseases. Nevertheless, a number of mutations are relatively common, such as the c.2299delG mutation in *USH2A* and companies may be incentivized by favorable pricing for rare indications, expedited regulatory review and approval pathways for orphan diseases, and clinical trial structures such as basket trials to test a prime editing program in multiple related indications.

Base editing can be applied to autosomal dominant diseases, where gene supplementation is not a suitable method due to the requirement for silencing or ablation of the dominant-negative mutant allele. 25–30% of the cases of retinitis pigmentosa (RP), the most common inherited retinal degeneration, are autosomal dominant. Mutations in at least 23 genes have been reported to cause adRP to date, including over 180 mutations in the *RHO* gene, which accounts for over 25% of adRP cases. Approximately half of the RHO-associated adRP cases are caused by the P23H mutation.[50] Targeted adenine or cytosine-mediated based editing would convert P23H into P23R (CAC→CGC) or P23T (CAC→TAT/C) amino-acid substitutions that may be milder than P23H, albeit likely still disease causing. A prime editing approach would allow for precise and targeted correction of the CAC mutation and thus presents the most attractive therapeutic strategy for this indication. Similarly, base editing may offer a novel therapeutic avenue in correction of heterozygous C > T mutations in *GUCY2D*, a leading cause of the autosomal dominant cone-rod dystrophy.

Two classes of DNA base editors have been described: cytosine base editors and adenine base editors.[51,52] DNA base editors encompass two key components: a Cas enzyme for programmable DNA binding and a single-stranded DNA modifying enzyme for targeted nucleotide alteration. Collectively, all four transition mutations (C→T, T→C, A→G, and G→A) can be installed with the available CRISPR-Cas base editor systems.[53] Recently, Kurt et al. described the engineering of two novel base editor architectures that can efficiently induce targeted C-to-G base transversions.[54] In addition, recent studies report dual base editor systems for combinatorial editing in human cells.[55–57] Together, these new base editors expand the range of DNA base editors to transversion mutations and may allow for targeting of more complex compound edits than are currently achievable by a single DNA base editor.

3.1 Base editor systems
3.1.1 Cytosine base editors
The first-generation base editor (CBE1) was developed by Liu and co-workers in 2016.[44,52] It was engineered by fusing a rat-derived cytosine deaminase Apolipoprotein B MRNA Editing Enzyme Catalytic Subunit 1 (APOBEC1) to the amino terminus of catalytically deficient, or "dead," Cas9 (dCas9). In a narrow window of the non-targeted strand, CBE1 deaminates cytosine to uracil. Uracil is then recognized by cell replication machinery as a thymine, resulting in a C-G to T-A transition. To improve base editing efficiency, a second-generation cytosine base editor (CBE2) was developed by fusing an uracil DNA glycosylase inhibitor (UGI) to the C-terminus of BE1, inhibiting the activity of Uracil-DNA glycosylase (UDG) enzyme that was found to reverse the U-G intermediate created by the base editor back to the C-G base pair. The inhibition of UNG by BE2 resulted in a threefold increase in editing efficiency in human cells.[52] To further improve editing efficacy, BE3 was developed by restoring histidine at position 840 (H840, HNH catalytic domain) in dCas9 to generate a base editor that uses Cas9 nickase (nCas9 D10A). This variant induces a nick in the G-containing strand of the U-G intermediate (non-edited DNA strand) to bias cellular repair of the intermediate towards a U-A outcome, further converted to T-A during DNA replication.

Tran and colleagues described the use of TARGET-AID (activation-induced cytidine deaminase) to target the AMD high-risk Y402H complement factor H variant and showed cytosine-to-thymine nucleotide correction in 21.5% of the total sequencing reads with minimal off target effects.[58] The authors demonstrated the feasibility of using a base editor-mediated approach for a common high-risk variant that is found to be strongly associated with AMD, although CBE has yet to be applied to the retina *in vivo*.

3.1.2 Adenine base editors

The cytosine base editor is limited to installing a C-G to T-A mutation, restricting the range of correctable disease-causing mutations. Importantly, methylated cytosines are vulnerable to high rates of spontaneous cytosine deamination,[59] and nearly half of all pathogenic point mutations in principle can be reversed using an ABE to convert an A-T base pair back into a G-C base pair. As such, base editing capabilities and study of genetic diseases were further expanded by the development of a new class of adenine base editors that could induce A to G conversions.[51] ABE-mediated DNA editing operates under a similar mechanism as CBE. The ABE-dCas9 fusion binds to a target DNA sequence in a guide RNA-programmed manner, and the deoxyadenosine deaminase domain catalyzes an adenine to inosine transition. In the context of DNA replication, inosine is interpreted as guanine, and the original A-T base pair may be replaced with a G-C base pair at the target site. Unlike cytosine deaminases, ssDNA adenosine deaminase enzymes do not occur in nature. David Liu and group overcame this limitation through extensive protein engineering and directed evolution of *Escherichia coli* tRNA adenosine deaminase, TadA (ecTadA). EcTadA converts adenine to inosine in the single-stranded anticodon loop of tRNA^{Arg}, and shares sequence similarity with the APOBEC family. The first-generation ABE (ABE1.2) was generated by fusing the evolved TadA variant (TadA*) to the N-terminus of nCas9 through XTEN (a 16 amino acid linked used in BE3), with the C terminal of nCas9 fused with a nuclear localization signal (TadA*-XTEN-nCas9-NLS) [13]. In comparison with cytosine base editing, adenine base editing by ABE yields a much cleaner product that has virtually no indels, and there are no reports of significant off-target (A-to-non-G) edits to date. In its native context, TadA acts as a homodimer, with one monomer catalyzing deamination and the other monomer enabling tRNA substrate binding.[60] To optimize ABEs, Gaudelli et al. engineered a single-chain heterodimer comprised of a wild-type non-catalytic TadA monomer and an evolved ecTadA monomer (TadA-TadA*).[51]

Recently, using an inherited retinal disease mouse model harboring a nonsense mutation in the *Rpe65* gene, Suh and colleagues show that lentivirus-mediated delivery of ABE corrected the mutation with an efficiency of up to 29% in the adult mouse eye with minimum detected indel or off-target mutations.[61] Lentivirus exhibits RPE-specific tropism when injected subretinally to the mouse eye, obviating the need for a tissue-specific promoter. The ABE-corrected eyes demonstrated restored RPE65 expression, retinoid isomerase activity, and retinal and visual function at near-normal levels. Although the ABE enabled precise correction of the target mutation with high precision and efficiency, and substantially higher than previous approaches utilizing HDR, it is yet to be seen whether the lentivirus-mediated base editing will translate to the clinic. Furthermore, the therapeutic potential of DNA base-editors to target premature termination codons, which frequently result in severe disease phenotypes due to complete loss in protein function, have been demonstrated *in vivo*. Ryu and colleagues utilized an AAV targeted ABE to correct a premature stop codon (Q871X) in the *Dmd* gene.[62] Targeted base-editing restored dystrophin expression in 17% of myofibers, a level which was sufficient to improve muscle function. Lee et al. developed CRISPR-pass, a targeted tool for bypassing premature termination codons using CRISPR-mediated adenine base-editors that showed systematic rescue of all possible cases of premature termination codons (PTCs).[63] This approach is applicable to 95.5% of clinically significant nonsense mutations in the ClinVar database. CRISPR-mediated base editing may be particularly well adapted in Usher syndrome, where nonsense mutations account for ~16% of all USH2A-mutant alleles (199 nonsense mutations of 1234 USH2A-mutant variants, https://portal.biobase-international.com/hgmd/pro/gene.php?gene=ush2a). Finally, CBEs and ABEs can disrupt donor/acceptor sites at AT/AA and GC/GG, respectively, a robust strategy for selectively skipping mutation-containing exons while keeping normally functioning variants in XL-RP and other IRDs.[53] Gapinske et al. demonstrated a novel method (CRISPR-SKIP) that utilizes cytidine deaminase single-base-editors to program exon skipping by mutating target DNA bases within splice acceptor sites.[64] The authors estimate that this approach could target ~63% (118,089 out of 187,636) of inner exons in protein coding transcripts. The interim readout from the Phase 1/2 trial of QR-421a, an RNA-based platform that utilizes exon skipping of exon 13 to restore functional Usherin protein, supports the clinical utility of a CRISPR-mediated exon skipping approach in Usher syndrome and other retinal degenerations.

3.1.3 Prime editors
Despite the profound capabilities of CBEs and ABEs to edit the genomic DNA, a major limitation of the current base editing technologies (until recently) has been the ability to generate precise base-edits beyond the four

transition mutations. Recently, a method to overcome these shortcomings, known as prime editing, has been described by Anzalone and colleagues.[46] As with CRISPR-mediated base editing, prime editing does not rely on DSBs. Prime editors use an engineered reverse transcriptase fused to Cas9 nickase and a prime editing guide RNA (pegRNA). Importantly, the pegRNA differs significantly from regular sgRNAs and plays a major role in the system's function. The pegRNA contains not only (i) the sequence complementary to the target sequence that directs nCas9 to its target site, but also (ii) an additional sequence spelling the desired sequence changes.

The first generation of PEs (PE1) was comprised of Moloney murine leukemia virus reverse transcriptase (M-MLV RT), linked to the c-terminus of nCas9 and pegRNA, which was expressed on a second plasmid. To further enhance the efficiency of the reverse transcriptase, Anzalone and colleagues tested different M-MLV RT variants that have been shown to enhance binding, enzyme processivity, and thermostability. The latest generation prime editor, designated PE3, performed all 12 possible transition and transversion mutations (24 single-nucleotide substitutions) with average editing efficiencies of 33% (±7.9%).[46] The number of off-target effects observed with PEs was greatly reduced, likely due to the need for complementation at Cas9 binding, PBS binding, and RT product complementation for flap resolution. Prime editing owns other advantages over previous CRISPR-mediated base editing approaches, including less stringent PAM requirements due to the varied length of the RT template and no "bystander" editing. Notwithstanding, prime editing is still in early stages of development, and its specificity and potential for off-target modifications remains to be studied. The latest generations of base editors are much better characterized, particularly *in vivo*, and offer higher efficiency rates and lower-indel formation. Thus, at this stage, these tools should be used over prime editors whenever possible. In any case, the prime editing system is an enormous milestone in the development of a universal method for genome editing, and its clinical adaptation towards the correction of known pathogenic mutations may prove tremendous.

4. RNA editing

Site-directed RNA editing is a base editing approach that targets the transcriptome rather than the genome. Post-transcriptional mRNA modifications occur naturally to a number of native transcripts in mammalian cells by a class of endogenous Adenosine Deaminase Acting on RNA (ADAR) enzymes. These deaminases convert adenosine bases to inosine, which is structurally similar to guanosine and read as such during splicing and translation.[65] This effectively creates an A>G edit in RNA which can be harnessed for the correction of G>A mutations. RNA editing strategies can be broadly separated into two approaches depending on what ADAR is used. Exogenous ADAR strategies deliver the full-length ADAR protein or the deaminase domain (ADAR$_{DD}$) fused to another RNA binding protein to recruit the target RNA. Endogenous approaches do not introduce ADAR into cells but rely on recruiting the endogenously expressed enzyme. A range of strategies have now been engineered to direct ADAR to edit targets of interest.[66]

4.1 ADAR physiology

Adenosine to inosine (A>I) editing in mammals is mediated by one of two catalytically active ADAR enzymes (ADAR1 and ADAR2) with a third (ADAR3) not known to be catalytically active.[65] ADAR1 is expressed ubiquitously from two promoters: one directs constitutive expression of the truncated ADAR-p110 isoform in the nucleus while the second directs inducible expression of the full-length ADAR-p150 isoform in both nucleus and cytoplasm in response to type I and II interferon.[66,67] ADAR2 is more selectively expressed, predominantly in the brain with nuclear localisation.[67,68] ADAR2 has high affinity for the Q/R site of *GRIA2* transcripts, where A-I editing is essential for the function of the subunit GluR2 in glutamate-activated ligand-gated ion channels (GluR).[69] The important physiological role of ADAR editing is illustrated by knockout models: knockout of *Adar1* is embryonically lethal in mice, associated with disruption of hematopoiesis and interferon signalling,[70] while *Adar2* knockout leads to seizures and early post-natal lethality in mice.[69]

ADARs edit at regions of double-stranded RNA (dsRNA) >15–20 nt in length.[71] In the endogenous context the double-stranded RNA binding domain (dsRBD) of the ADAR binds dsRNA regions formed by complementary sequences such as found in introns[72] or *Alu* repeats.[73] For site-directed RNA editing, all strategies employ a mechanism to create double-stranded RNA at the site of interest through the introduction of complementary RNA sequences that hybridize to the target RNA, leading to hydrolytic deamination of the adenosine to inosine.

4.2 RNA editing tools

Common to both endogenous and exogenous ADAR strategies is the use of a guide RNA (gRNA; also known as an ADAR guiding RNA, adRNA). This is an RNA sequence with a region of varying length complementary to the target that can interact with the relevant RNA binding protein and create a dsRNA region for ADAR$_{DD}$ activity. ADAR editing preferentially occurs at adenosines that are mismatched with a cytidine opposite the target adenosine (rather than a complimentary thymidine),[74] and this A-C mismatch is commonly encoded in the gRNA to specify ADAR editing to the target adenosine and improve editing efficiency at the target site.

Engineering of the ADAR deaminase has produced mutants that alter editing efficiency and specificity. The ADAR2-E488Q mutation markedly increases editing activity through increasing base flipping stability and deamination rates,[75,76] however is known to increase off-target editing throughout the transcriptome.[77,78] An ADAR2-E488Q/T375G double mutant has demonstrated promise for much greater specificity although with some loss of on-target efficiency.[77] In ADAR1, the E1008Q variant demonstrates similar hyperactivity like that seen for ADAR2-E488Q.[79] Recently, mutagenesis studies have been used to introduce 16 mutations to ADAR2$_{DD}$ to create a system capable of C-to-U editing in addition to A-to-I editing when paired with a CRISPR-Cas13 construct (see below).[80] These deaminase mutants are used widely with exogenous ADAR strategies.[66]

4.3 Exogenous ADAR approaches
4.3.1 CRISPR-Cas13

Cas13 is a type VI CRISPR nuclease that selectively binds to single-stranded RNA for bacterial immunity against foreign RNA and has been repurposed for RNA targeting in mammalian cells. Native Cas13 acts as an RNAse with cleavage activity both for the target and bystander transcripts. Selective mutation of the nuclease domains yields a catalytically deactivated RNA binding protein (dCas13) capable of specific RNA targeting directed by a gRNA. Many Cas13 proteins (Cas13a,b,d) with varying properties have been studied, and a selection of these tested for RNA editing by fusing the ADAR$_{DD}$ to dCas13.[78,80] Unlike Cas9 where gRNA binding is limited by the availability of local PAM sequences, the dCas13 enzyme appears to have few targeting constraints for RNA editing, resulting in a broad scope for mutation targeting.[78]

The REPAIR system (RNA Editing for Programmable A to I Replacement) was developed for A-I editing using dPspCas13b derived from *Prevotella Sp.*[78] Two systems have been developed: REPAIRv1 uses an ADAR$_{DD}$(E488Q) and demonstrates high editing efficiency but also widespread off-target editing, while REPAIRv2 was developed with the ADAR$_{DD}$(E488Q/T375G), and these modifications produced a 900-fold reduction in off-target editing and a twofold reduction in on-target efficiency.[78]

The RESCUE system (RNA Editing for Specific C to U Exchange) employs the engineered RESCUE ADAR$_{DD}$ variant to enable C—U editing fused to dRanCas13b derived from *Riemerella anatipestifer*. As this allows both C—U and A-I editing, it is possible to used multiplexed gRNAs to create a range of different edits with the same construct. A-I off-targets and C—U off-targets (to a lesser extent) occur with RESCUE. To address this a more specific variant with a S375A mutation ADAR$_{DD}$(C—U) termed RESCUE-S has also been developed. This yielded an approximately 12-fold reduction in A-I off-targets and ~1.8-fold reduction in C—U off-targets while maintaining specificity.[80] Both systems have shown promise *in vitro* however have yet to be tested *in vivo*.

4.3.2 BoxB-λN-ADAR

The bacteriophage-derived λN peptide binds short stem-loop RNA structures called BoxBs.[81] To facilitate RNA editing, the ADAR2$_{DD}$ is fused to the λN peptide, and the BoxB binding motifs are incorporated into the middle of the gRNA sequence. The gRNA complementary region binds the target sequence, and the λN-ADAR2$_{DD}$ fusion is recruited to bind the BoxB motifs at the target site.[81] This system has been optimized to increase editing rates, using the ADAR2$_{DD}$(E488Q), four copies of the λN peptide and including multiple copies of the BoxB hairpin in the gRNA.[80] This minimal system is able to be delivered with AAV, and has been used to edit *Mecp2* in primary murine neurons.[82,83] Off-target events seem to be particularly prevalent with the BoxB-λN-ADAR system, although localizing expression to the nucleus may improve these.[84]

4.3.3 Snap-ADAR

SNAP-tags are enzymes that form covalent linkages with a O6-benzylguanine (BG) substrate. SNAP-tags fused to ADAR$_{DD}$ can bind to gRNAs conjugated with BG that then bind an RNA target.[85] As the gRNAs must be chemically modified for the attachment of the BG and for nuclease-protection, these guides are delivered separately from the SNAP-ADAR construct, similarly to anti-sense oligonucleotides.[85,86] To date these have shown high rates of editing efficiency in immortalized cell lines using both ADAR deaminases and their hyperactive mutants,[87] but have yet to be used in primary cells or delivered *in vivo*.

4.3.4 MS2-MCP-ADAR

The MS2 bacteriophage coat protein (MCP) binds the MS2 RNA stem-loop from its genome. For RNA editing, MS2 loops can be attached to gRNAs, and MCP-ADAR$_{DD}$ fusions are recruited to bind the MS2 loop and directed to the editing site.[88–90] MCP-ADAR1$_{DD}$(E1008Q) and MCP-ADAR2$_{DD}$(E488Q) constructs have not only demonstrated efficacy *in vitro*, but have been delivered by AAV *in vivo*. An AAV8-MCP-ADAR1$_{DD}$(E1008Q) construct delivered with an MS2 gRNA demonstrated 2% on-target efficiency and partial restoration of dystrophin expression in an *mdx* mouse model of muscular dystrophy.[89] Similarly to the λN-BoxB system, use of the hyperactive mutants and cytosolic rather than nuclear localization of the editing construct is associated with higher editing rates and concomitant higher off-target rates.[89]

4.3.5 GluR2-ADAR

The Q/R motif in the *GRIA2* transcript is a natural target for the binding of the dsRBD domains of full-length ADAR2, as discussed earlier. This has been harnessed for RNA editing, by attaching an optimized Q/R hairpin motif to gRNA sequences to recruit full-length ADAR2. Overexpression of ADAR2(E488Q) has been used to target endogenous transcripts in immortalized cells with the GluR2 system.[89,91] Furthermore, Katrekar and colleagues delivered the 2.1 kb ADAR2 and ADAR2(E488Q) sequences with GluR2-gRNAs *via* AAV8 intramuscularly to the *mdx* mouse and systemically to the sparse fur ash (*spf$_{ash}$*) mouse model of ornithine transcarbamylase (OTC) deficiency.[90] This resulted in the low-level correction of a stop codon and splice defect in each model respectively, with resultant rescue of protein expression.

4.3.6 Synthetic RNA targeting systems

In order to develop an entirely human-derived editing system to overcome fears of immunogenicity from bacterial derived effectors, the synthetic CRISPR-Cas Inspired RNA targeting system (CIRTS) was developed. Mimicking Cas13 ssRNA recognition, binding and gRNA targeting functions, a combination of human-derived elements were assembled and fused to $ADAR_{DD}$(E488Q).[92] While only tested in immortalized cells, this has demonstrated the potential for the development of engineered proteins for RNA editing, although the immune significance is still less clear.

4.4 Endogenous ADAR approaches

4.4.1 Leaper

To recruit endogenous ADAR, the LEAPER (Leveraging Endogenous ADAR for Programmable Editing of RNA) system uses long gRNAs (71–191 nts) to create long dsRNA regions for the binding of naturally expressed ADAR1, without the need for any recruitment domains.[93] These have achieved high editing rates in the 5'UTR of endogenous genes, and successful editing at lower efficiencies in coding regions. Chemical modification of the guides to enhance stability increased editing and demonstrated correction of mutations in Hurler syndrome patient-derived fibroblasts. Low editing rates were seen from lentiviral delivery of the long gRNAs.[93] These findings were supported by a separate study demonstrating that 100 nt guides with no recruitment domains could facilitate ADAR editing *in vitro* to edit endogenous transcripts.[90]

4.4.2 RESTORE and GluR2 recruitment

Building on observations that GluR2 R/G domains can recruit overexpressed ADAR, it has also been shown that GluR2 gRNAs alone are able to recruit endogenous ADAR for site-directed editing. RESTORE (Recruiting Endogenous ADAR to Specific Transcripts for Oligonucleotide-mediated RNA Editing) uses shorter (20–40 nt) chemically modified oligonucleotides with the R/G domain to recruit ADAR1 and edit both immortalized and primary cells *in vitro*, including in RPE.[94] Notably, the induction of ADAR1p150 expression with interferon increased editing rates due to induction of antiviral interferon-stimulated genes and may enable a long-term therapeutic approach to IRDs.

Katrekar and colleagues found that short 20 nt guides with a single GluR2 domain delivered alone *via* AAV8 to the *spf$_{ash}$* mouse without overexpression of ADAR2 were able to mediate very low (0.6%) editing rates in the liver.[90] Given the liver expresses ADAR2 at a much lower level relative to neuronal tissue, it is of interest to assess whether this rate might be higher in tissues with higher ADAR2 expression.

4.5 Clinical applications and challenges for RNA editing treatment of inherited retinal disease

4.5.1 Clinical targets

Currently there have been no reports of pre-clinical testing of RNA editing constructs for inherited retinal disease models. Like other base editing strategies, RNA editing has significant promise for the treatment of recessive disorders. For common recessive genes with long coding sequences not packageable in a single AAV for gene therapy such as *USH2A, ABCA4, MYO7A, EYS, CDH23* and *CEP290*, RNA editing could be used to treat common mutations.[66] As transition mutations (including G > A and T > C) are the most common single base mutations, a significant proportion of all mutations in these genes should be treatable.[66]

Termination codons and splice site mutations may be particularly amenable to RNA editing. Premature termination codons TGA, TAG and TAA are all editable by re-programming to tryptophan (TGG) codons with A > I editing. For Trp > Ter mutations this would restore the normal amino acid sequence, while in other cases installing a tryptophan rather than the native amino acid may not unduly affect the protein function and allow for restoration of protein expression and activity. As guanosines occupy the +1 and −1 positions of the 5′ and 3′ canonical splice sites respectively, G > A mutations affecting splice sites can be targeted. Furthermore, the −2 A in the 3′ splice site might be targeted to allow for exon skipping, however this might also be achieved using other strategies such as with dCasRx or antisense oligonucleotides.[95,96]

4.5.2 Delivery and approach

As RNA molecules are transient, any delivery strategy to the retina for repair of mutations will require RNA editing effectors to be continuously expressed or administered. The extensive experience in retinal gene therapy with the safe and stable expression of transgenes from AAV vectors has shown that stable expression of an RNA editor in the retina may be possible, especially given photoreceptors are post-mitotic and amenable to AAV transduction. This makes the retina a promising site for proof-of-principle *in vivo* RNA editing. The BoxB-λN, MS2-MCP and GluR2 systems have all been delivered with AAV vectors, while the truncated REPAIR-Cas13 system and the endogenous editing strategies could be delivered by AAV. Strategies which use chemically modified gRNAs would require these to be delivered separately, however reports that modified antisense oligonucleotides can be delivered to photoreceptors and RNA splicing suggest that this might be feasible.[97]

Studies demonstrating that endogenous ADAR can be harnessed for RNA editing are encouraging. Endogenous strategies do not suffer from concerns regarding immunogenicity that surround exogenous strategies that over-express bacterial or viral derived proteins, although these immunogenic events have yet to be observed. Furthermore, off-targeting rates appear much lower for endogenous editing. Broadly, lower editing rates are observed with endogenous strategies, likely because exogenous strategies can use the hyperactive ADAR mutants such as ADAR(E488Q). For endogenous editing in the retina, further work to understand the expression of ADAR1 and ADAR2 is required as this will influence the choice of endogenous strategies going forward. RNA editing is observed endogenously in the inner retina,[98] however, little is known about photoreceptors and RPE.

4.5.3 Off-target concerns

Although off-target editing remains a concern in RNA editing approaches, the significant advantage of targeting the transcriptome over the genome is that off-target edits are by nature transient, and although they occur, they do not accumulate as they might following DNA editing. The transient nature of the edits also opens the field to future innovations that might allow for additional safety features to be built in. For example, molecular "off-switches" could be installed in RNA editing constructs in case of adverse events, or mechanisms might be employed to adjust RNA editing rates, for example by controlling the expression of the gRNA or effector. It is important to assess and optimize off-target editing, and strategies such as using high specificity ADAR mutants,[78,80] localization of editing to the nucleus,[84,90] and design of guide RNAs to install mismatched guanosines at common editing sites within the gRNA binding region all represent advances that can improve off-target rates.[99]

While RNA editing strategy developments are still in their infancy, it is clear they offer great potential and scope for providing a viable therapeutic intervention in the future for common G > A mutations that lead to inherited retinal diseases.

5. Epigenetic editing

CRISPRi interference (CRISPRi) and CRISPR activation (CRISPRa) are techniques derived from the CRISPR-Cas9 system that are used to down- and up-regulate expression of a target gene by influencing its transcription. Transcription in eukaryotes is determined by interactions between DNA elements and regulatory proteins. Briefly, RNA polymerase II forms a complex with a range of transcription factors, which determine its ability to bind to a gene's promoter. Initiation of transcription is then determined by interactions with distal regulatory elements. Transcription factors contain specific protein domains which decrease or increase gene expression (called repressor and activator domains, respectively) through interactions with other cellular proteins.[100–102]

The CRISPRi and CRISPRa systems typically consist of two components. First, a gRNA which targets a regulatory region of a gene (usually the promoter but distal regulatory regions can also be targeted). Second, a fusion protein of catalytically inactive Cas9 and transcriptional activator or repressor domains.[23] The dCas9 protein binds to DNA *via* the gRNA but cannot cleave DNA as it is an inactive nuclease. Here the repressor or activator domains act as transcription factors, influencing the transcription rate by triggering native regulatory pathways. When targeting the promoter, the optimal gRNA binding site relative to the transcriptional start site is roughly −50 to +250 bp for CRISPRi and −40 to +400 bp for CRISPRa.[103,104]

While dCas9 alone can be used to drive CRISPRi, fusion of a repressor domain significantly improves knock down. One of the most commonly used repressors is the Krüppel-associated box (KRAB) repressor which alters the histone methylation state of the gene locus.[105] Eukaryotic DNA is wound around histone proteins that can be chemically modified, influencing how closely they associate.[106] The KRAB repressor interacts with histone modifying proteins to create a tightly wound histone state which is inaccessible to RNA polymerase II and therefore reduces gene transcription. dCas9-KRAB has driven clinically significant gene repression *in vivo* in a range of disease models, including retinitis pigmentosa, familial hypercholesterolemia, and obesity.[107–109]

Before CRISPRa, increasing expression of a target gene relied on gene augmentation: delivering the gene of interest into the cell as a transgene. Overexpression of transgenes by gene augmentation may be associated with toxic effects to retinal cells.[110–112] The level of gene activation with CRISPRa has the potential to be tailored by varying the number of activators recruited to a particular site, ensuring the level of expression is non-toxic.[113,114] CRISPRa systems are varied but most involve the transcriptional activator VP64. VP64 is a fusion of four copies of the activator domain of herpes simplex virus transcription factor VP16. Initially, this was fused to dCas9 to create dCas9-VP64 but this drove only modest levels of gene activation.[115,116] Increasing gene activation requires the recruitment of multiple activators to a single gene promoter. This is achieved through the delivery of multiple gRNAs or by creating more complex fusion proteins. A comparison of CRISPRa constructs in 2016 revealed that the SunTag, VPR, and SAM systems typically drive the highest activation.[117] In the SunTag system, dCas9 is fused with a peptide chain containing 10 binding sites for the single-chain fragment variable(scFv)-GCN4 scFv intrabody, and VP64 is fused with scFv-GCN4. This allows the recruitment of multiple VP64 copies to a promoter with a single gRNA.[118] The dCas9-VPR system uses a quadruple fusion protein of dCas9 to three transcriptional activators: VP64, p65, and Rta.[117] The SAM system delivers dCas9-VP64 and a modified gRNA which contains a binding site for the fusion protein MPH. MPH includes activator domains p65 and heat shock factor 1, allowing the recruitment of these plus VP64 to the target site.[96] Finally, dCas9 can be fused to proteins to deliver heritable changes to gene activity. These strategies can include modifying the methylation status of promoters, or modifying chemical tags on histone molecules.[119,120]

5.1 Application to retinal disease: Targeting pathogenic genes in dominant diseases

CRISPRi can be used to suppress the expression of the pathogenic allele, thus reducing the level of the mutant protein.[121,122] This is a potentially useful therapy for gain-of-function mutations and dominant-negative mutations. For allele-specific disruption, CRISPRi must be targeted to a unique region present only on the mutant allele, such as the pathogenic mutation or a single nucleotide polymorphism. This will create a mismatch between the gRNA and the wild-type DNA allele which may prevent binding. As CRISPRi is highly sensitive to mismatches between its gRNA sequence and the target DNA sequence, even a single base-pair mismatch may strongly affect CRISPRi activity.[103,123] Allele-specific disruption of the pathogenic gene *Rho* (using CRISPR-Cas9 gene disruption) has been shown to improve clinical outcomes in two models of autosomal dominant retinitis pigmentosa.[26,41]

Dominant-negative mutations and haploinsufficiency mutations cause disease due to a disruption and deficiency of wild-type protein function, respectively. Over-expression of the wild-type allele through gene augmentation has shown a therapeutic effect in treating retinitis pigmentosa caused by the dominant-negative P23H mutation in rhodopsin.[124] CRISPRa can be used in a similar way to upregulate the expression of the wild-type allele. This was demonstrated with dCas9-VP64 in the

Sim1 and *Mc4r* heterozygous knock out mouse models of obesity. *SIM1* and *MC4R* are involved in regulation of energy homeostasis, and haploinsufficiency mutations in these genes are common causes of monogenic obesity in humans. The AAV.dCas9-VP64 system was targeted to either *Sim1* or *Mc4r* and delivered *via* stereotactic injection. There was a significant increase in target gene expression in the brain, and body weights of the treated animals were significantly reduced, indicating a rescued phenotype.[125] Use of CRISPRa systems in the retina have yet to be published, but this recent study indicates the potential of the dCas9-VP64 system.

5.2 Targeting disease pathways

Targeting the pathogenic gene directly is not always an ideal strategy. Some diseases, such as age-related macular degeneration (AMD), are polygenic, meaning that the accumulative effect of multiple genes determines an individual's risk factor.[126] Within monogenic diseases, many retinal dystrophies can be caused by a number of genes, for example, mutations in 63 genes are known to cause retinitis pigmentosa (https://sph.uth.edu/retnet/home.htm). Rarer mutations can be limited to a single family, and therefore the pathogenic mutation is not a practical target.[127] Finally, despite genome sequencing, many patient's pathogenic gene remains unknown. For these reasons, it can be more useful to target genes involved in the pathogenesis of the disease rather than the pathogenic gene or pathogenic mutation itself.

CRISPRi and CRISPRa can be delivered *via* AAV injection to the target site to function *in situ*, preserving existing retinal cells.[107,108,119,125,128–130] This treatment would be most beneficial in early-stage patients prior to severe retinal degeneration. CRISPRi has demonstrated success in treating the *Rd10* mouse model of autosomal recessive retinitis pigmentosa.[107] In retinitis pigmentosa, mutations in rod cell-specific genes cause rod cell death. The authors used a dual AAV-delivered dCas9-KRAB to repress *Nrl*, the transcription factor dictating a rod cell phenotype in photoreceptors. The treated rod cells developed a more cone-like phenotype and were therefore less susceptible to the pathogenic mutation. Treated mice had an increased photoreceptor cell layer and visual acuity.[107] CRISPRa has also been delivered *in vivo* to improve outcomes in a range of diseases, including cancers, epilepsy, kidney damage, type I diabetes, and Duchenne muscular dystrophy. In all cases there was an increased expression of the target gene and subsequently improved measures of health in the treated animals.[125,129–131]

CRISPRi and CRISPRa have also shown promise in modifying expression of key genes in stem cell-derived cells before autologous transplantation. This treatment has potential benefits for late stage patients by converting induced pluripotent stem cells (iPSCs) into retinal progenitor cells, photoreceptors or RPE, for surgical transplantation into the subretinal space to replace the lost tissue.[35,37,38] Although not yet applied in retinal diseases, CRISPRi and CRISPRa have been used simultaneously in transplanted stem cell-derived engineered cartilage to treat skull injury.[132] This CRISPRai system is similar to the SAM system in that it uses modified gRNAs to recruit activator and repressor domains. The CRISPRai system was delivered to bone marrow-derived stem cells *via* baculovirus transduction where they inhibited PPAR-γ and activated Sox9, which are involved in the inhibition and promotion of cartilage formation, respectively. Following implantation into calvarial bone defects in rats, the treated cartilage grew over 23–29% of the defect, whereas the untreated cartilage showed negligible growth.[132]

5.3 Comparisons to other methods

The most common methods of reducing expression of a gene are RNA interference (RNAi), CRISPR-Cas9 gene disruption, and CRISPRi. In RNAi, a short double stranded piece of RNA is designed which is homologous, and binds to, the mRNA of the target gene. This triggers the native RNAi pathway ultimately leading to mRNA degradation or decreased translation.[133] CRISPR-Cas9 introduces indels into the target gene coding sequencing, creating non-functional mRNA and therefore reducing protein production. The optimal target site is distinct between these techniques: regulatory DNA regions for CRISPRi, coding DNA for CRISPR-Cas9, and mRNA for RNAi. This means a given target site may be more suited to one technique over another.

While RNAi and CRISPRi have comparable repression rates,[123,134] CRISPR-Cas9 typically outperforms CRISPRi, although this is highly variable across targets.[135–137] The main advantage of the CRISPRi system is its high specificity: multiple CRISPRi studies have reported no off-target effects detectable with RNA-seq.[103,105,123,136,138] RNAi on the other hand, commonly suffers from off-target effects.[134,135,138,139] CRISPR-Cas9 has variable specificity, with indel activity often reported at unexpected sites with a 5 bp discrepancy between the gRNA and DNA sequence.[23,140–143] This has been improved with the development of "high fidelity" Cas9 species, which are engineered to reduce off-target activity, and in some cases have reported no detectable off target effects.[144,145] Another advantage of CRISPRi over CRISPR-Cas9 is its reversibility. Gene repression by CRISPRi is only possible while dCas9 and gRNA are present. By expressing these from inducible promoters, the effects of CRISPRi can be controlled in a time, and dose-dependent manner, although these tools are still early in their development and substantial optimisations are required before their adaptation in a clinically relevant manner.[103,135]

5.4 Challenges

The first challenge for CRISPRi and CRISPRa is identification of an efficient target site. Both CRISPRi and CRISPRa efficiency are affected by two factors: the first is the binding of the gRNA to the DNA, and the second is the ability of the binding site to influence transcription rates.[103] gRNA binding efficiency has been studied at length and various bioinformatics programmes exist to identify the most promising gRNAs.[146] Identifying binding sites that can influence transcription rates is more challenging. CRISPRi and CRISPRa are often most effective when bound to the promoter of a gene, and large screens have identified regions most likely to produce knock down.[103,104] As promoter regions are often mis-annotated, one study found that using the FANTOM5/CAGE database to annotate regulatory elements significantly improved CRISPRi-mediated gene repression.[104] Even when correctly annotated, gRNAs in the optimal target regions can have vastly different effects, with one study finding only 54% of gRNAs in the CRISPRi target region produced moderate gene repression.[104] CRISPRa can also be targeted to distal enhancer elements, with one paper targeting an enhancer 270 kB from the coding region.[125] These enhancers are not included in the recommended target region and their effectiveness is likely to vary greatly between genes.

The second challenge of this technology is the size of the constructs. Cas9 is a relatively large protein, with the most commonly used variant, SpCas9, at 4104 bp.[147] As CRISPRi and CRISPRa involve the fusion of additional domains, their total size frequently exceeds the packaging constrains of AAV. This is particularly an issue for CRISPRa as it can require delivery of multiple, large elements. To overcome delivery limitations, some studies have adopted a dual vector approach, delivering the components in separate vectors, but this has added complications for dosing.[125,131,147] Multiple shorter variants of Cas9 have been identified which may be packaged into a single AAV for CRISPRi. In particular, Cas9 species from *Staphylococcus aureus*, *Streptococcus thermophiles*, *Neisseria meningitides*, and *Campylobacter jejuni* are all under 3.5 kb, and have been used in CRISPRi or CRISPRa experiments.[107,125,148–151]

6. Toward clinical use of CRISPR genome editing for the treatment of retinal disease

With an FDA-approved gene therapy treatment for *RPE65*-associated Leber congenital amaurosis already available,[152] similar gene augmentation therapies for other inherited retinal diseases, such as choroideremia[9] and *RPGR*-related retinitis pigmentosa,[10] are fast approaching clinical approval. However, these AAV vector-based gene augmentation therapies are generally limited by the size of the transgene and to the treatment of autosomal recessive and X-linked diseases. Thus, genome editing systems such as CRISPR-Cas technology, capable of directly correcting disease-causing mutations in the human genome and of targeted knock-down of disease alleles, show great potential as a complementary tool for the treatment of these blinding conditions. In addition, introduction of DNA base editing, prime editing and mRNA editing methodologies is expected to make gene editing safer and adaptable for a broader range of genetic pathologies, including applications in both dividing and terminally differentiated cell types. In principle, over 25% of pathogenic SNPs may be editable by targeting the four transition mutations, with prime editing poised to correct up to 89% of known genetic variants associated with human disease.[44,46] For retinal degenerative disease, gene editing techniques could be particularly well suited for the correction of large genes and treatment of the dominant diseases requiring silencing or ablation of the defective gene in addition to AAV gene augmentation therapy (Fig. 2).

The expanding range of gene editing systems, their versatility and adaptability make them attractive for clinical development pathways, especially given the heterogeneous nature of inherited retinal diseases. With relatively small changes to the guide RNA sequence, while keeping the rest of the medicinal product constant, a large number of pathogenic mutations may be treated. However, the sequence-specific nature of guide RNAs will require extensive safety and efficacy pre-clinical evaluations in human tissue (*ex-vivo* human retinal organoids or retinal explants) and *in vivo* with non-human primate (NHP) models. Nonetheless, this rapidly developing field has already resulted in the first potential therapeutic for retinal disease,

EDIT-101 (Editas Medicine Inc., Cambridge, MA, USA), which has now entered phase I clinical trial (ClinicalTrials.gov, NCT03872479). This is for the retinal condition Leber congenital amaurosis type 10 and specifically the form associated with a deep-intronic c.2991 + 1655A > G single nucleotide transition in the *CEP290* gene. This mutation, the most common in *CEP290*, accounts for 10% of patients with Leber congenital amaurosis and creates a cryptic splice donor site, which leads to the insertion of an aberrant pseudoexon (exon X) containing a premature stop-codon between exon 26 and 27. The resultant aberrant splicing in *CEP290*, which plays a critical role in photoreceptor ciliogenesis, leads to blindness in early infancy. The EDIT-101 therapeutic uses a CRISPR-SaCas9 editing system to excise a segment of intron 26 that contains the mutation thus restoring normal splicing between exon 26 and exon 27. FDA-approval of the clinical trial followed pre-clinical testing in a mouse model which demonstrated efficacious genome editing reaching the therapeutic threshold of 10%.[28] In addition, guide RNA expression was maintained for the duration of the study (40 weeks), the Cas9 expression driven by the human rhodopsin kinase promoter was photoreceptor-specific and there was limited drug-induced immunogenicity. The study was translatable to non-human primates, demonstrating the therapeutic threshold without evidence of significant off-target effects. These encouraging results are likely to pave the way for other gene editing systems to enter human trials and the results of the EDIT-101 trial are eagerly awaited. However, despite the encouraging pre-clinical data, there are still concerns and risks associated with introducing CRISPR-Cas systems into humans. The ongoing EDIT-101 trial will enlighten some of these concerns, while others continue to require pre-clinical investigation. Of particular relevance are the risks of immune responses to the expressed Cas proteins, the rates of on-target efficacy and the consequences of undesired editing in the retina.

6.1 Immunogenic concerns

Being of bacterial origin, expression of CRISPR-Cas components in the eye raises the possibility of intraocular immune responses, which could lead to cell immune cell infiltration, structural damage and reduced clinical efficacy. Pre-existing antibodies to the current most commonly used SaCas9 and SpCas9 have been detected in 78% and 58% of humans, respectively[153,154] as well as NHPs.[28] Despite this, no significant inflammatory response was observed in NHPs following subretinal delivery of AAV5 vectors carrying a CRISPR-SaCas9 transgene for editing of *GUCY2D*, mutations in which cause cone-rod dystrophy.[43] It was observed that levels of AAV neutralizing antibodies and Cas9-induced T-cell responses correlated with the number of subretinal blebs performed, yet no significant intraocular inflammation was observed in treated animals. Interestingly, the NHP with the greatest pre-existing T-cell response to SaCas9 provided the highest achieved rate of *GUCY2D* editing. A similar finding was observed in the pre-clinical NHP study performed using the EDIT-101 AAV5 vector.[28]

The absence of significant signs of inflammation despite a pre-existing response to Cas9 is encouraging, but should nevertheless be taken cautiously due to the timeframes of the studies. Testing in humans will continue for more extended periods and negative consequences may not be evident in a short term follow up. It will be important for any clinical trial to assess each patient for pre-existing antibodies to the CRISPR-Cas system being administered in order to best anticipate and interpret subsequent events.

6.2 Undesired genome editing

The development of CRISPR-Cas9 genome editing tools has simultaneously created great excitement in addition to many technical and ethical issues, which are yet to be resolved. For some concerns, the consequences of use in humans remain unknown and will be enlightened by current trials, but others can be anticipated, and the technology is rapidly evolving to create safer editing tools. These include the systems discussed throughout this chapter that do not directly alter the genome as well as strategies to control Cas9 activity and mitigate the risks of off-target editing. RNA editing rather than DNA editing may circumvent some risks associated with genome editing as off-target events would not result in permanent mutations. Base and prime editing approaches also provide potentially lower risks than Cas9 genome editing. Compared with base editing, prime editing appears to be associated with lower off-target mutagenesis in human cell lines, although additional concerns may emerge as the system is studied further. Indeed, due to the infancy of CRISPR reagents, additional research is needed to further characterize and improve base and prime editing in a broad range of cell types and organisms before the full potential of these platforms can be realized for clinical use.

While there is concern about the lack of control of on-target editing with CRISPR-Cas systems, particularly those that employ NHEJ mechanisms, the control of off-target editing is also an important consideration.

Off-target mutations have the potential to alter expression of other genes, disrupt tumor suppressor genes, cause chromosomal translocations or integration of the viral vector genome.[143,155] Prior to human trials, the potential off-target activities of gene-editing therapies must be assessed in a variety of primate-specific model systems, ranging from *in silico* predictions, to human cell lines, iPSC-derived retinal organoids, *ex vivo* tissue and *in vivo* testing in non-human primates. Computer algorithms may be used to predict off-target sites based on the sgRNA, PAM preference, chromatin accessibility, and machine learning from previous experimental data, but their accuracies remain limited.[156] *In vitro* assays for unbiased detection of off-targets include GUIDE-seq (genome-wide unbiased identification of DSBs enabled by sequencing),[157] Digenome-Seq,[158] and CIRCLE-Seq (circularization for *in vitro* reporting of cleavage effects by sequencing).[159] Assessment of off-target mutations *in vivo* is more challenging and relies on DISCOVER-Seq (discovery of *in situ* Cas off-targets and verification by sequencing). This approach utilized chromatin immunoprecipitation sequencing (ChIP-Seq) to identify the positions of Cas9-cut DNA ends bound by MRE11, a DNA repair factor.[160] Importantly, DISCOVER-Seq could only detect DSBs at the time of the experiment and is not suitable for assessing off-targets in base editing or prime editing therapies. For instance, cytidine or adenine deaminases within DNA and RNA base editing systems could potentially cause off-target deamination of bystander C or A within the edit window or elsewhere.[161] The reverse transcriptase within prime editing systems may create off-target mutations by extending a short way into the pegRNA scaffold beyond its primer sequence.[46] In order to directly assess the accumulated mutations following genome editing therapies, targeted, exome or whole genome sequencing needs to be performed on treated *versus* untreated tissues. In this respect, pre-clinical assessments of CRISPR genome engineering systems in the eye are fortunate to have the paired eye available as an untreated control.

6.3 Improving safety of CRISPR-Cas systems

While genotoxicity may be reduced by the use of high-fidelity Cas variants, even a very low level of off-target activity may become significant if the Cas expression is sustained *in vivo*. Several studies have described Cas9 variants evolved to increase the targeting specificity.[162–164] In addition to engineering approaches of the Cas9 protein, Kocak and colleagues demonstrated that engineering a hairpin secondary structure onto the sgRNA guiding sequence can enhance the targeting specificity by several orders of magnitude.[165] Finally, increasing or decreasing the length of the sgRNA guiding sequence by a few base pairs can enhance the targeting specificity.[166,167] Appropriate and thorough optimisation of construct design should therefore be employed throughout pre-clinical testing phases of a given CRISPR-Cas system, as considered in the relevant sections of this chapter.

Temporal and spatial control of Cas9 endonuclease expression *in vivo* maybe achieved *via* a number of strategies. Viral vectors, especially AAV, remain the most attractive mode of delivering CRISPR-Cas systems to the retina due to their proven clinical efficacy and safety.[168] Most retinal gene augmentation therapies use ubiquitous (*e.g.*, chicken beta-actin) or tissue-specific promoters (*e.g.*, human rhodopsin kinase promoter) to drive transgene expression within target retinal cell groups. However, with the shift in clinical need from long-term transgene expression to transient Cas9/sgRNA expression, the choice of promoters in viral vector constructs needs to be reconsidered. Temporal control of Cas9 expression may be achieved using drug-inducible promoters. For instance, the steroid-inducible *GRE5* promoter may be used to drive Cas9 expression as corticosteroid drugs are readily-available for ocular use.[169] However, basal levels of corticosteroids in the body may cause leaky expression and thus low level off-target activity, and it is possible that the increased ocular pressure caused by these steroids would limit use in patients with retinitis. The doxycycline (or tetracycline)-inducible Tet-ON expression system may afford tighter control of Cas9 expression.[170–172] Alternatively, temporal control may be achieved through protein perturbation techniques. Insertion of a 4-hydroxytamoxifen (4-HT)-responsive intein domain within Cas9 has been shown to enable internal protein splicing to generate active Cas9 in the presence of tamoxifen.[173] Fusion of the estrogen receptor-α (hERα) ligand binding domain or the destabilization domain of *E. coli* dihydrofolate reductase (DHFR-DD) to Cas9 could enable inducible Cas9 stabilization by tamoxifen or trimethoprim, respectively.[174,175] Of the exogenous regulation systems covered, tetracycline-regulated systems have been refined sufficiently such that they are on the cusp of being applied clinically. While low basal transgene expression is of concern, in a clinical setting these levels may not be sufficient to elicit a phenotypic response.[176] One of the drawbacks of modifying the Cas nuclease is often reduced editing efficiency. In addition, the introduction of inducible transcriptional regulatory elements, intein or fusion domains on the whole results in a construct that exceeds the capacity of a single AAV vector.

One approach to temporal control with minimal increase in construct size is a self-destructing CRISPR system, which contains a self-targeting sgRNA.[177–179] The Self-Limiting Cas9 circuit for Enhanced Safety and specificity (SLiCES) system consists of an expression SpCas9, a self-targeting sgRNA and a second sgRNA targeting a chosen genomic locus. Upon delivery into target cells using lentivirus, the system exploits the efficiency of viral-based delivery and simultaneously limiting its expression post transduction through SpCas9 inactivation.[179] It remains to be seen whether such a self-limiting system could provide complete inactivation of Cas9 activity *in vivo* and how it compares against transcriptional or post-translational regulatory systems.

Finally, non-viral delivery of CRISPR-Cas elements could provide transient editing activity, avoid accumulation of off-target effects, and deliver bulky constructs. These are considered further below.

6.4 Delivery of CRISPR reagents to the retina

One of the major challenges facing clinical implementation of CRISPR-Cas gene editing technologies is the safe and efficient intraocular and intracellular delivery of gene editing system components. Gene delivery systems are typically classified as viral or non-viral systems where the latter can be further divided into physical and chemical approaches.[180] While AAV is the viral vector of choice for current retinal delivery strategies, some of the safety concerns relating to immunogenic risks and undesired genome editing may be possible to mitigate by use of non-viral delivery methods. Potential advantages of using non-viral delivery systems is their transient nature (thus relative biosafety), lower cost and ease of production. However, non-viral carriers encapsulating gene editing tools must overcome extracellular barriers, be endocytosed by the cells of interest, unpackage the payload in the cytoplasm or nucleus to achieve their objectives. So far, most non-viral gene delivery approaches have demonstrated low efficiency and thus low transient expression of the transgenes. Access to the photoreceptor cell bodies from the subretinal space by transfection is likely to be particularly challenging due to the presence of stacks of outer segment discs (Fig. 1).

Chemical and physical methods for delivering CRISPR-Cas9 components into target tissues are under investigation, including lipid/gold/polymer/exosome nanoparticles, electroporation and cell-penetrating peptides (CPP).[181] Delivery of Cas9-sgRNA ribonucleoproteins either in suspension or *via* synthetic gold or lipid nanoparticles, which naturally degrade after approximately 3 days, could provide transient genome editing while minimizing off-target effects. Notably, cationic lipid nanoparticles have been chemically engineered to deliver CRISPR-Cas9 ribonucleoproteins (RNPs) via intravenous administration to elicit multiplex editing of lung and liver tissues in mouse models.[182] Furthermore, the addition of multiple NLS sequences to SpCas9 facilitates penetration of RNP proteins in suspension to edit neurons when injected to the brain.[183] PEGylated nanoparticles based on a cationic α-helical polypeptide have been used to deliver Cas9 plasmids and sgRNA into mouse tumor tissue to induce 35% target gene deletion in vivo.[155] While CRISPR-Cas systems have yet to be reported as being successfully delivered by non-viral delivery to outer retinal cells in vivo, use of such methods for delivering genetic material has been tested in a range of settings.[184] Johnson and colleagues described the use of a novel peptide delivery for small and large molecules across the plasma membrane of ocular tissue.[185] The authors delivered fluorophores, siRNA, DNA and quantum dots, achieving robust silencing of transgene expression and penetration of the RPE, photoreceptors and ganglion cells.[185] Similarly, intravitreal delivery of fluorescently labeled peptides were found to localize to the photoreceptors and RPE.[186] Other studies have used compacted DNA nanoparticles (NPs) formulated with polyethylene glycol-substituted polylysine (CK30PEG) to subretinally deliver the *Rds* coding sequence into mice carrying a haploinsufficiency mutation in the retinal degeneration slow *(rds$^{+/-}$)* gene. Elevated mRNA levels were demonstrated at the latest time point examined (PI-120) and modest but statistically significant improvements in rod function were observed.[187] Similar approaches have been shown to mediate improved structural changes in the retina of several ocular disease models, including the *Abca4* knockout model of Stargardt disease and the rhodopsin knockout and P23H models of rhodopsin associated RP.[188–191]

Recently, it was demonstrated that gold nanoparticles conjugated to DNA and complexed with cationic endosomal disruptive polymers could deliver both Cas9 and Cpf1 ribonucleoprotein and donor DNA into a wide variety of cell types.[192,193] The authors demonstrated efficient correction of a DNA mutation that causes Duchenne muscular dystrophy and reduction of local mGluR5 levels in the striatum. A similar approach may present an attractive approach for the delivery of CRISPR-genome editing reagents to the retina.

For CRISPR-Cas systems that rely on fusion proteins consisting of a Cas protein plus additional effector elements, development of smaller variants could be critical to the future translation of such systems into human trials. Future research directions are likely to focus on strategies that maximize the cargo capacity of AAV, including increasing efficacy of dual vector approaches to deliver the necessary genetic cargo. Minicircles offer an alternative option for transgene delivery, these are plasmid derivatives devoid of bacterial backbone elements and are known to increase transfection efficiencies likely due to their smaller size and, thus, are more adept to achieve therapeutic gene expression.[194–196] Minicircles have proven to be a reliable tool for efficient transgene expression in eukaryotic cells both *in vitro* and *in vivo*, as well as for *ex vivo* modification in a range of cell and gene therapy applications, including in retinal cells.[196,197] In transfected photoreceptor cells, survival of minicircle vectors extended to 3 months post-injection,[197] they therefore represent a further option for achieving retinal gene editing by delivery of CRISPR-Cas coding transgenes.

7. Challenges and future perspectives

Since the foundational and subsequently Nobel-prize winning report of the first programmable genome editing CRISPR-Cas system was published in 2012, many areas of research have been revolutionized by the potential of the molecular system. With a retinal disease having the first approved gene therapy treatment and a growing body of data confirming safety and efficacy from multiple clinical trials, it was perhaps inevitable that one of the first direct uses of a CRISPR-Cas genome engineering system would be for the treatment of an inherited retinal disease. The success of retinal gene therapy clinical trials over the past decade was always likely to be curbed by the diversity of inherited retinal diseases, which are genetically highly heterogenous. Gene augmentation strategies are, by their nature, largely limited to treating autosomal recessive and X-linked, single gene disorders. The advent of CRISPR has dramatically altered future treatment options for the numerous forms of inherited retinal disease not treatable by other means. In this chapter, we have discussed the current scope of CRISPR systems available and their varying applications for the treatment of inherited retinal diseases (Table 1), which is dependent on the nature of the causative mutation. This is a rapidly developing field and, importantly, there are still many challenges to consider, not least the impact of expressing these bacterial molecular systems in the human retina and the potential consequences of undesired genome editing activity. Progress towards these challenges appears to be occurring at an unprecedented rate and with a CRISPR

Table 1 A summary of CRISPR studies relevant to inherited retinal disease to date.

Genome editing strategy	Delivery method	Model	Gene Target	Relevant mutation	Model of inheritance retinal disease	Summarized outcomes	References
Genome editing							
SpCas9: HDR	Nucleofection	Patient-derived iPSCs	*MAK*	c.1513ins353	Autosomal recessive RP	Restoration of expression of the wild-type allele	38
SpCas9: HDR	Plasmid transfection	Patient-derived iPSCs	*NR2F3*	c.119-2A>C; c.219G>C (p.Arg73Ser)	Enhanced S-cone syndrome	Restoration of expression of the wild-type allele	36
SpCas9: HDR	Plasmid electroporation	*rd10* mice	*Pde6b*	c.1678C>T (p.Arg560Cys)	Autosomal recessive RP	Correction of mutation with improved visual responses up to 3 months post-treatment	27
SpCas9/RecA: HDR	Plasmid electroporation	*rd1* mice	*Pde6b*	c.1041C>A (p.Tyr347ter)	Autosomal recessive RP	Correction of mutation promoted photoreceptor cell survival and improved visual responses	24
SpCas9: HDR	Plasmid transfection	Patient-derived iPSCs	*RPGR*	c.3070G>T (p.Glu1024X)	X-linked RP	13% correction of the mutant allele by conversion to wild type	35
SpCas9: HDR	Plasmid electroporation	Patient-derived iPSCs and retinal organoids	*RPGR*	c.1685_1686delAT c.2234_2235delGA	X-linked RP	Correction of mutation in iPSCs with development into retinal organoids revealing correction of photoreceptor defects	37

SpCas9 & SaCas9: NHEJ & HDR	Nucleofection & electroporation	Patient-derived iPSCs	CEP290	c.2991 +1665A>G	Leber congenital amaurosis	Restoration of expression of the wild-type allele	38
SaCas9: NHEJ	AAV5	Retinal explants, IVS26 knock-in mice and non-human primates	CEP290	c.2991 +1665A>G	Leber congenital amaurosis	16% functional editing in retinal explants, 21% in knock-in mice, 28% in non-human primates with evidence of safety and efficacy	28
SaCas9: NHEJ	AAV5	GC knockout mice and macaques	Gucy2e/ GUCY2D	Wild type	Autosomal dominant CORD	Editing achieved in 44% of transduced mouse photoreceptor cells and 6–20% in macaques	43
SaCas9 & SpCas9: NHEJ & HDR	Transfection & AAV5	Patient-derived iPSCs & P23H knock-in pigs	RHO	c.163C>A (Pro23His)	Autosomal dominant RP	Repression of the mutant variant	38
SpCas9: NHEJ	Plasmid electroporation	Rho^{S334} rat	Rho	S334ter	Autosomal dominant RP	Mutant allele-specific disruption with subsequent prevention in retinal degeneration evident by 9-fold increase in photoreceptor nuclei	41
SpCas9-VQR: NHEJ	Plasmid electroporation	P23H knock-in mice	Rho	c.163C>A (Pro23His)	Autosomal dominant RP	Mutant allele-specific disruption with a subsequent 20% increase in wild-type mRNA with delayed retinal degeneration	26

Continued

Table 1 A summary of CRISPR studies relevant to inherited retinal disease to date.—cont'd

Genome editing strategy	Delivery method	Model	Gene Target	Relevant mutation	Model of inheritance retinal disease	Summarized outcomes	References
SpCas9: NHEJ	Plasmid electroporation	P23H knock-in mice	*Rho*	*c.163C>A (Pro23His)*	Autosomal dominant RP	Reduction in mutant protein	198
SpCas9: NHEJ & supplementation	Dual AAV8 Y733F	D190N knock-in mice P23H knock-in mice	*Rho*	*c.163C>A (Pro23His) c.568G>A (Asp190Asn)*	Autosomal dominant RP	Mutation-specific gene destruction with provision of wild-type gene for functional protein and improved retinal function	42
SpCas9-VQR: NHEJ	Plasmid electroporation & AAV9-PHP·B	P23H knock-in mice	*Rho*	*c.163C>A (Pro23His)*	Autosomal dominant RP	10% editing of photoreceptor cells achieved *in vivo*	25
CjCas9 - NHEJ	AAV9	Wild-type mice	*Vegfa & Hif1a*	*Wild type*	Choroidal neovascularisation	Knockdown of *Vegfa* and *Hif1a* with associated reduction in the area of laser induced choroidal neovascularisation	148
SpCas9 - HITI	Dual AAV8 or AAV9	Royal College of surgeons rat	*Mertk*	*1.9 kb mutation*	Autosomal recessive RP	Expression of Mertk with associated improved photoreceptor layer thickening and visual responses	40

Base editing					
SpCas9-CBE: TARGET-AID	Plasmid transfection	HEK293A-CFH	*CFH*	*c.1204C>T (Tyr402His)*	Age-related macular degeneration	Cytosine to thymine correction in 21% of sequencing reads with minimal off target edits	199
SpCas9-ABE	Lentivirus	*rd12* mice	*Rpe65*	*c.130C>T (p.R44X)*	Autosomal recessive RP	Mutation correction in 29% of assessed genetic material extracted from treated eyes with associated restoration of visual function	61
Epigenetic editing					
SpCas9-KRAB: CRISPRi	Dual AAV8	*Rho* knockout and *rd10* mice	*Nrl*	*Wild type*	Autosomal recessive RP	Reprogramming of rod photoreceptors to achieve 25% increase in cone preservation with associated improved visual acuity	107

clinical trial for an inherited retinal disease already ongoing, our understanding of the impact of such administration will soon be further advanced. There is no doubt that CRISPR genome engineering in the retina will be at the forefront of future developments and use in humans. Researchers now have the molecular tools to potentially treat patients destined to go blind, which provides an exciting future for scientists, clinicians and, most importantly, patients.

References

1. Cremers FPM, Boon CJF, Bujakowska K, Zeitz C. Special issue introduction: inherited retinal disease: novel candidate genes, genotype-phenotype correlations, and inheritance models. *Genes (Basel)*. 2018;9(4):215.
2. Cehajic-Kapetanovic J, Birtel J, McClements ME, et al. Clinical and molecular characterization of PROM1-related retinal degeneration. *JAMA Netw Open*. 2019;2(6): e195752.
3. Nanda A, McClements ME, Clouston P, Shanks ME, MacLaren RE. The location of exon 4 mutations in RP1 raises challenges for genetic counseling and gene therapy. *Am J Ophthalmol*. 2019;202:23–29.
4. Thompson JA, De Roach JN, McLaren TL, Lamey TM. A mini-review: leber congenital amaurosis: identification of disease-causing variants and personalised therapies. *Adv Exp Med Biol*. 2018;1074:265–271.
5. Jolly JK, Bridge H, MacLaren RE. Outcome measures used in ocular gene therapy trials: a scoping review of current practice. *Front Pharmacol*. 2019;10:1076.
6. Naylor A, Hopkins A, Hudson N, Campbell M. Tight junctions of the outer blood retina barrier. *Int J Mol Sci*. 2019;21(1):211.
7. Bainbridge JWB, Mehat MS, Sundaram V, et al. Long-term effect of gene therapy on Leber's congenital amaurosis. *N Engl J Med*. 2015;372(20):1887–1897.
8. Russell S, Bennett J, Wellman JA, et al. Efficacy and safety of voretigene neparvovec (AAV2-hRPE65v2) in patients with RPE65-mediated inherited retinal dystrophy: a randomised, controlled, open-label, phase 3 trial. *Lancet*. 2017;390(10097):849–860.
9. Xue K, Jolly JK, Barnard AR, et al. Beneficial effects on vision in patients undergoing retinal gene therapy for choroideremia. *Nat Med*. 2018;24(10):1507–1512.
10. Cehajic-Kapetanovic J, Xue K, de la Camara CM, et al. Initial results from a first-in-human gene therapy trial on X-linked retinitis pigmentosa caused by mutations in RPGR. *Nat Med*. 2020;26(3):354–359.
11. Ochakovski GA, Bartz-Schmidt KU, Fischer MD. Retinal gene therapy: surgical vector delivery in the translation to clinical trials. *Front Neurosci*. 2017;11:174.
12. Xue K, Groppe M, Salvetti AP, MacLaren RE. Technique of retinal gene therapy: delivery of viral vector into the subretinal space. *Eye (Lond)*. 2017;31(9):1308–1316.
13. McClements ME, Barnard AR, Singh MS, et al. An AAV dual vector strategy ameliorates the stargardt phenotype in adult Abca4(−/−) mice. *Hum Gene Ther*. 2019;30 (5):590–600.
14. McClements ME, Barnard AR, Charbel Issa P, MacLaren RE. Assessment of AAV dual vector safety in the Abca4−/− mouse model of Stargardt disease. *Transl Vis Sci Technol*. 2020;9(7):20.
15. Binley K, Widdowson P, Loader J, et al. Transduction of photoreceptors with equine infectious anemia virus lentiviral vectors: safety and biodistribution of StarGen for Stargardt disease. *Invest Ophthalmol Vis Sci*. 2013;54(6):4061–4071.

16. Kong J, Kim SR, Binley K, et al. Correction of the disease phenotype in the mouse model of Stargardt disease by lentiviral gene therapy. *Gene Ther*. 2008;15(19):1311–1320.
17. Wheway G, Douglas A, Baralle D, Guillot E. Mutation spectrum of PRPF31, genotype-phenotype correlation in retinitis pigmentosa, and opportunities for therapy. *Exp Eye Res*. 2020;192:107950.
18. Price BA, Sandoval IM, Chan F, et al. Rhodopsin gene expression determines rod outer segment size and rod cell resistance to a dominant-negative neurodegeneration mutant. *PLoS One*. 2012;7(11):e49889.
19. Haeri M, Knox BE. Rhodopsin mutant P23H destabilizes rod photoreceptor disk membranes. *PLoS One*. 2012;7(1):e30101.
20. Orlans HO, Barnard AR, Patrício MI, McClements ME, MacLaren RE. Effect of AAV-mediated rhodopsin gene augmentation on retinal degeneration caused by the dominant P23H rhodopsin mutation in a knock-in murine model. *Hum Gene Ther*. 2020;31(13–14):730–742.
21. Cideciyan AV, Sudharsan R, Dufour VL, et al. Mutation-independent rhodopsin gene therapy by knockdown and replacement with a single AAV vector. *Proc Natl Acad Sci U S A*. 2018;115(36):E8547–e8556.
22. Cremers FPM, Lee W, Collin RWJ, Allikmets R. Clinical spectrum, genetic complexity and therapeutic approaches for retinal disease caused by ABCA4 mutations. *Prog Retin Eye Res*. 2020;79:100861.
23. Jinek M, Chylinski K, Fonfara I, Hauer M, Doudna JA, Charpentier E. A programmable dual-RNA-guided DNA endonuclease in adaptive bacterial immunity. *Science*. 2012;337(6096):816–821.
24. Cai Y, Cheng T, Yao Y, et al. In vivo genome editing rescues photoreceptor degeneration via a Cas9/RecA-mediated homology-directed repair pathway. *Sci Adv*. 2019;5(4):eaav3335.
25. Giannelli SG, Luoni M, Castoldi V, et al. Cas9/sgRNA selective targeting of the P23H rhodopsin mutant allele for treating retinitis pigmentosa by intravitreal AAV9.PHP.B-based delivery. *Hum Mol Genet*. 2018;27(5):761–779.
26. Li P, Kleinstiver BP, Leon MY, et al. Allele-specific CRISPR-Cas9 genome editing of the single-base P23H mutation for rhodopsin-associated dominant retinitis pigmentosa. *CRISPR J*. 2018;1(1):55–64.
27. Vagni P, Perlini LE, Chenais NAL, et al. Gene editing preserves visual functions in a mouse model of retinal degeneration. *Front Neurosci*. 2019;13:945.
28. Maeder ML, Stefanidakis M, Wilson CJ, et al. Development of a gene-editing approach to restore vision loss in Leber congenital amaurosis type 10. *Nat Med*. 2019;25(2):229–233.
29. First CRISPR therapy dosed. *Nat Biotechnol*. 2020;38(4):382.
30. Sander JD, Joung JK. CRISPR-Cas systems for editing, regulating and targeting genomes. *Nat Biotechnol*. 2014;32(4):347–355.
31. Hidalgo-Cantabrana C, Barrangou R. Characterization and applications of type I CRISPR-Cas systems. *Biochem Soc Trans*. 2020;48(1):15–23.
32. Gallagher DN, Haber JE. Repair of a site-specific DNA cleavage: old-school lessons for Cas9-mediated gene editing. *ACS Chem Biol*. 2018;13(2):397–405.
33. Mao Z, Bozzella M, Seluanov A, Gorbunova V. Comparison of nonhomologous end joining and homologous recombination in human cells. *DNA Repair*. 2008;7(10):1765–1771.
34. Nami F, Basiri M, Satarian L, Curtiss C, Baharvand H, Verfaillie C. Strategies for in vivo genome editing in nondividing cells. *Trends Biotechnol*. 2018;36(8):770–786.
35. Bassuk AG, Zheng A, Li Y, Tsang SH, Mahajan VB. Precision medicine: genetic repair of retinitis pigmentosa in patient-derived stem cells. *Sci Rep*. 2016;6:19969.

36. Bohrer LR, Wiley LA, Burnight ER, et al. Correction of NR2E3 associated enhanced S-cone syndrome patient-Specific iPSCs using CRISPR-Cas9. *Genes (Basel)*. 2019; 10(4):278.
37. Deng WL, Gao ML, Lei XL, et al. Gene correction reverses ciliopathy and photoreceptor loss in iPSC-derived retinal organoids from retinitis pigmentosa patients. *Stem Cell Rep*. 2018;10(4):1267–1281.
38. Burnight ER, Gupta M, Wiley LA, et al. Using CRISPR-Cas9 to generate gene-corrected autologous iPSCs for the treatment of inherited retinal degeneration. *Mol Ther*. 2017;25(9):1999–2013.
39. Sanjurjo-Soriano C, Erkilic N, Baux D, et al. Genome editing in patient iPSCs corrects the most prevalent USH2A mutations and reveals intriguing mutant mRNA expression profiles. *Mol Ther Methods Clin Dev*. 2020;17:156–173.
40. Suzuki K, Tsunekawa Y, Hernandez-Benitez R, et al. In vivo genome editing via CRISPR/Cas9 mediated homology-independent targeted integration. *Nature*. 2016;540(7631):144–149.
41. Bakondi B, Lv W, Lu B, et al. In vivo CRISPR/Cas9 gene editing corrects retinal dystrophy in the S334ter-3 rat model of autosomal dominant retinitis pigmentosa. *Mol Ther*. 2016;24(3):556–563.
42. Tsai YT, Wu WH, Lee TT, et al. Clustered regularly interspaced short palindromic repeats-based genome surgery for the treatment of autosomal dominant retinitis pigmentosa. *Ophthalmology*. 2018;125(9):1421–1430.
43. McCullough KT, Boye SL, Fajardo D, et al. Somatic gene editing of GUCY2D by AAV-CRISPR/Cas9 alters retinal structure and function in mouse and macaque. *Hum Gene Ther*. 2019;30(5):571–589.
44. Komor AC, Badran AH, Liu DR. CRISPR-based technologies for the manipulation of eukaryotic genomes. *Cell*. 2017;168(1–2):20–36.
45. Nishida K, Arazoe T, Yachie N, et al. Targeted nucleotide editing using hybrid prokaryotic and vertebrate adaptive immune systems. *Science*. 2016;353(6305):aaf8729.
46. Anzalone AV, Randolph PB, Davis JR, et al. Search-and-replace genome editing without double-strand breaks or donor DNA. *Nature*. 2019;576(7785):149–157.
47. Yeh W-H, Chiang H, Rees HA, Edge AS, Liu DR. In vivo base editing of post-mitotic sensory cells. *Nat Commun*. 2018;9(1):1–10.
48. Chamberlain K, Riyad JM, Weber T. Expressing transgenes that exceed the packaging capacity of adeno-associated virus capsids. *Hum Gene Ther Methods*. 2016; 27(1):1–12.
49. Stone EM, Andorf JL, Whitmore SS, et al. Clinically focused molecular investigation of 1000 consecutive families with inherited retinal disease. *Ophthalmology*. 2017; 124(9):1314–1331.
50. Dryja TP, McGee TL, Hahn LB, et al. Mutations within the rhodopsin gene in patients with autosomal dominant retinitis pigmentosa. *N Engl J Med*. 1990;323(19): 1302–1307.
51. Gaudelli NM, Komor AC, Rees HA, et al. Programmable base editing of A•T to G•C in genomic DNA without DNA cleavage. *Nature*. 2017;551(7681):464–471.
52. Komor AC, Kim YB, Packer MS, Zuris JA, Liu DR. Programmable editing of a target base in genomic DNA without double-stranded DNA cleavage. *Nature*. 2016;533 (7603):420–424.
53. Kantor A, McClements ME, MacLaren RE. CRISPR-Cas9 DNA base-editing and prime-editing. *Int J Mol Sci*. 2020;21(17):6240.
54. Kurt IC, Zhou R, Iyer S, et al. CRISPR C-to-G base editors for inducing targeted DNA transversions in human cells. *Nat Biotechnol*. 2020;39:1–6.
55. Zhang X, Zhu B, Chen L, et al. Dual base editor catalyzes both cytosine and adenine base conversions in human cells. *Nat Biotechnol*. 2020;38:1–5.

56. Sakata RC, Ishiguro S, Mori H, et al. Base editors for simultaneous introduction of C-to-T and A-to-G mutations. *Nat Biotechnol.* 2020;1–5.
57. Grünewald J, Zhou R, Lareau CA, et al. A dual-deaminase CRISPR base editor enables concurrent adenine and cytosine editing. *Nat Biotechnol.* 2020;38:1–4.
58. Tran MTN, Khalid M, Pébay A, et al. Screening of CRISPR/Cas base editors to target the AMD high-risk Y402H complement factor H variant. *Mol Vis.* 2019; 25:174–182.
59. Alsøe L, Sarno A, Carracedo S, et al. Uracil accumulation and mutagenesis dominated by cytosine deamination in CpG dinucleotides in mice lacking UNG and SMUG1. *Sci Rep.* 2017;7(1):1–14.
60. Losey HC, Ruthenburg AJ, Verdine GL. Crystal structure of *Staphylococcus aureus* tRNA adenosine deaminase TadA in complex with RNA. *Nat Struct Mol Biol.* 2006;13(2):153–159.
61. Suh S, Choi EH, Leinonen H, et al. Restoration of visual function in adult mice with an inherited retinal disease via adenine base editing. *Nat Biomed Eng.* 2020.
62. Ryu S-M, Koo T, Kim K, et al. Adenine base editing in mouse embryos and an adult mouse model of Duchenne muscular dystrophy. *Nat Biotechnol.* 2018;36(6):536–539.
63. Lee C, Jo DH, Hwang G-H, et al. CRISPR-pass: gene rescue of nonsense mutations using adenine base editors. *Mol Ther.* 2019;27(8):1364–1371.
64. Gapinske M, Luu A, Winter J, et al. CRISPR-SKIP: programmable gene splicing with single base editors. *Genome Biol.* 2018;19(1):107.
65. Nishikura K. A-to-I editing of coding and non-coding RNAs by ADARs. *Nat Rev Mol Cell Biol.* 2016;17(2):83–96.
66. Fry LE, Peddle CF, Barnard AR, McClements ME, MacLaren RE. RNA editing as a therapeutic approach for retinal gene therapy requiring long coding sequences. *Int J Mol Sci.* 2020;21(3):777.
67. Kim U, Wang Y, Sanford T, Zeng Y, Nishikura K. Molecular cloning of cDNA for double-stranded RNA adenosine deaminase, a candidate enzyme for nuclear RNA editing. *Proc Natl Acad Sci U S A.* 1994;91(24):11457–11461.
68. Desterro JM, Keegan LP, Lafarga M, Berciano MT, O'Connell M, Carmo-Fonseca M. Dynamic association of RNA-editing enzymes with the nucleolus. *J Cell Sci.* 2003;116(Pt. 9):1805–1818.
69. Hartner JC, Walkley CR, Lu J, Orkin SH. ADAR1 is essential for the maintenance of hematopoiesis and suppression of interferon signaling. *Nat Immunol.* 2009;10(1): 109–115.
70. Melcher T, Maas S, Herb A, Sprengel R, Seeburg PH, Higuchi M. A mammalian RNA editing enzyme. *Nature.* 1996;379(6564):460–464.
71. Higuchi M, Maas S, Single FN, et al. Point mutation in an AMPA receptor gene rescues lethality in mice deficient in the RNA-editing enzyme ADAR2. *Nature.* 2000;406 (6791):78–81.
72. Nishikura K, Yoo C, Kim U, et al. Substrate specificity of the dsRNA unwinding/modifying activity. *EMBO J.* 1991;10(11):3523–3532.
73. Higuchi M, Single FN, Köhler M, Sommer B, Sprengel R, Seeburg PH. RNA editing of AMPA receptor subunit GluR-B: a base-paired intron-exon structure determines position and efficiency. *Cell.* 1993;75(7):1361–1370.
74. Eggington JM, Greene T, Bass BL. Predicting sites of ADAR editing in double-stranded RNA. *Nat Commun.* 2011;2:319.
75. Bazak L, Haviv A, Barak M, et al. A-to-I RNA editing occurs at over a hundred million genomic sites, located in a majority of human genes. *Genome Res.* 2014;24(3):365–376.
76. Matthews MM, Thomas JM, Zheng Y, et al. Structures of human ADAR2 bound to dsRNA reveal base-flipping mechanism and basis for site selectivity. *Nat Struct Mol Biol.* 2016;23(5):426–433.

77. Wong SK, Sato S, Lazinski DW. Substrate recognition by ADAR1 and ADAR2. *RNA.* 2001;7(6):846–858.
78. Cox DBT, Gootenberg JS, Abudayyeh OO, et al. RNA editing with CRISPR-Cas13. *Science.* 2017;358(6366):1019–1027.
79. Wang Y, Havel J, Beal PA. A phenotypic screen for functional mutants of human adenosine deaminase acting on RNA 1. *ACS Chem Biol.* 2015;10(11):2512–2519.
80. Abudayyeh OO, Gootenberg JS, Franklin B, et al. A cytosine deaminase for programmable single-base RNA editing. *Science.* 2019;365(6451):382–386.
81. O'Connell MR. Molecular mechanisms of RNA targeting by Cas13-containing type VI CRISPR-Cas systems. *J Mol Biol.* 2019;431(1):66–87.
82. Montiel-Gonzalez MF, Vallecillo-Viejo I, Yudowski GA, Rosenthal JJ. Correction of mutations within the cystic fibrosis transmembrane conductance regulator by site-directed RNA editing. *Proc Natl Acad Sci U S A.* 2013;110(45):18285–18290.
83. Baron-Benhamou J, Gehring NH, Kulozik AE, Hentze MW. Using the lambdaN peptide to tether proteins to RNAs. *Methods Mol Biol.* 2004;257:135–154.
84. Vallecillo-Viejo IC, Liscovitch-Brauer N, Montiel-Gonzalez MF, Eisenberg E, Rosenthal JJC. Abundant off-target edits from site-directed RNA editing can be reduced by nuclear localization of the editing enzyme. *RNA Biol.* 2018;15(1):104–114.
85. Montiel-González MF, Vallecillo-Viejo IC, Rosenthal JJ. An efficient system for selectively altering genetic information within mRNAs. *Nucleic Acids Res.* 2016;44(21):e157.
86. Sinnamon JR, Kim SY, Corson GM, et al. Site-directed RNA repair of endogenous Mecp2 RNA in neurons. *Proc Natl Acad Sci U S A.* 2017;114(44):E9395–e9402.
87. Stafforst T, Schneider MF. An RNA-deaminase conjugate selectively repairs point mutations. *Angew Chem Int Ed Engl.* 2012;51(44):11166–11169.
88. Vogel P, Moschref M, Li Q, et al. Efficient and precise editing of endogenous transcripts with SNAP-tagged ADARs. *Nat Methods.* 2018;15(7):535–538.
89. Vogel P, Schneider MF, Wettengel J, Stafforst T. Improving site-directed RNA editing in vitro and in cell culture by chemical modification of the guideRNA. *Angew Chem Int Ed Engl.* 2014;53(24):6267–6271.
90. Katrekar D, Chen G, Meluzzi D, et al. In vivo RNA editing of point mutations via RNA-guided adenosine deaminases. *Nat Methods.* 2019;16(3):239–242.
91. Azad MTA, Qulsum U, Tsukahara T. Comparative activity of adenosine deaminase acting on RNA (ADARs) isoforms for correction of genetic code in gene therapy. *Curr Gene Ther.* 2019;19(1):31–39.
92. Rauch S, He E, Srienc M, Zhou H, Zhang Z, Dickinson BC. Programmable RNA-guided RNA effector proteins built from human parts. *Cell.* 2019;178(1):122–134.e112.
93. Qu L, Yi Z, Zhu S, et al. Programmable RNA editing by recruiting endogenous ADAR using engineered RNAs. *Nat Biotechnol.* 2019;37(9):1059–1069.
94. Merkle T, Merz S, Reautschnig P, et al. Precise RNA editing by recruiting endogenous ADARs with antisense oligonucleotides. *Nat Biotechnol.* 2019;37(2):133–138.
95. Xue K, MacLaren RE. Antisense oligonucleotide therapeutics in clinical trials for the treatment of inherited retinal diseases. *Expert Opin Investig Drugs.* 2020;29(10):1163–1170.
96. Konermann S, Lotfy P, Brideau NJ, Oki J, Shokhirev MN, Hsu PD. Transcriptome engineering with RNA-targeting type VI-D CRISPR effectors. *Cell.* 2018;173(3):665–676.e614.
97. Cideciyan AV, Jacobson SG, Drack AV, et al. Effect of an intravitreal antisense oligonucleotide on vision in Leber congenital amaurosis due to a photoreceptor cilium defect. *Nat Med.* 2019;25(2):225–228.

98. Wang AL, Carroll RC, Nawy S. Down-regulation of the RNA editing enzyme ADAR2 contributes to RGC death in a mouse model of glaucoma. *PLoS One.* 2014;9(3):e91288.
99. Naeem M, Majeed S, Hoque MZ, Ahmad I. Latest developed strategies to minimize the off-target effects in CRISPR-Cas-mediated genome editing. *Cell.* 2020;9(7):1608.
100. Sainsbury S, Bernecky C, Cramer P. Structural basis of transcription initiation by RNA polymerase II. *Nat Rev Mol Cell Biol.* 2015;16(3):129–143.
101. Yao L, Berman BP, Farnham PJ. Demystifying the secret mission of enhancers: linking distal regulatory elements to target genes. *Crit Rev Biochem Mol Biol.* 2015;50(6):550–573.
102. Heintzman ND, Ren B. Finding distal regulatory elements in the human genome. *Curr Opin Genet Dev.* 2009;19(6):541–549.
103. Gilbert LA, Horlbeck MA, Adamson B, et al. Genome-scale CRISPR-mediated control of gene repression and activation. *Cell.* 2014;159(3):647–661.
104. Radzisheuskaya A, Shlyueva D, Müller I, Helin K. Optimizing sgRNA position markedly improves the efficiency of CRISPR/dCas9-mediated transcriptional repression. *Nucleic Acids Res.* 2016;44(18):e141.
105. Gilbert LA, Larson MH, Morsut L, et al. CRISPR-mediated modular RNA-guided regulation of transcription in eukaryotes. *Cell.* 2013;154(2):442–451.
106. Lawrence M, Daujat S, Schneider R. Lateral thinking: how histone modifications regulate gene expression. *Trends Genet.* 2016;32(1):42–56.
107. Moreno AM, Fu X, Zhu J, et al. In situ gene therapy via AAV-CRISPR-Cas9-mediated targeted gene regulation. *Mol Ther.* 2018;26(7):1818–1827.
108. Chung JY, Ain QU, Song Y, Yong SB, Kim YH. Targeted delivery of CRISPR interference system against Fabp4 to white adipocytes ameliorates obesity, inflammation, hepatic steatosis, and insulin resistance. *Genome Res.* 2019;29(9):1442–1452.
109. Thakore PI, Kwon JB, Nelson CE, et al. RNA-guided transcriptional silencing in vivo with *S. aureus* CRISPR-Cas9 repressors. *Nat Commun.* 2018;9(1):1674.
110. Burnight ER, Wiley LA, Drack AV, et al. CEP290 gene transfer rescues Leber congenital amaurosis cellular phenotype. *Gene Ther.* 2014;21(7):662–672.
111. Seo S, Mullins RF, Dumitrescu AV, et al. Subretinal gene therapy of mice with Bardet-Biedl syndrome type 1. *Invest Ophthalmol Vis Sci.* 2013;54(9):6118–6132.
112. Olsson JE, Gordon JW, Pawlyk BS, et al. Transgenic mice with a rhodopsin mutation (Pro23His): a mouse model of autosomal dominant retinitis pigmentosa. *Neuron.* 1992;9(5):815–830.
113. Hsu MN, Chang YH, Truong VA, Lai PL, Nguyen TKN, Hu YC. CRISPR technologies for stem cell engineering and regenerative medicine. *Biotechnol Adv.* 2019;37(8):107447.
114. Larouche J, Aguilar CA. New technologies to enhance in vivo reprogramming for regenerative medicine. *Trends Biotechnol.* 2019;37(6):604–617.
115. Maeder ML, Linder SJ, Cascio VM, Fu Y, Ho QH, Joung JK. CRISPR RNA-guided activation of endogenous human genes. *Nat Methods.* 2013;10(10):977–979.
116. Balboa D, Weltner J, Eurola S, Trokovic R, Wartiovaara K, Otonkoski T. Conditionally stabilized dCas9 activator for controlling gene expression in human cell reprogramming and differentiation. *Stem Cell Rep.* 2015;5(3):448–459.
117. Chavez A, Tuttle M, Pruitt BW, et al. Comparison of Cas9 activators in multiple species. *Nat Methods.* 2016;13(7):563–567.
118. Tanenbaum ME, Gilbert LA, Qi LS, Weissman JS, Vale RD. A protein-tagging system for signal amplification in gene expression and fluorescence imaging. *Cell.* 2014;159(3):635–646.
119. Lau CH, Suh Y. In vivo epigenome editing and transcriptional modulation using CRISPR technology. *Transgenic Res.* 2018;27(6):489–509.

120. Liu XS, Wu H, Ji X, et al. Editing DNA methylation in the mammalian genome. *Cell.* 2016;167(1):233–247.e217.
121. Peddle CF, Fry LE, McClements ME, MacLaren RE. CRISPR interference-potential application in retinal disease. *Int J Mol Sci.* 2020;21(7):2329.
122. Peddle CF, MacLaren RE. The application of CRISPR/Cas9 for the treatment of retinal diseases. *Yale J Biol Med.* 2017;90(4):533–541.
123. Zheng Y, Shen W, Zhang J, et al. CRISPR interference-based specific and efficient gene inactivation in the brain. *Nat Neurosci.* 2018;21(3):447–454.
124. Mao H, James Jr T, Schwein A, et al. AAV delivery of wild-type rhodopsin preserves retinal function in a mouse model of autosomal dominant retinitis pigmentosa. *Hum Gene Ther.* 2011;22(5):567–575.
125. Matharu N, Rattanasopha S, Tamura S, et al. CRISPR-mediated activation of a promoter or enhancer rescues obesity caused by haploinsufficiency. *Science.* 2019;363(6424):eaau0629.
126. Cooke Bailey JN, Hoffman JD, Sardell RJ, Scott WK, Pericak-Vance MA, Haines JL. The application of genetic risk scores in age-related macular degeneration: a review. *J Clin Med.* 2016;5(3):31.
127. Daiger SP, Bowne SJ, Sullivan LS. Genes and mutations causing autosomal dominant retinitis pigmentosa. *Cold Spring Harb Perspect Med.* 2014;5(10):a017129.
128. Thakore PI, D'Ippolito AM, Song L, et al. Highly specific epigenome editing by CRISPR-Cas9 repressors for silencing of distal regulatory elements. *Nat Methods.* 2015;12(12):1143–1149.
129. Wang G, Chow RD, Bai Z, et al. Multiplexed activation of endogenous genes by CRISPRa elicits potent antitumor immunity. *Nat Immunol.* 2019;20(11):1494–1505.
130. Colasante G, Qiu Y, Massimino L, et al. In vivo CRISPRa decreases seizures and rescues cognitive deficits in a rodent model of epilepsy. *Brain.* 2020;143(3):891–905.
131. Liao HK, Hatanaka F, Araoka T, et al. In vivo target gene activation via CRISPR/Cas9-mediated trans-epigenetic modulation. *Cell.* 2017;171(7):1495–1507.e1415.
132. Truong VA, Hsu MN, Kieu Nguyen NT, et al. CRISPRai for simultaneous gene activation and inhibition to promote stem cell chondrogenesis and calvarial bone regeneration. *Nucleic Acids Res.* 2019;47(13):e74.
133. Bobbin ML, Rossi JJ. RNA interference (RNAi)-based therapeutics: delivering on the promise? *Annu Rev Pharmacol Toxicol.* 2016;56:103–122.
134. Smith I, Greenside PG, Natoli T, et al. Evaluation of RNAi and CRISPR technologies by large-scale gene expression profiling in the connectivity map. *PLoS Biol.* 2017;15(11):e2003213.
135. Qi LS, Larson MH, Gilbert LA, et al. Repurposing CRISPR as an RNA-guided platform for sequence-specific control of gene expression. *Cell.* 2013;152(5):1173–1183.
136. Yeo NC, Chavez A, Lance-Byrne A, et al. An enhanced CRISPR repressor for targeted mammalian gene regulation. *Nat Methods.* 2018;15(8):611–616.
137. Evers B, Jastrzebski K, Heijmans JP, Grernrum W, Beijersbergen RL, Bernards R. CRISPR knockout screening outperforms shRNA and CRISPRi in identifying essential genes. *Nat Biotechnol.* 2016;34(6):631–633.
138. Stojic L, Lun ATL, Mangei J, et al. Specificity of RNAi, LNA and CRISPRi as loss-of-function methods in transcriptional analysis. *Nucleic Acids Res.* 2018;46(12):5950–5966.
139. Svoboda P. Off-targeting and other non-specific effects of RNAi experiments in mammalian cells. *Curr Opin Mol Ther.* 2007;9(3):248–257.
140. Cong L, Ran FA, Cox D, et al. Multiplex genome engineering using CRISPR/Cas systems. *Science.* 2013;339(6121):819–823.
141. Hsu PD, Scott DA, Weinstein JA, et al. DNA targeting specificity of RNA-guided Cas9 nucleases. *Nat Biotechnol.* 2013;31(9):827–832.

142. Jiang W, Bikard D, Cox D, Zhang F, Marraffini LA. RNA-guided editing of bacterial genomes using CRISPR-Cas systems. *Nat Biotechnol.* 2013;31(3):233–239.
143. Fu Y, Foden JA, Khayter C, et al. High-frequency off-target mutagenesis induced by CRISPR-Cas nucleases in human cells. *Nat Biotechnol.* 2013;31(9):822–826.
144. Casini A, Olivieri M, Petris G, et al. A highly specific SpCas9 variant is identified by in vivo screening in yeast. *Nat Biotechnol.* 2018;36(3):265–271.
145. Kleinstiver BP, Pattanayak V, Prew MS, et al. High-fidelity CRISPR-Cas9 nucleases with no detectable genome-wide off-target effects. *Nature.* 2016;529(7587):490–495.
146. Zischewski J, Fischer R, Bortesi L. Detection of on-target and off-target mutations generated by CRISPR/Cas9 and other sequence-specific nucleases. *Biotechnol Adv.* 2017;35(1):95–104.
147. Friedland AE, Baral R, Singhal P, et al. Characterization of *Staphylococcus aureus* Cas9: a smaller Cas9 for all-in-one adeno-associated virus delivery and paired nickase applications. *Genome Biol.* 2015;16:257.
148. Kim E, Koo T, Park SW, et al. In vivo genome editing with a small Cas9 orthologue derived from campylobacter jejuni. *Nat Commun.* 2017;8:14500.
149. Hou Z, Zhang Y, Propson NE, et al. Efficient genome engineering in human pluripotent stem cells using Cas9 from Neisseria meningitidis. *Proc Natl Acad Sci U S A.* 2013;110(39):15644–15649.
150. Murovec J, Pirc Ž, Yang B. New variants of CRISPR RNA-guided genome editing enzymes. *Plant Biotechnol J.* 2017;15(8):917–926.
151. Ran FA, Cong L, Yan WX, et al. In vivo genome editing using *Staphylococcus aureus* Cas9. *Nature.* 2015;520(7546):186–191.
152. FDA approves hereditary blindness gene therapy. *Nat Biotechnol.* 2018;36(1):6.
153. Simhadri VL, McGill J, McMahon S, Wang J, Jiang H, Sauna ZE. Prevalence of pre-existing antibodies to CRISPR-associated nuclease Cas9 in the USA population. *Mol Ther Methods Clin Dev.* 2018;10:105–112.
154. Charlesworth CT, Deshpande PS, Dever DP, et al. Identification of preexisting adaptive immunity to Cas9 proteins in humans. *Nat Med.* 2019;25(2):249–254.
155. Wang HX, Song Z, Lao YH, et al. Nonviral gene editing via CRISPR/Cas9 delivery by membrane-disruptive and endosomolytic helical polypeptide. *Proc Natl Acad Sci U S A.* 2018;115(19):4903–4908.
156. Wang Y, Wang M, Zheng T, et al. Specificity profiling of CRISPR system reveals greatly enhanced off-target gene editing. *Sci Rep.* 2020;10(1):2269.
157. Tsai SQ, Zheng Z, Nguyen NT, et al. GUIDE-seq enables genome-wide profiling of off-target cleavage by CRISPR-Cas nucleases. *Nat Biotechnol.* 2015;33(2):187–197.
158. Kim D, Kim S, Kim S, Park J, Kim JS. Genome-wide target specificities of CRISPR-Cas9 nucleases revealed by multiplex Digenome-seq. *Genome Res.* 2016;26(3):406–415.
159. Tsai SQ, Nguyen NT, Malagon-Lopez J, Topkar VV, Aryee MJ, Joung JK. CIRCLE-seq: a highly sensitive in vitro screen for genome-wide CRISPR-Cas9 nuclease off-targets. *Nat Methods.* 2017;14(6):607–614.
160. Wienert B, Wyman SK, Richardson CD, et al. Unbiased detection of CRISPR off-targets in vivo using DISCOVER-Seq. *Science.* 2019;364(6437):286–289.
161. McGrath E, Shin H, Zhang L, et al. Targeting specificity of APOBEC-based cytosine base editor in human iPSCs determined by whole genome sequencing. *Nat Commun.* 2019;10(1):5353.
162. Vakulskas CA, Dever DP, Rettig GR, et al. A high-fidelity Cas9 mutant delivered as a ribonucleoprotein complex enables efficient gene editing in human hematopoietic stem and progenitor cells. *Nat Med.* 2018;24(8):1216–1224.
163. Lee JK, Jeong E, Lee J, et al. Directed evolution of CRISPR-Cas9 to increase its specificity. *Nat Commun.* 2018;9(1):3048.

164. Kim D, Luk K, Wolfe SA, Kim JS. Evaluating and enhancing target specificity of gene-editing nucleases and deaminases. *Annu Rev Biochem.* 2019;88:191–220.
165. Kocak DD, Josephs EA, Bhandarkar V, Adkar SS, Kwon JB, Gersbach CA. Increasing the specificity of CRISPR systems with engineered RNA secondary structures. *Nat Biotechnol.* 2019;37(6):657–666.
166. Cho SW, Kim S, Kim Y, et al. Analysis of off-target effects of CRISPR/Cas-derived RNA-guided endonucleases and nickases. *Genome Res.* 2014;24(1):132–141.
167. Fu Y, Sander JD, Reyon D, Cascio VM, Joung JK. Improving CRISPR-Cas nuclease specificity using truncated guide RNAs. *Nat Biotechnol.* 2014;32(3):279–284.
168. Nuzbrokh Y, Kassotis AS, Ragi SD, Jauregui R, Tsang SH. Treatment-emergent adverse events in gene therapy trials for inherited retinal diseases: a narrative review. *Ophthalmol Therapy.* 2020;9(4):709–724.
169. Mader S, White JH. A steroid-inducible promoter for the controlled overexpression of cloned genes in eukaryotic cells. *Proc Natl Acad Sci U S A.* 1993;90(12):5603–5607.
170. Gangopadhyay SA, Cox KJ, Manna D, et al. Precision control of CRISPR-Cas9 using small molecules and light. *Biochemistry.* 2019;58(4):234–244.
171. Aubrey BJ, Kelly GL, Kueh AJ, et al. An inducible lentiviral guide RNA platform enables the identification of tumor-essential genes and tumor-promoting mutations in vivo. *Cell Rep.* 2015;10(8):1422–1432.
172. Dow LE, Fisher J, O'Rourke KP, et al. Inducible in vivo genome editing with CRISPR-Cas9. *Nat Biotechnol.* 2015;33(4):390–394.
173. Davis KM, Pattanayak V, Thompson DB, Zuris JA, Liu DR. Small molecule-triggered Cas9 protein with improved genome-editing specificity. *Nat Chem Biol.* 2015;11 (5):316–318.
174. Oakes BL, Nadler DC, Flamholz A, et al. Profiling of engineering hotspots identifies an allosteric CRISPR-Cas9 switch. *Nat Biotechnol.* 2016;34(6):646–651.
175. Zhang J, Chen L, Zhang J, Wang Y. Drug inducible CRISPR/Cas systems. *Comput Struct Biotechnol J.* 2019;17:1171–1177.
176. Naidoo J, Young D. Gene regulation systems for gene therapy applications in the central nervous system. *Neurol Res Int.* 2012;2012:595410.
177. Li A, Lee CM, Hurley AE, et al. A self-deleting AAV-CRISPR system for in vivo genome editing. *Mol Ther Methods Clin Dev.* 2019;12:111–122.
178. Merienne N, Vachey G, de Longprez L, et al. The self-inactivating KamiCas9 system for the editing of CNS disease genes. *Cell Rep.* 2017;20(12):2980–2991.
179. Petris G, Casini A, Montagna C, et al. Hit and go CAS9 delivered through a lentiviral based self-limiting circuit. *Nat Commun.* 2017;8:15334.
180. Bordet T, Behar-Cohen F. Ocular gene therapies in clinical practice: viral vectors and nonviral alternatives. *Drug Discov Today.* 2019;24(8):1685–1693.
181. Eoh J, Gu L. Biomaterials as vectors for the delivery of CRISPR-Cas9. *Biomater Sci.* 2019;7(4):1240–1261.
182. Wei T, Cheng Q, Min YL, Olson EN, Siegwart DJ. Systemic nanoparticle delivery of CRISPR-Cas9 ribonucleoproteins for effective tissue specific genome editing. *Nat Commun.* 2020;11(1):3232.
183. Xu X, Wan T, Xin H, et al. Delivery of CRISPR/Cas9 for therapeutic genome editing. *J Gene Med.* 2019;21(7):e3107.
184. Charbel Issa P, MacLaren RE. Non-viral retinal gene therapy: a review. *Clin Experiment Ophthalmol.* 2012;40(1):39–47.
185. Johnson LN, Cashman SM, Read SP, Kumar-Singh R. Cell penetrating peptide POD mediates delivery of recombinant proteins to retina, cornea and skin. *Vision Res.* 2010;50(7):686–697.

186. Leaderer D, Cashman SM, Kumar-Singh R. G-quartet oligonucleotide mediated delivery of proteins into photoreceptors and retinal pigment epithelium via intravitreal injection. *Exp Eye Res.* 2016;145:380–392.
187. Cai X, Nash Z, Conley SM, Fliesler SJ, Cooper MJ, Naash MI. A partial structural and functional rescue of a retinitis pigmentosa model with compacted DNA nanoparticles. *PLoS One.* 2009;4(4):e5290.
188. Cai X, Conley SM, Nash Z, Fliesler SJ, Cooper MJ, Naash MI. Gene delivery to mitotic and postmitotic photoreceptors via compacted DNA nanoparticles results in improved phenotype in a mouse model of retinitis pigmentosa. *FASEB J.* 2010;24(4):1178–1191.
189. Ding XQ, Quiambao AB, Fitzgerald JB, Cooper MJ, Conley SM, Naash MI. Ocular delivery of compacted DNA-nanoparticles does not elicit toxicity in the mouse retina. *PLoS One.* 2009;4(10):e7410.
190. Han Z, Conley SM, Makkia RS, Cooper MJ, Naash MI. DNA nanoparticle-mediated ABCA4 delivery rescues Stargardt dystrophy in mice. *J Clin Invest.* 2012;122(9):3221–3226.
191. Han Z, Banworth MJ, Makkia R, et al. Genomic DNA nanoparticles rescue rhodopsin-associated retinitis pigmentosa phenotype. *FASEB J.* 2015;29(6):2535–2544.
192. Lee K, Conboy M, Park HM, et al. Nanoparticle delivery of Cas9 ribonucleoprotein and donor DNA in vivo induces homology-directed DNA repair. *Nat Biomed Eng.* 2017;1:889–901.
193. Lee B, Lee K, Panda S, et al. Nanoparticle delivery of CRISPR into the brain rescues a mouse model of fragile X syndrome from exaggerated repetitive behaviours. *Nat Biomed Eng.* 2018;2(7):497–507.
194. Munye MM, Tagalakis AD, Barnes JL, et al. Minicircle DNA provides enhanced and prolonged transgene expression following airway gene transfer. *Sci Rep.* 2016;6:23125.
195. Chabot S, Orio J, Schmeer M, Schleef M, Golzio M, Teissié J. Minicircle DNA electrotransfer for efficient tissue-targeted gene delivery. *Gene Ther.* 2013;20(1):62–68.
196. Gallego I, Villate-Beitia I, Martínez-Navarrete G, et al. Non-viral vectors based on cationic niosomes and minicircle DNA technology enhance gene delivery efficiency for biomedical applications in retinal disorders. *Nanomedicine.* 2019;17:308–318.
197. Barnea-Cramer AO, Singh M, Fischer D, et al. Repair of retinal degeneration following ex vivo minicircle DNA gene therapy and transplantation of corrected photoreceptor progenitors. *Mol Ther.* 2020;28(3):830–844.
198. Latella MC, Di Salvo MT, Cocchiarella F, et al. In vivo editing of the human mutant rhodopsin gene by electroporation of plasmid-based CRISPR/Cas9 in the mouse retina. *Mol Ther Nucleic Acids.* 2016;5(11):e389.
199. Nguyen Tran MT, Mohd Khalid MKN, Wang Q, et al. Engineering domain-inlaid SaCas9 adenine base editors with reduced RNA off-targets and increased on-target DNA editing. *Nat Commun.* 2020;11(1):4871.

CHAPTER THREE

Advances in gene editing strategies for epidermolysis bullosa

Thomas Kocher and Ulrich Koller*

EB House Austria, Research Program for Molecular Therapy of Genodermatoses, Department of Dermatology and Allergology, University Hospital of the Paracelsus Medical University Salzburg, Salzburg, Austria
*Corresponding author: e-mail address: u.koller@salk.at

Contents

1. The blistering skin disease epidermolysis bullosa	82
1.1 Epidermolysis bullosa simplex	84
1.2 Junctional epidermolysis bullosa	85
1.3 Dystrophic epidermolysis bullosa	85
2. Gene therapeutic applications for EB	86
3. Gene editing development for the treatment of genodermatoses	87
3.1 Inactivation of dominant-negative alleles *via* gene depletion	91
3.2 Genome editing-mediated reading frame restoration of pathogenic alleles	92
3.3 Footprint-less correction of pathogenic alleles *via* homology-dependent repair	95
3.4 Alternative footprint-less correction strategies for pathogenic alleles	98
4. Conclusion and considerations for future gene editing applications in EB	101
Acknowledgment	103
References	104

Abstract

Epidermolysis bullosa represents a monogenetic disease comprising a variety of heterogeneous mutations in at least 16 genes encoding structural proteins crucial for skin integrity. Due to well-defined mutations but still lacking causal treatment options for the disease, epidermolysis bullosa represents an ideal candidate for gene therapeutic interventions. Recent developments and improvements in the genome editing field have paved the way for the translation of various gene repair strategies into the clinic. With the ability to accurately predict and monitor targeting events within the human genome, the translation might soon be possible. Here, we describe current advancements in the genome editing field for epidermolysis bullosa, along with a discussion of aspects and strategies for precise and personalized gene editing-based medicine, in order to develop efficient and safe *ex vivo* as well as *in vivo* genome editing therapies for epidermolysis bullosa patients in the future.

Abbreviations

AAV	adeno-associated virus
AON	antisense oligonucleotide
BMZ	basal membrane zone
C7	type VII collagen
Cas9	CRISPR-associated protein 9
CRISPR	clustered regularly interspaced short palindromic repeats
DDEB	dominant dystrophic epidermolysis bullosa
DEB	dystrophic epidermolysis bullosa
DEJ	dermal epidermal junction
DSB	double-strand break
dsDNA	double-stranded DNA
EB	epidermolysis bullosa
EBS	epidermolysis bullosa simplex
EI	epidermolytic ichthyosis
EJ	end-joining
GFP	green fluorescent protein
GMP	Good Manufacturing Practice
HDR	homology-directed repair
HR	homologous recombination
HSC	hematopoietic stem cell
HSE	human skin equivalent
IF	intermediate filament
indels	insertions and deletions
iPSC	induced pluripotent stem cell
JEB	junctional epidermolysis bullosa
MH	microhomology
NMD	nonsense-mediated RNA decay
PBS	primer binding site
PC	pachyonychia congenita
PE	prime editor
pegRNA	prime editing guide RNA
PTC	premature termination codon
RDEB	recessive dystrophic epidermolysis bullosa
RNP	ribonucleoprotein
RT	reverse transcriptase
SCC	squamous cell carcinoma
sgRNA	single guide RNA
ssDNA	single-stranded DNA
ssOligo	single-stranded oligonucleotide
TALEN	transcription activator-like effector nuclease
ZFN	zinc-finger-nuclease

1. The blistering skin disease epidermolysis bullosa

Genodermatoses comprise a group of monogenetic skin disorders with high heterogeneity. Clinical manifestations include superficial epidermis, mucosal involvement, inherited tumorigenesis, increased photosensitivity and deep dermal trauma.[1-3] Therapeutic targets comprise structural epidermal genes such as keratins, filaggrin, laminins, collagens, integrins or components of the DNA repair pathway.[4,5] The high diversity of associated genes and disease-causing mutations result in a variety of phenotypic severities and clinical outcomes.[6,7] The inherited blistering skin disease epidermolysis bullosa (EB) is caused by mutations in genes involved in skin stability. Distinctive for this disease is the formation of extended blisters and lesions on the skin and mucous membranes upon minimal mechanical trauma.[6] The genetic inheritance can occur in a dominant or recessive fashion depending on the respective EB subform. In general, affected genes encode structural proteins within the basal membrane zone (BMZ) of the skin. Absence or functional loss of one of these proteins is accompanied by a lack of stability of the microarchitecture and the connection between dermis and epidermis, resulting in a loss of coherence.[6] Currently, EB can be divided into 4 classical types: EB simplex (EBS), junctional EB (JEB), dystrophic EB (DEB) and Kindler syndrome (Fig. 1).[7] The prevalence of the EB types, displayed at www.orpha.net, may vary between the countries. 85% of EB patients are affected by EBS, the least severe form, associated with more superficial blisters due to the restricted epidermal localization of the affected proteins keratin 5 and keratin 14.[6-10] In JEB and DEB manifestations include more severe clinical complications and deeper blister formations. JEB, characterized by blistering within the lamina lucida of the basement membrane, is caused by mutations in genes encoding laminin-332, type XVII collagen and integrin-α6β4. Mutations within the gene encoding type VII collagen (C7) are responsible for DEB.[6,7] Kindler

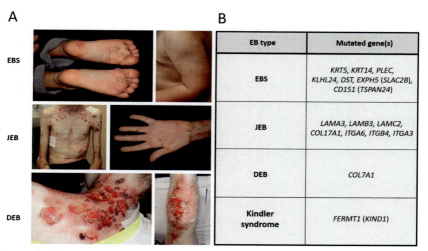

Fig. 1 The blistering skin disease epidermolysis bullosa. (A) Typical skin appearance in EBS, JEB and DEB. (B) EB types and affected genes. *Images courtesy of Rudolf Hametner.*

syndrome is caused by recessive mutations within the *KIND1* gene leading to the loss of the cytoskeletal linker protein Kindlin-1 that connects actin filaments to the extracellular matrix.[6]

Currently, the treatment of EB is restricted to wound management. Thus, the economic burden of the disease for affected families is high. The average estimated treatment cost is €31,390 annually across different EB subtypes.[11] Therefore, the development of novel therapies for EB is mandatory.[12] The direct accessibility of the skin, the facilitated cultivation of skin cells and the possible *in vitro* reconstitution of the organ simplify the research in this field. However, scientists have yet to overcome various skin-associated challenges including the large body area to treat, the avascular epidermis and the general heterogeneity of the skin.[13,14]

1.1 Epidermolysis bullosa simplex

The keratinopathies, EBS, epidermolytic ichthyosis (EI) and pachyonychia congenita (PC), represent a group of inherited skin disorders mainly caused by dominant-negative mutations in genes encoding keratins. As cytoskeletal proteins, keratins generate a heteropolymeric intermediate filament (IF) network, which protects the cell from mechanical stress induced consequences. In EBS, mutations in the keratin 5 (*KRT5*) and keratin 14 (*KRT14*) genes, result in intraepidermal blistering after mechanical trauma.[15] In EBS most mutations are dominantly inherited and are either missense mutations or small in-frame insertions or deletions (indels).[15] The type II IF protein keratin 5 and its binding partner keratin 14, a type I IF protein, are both expressed in the basal proliferative keratinocyte layer of the epidermis.[16] The phenotypic severity depends on the location of the disease-associated mutation within the affected genes *KRT5* or *KRT14*.[17,18] Mutant keratins polymerize with wild-type keratins into abnormal filaments that collapse upon minimal mechanical stress induction. Aggregate formation, cytolysis and skin blistering are direct consequences.[19,20] Cell culture experiments with EBS patient keratinocytes revealed the impairment of IF structure and aggregate formation as a stress response.[21–25] Other less common EBS subtypes are caused by mutations in genes encoding proteins that directly interact with the IF, such as plectin,[26] or are involved in keratin turnover.[27,28] So far, for EBS or keratinopathies in general, there is no cure available. The alleviation of symptoms and wound managements are the only current treatment options to improve the quality of a patient's life. In particular, therapies for the dominantly inherited EBS subtype are limited.

The dominant negative influence of mutant keratins on the IF cytoskeleton cannot be resolved by conventional gene-replacement strategies, that have been successfully applied for the correction of the junctional form of EB.[29] Besides the provision of the wild-type gene, the downregulation or knock-out of the disease-causing allele is mandatory. Gene editing can mediate an efficient gene knock-out,[24,25] whereas downregulation can be achieved by RNA-based approaches such as spliceosome-mediated RNA *trans*-splicing.[30–32]

1.2 Junctional epidermolysis bullosa

In JEB, the skin separation takes place within the lamina lucida of the BMZ. Common autosomal recessive JEB subtypes comprise generalized and localized forms. Both are the result of mutations in *COL17A1, ITGA6, ITGB4, LAMA3, LAMB3* and *LAMC2*, respectively, all of which encode hemidesmosome function related proteins that interact with each other. *COL17A1* encodes type XVII collagen, a transmembrane protein necessary for connections between the plasma membrane of the basal keratinocytes and the lamina densa. Homozygous nonsense, missense or in-frame deletion mutations within *COL17A1* result in reduced or absent expression in JEB patients.[6,7] Defects in laminin-332, the major adhesion ligand of basal epithelial cells expressed in the BMZ, are the cause of the majority of JEB cases. Laminin-332 is a multi-domain glycoprotein composed of 3 subunits; $\alpha 3$, $\beta 3$, and $\gamma 2$, encoded by *LAMA3, LAMB3*, and *LAMC2* genes, respectively.[6,33] Recently, promising gene replacement strategies led to a complete re-establishment of the adhesion property of keratinocytes *in vitro*[34,35] and *in vivo*.[29,36,37] However, the scarcity of available therapies for JEB restricts treatment to palliative wound and pain management.

1.3 Dystrophic epidermolysis bullosa

Patients, harboring mutations in the *COL7A1* gene, suffer from the severe recessive form of EB associated with the development of aggressive squamous cell carcinoma (SCC) in early adulthood. In case of SCC progression neither chemo- nor radiotherapies represent useful treatment options, resulting in the high mortality rate observed in patients with the severe recessive form of dystrophic EB (RDEB).[6,38] The majority of *COL7A1* mutations, from more than 800 reported,[39] are loss-of-function, although the severity of the RDEB phenotype is dependent upon the mutation type and its location within the gene. As a result, diagnosis, course of disease and the choice of wound therapy vary significantly. Blister formation is observed as generalized or restricted to palms and soles. Dystrophic EB (DEB) can be classified into two main subtypes according to the mode of inheritance, dominant or recessive.[6,7] *COL7A1* encodes type VII collagen, which builds up the anchoring fibrils needed for the firm adhesion of the dermis and epidermis.[40] Depending on the underlying pathogenic mutation the fibrils are morphologically abnormal, reduced in number or completely absent in DEB patient's skin.[6,40,41] Current experimental therapies for DEB focus on allogeneic bone marrow transplantation,[42] fibroblast injections,[43] gene therapy,[44,45] protein therapy[46] or systemic application of allogeneic mesenchymal stem cells.[47]

2. Gene therapeutic applications for EB

As causal mutations for most cases are well-defined, EB represents an ideal target disease for gene therapeutic applications. Two promising forms of gene therapy are potentially feasible for EB: gene replacement and genome editing. A stark difference between these two approaches is the reliance of replacement therapies on viral transfer vectors (*e.g.,* murine leukemia virus, MLV-derived vectors). As gene replacement therapy aims to maintain stable exogenous expression cassettes, efficient virus-mediated nuclear delivery and integration of the transgene (including viral regulatory elements) into the host cell genome is compulsory. Controlled integration and activity of the transgene is limited and therefore potentially associated with genomic toxicity due to insertional mutagenesis. However, 4 years after successful transplantation of 80% of diseased skin in a JEB patient no such event has occurred (personal communication Prof. Johann Bauer).[29]

In contrast, genome editing aims to target the mutant locus specifically. These approaches preferably rely on robust yet transient expression of gene editing molecules that enable permanent repair of mutations, thereby restoring normal endogenous gene expression, without the need for viral integration.

While gene editing for genodermatoses remains at the preclinical stage,[48,49] gene replacement approaches are far more clinically advanced and have already been successfully applied to the treatment of *LAMB3*-deficient junctional EB.[29,36,37] The most recent application[29] was life-saving, suggesting the use of this strategy in clinical studies for other EB-associated genes. However, similar strategies to stably introduce *COL7A1* cDNA into RDEB patient cells have yielded limited therapeutic applicability so far.[45] This can be due to many different factors, including low transduction efficiencies, the large size of *COL7A1* cDNA (9.3 kb), random integration of the transgene flanked by viral sequences, or post-transcriptional deregulation of target genes by aberrant splicing.[50,51] These observations demonstrate that gene replacement is not a universal treatment option for EB, and that varying therapeutic efficacies can be expected depending on the affected gene, of which there are ~16 implicated in EB.[7] Taking these aspects into consideration, achieving a permanent and potentially traceless repair of a disease-causing mutation *via* genome editing is one preferable aspect of successful gene correction. In addition, this strategy could be easily adapted and applied for any disease-associated gene.

3. Gene editing development for the treatment of genodermatoses

Currently, gene editing comprises a suite of technologies that can be custom-tailored to almost any target sequence and specific application required. Essentially, gene editing tools are comprised of a programmable DNA-binding domain and a nuclease. Following specific localization of the DNA-binding domain to a target site, the guided nuclease induces a double-strand break (DSB), triggering endogenous cellular repair machineries to repair the lesion in order to maintain genomic integrity[52] (Fig. 2).

These repair mechanisms are co-opted for gene editing,[53] with the predominant mode of repair often dictated by the cell cycle.[52] Currently, homology-directed repair (HDR) comprises the best strategy for a perfect, traceless repair, but is hindered by low efficiencies due to the additional requirement of an exogenously-provided DNA template for the repair reaction[33,54,55] (Fig. 2). In comparison to HDR, end-joining (EJ) repair pathways exhibit higher efficiencies, regardless whether occurring through non-homologous end joining mechanisms or driven by small areas of microhomology (MH) on each end of the double strand break. EJ pathways typically result in indel mutations and imperfect repair outcomes.[56,57] Naturally, these pathways have been co-opted in applications aimed at disrupting the target locus, which we have successfully exploited for the highly efficient knock-out of alleles carrying dominant-negative disease-causing mutations in genodermatoses[24,25] (Fig. 3) (Table 1). A significant development in the field that has broadened the therapeutic application of EJ-mediated repair came with the ability to predict repair outcomes simply based upon the sequence context at the clustered regularly interspaced short

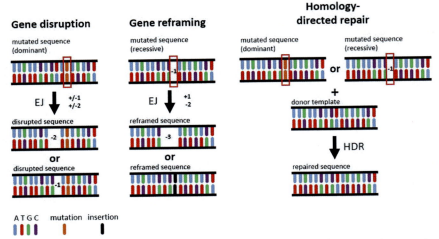

Fig. 2 Gene editing strategies for epidermolysis bullosa. EJ-based gene repair strategies can be used to disrupt a gene, carrying a dominant-negative mutation, or to reframe a gene with a pathogenic insertion or deletion (indel). For HDR a donor template is required to repair the mutation in a potentially traceless manner.

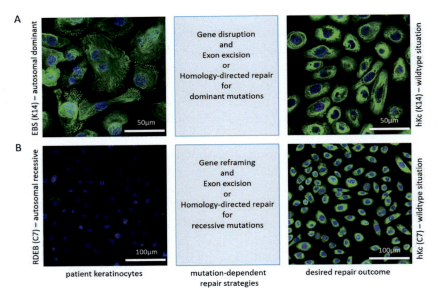

Fig. 3 Summary of gene editing strategies considered for the correction of EB causing mutations. (A) K14 IF staining of EBS (c.374G > A; exon 1) and wild-type keratinocytes. The expression of mutant K14 and the formation of K14/K5 heterodimers results in intermediate filament instability and the characteristic presence of cytoplasmic aggregates containing mutant keratin. Due to the fact, that the vast majority of mutations are either missense or small in-frame insertions or deletions (indels) gene disruption and exon excision, thereby knocking out the mutant allele, might be the most efficient and preferred treatment strategies to reach the desired repair outcome. (B) C7 IF staining of DEB (c.6081delC; exon 73) and wild-type keratinocytes. The most efficient repair strategies to achieve restoration and re-expression of C7 in patients with *COL7A1* frameshift indels are gene reframing as well as whole exon excision strategies. For autosomal recessive missense mutations, application of HDR-dependent repair strategies offers the most commonly used treatment option.

Table 1 Gene editing strategies for epidermolysis bullosa.

Targeting strategy (nuclease)	Gene (mutation)	Dominant recessive	EB Type	References
Gene disruption (TALEN, CRISPR/Cas9)	COL7A1 (c.8068_8084delinsGA)	Dominant	DDEB	65
Gene disruption (TALEN)	KRT5 (c.556G>T, c.1424A>G)	Dominant, Dominant	EBS	24
Gene reframing (TALEN)	COL7A1 (c.6527insC)	Recessive	RDEB	68
Gene reframing (TALEN)	COL7A1 (c.6527insC)	Recessive	RDEB	70
Gene reframing (CRISPR/Cas9)	COL7A1 (c.5819delC)	Recessive	RDEB	66
Gene reframing (CRISPR/Cas9)	COL7A1 (c.6081delC)	Recessive	RDEB	57
Gene reframing (Dual CRISPR/Cas9)	COL7A1 (c.6485G>A)	Recessive	RDEB	69
Gene reframing (Dual CRISPR/Cas9)	COL7A1 (c.6527insC)	Recessive	RDEB	71
Homology-directed repair (TALEN)	COL7A1 (c.1837C>T)	Recessive	RDEB	54
Homology-directed repair (ZFN, TALEN, CRISPR/Cas9)	COL7A1 (c.356_357delCA, c.90delC)	Recessive	RDEB	44
Homology-directed repair (TALEN)	COL7A1 (c.6527insC)	Recessive	RDEB	68
Homology-directed repair (CRISPR/Cas9)	COL7A1 (c.4317delC)	Recessive	RDEB	79
Homology-directed repair (CRISPR/Cas9)	COL7A1 (c.6527insC)	Recessive	RDEB	82
Homology-directed repair (CRISPR/Cas9)	LAMB3 (c.1945dupG, c.1903C>T)	Recessive	JEB	33

Continued

Table 1 Gene editing strategies for epidermolysis bullosa.—cont'd

Targeting strategy (nuclease)	Gene (mutation)	Dominant recessive	EB Type	References
Homology-directed repair (Meganuclease)	COL7A1 (c.189delG, c.425A > G)	Recessive, Recessive	RDEB	80
Homology-directed repair (Dual CRISPR/Cas9 D10A nicking)	KRT14 (c.1231G > A)	Dominant	EBS	60
Homology-directed repair (CRISPR/Cas9)	COL7A1 (c.2470insG, c.3948insT)	Recessive	RDEB	83
Homology-directed repair (CRISPR/Cas9)	COL7A1 (c.189delG)	Recessive	RDEB	81
Homology-directed repair (CRISPR/Cas9)	COL7A1 (c.2005C > T, c.8569G > T)	Recessive	RDEB	84
Homology-directed repair (Dual CRISPR/Cas9 D10A nicking)	COL7A1 (c.425A > G)	Recessive	RDEB	55
Base editing (ABE)	COL7A1 (c.553C > T, c.1573C > T)	Recessive, Recessive	RDEB	87

palindromic repeats (CRISPR)/CRISPR-associated protein 9 (Cas9) target site accurately.[58] Using this algorithm to predict the most predominant EJ-based repair outcome, we were able to design a COL7A1 reframing strategy to correct a single nucleotide deletion mutation, resulting in C7 protein restoration in >70% of primary RDEB keratinocytes following an *in vitro* single treatment.[57] While these strategies lead to more homogenous repair outcomes, suggesting their consideration for direct *in vivo* therapeutic applications, they also comprise disadvantages. Reframing strategies are only applicable for a small fraction of disease-causing mutations in EB (Fig. 3) (Table 1). They are not suitable for restoring the wild-type sequence of the target gene in the context of missense mutations. Furthermore, the inevitable heterogeneity of gene editing outcomes, albeit in low frequency, will yet produce C7 variants in addition to the predicted, desired repair outcomes. The impact of these variants on functional fibril formation will need to be fully analyzed. Therefore, a highly efficient and safe method to achieve perfect wild-type sequence restoration that can be applied to a significant fraction of disease-causing mutations still remains the ultimate goal of gene editing. Steps to improve HDR efficiencies and safety profile are largely based on the use of the D10A Cas9 nickase, which preferably induces single-strand breaks (nicks) within DNA. By combining two nickases in a double-nicking configuration both an increase in HR rates and a reduction in the frequency of off-target events could be observed.[55,59,60] Nevertheless, the reliance on a selection-based protocol and possible donor template integrations remain the main hurdles for future clinical applications.[55] The recent development of prime editing[61] appears to represent a significant and highly promising milestone towards the goal of traceless gene editing. Prime editing directly writes new genetic information into a specified DNA site using a Cas9 nickase fused to an engineered reverse transcriptase (RT) domain.

3.1 Inactivation of dominant-negative alleles *via* gene depletion

Dominant-negative skin diseases are suitable targets for gene depletion approaches aiming at the disruption of pathogenic alleles[24,25] (Table 1). In general, this can be achieved *via* error-prone EJ-based repair upon nuclease-mediated DSB generation. Introduced insertions and deletions (indels) generate frameshifts within the target allele. Thus, premature termination codons (PTCs) are generated leading in the majority of cases to nonsense-mediated RNA decay (NMD).[62] Typically, two thirds of nuclease-induced indels are expected to lead to PTCs. Therefore, this simple approach can have highly efficient outcomes regarding gene disruption, without the need for extensive screening.[25] Dominant negative mutations are common in keratinopathies, in which keratins, expressed within specific epidermal layers of the skin, are the causal factors.[16] They build up the heteropolymeric IF network in epidermal keratinocytes, which becomes destabilized when mutant keratins are incorporated, resulting in cytoskeletal fragility of epidermal cells.[6,63] Recently, Aushev *et al.* published an EJ-based approach to inactivate the *KRT5* gene carrying a dominant-negative mutation.[24] Edited EBS keratinocyte clones, screened and isolated without selection pressure, demonstrated phenotypic reversion *in vitro*. This non-allele specific gene editing protocol should be suitable for many EBS patients with pathogenic *KRT5* mutations. Although an allele-specific strategy would avoid targeting of the wild-type allele, this strategy is difficult to implement for clinical translation as for each dominant-negative mutation an individual nuclease requires engineering. This approach would be a time consuming and cost intensive endeavor. March et al. followed a similar approach for EI.[25] Using the transcription activator-like effector nuclease (TALEN) technology, the authors disrupted the *KRT10* gene upstream of a recently described PTC[64] leading to NMD. A normalization of *KRT10* expression at protein level was detectable *via* Western blot analysis. Further, a gene editing efficiency of over 20% was achieved in treated primary EI keratinocytes.[25] Besides EBS, dominant-negative mutations are common in DEB.[6,7] In 2016, Shinkuma and co-authors described an allele-specific gene editing approach aiming to disrupt the *COL7A1* gene, in which an underlying mutation interferes with the collagen triple helix formation in a dominant manner.[65] Patient fibroblasts-derived induced pluripotent stem cells (iPSCs) were targeted with TALEN and CRISPR/Cas9 nucleases, resulting in indels leading to PTCs not associated with NMD. The analysis of fibroblasts and keratinocytes, differentiated from treated iPSCs, revealed the presence of truncated C7 proteins. This underlined the importance of prior characterization of target PTCs in gene depletion approaches.[25] The prediction of predominant DSB repair outcomes, induced by CRISPR/Cas9 nucleases, would further increase the gene correction efficiencies, as recently shown in a complementary approach dealing with the EJ-based reframing of *COL7A1*.[57] Keratinopathies represent the majority of dominant genodermatoses and are therefore the most suitable disease group for gene disruption strategies. In contrast, many other genodermatoses are inherited recessively, and thus unsuitable for this approach. However, for the phenotypic correction of dominant DEB (DDEB), targeting of an effective NMD-inducing PTC in *COL7A1* could represent the strategy of choice.

3.2 Genome editing-mediated reading frame restoration of pathogenic alleles

Identical and similar strategies to gene disruption approaches can be used to restore the reading frame of genes destructed by pathogenic frameshift mutations. However, in contrast to gene disruption, exon reframing is mainly suitable for the correction of recessive disease forms and frameshift-inducing indel mutations, which represent the only feasible targets for this gene editing strategy (Table 1). Importantly, only gene targets tolerating amino acid divergences from wild-type sequence should be considered.[66] Respective changes in the amino acid sequence of restored protein variants should have no negative impact on whose functionality. Thus, the restored and altered protein must be fully characterized for its functionality, immunogenicity and stability. The expression of altered protein variants within EB patients and the reduced severity of dominant-negative forms of EB[6,67] already indicate potential therapeutic alleviation could be achieved *via* reframing strategies. Currently, all gene editing-based reframing approaches for genodermatoses have been performed for recessive DEB (RDEB).[57,66,68–71] This might be due to the proven amenability of *COL7A1* to truncation and reading frame restoration within the collagenous domains.[14] Exons encoding the triple helix-forming region, essential for the structural function of C7, are small, in frame and encode Gly-X-Y repeats.[70,72] Antisense oligonucleotide (AON)-based exon-skipping approaches have been described for DEB[73–76] confirming the functionality of internally deleted C7 variants lacking sequences encoded by these specific collagenous domain exons *in vitro* and *in vivo*. In these studies truncated C7 variants, lacking sequences encoded by exons 70,[73] 73,[74,76] 80[74] and 105,[75] have been shown to retain the functions of the full-length protein.

Chamorro *et al.* described the first gene editing-based permanent exon reframing approach in genodermatoses. This *ex vivo* approach involved adenoviral delivery of a TALEN pair, targeting the c.6527insC mutation in exon 80 of the *COL7A1* gene, which is highly prevalent in the Spanish RDEB patient population.[68] Since than many studies based upon CRISPR/Cas9-mediated reframing have been performed to restore C7 expression *in vitro*,[57] *ex vivo*[66,68,70,71] and *in vivo*.[69] In addition to strategies using a single nuclease and EJ-based repair of the resultant DSB, which often generates small indels and requires screening for low-frequency desired edits, several studies have endeavored to establish more efficient strategies to achieve more precise, predictable and therefore homogenous repair outcomes. Recent developments, in particular the introduction of ribonucleoproteins (RNPs) have revealed more efficient, homogenous, precise and predictable outcomes, following single as well as dual targeting *via* CRISPR/Cas9 molecules.[57,71] Dual CRISPR/Cas9 targeting can be used for highly efficient excision of an in-frame exon and to effectively excise any mutation type from mutant alleles, while maintaining or restoring the wild-type reading frame. This was shown recently *ex vivo*[70,71] as well as *in vivo*.[69] Wu *et al.* described the first gene editing-based exon excision approach for genodermatoses *in vivo*.[69] The intradermal injection of dual CRISPR/Cas9 RNPs into the tail skin of an RDEB mouse model and subsequent electroporation resulted in moderate targeting efficiency (2% of epidermal cells), but in an enrichment of BMZ-localized C7 and increased stability of the dermal epidermal junction (DEJ), which was indicated by the measurement of epidermal detachment within a given area.[69] A similar dual CRISPR/Cas9 RNP electroporation strategy for the excision of exon 80 has been applied to primary human keratinocytes by Bonafont et al., in a proposed *ex vivo* therapeutic approach.[71] They reached excision efficiencies of over 80% in bulk-treated primary keratinocytes, which was confirmed subsequently *via* restored C7 transcription and translation, as well as a long-term engraftment of bulk-treated samples, resulting in highly efficient phenotypic restoration.[71] Another recent and significant development in the field, which revealed more precise, predictable outcomes, following single targeting *via* CRISPR/Cas9, at many loci,[58,77] broadened the therapeutic application of EJ-mediated repair. This method is simply based upon the sequence context of the protospacer-adjacent motif (PAM) site, enabling rapid and efficient development of precise CRISPR/Cas9-based strategies. Recently, we presented a promising CRISPR/Cas9-based targeting strategy to reframe *COL7A1* in primary keratinocytes, carrying a homozygous frameshift mutation (c.6081delC) within exon 73.[57] We delivered a single CRISPR/Cas9 RNP into recessive DEB keratinocytes to introduce a precise predictable single adenine sense-strand insertion at the target site, achieving C7 restoration in >70% of RNP-treated RDEB keratinocytes. In contrast to the exon excision approaches described previously, this precise adenine sense-strand insertion at the target site only resulted in a single Arg→Glu amino acid divergence from the wild-type coding sequence, representing the exchange of polar residues within the collagenous domain.[57] In theory it can be possible to reach perfect repair and restoration of wild-type C7 *via* this strategy, at least for some patients. This is largely dependent upon the size of the frameshift mutation and the sequence context of the PAM site. A possible predictable and perfect repair can only be achieved for ±1 nucleotide insertions or deletions, respectively. Efficient reframing approaches would obviate the need for single cell expansion, mitigating issues with genetic instability, culture stress and replicative senescence.[78] However, the functionality of reframed proteins, the preservation of open reading frames and amino acid-encoding repeats must be accounted for in the design of future approaches. These advanced gene editing strategies and studies provide important implications for model generation and therapeutic applications for DEB and in the field of genome editing for genodermatoses.

3.3 Footprint-less correction of pathogenic alleles *via* homology-dependent repair

Although EJ-based approaches have a clear clinical translation potential for EB patients carrying mutations in small exons within collagenous domains, HDR-based protocols offer the only option to perfectly correct any kind of mutation. Homology-dependent repair pathways are only activated following the failure of alternative EJ pathways, and in the presence of DNA repair templates bearing homology to the area of the DSB. This repair pathway occurs exclusively during the S and G2 phases of the cell cycle in proliferating cells, when intact sister chromatids are available for use as repair templates.[56] As a result, frequencies of homology-dependent repair events are often lower due to the dominance of other pathways, such as classical EJ. However, in addition and in contrast to EJ-based strategies, a single repair template could allow for the correction of a wider number of mutations within different intronic and exonic regions, as long as they are within the length of the designed donor template. This strategy carries potential for the development of therapies for a larger cohort of EB patients. However, homologous recombination (HR)-based strategies are typically only efficient in a patient-personalized manner. Recently, Benati *et al.* described a very promising approach, which theoretically enables the development of single non-personalized approaches to treat patients with several distinct mutations in genes.[33] Delivering an adenovector, carrying Cas9/single guide RNA (sgRNA) tailored to the intron 2 of *LAMB3* gene and an integration defective lentiviral vector bearing a *LAMB3* cDNA cassette starting from exon 3 and flanked by a splice acceptor, into JEB keratinocytes, they were able to safely *in situ* integrate a therapeutic *LAMB3* cDNA *via* CRISPR-mediated HR. The outcome of this study and barely detectable 0.48% correct HR efficiency, confirmed the assumption that a repair template, including a longer homology wild-type sequence, would result in several different unpredictable HR events, lowering the already limited HR efficiencies. However, taking advantage of *in vitro* adhesion properties of *LAMB3*-expressing cells,[35] they were able to enrich and select for *LAMB3*-corrected keratinocytes in treated bulk populations *via* cell detachment assay.[33] Importantly, several recent molecular tool advances have increased the efficiencies of personalized homology-based gene correction strategies,[33,44,54,55,60,68,79–84] which still makes it a further interesting strategy option for *ex vivo* gene editing approaches. The vast majority of homology-dependent approaches reported to date have been performed in DEB targeting *COL7A1*,[44,54,55,68,79–84] with the exception of two studies, one targeting *KRT14* in EBS[60] and the above-described study targeting the *LAMB3* gene in JEB.[33] In a first proof of concept study for precise homology-dependent repair, Coluccio et al. targeted a "safe harbor" locus, in a human keratinocyte cell line and in primary keratinocytes using zinc-finger-nuclease (ZFN).[85] Their data indicated poor induction of the HR-dependent DNA repair pathways, especially in primary keratinocytes. These relatively low HR efficiencies already suggested use of selection-based systems and expansion of corrected single-cell clones for future studies. Still, there have been some attempts to efficiently target and correct *COL7A1* in primary RDEB keratinocytes and RDEB fibroblasts *via* HR, based on the delivery of meganucleases or CRISPR/Cas9 together with lentiviral vectors, without any antibiotic- or fluorescence-based selection.[80,81] Izmyrian and colleagues achieved *COL7A1* transcript correction efficiencies of 11% and 15.7% in transduced primary keratinocytes and fibroblasts, respectively.[81] Higher HR efficiencies could be reached *via* introduction of antibiotic/fluorescence selection cassettes in between the homology repair arms, flanked by loxP sites, allowing for selection of HR events, or the addition of antibiotic/fluorescence cassettes within the donor plasmid.[44,54,55,60,68,79,82–84] With these selection-based strategies, remarkably high HR frequencies of up to 89% and C7 restoration of up to 77% could be reached in immortalized patient keratinocytes.[55] Especially double-nicking, a method using a pair of the mutant Cas9 D10A nickase in close proximity to each other and targeting both DNA strands *via* single nicks,[59] was shown to have beneficial effects on HR efficiency and safety. Comparing HR efficiencies of either spCas9-, single-nicking-, or double-nicking-based strategies for the correction of the respective mutation highlighted, that double-nicking consistently outperformed all other approaches.[55,60] Interestingly, strategic nicking of repair templates could further improve HR efficiencies in treated RDEB patient keratinocytes. Two obvious pitfalls of these very efficient selection-based strategies are remaining loxP sites in the genome even after Cre-recombinase expression and the possibility of targeted or random integration of the entire donor plasmid or large parts thereof.[55,82] As a result, they cannot be considered as traceless, impeding CRISPR/Cas9's translation into the clinic. Beside this, it is still difficult to induce HR events in somatic cells, which are further difficult to maintain due to limited cell division. A possible solution for this could be using iPSCs. iPSCs have been widely used for HR-based gene therapies for inherited disorders[86] and more recently, several studies have also demonstrated CRISPR/Cas9-based gene therapy in EB-specific iPSCs.[44,54,65,79,83,86,87] iPSCs have unlimited proliferation potential and their pluripotent property allows for differentiation into different cell lineages, like fibroblasts and keratinocytes, for therapeutic applications.[65,83,88–90] Itoh and colleagues recently presented a possible footprint-free HR approach for the correction of iPSCs derived from RDEB, combining the CRISPR/Cas9 system together with the piggyBac transposon.[84] They demonstrated correction of the pathogenic mutation in RDEB-specific iPSCs, removing the selection cassette by transposon system in iPSCs without leaving footprints, and maintenance of their "stemness". In addition, they showed, when differentiating iPSCs into keratinocytes, that the expression of C7 could be restored.[84] Jackow et al. presented a distinct approach for a possible traceless correction of RDEB using iPSCs.[83] They optimized their correction strategy by moving from a plasmid-based system to a protein-based RNP delivery system. Electroporation of RNPs together with single-stranded oligonucleotides (ssOligos) allowed for the correction of a homozygous mutation in exon 19 and the simultaneous correction of patient derived iPSCs harboring mutations in exon 19 and exon 32. The co-transfection of HR repair molecules together with a green fluorescent protein (GFP) plasmid allowed them to apply a FACS strategy to sort the possible gene-corrected clones out of the bulk population without the necessity of using a drug selection cassette. Via this strategy, they were able to increase the gene correction efficiency in positive iPSC colonies from 10% to 58% for biallelic correction. Three-dimensional human skin equivalents (HSEs), generated from gene-corrected iPSCs and further differentiated into keratinocytes and fibroblasts, were grafted onto immune-deficient mice, and showed normal expression of C7 at the BMZ, as well as restored anchoring fibrils 2 months post grafting. Since there is evidence in RDEB, that C7 expression from both major C7-expressing cell compartments, namely the fibroblasts and keratinocytes, are required for the formation of proper anchoring fibrils within the BMZ,[91] iPSC approaches should be definitively considered for future applications and in preparation for clinical investigation. Already highlighting the advantages and improvements of RNP usage together with single stranded repair templates for efficient gene targeting via electroporation, Osborn et al. tried to employ a different HR-dependent and CRISPR/Cas9-based strategy, aiming for distinct cell types as a source of missing C7 expression in RDEB.[92] They reached highly efficient gene targeting and HR induction (>60%) in hematopoietic stem cells (HSCs) and T-cells via adeno-associated virus 6 (AAV6) infection, delivering the respective donor template, and subsequent electroporation of Cas9 RNP complexes targeting 164 bp upstream of the *COL7A1* start codon. The very efficient HR-mediated insertion of the transcriptional promoting element termed UMET, allowed high C7 levels in HSCs and T-cells. Their strategy is reliant on AAV-based delivery of the donor template and merges the specificity of gene editing tools with a high level of sustained gene expression, which represents a facile cellular engineering platform for promoting gene expression in cells with high translational impact.[92]

In summary, numerous aspects mentioned above have an influence on HR efficiencies, including the delivery method, the nature (mRNA, DNA or Protein) and modus operandi of DNA-modifying agents, the nature of the repair template (double-stranded or single-stranded DNA), the cell type to be corrected and the cell cycle stage. High targeting efficiency, precision, and seamless repair all need to be carefully balanced in future *ex vivo* therapies to circumvent the need for antibiotic selection and clonal expansion of corrected cells.

3.4 Alternative footprint-less correction strategies for pathogenic alleles

In the near future, CRISPR/Cas9 developments will lead to rapid advances in the applicability and possible translation of this technology into the clinics. Delivery of CRISPR/Cas9 into target cells *via* electroporation of RNPs has been shown to result in high on-target and correction efficiencies,[57,71] and low off-target activity.[71] Cas9 protein, sgRNA and nucleofectors for high scale cell numbers are already available at Good Manufacturing Practice (GMP)-grade. In addition, CRISPR/Cas9 studies in RDEB have recently revealed more precise, predictable outcomes, following targeting *via* single or dual CRISPR/Cas9 nucleases, resulting in predictable homogenous DSB repair outcomes for reframing and knock-out approaches.[57,71] Despite the relative homogenous and predictable repair outcomes, they will still be accompanied with undesired indels and in most cases, the resulting C7 will be modified or even truncated compared to wild-type C7. So far, there have been several studies aiming for perfect and seamless repair of EB associated mutations *via* homology-dependent repair. However, efficiency in therapeutically relevant cells was very low, and DSB repair resulted in an excess of EJ indels accompanying the desired repair product. These challenges motivated the exploration of alternative precision genome editing strategies not based on exogenous donor DNA repair templates. One of these is base editing, which can be used to efficiently induce four transition mutations

(C→T, G→A, A→G, and T→C), without DSB formation or repair templates, in many cell types and organisms.[93–95] Two types of base editors have been developed, both chimeric fusion proteins utilizing the sgRNA-targeting and DNA strand-nicking activity of Cas9 D10A nickase.[94,95] However, base editors cannot currently be used to perform the following eight transversion mutations (C→A, C→G, G→C, G→T, A→C, A→T, T→A, and T→G). Recently, Osborn et al. presented the first base editing approach for RDEB targeting clinically relevant cell populations (fibroblasts and iPSCs) via electroporation of sgRNAs and mRNA of the current optimized version of the adenine base editor, ABEmax.[87,96] Targeting two different mutations, one homozygous c.553C > T (R185X) mutation and one compound heterozygous with c.1573C > T (R525X) and c.2005C > T (R669X) mutations, they achieved average T > C mutation correction rates of up to 23.8% in genomic DNA and increased presence of corrected transcripts in up to 45% of analyzed patient cells, respectively. Both 3D culture of corrected cells and the injection of corrected iPSC in mice showed normalized epithelial layer attachment in vitro and contiguous C7 deposition at the DEJ in vivo, respectively. Whereas targeting of the homozygous c.553C > T mutation resulted in efficient editing with little or no bystander events, the compound heterozygous c.1573 target had two bystander nucleotides within the target sequence, one edited at a much greater efficiency than the disease causing mutation itself. Collectively, the on- and off-target data not only highlights the benefits of base editors, it also highlights the non-specific nature of base editing within the target region, which necessitates efficient strategy design and precludes the use of this technology in repetitive target sites.[87] Building upon base editing, the recent development of prime editing[61] seems to represent a significant and highly promising milestone towards the goal of traceless gene editing. Prime editing directly writes new genetic information into a specified DNA site using a Cas9 nickase fused to an engineered RT domain. This "prime editor" is programmed with a prime editing guide RNA (pegRNA) that is both gRNA, specifying the target site, and RT template, encoding the desired edit. The pegRNA directs the prime editor (PE) to its target DNA where the Cas9 nickase makes a nick in the PAM-containing strand. The nicked genomic strand hybridizes to a complementary primer binding site (PBS) sequence and serves as the initiation point for reverse transcription of new DNA containing the desired edit using the RT template of the pegRNA. Anzalone et al. not only demonstrated the correction of the primary genetic causes of two genetic diseases, they also stated that prime editing is suitable for the correction of ~89% of all disease-associated mutations.[61] Their study concluded, that prime editing features complementary strengths and weaknesses compared to base editors, and higher or similar correction efficiency compared to Cas9-initiated HDR. Interestingly, prime editing has the potential to target all kinds of mutations without requiring DSBs or donor DNA templates, which would allow for the correction of the vast majority of pathogenic alleles. Anzalone and colleagues started with prime editors (PEs), initially exemplified by PE1, which uses a RT fused to an RNA-programmable nickase and a pegRNA. Due to low efficiencies they further moved on to PE2, where they used an engineered RT to improve DNA synthesis during prime editing and thereby increasing editing efficiencies. The usage of PE2 led to a 1.6- to 5.1-fold improvement in the efficiency of prime editing point mutations over PE1. Since nicking of the non-edited strand has been previously shown to optimize base editing,[93–95] they added the Cas9 (H840A) nickase, already present in PE2, and a simple sgRNA directing DNA repair to that strand using the edited strand as a template, to establish the PE3 system. In four of five tested DNA loci, nicking the non-edited strand increased editing efficiency by 1.5- to 4.2-fold compared to PE2, to as high as 55%. However, additional nicking of the non-edited strand also resulted in bystander indel events at the target site. To further minimize the formation of DSBs and indels, they designed sgRNAs with spacers, denoted PE3b, that matched the edited strand, but not the original allele.[61] Interestingly, the very efficient prime editor 3 (PE3) system is analogical and similar to the double-nicking setting, which we applied for *KRT14* and *COL7A1* correction,[55,60] where we reached HR frequencies of up to 89%, while keeping unwanted repair outcomes, such as non-homologous end joining (NHEJ), at a minimum (11%). Taken together, prime editing provides a new search-and-replace strategy, which is not dependent on DSBs, donor DNA templates or HDR, highlighting it as an additional interesting candidate approach for the treatment of several genetic diseases, such as EB.[61] Still, additional knowledge and research is required to better understand and improve prime editing in a broad range of cell types such as skin cells. The assessment of off-target prime editing in a genome-wide manner, and the further characterization of the extent to which prime editors and especially the presence of a RT within a cells cytoplasm and nuclei might affect cells are essential for future clinical applications.

4. Conclusion and considerations for future gene editing applications in EB

Genome editing and especially the CRISPR/Cas9 system represents a transformative and rapidly advancing platform for generating future clinical medicines. It enables the engineering and manipulation of the genome at a level of precision unimaginable few years ago. The overall efficiencies of CRISPR/Cas9-based approaches for *ex vivo* applications appear to have reached a plateau of editing efficiency, with targeting efficiencies routinely exceeding 80% without further selection.[55,71] This was only possible, due to the introduction of CRISPR/Cas9 RNPs, the electroporation of which can be effective and relatively nontoxic. This strategy is of utmost importance for the treatment of primary cells, which have an intact antiviral, cytoplasmic DNA-sensing mechanism.[97] Importantly, GMP-grade Cas9 proteins, sgRNAs and nucleofectors for high cell amounts and possible future clinical applications are commercially available. However, primary cells have limited growth potential to undergo large-scale expansion before treatment and to endure genetic manipulation thereafter. This limitation could be circumvented by using iPSCs, which can be reprogrammed from different cell sources and expanded to large cell numbers of the desired cell types.[98] In a proof-of-concept study, Jackow and colleagues could show that induced fibroblasts and keratinocytes derived from CRISPR/Cas9-treated RDEB iPSCs have the therapeutic potential to improve RDEB and to further move on toward clinical *ex vivo* implementation.[83] Just recently, Supp et al. presented data especially important in the context of RDEB *ex vivo* treatment.[91] The inclusion of both, keratinocytes and fibroblasts, corrected skin cell types as sources for required C7 in skin equivalents is likely to be fundamental to create proper C7 anchoring fibrils within the BMZ and to improve long-term functionality and stability. Beside other explanations for the relatively unsuccessful efforts to engraft gene-replaced keratinocytes long-term,[45] the above-described issue might be a further one to be considered in future *ex vivo* therapies for RDEB.

Although the CRISPR/Cas9 technology advanced and is quite promising regarding high correction and targeting efficiencies, many hurdles to its clinical translation still have to be overcome. One major concern is the risk of off-target activity, DSB inductions at unintended areas of the cells genome, which could lead to loss-of-function mutations within tumor suppressor genes, gain-of-function mutations within oncogenes and to possible genomic instability. Although the accessible nature of the skin theoretically enables careful monitoring of edited tissues in genodermatoses patients, possible malignant transformations may not be seen immediately and have to be observed as long-term serious adverse events. Therefore, *in silico* analysis with *in vitro* and/or cell-based analytical methods have to be considered. Recent advances in off-target analysis have been made by the introduction of several methods. Keith Joung's lab has established two highly sensitive, sequencing-efficient *in vitro* screening strategies, GUIDE-seq and CIRCLE-seq, to identify CRISPR/Cas9 genome-wide off-target events.[99,100] Additionally Keith Joung and colleagues could observe translocations that can occur between CRISPR/Cas9-induced on-target and off-target DSBs as well as nuclease-independent DSB hotspots. Together with a recently published LAM-PCR-based approach[101] for the detection of engineered nuclease-induced chromosomal translocations, these techniques are likely to become important new approaches to assess the safety issues of genome editing. However, it might also raise the concern of DSB-induced translocations associated with cancer that might result in an increased safety risk.[102] Attempts to reduce the likelihood of unwanted off-target events included the decrease of duration of nuclease expression by delivering Cas9–gRNA as a RNP complex, improvements of the specificity *via* changing the binding and catalytic activity of the nuclease[100,103–105] and the inactivation of one of the two catalytic domains of the spCas9 protein.[106] Especially the application of a Cas9 nickase, only differing in an aspartate-to-alanine (D10A) mutation in the RuvC catalytic domain, thereby nicking rather than cleaving DNA, has been already shown to improve safety profiles compared to its wild-type version in several EB studies.[55,60,82] Furthermore, the combination of two Cas9 nickases binding in close proximity to each other (double-nicking), the simultaneous nicking of both opposite strands and the addition of a repair template in selection-based approaches, resulted in increased correction and HR efficiencies compared to wild-type spCas9.[55,60] Nickases are the basis for recently published strategies like base editing and prime editing. In contrast to HR approaches, they are not dependent on the simultaneous addition of exogenous sequences to repair the disease causing mutation. Osborn *et al.* described the first base editing approach for RDEB, correction rates of over 20% and over 40% at the DNA and RNA level respectively.[87] However, the non-specific nature of base editing precludes the use of this technology in repetitive target sites and the off-target potential of base editors requires further elucidation. Prime editing might be a good alternative option to HDR, but has not been tested in EB yet. The possible development of prime editing RNPs together with the high transfection efficiency reached *via* electroporation might be the key to highly efficient seamless repair in the near future. Beside this, HR approaches dealing with AAV delivery of single-stranded DNA (ssDNA) repair templates in combination with electroporation of the double-nicking RNPs could be another efficient and safer traceless repair strategy for EB. Similar strategies based on DSB induction have already been shown to achieve 30–70% HR rates in primary blood cells without the necessity of selection.[92,107] For the already highly efficient EJ-based reframing strategies, double-nicking could improve safety issues and also lead to efficient but less homogenous repair outcomes, at least on DNA level. While the genome editing field is advancing exceptionally, especially for *ex vivo* applications, key issues regarding development of appropriate preclinical assays to evaluate off-target effects and possible resulting cytotoxicity, genotoxicity and tumorogenicity still remain. Although *in vitro* and *in silico* models are important for the development of new approaches, animal models have always been used to more definitively explore safety issues, before moving into clinics. Animal studies would allow for assessment of the viability and functionality of edited cells in an environment in which they will compete with unmodified cells. Animal models would also be highly important to estimate immunological aspects, like possible tumor development, and the general success of *ex vivo* approaches. Furthermore, humanized animal models could serve as tools to establish and further develop preclinical local *in vivo* delivery approaches for gene editing, since efficient delivery and targeting *in vivo* are still big hurdles to become clinically relevant. Keeping all advantages, developments and safety considerations of each single gene editing strategy in mind it should be possible to develop a personalized reagent that is ready for clinical use within few months from diagnosis. The development of a common well-defined pipeline for translation of a gene therapeutic agent might be the ultimate path for precision medicine not only for EB but also for several other genetic diseases.

Acknowledgment

The authors thank Oliver P. March and Johann W. Bauer for proofreading this book chapter and DEBRA Austria for funding.

References

1. Ishida-Yamamoto A, Igawa S. Genetic skin diseases related to desmosomes and corneodesmosomes. *J Dermatol Sci.* 2014;74(2):99–105.
2. Castori M, Morrone A, Kanitakis J, Grammatico P. Genetic skin diseases predisposing to basal cell carcinoma. *Eur J Dermatol.* 2012;22(3):299–309.
3. Has C, Nystrom A. Epidermal basement membrane in health and disease. *Curr Top Membr.* 2015;76:117–170.
4. Feramisco JD, Sadreyev RI, Murray ML, Grishin NV, Tsao H. Phenotypic and genotypic analyses of genetic skin disease through the Online Mendelian Inheritance in Man (OMIM) database. *J Invest Dermatol.* 2009;129(11):2628–2636.
5. Lehmann J, Schubert S, Emmert S. Xeroderma pigmentosum: diagnostic procedures, interdisciplinary patient care, and novel therapeutic approaches. *J Dtsch Dermatol Ges.* 2014;12(10):867–872.
6. Fine JD, Bruckner-Tuderman L, Eady RAJ, et al. Inherited epidermolysis bullosa: updated recommendations on diagnosis and classification. *J Am Acad Dermatol.* 2014;70(6):1103–1126.
7. Has C, Bauer JW, Bodemer C, et al. Consensus reclassification of inherited epidermolysis bullosa and other disorders with skin fragility. *Br J Dermatol.* 2020;183(4):614–627.
8. Coulombe PA, Kerns ML, Fuchs E. Epidermolysis bullosa simplex: a paradigm for disorders of tissue fragility. *J Clin Invest.* 2009;119(7):1784–1793.
9. Horn HM, Tidman MJ. The clinical spectrum of epidermolysis bullosa simplex. *Br J Dermatol.* 2000;142(3):468–472.
10. De Rosa L, Koller U, Bauer JW, De Luca M, Reichelt J. Advances on potential therapeutic options for epidermolysis bullosa. *Expert Opin Orphan Drugs.* 2018;6(4):283–293.
11. Angelis A, Kanavos P, Lopez-Bastida J, et al. Social/economic costs and health-related quality of life in patients with epidermolysis bullosa in Europe. *Eur J Health Econ.* 2016;17(Suppl. 1):31–42.
12. Has C, South A, Uitto J. Molecular therapeutics in development for epidermolysis bullosa: update 2020. *Mol Diagn Ther.* 2020;24(3):299–309.
13. Sriram G, Bigliardi PL, Bigliardi-Qi M. Fibroblast heterogeneity and its implications for engineering organotypic skin models in vitro. *Eur J Cell Biol.* 2015;94(11):483–512.
14. Bornert O, Peking P, Bremer J, et al. RNA-based therapies for genodermatoses. *Exp Dermatol.* 2017;26(1):3–10.
15. Szeverenyi I, Cassidy AJ, Chung CW, et al. The Human Intermediate Filament Database: comprehensive information on a gene family involved in many human diseases. *Hum Mutat.* 2008;29(3):351–360.
16. Fuchs E. Keratins and the skin. *Annu Rev Cell Dev Biol.* 1995;11:123–153.
17. Letai A, Coulombe PA, McCormick MB, Yu QC, Hutton E, Fuchs E. Disease severity correlates with position of keratin point mutations in patients with epidermolysis bullosa simplex. *Proc Natl Acad Sci U S A.* 1993;90(8):3197–3201.
18. Sorensen CB, Ladekjaer-Mikkelsen AS, Andresen BS, et al. Identification of novel and known mutations in the genes for keratin 5 and 14 in Danish patients with epidermolysis bullosa simplex: correlation between genotype and phenotype. *J Invest Dermatol.* 1999;112(2):184–190.
19. Ishida-Yamamoto A, McGrath JA, Chapman SJ, Leigh IM, Lane EB, Eady RA. Epidermolysis bullosa simplex (Dowling-Meara type) is a genetic disease characterized by an abnormal keratin-filament network involving keratins K5 and K14. *J Invest Dermatol.* 1991;97(6):959–968.

20. Kitajima Y, Inoue S, Yaoita H. Abnormal organization of keratin intermediate filaments in cultured keratinocytes of epidermolysis bullosa simplex. *Arch Dermatol Res.* 1989;281(1):5–10.
21. Chamcheu JC, Lorie EP, Akgul B, et al. Characterization of immortalized human epidermolysis bullosa simplex (KRT5) cell lines: trimethylamine N-oxide protects the keratin cytoskeleton against disruptive stress condition. *J Dermatol Sci.* 2009;53(3):198–206.
22. Chamcheu JC, Virtanen M, Navsaria H, Bowden PE, Vahlquist A, Torma H. Epidermolysis bullosa simplex due to KRT5 mutations: mutation-related differences in cellular fragility and the protective effects of trimethylamine N-oxide in cultured primary keratinocytes. *Br J Dermatol.* 2010;162(5):980–989.
23. Chamcheu JC, Navsaria H, Pihl-Lundin I, Liovic M, Vahlquist A, Torma H. Chemical chaperones protect epidermolysis bullosa simplex keratinocytes from heat stress-induced keratin aggregation: involvement of heat shock proteins and MAP kinases. *J Invest Dermatol.* 2011;131(8):1684–1691.
24. Aushev M, Koller U, Mussolino C, Cathomen T, Reichelt J. Traceless targeting and isolation of gene-edited immortalized keratinocytes from epidermolysis bullosa simplex patients. *Mol Ther Methods Clin Dev.* 2017;6:112–123.
25. March OP, Lettner T, Klausegger A, et al. Gene editing-mediated disruption of epidermolytic ichthyosis-associated KRT10 alleles restores filament stability in keratinocytes. *J Invest Dermatol.* 2019;139(8):1699–1710. e1696.
26. Chavanas S, Pulkkinen L, Gache Y, et al. A homozygous nonsense mutation in the PLEC1 gene in patients with epidermolysis bullosa simplex with muscular dystrophy. *J Clin Invest.* 1996;98(10):2196–2200.
27. He Y, Maier K, Leppert J, et al. Monoallelic mutations in the translation initiation codon of KLHL24 cause skin fragility. *Am J Hum Genet.* 2016;99(6):1395–1404.
28. Lin Z, Li S, Feng C, et al. Stabilizing mutations of KLHL24 ubiquitin ligase cause loss of keratin 14 and human skin fragility. *Nat Genet.* 2016;48(12):1508–1516.
29. Hirsch T, Rothoeft T, Teig N, et al. Regeneration of the entire human epidermis using transgenic stem cells. *Nature.* 2017;551(7680):327–332.
30. Peking P, Breitenbach JS, Ablinger M, et al. An ex vivo RNA trans-splicing strategy to correct human generalized severe epidermolysis bullosa simplex. *Br J Dermatol.* 2019;160(1):141–148.
31. Liemberger B, Pinon Hofbauer J, Wally V, et al. RNA trans-splicing modulation via antisense molecule interference. *Int J Mol Sci.* 2018;19(3):762.
32. Wally V, Brunner M, Lettner T, et al. K14 mRNA reprogramming for dominant epidermolysis bullosa simplex. *Hum Mol Genet.* 2010;19(23):4715–4725.
33. Benati D, Miselli F, Cocchiarella F, et al. CRISPR/Cas9-mediated in situ correction of LAMB3 gene in keratinocytes derived from a junctional epidermolysis bullosa patient. *Mol Ther.* 2018;26(11):2592–2603.
34. Melo SP, Lisowski L, Bashkirova E, et al. Somatic correction of junctional epidermolysis bullosa by a highly recombinogenic AAV variant. *Mol Ther.* 2014;22(4):725–733.
35. Dellambra E, Vailly J, Pellegrini G, et al. Corrective transduction of human epidermal stem cells in laminin-5-dependent junctional epidermolysis bullosa. *Hum Gene Ther.* 1998;9(9):1359–1370.
36. Bauer JW, Koller J, Murauer EM, et al. Closure of a large chronic wound through transplantation of gene-corrected epidermal stem cells. *J Invest Dermatol.* 2017;137(3):778–781.
37. Mavilio F, Pellegrini G, Ferrari S, et al. Correction of junctional epidermolysis bullosa by transplantation of genetically modified epidermal stem cells. *Nat Med.* 2006;12(12):1397–1402.

38. Gruber C, Gratz IK, Murauer EM, et al. Spliceosome-mediated RNA trans-splicing facilitates targeted delivery of suicide genes to cancer cells. *Mol Cancer Ther*. 2011;10 (2):233–241.
39. Wertheim-Tysarowska K, Sobczynska-Tomaszewska A, Kowalewski C, et al. The COL7A1 mutation database. *Hum Mutat*. 2012;33(2):327–331.
40. Lin AN, Carter DM. Epidermolysis bullosa. *Annu Rev Med*. 1993;44:189–199.
41. Tockner B, Kocher T, Hainzl S, et al. Construction and validation of a RNA trans-splicing molecule suitable to repair a large number of COL7A1 mutations. *Gene Ther*. 2016;23(11):775–784.
42. Wagner JE, Ishida-Yamamoto A, McGrath JA, et al. Bone marrow transplantation for recessive dystrophic epidermolysis bullosa. *N Engl J Med*. 2010;363(7):629–639.
43. Wong T, Gammon L, Liu L, et al. Potential of fibroblast cell therapy for recessive dystrophic epidermolysis bullosa. *J Invest Dermatol*. 2008;128(9):2179–2189.
44. Sebastiano V, Zhen HH, Haddad B, et al. Human COL7A1-corrected induced pluripotent stem cells for the treatment of recessive dystrophic epidermolysis bullosa. *Sci Transl Med*. 2014;6(264), 264ra163.
45. Siprashvili Z, Nguyen NT, Gorell ES, et al. Safety and wound outcomes following genetically corrected autologous epidermal grafts in patients with recessive dystrophic epidermolysis bullosa. *JAMA*. 2016;316(17):1808–1817.
46. Woodley DT, Wang X, Amir M, et al. Intravenously injected recombinant human type VII collagen homes to skin wounds and restores skin integrity of dystrophic epidermolysis bullosa. *J Invest Dermatol*. 2013;133(7):1910–1913.
47. Petrof G, Lwin SM, Martinez-Queipo M, et al. Potential of systemic allogeneic mesenchymal stromal cell therapy for children with recessive dystrophic epidermolysis bullosa. *J Invest Dermatol*. 2015;135(9):2319–2321.
48. March OP, Kocher T, Koller U. Context-dependent strategies for enhanced genome editing of genodermatoses. *Cell*. 2020;9(1):112.
49. March OP, Reichelt J, Koller U. Gene editing for skin diseases: designer nucleases as tools for gene therapy of skin fragility disorders. *Exp Physiol*. 2017;103(4):449–455.
50. Titeux M, Pendaries V, Zanta-Boussif MA, et al. SIN retroviral vectors expressing COL7A1 under human promoters for ex vivo gene therapy of recessive dystrophic epidermolysis bullosa. *Mol Ther*. 2010;18(8):1509–1518.
51. Montini E, Cesana D, Schmidt M, et al. The genotoxic potential of retroviral vectors is strongly modulated by vector design and integration site selection in a mouse model of HSC gene therapy. *J Clin Invest*. 2009;119(4):964–975.
52. Sfeir A, Symington LS. Microhomology-mediated end joining: a back-up survival mechanism or dedicated pathway? *Trends Biochem Sci*. 2015;40(11):701–714.
53. Danner E, Bashir S, Yumlu S, Wurst W, Wefers B, Kuhn R. Control of gene editing by manipulation of DNA repair mechanisms. *Mamm Genome*. 2017;28(7–8):262–274.
54. Osborn MJ, Starker CG, McElroy AN, et al. TALEN-based gene correction for epidermolysis bullosa. *Mol Ther*. 2013;21(6):1151–1159.
55. Kocher T, Wagner RN, Klausegger A, et al. Improved double-nicking strategies for COL7A1 editing by homologous recombination. *Mol Ther Nucleic Acids*. 2019;18:496–507.
56. Rodgers K, McVey M. Error-prone repair of DNA double-strand breaks. *J Cell Physiol*. 2016;231(1):15–24.
57. Kocher T, March OP, Bischof J, et al. Predictable CRISPR/Cas9-mediated COL7A1 reframing for dystrophic epidermolysis bullosa. *J Invest Dermatol*. 2020;140 (10):1985–1993.
58. Chakrabarti AM, Henser-Brownhill T, Monserrat J, Poetsch AR, Luscombe NM, Scaffidi P. Target-specific precision of CRISPR-mediated genome editing. *Mol Cell*. 2019;73(4):699–713. e696.

59. Ran FA, Hsu PD, Lin CY, et al. Double nicking by RNA-guided CRISPR Cas9 for enhanced genome editing specificity. *Cell.* 2013;154(6):1380–1389.
60. Kocher T, Peking P, Klausegger A, et al. Cut and paste: efficient homology-directed repair of a dominant negative KRT14 mutation via CRISPR/Cas9 nickases. *Mol Ther.* 2017;25(11):2585–2598.
61. Anzalone AV, Randolph PB, Davis JR, et al. Search-and-replace genome editing without double-strand breaks or donor DNA. *Nature.* 2019;576(7785):149–157.
62. Santiago Y, Chan E, Liu PQ, et al. Targeted gene knockout in mammalian cells by using engineered zinc-finger nucleases. *Proc Natl Acad Sci U S A.* 2008;105(15):5809–5814.
63. Lee CH, Coulombe PA. Self-organization of keratin intermediate filaments into cross-linked networks. *J Cell Biol.* 2009;186(3):409–421.
64. Terheyden P, Grimberg G, Hausser I, et al. Recessive epidermolytic hyperkeratosis caused by a previously unreported termination codon mutation in the keratin 10 gene. *J Invest Dermatol.* 2009;129(11):2721–2723.
65. Shinkuma S, Guo Z, Christiano AM. Site-specific genome editing for correction of induced pluripotent stem cells derived from dominant dystrophic epidermolysis bullosa. *Proc Natl Acad Sci U S A.* 2016;113(20):5676–5681.
66. Takashima S, Shinkuma S, Fujita Y, et al. Efficient gene reframing therapy for recessive dystrophic epidermolysis bullosa with CRISPR/Cas9. *J Invest Dermatol.* 2019;139(8):1711–1721. e1714.
67. Escamez MJ, Garcia M, Cuadrado-Corrales N, et al. The first COL7A1 mutation survey in a large Spanish dystrophic epidermolysis bullosa cohort: c.6527insC disclosed as an unusually recurrent mutation. *Br J Dermatol.* 2010;163(1):155–161.
68. Chamorro C, Mencia A, Almarza D, et al. Gene editing for the efficient correction of a recurrent COL7A1 mutation in recessive dystrophic epidermolysis bullosa keratinocytes. *Mol Ther Nucleic Acids.* 2016;5: e307.
69. Wu W, Lu Z, Li F, et al. Efficient in vivo gene editing using ribonucleoproteins in skin stem cells of recessive dystrophic epidermolysis bullosa mouse model. *Proc Natl Acad Sci U S A.* 2017;114(7):1660–1665.
70. Mencia A, Chamorro C, Bonafont J, et al. Deletion of a pathogenic mutation-containing exon of COL7A1 allows clonal gene editing correction of RDEB patient epidermal stem cells. *Mol Ther Nucleic Acids.* 2018;11:68–78.
71. Bonafont J, Mencia A, Garcia M, et al. Clinically relevant correction of recessive dystrophic epidermolysis bullosa by dual sgRNA CRISPR/Cas9-mediated gene editing. *Mol Ther.* 2019;27(5):986–998.
72. Woodley DT, Hou Y, Martin S, Li W, Chen M. Characterization of molecular mechanisms underlying mutations in dystrophic epidermolysis bullosa using site-directed mutagenesis. *J Biol Chem.* 2008;283(26):17838–17845.
73. Goto M, Sawamura D, Nishie W, et al. Targeted skipping of a single exon harboring a premature termination codon mutation: implications and potential for gene correction therapy for selective dystrophic epidermolysis bullosa patients. *J Invest Dermatol.* 2006;126(12):2614–2620.
74. Turczynski S, Titeux M, Tonasso L, Decha A, Ishida-Yamamoto A, Hovnanian A. Targeted exon skipping restores type VII collagen expression and anchoring fibril formation in an in vivo RDEB model. *J Invest Dermatol.* 2016;136(12):2387–2395.
75. Bremer J, Bornert O, Nystrom A, et al. Antisense oligonucleotide-mediated exon skipping as a systemic therapeutic approach for recessive dystrophic epidermolysis bullosa. *Mol Ther Nucleic Acids.* 2016;5(10): e379.
76. Bornert O, Hogervorst M, Nauroy P, et al. QR-313, an antisense oligonucleotide, shows therapeutic efficacy for treatment of dominant and recessive dystrophic

epidermolysis bullosa: a preclinical study. *J Invest Dermatol.* 2020. S0022-202X(20) 32061-3.
77. Shen MW, Arbab M, Hsu JY, et al. Predictable and precise template-free CRISPR editing of pathogenic variants. *Nature.* 2018;563(7733):646–651.
78. McKee C, Chaudhry GR. Advances and challenges in stem cell culture. *Colloids Surf B Biointerfaces.* 2017;159:62–77.
79. Webber BR, Osborn M, McElroy A, et al. CRISPR/Cas9-based genetic correction for recessive dystrophic epidermolysis bullosa. *Regen Med.* 2016;1:16014.
80. Izmiryan A, Danos O, Hovnanian A. Meganuclease-mediated COL7A1 gene correction for recessive dystrophic epidermolysis bullosa. *J Invest Dermatol.* 2016;136 (4):872–875.
81. Izmiryan A, Ganier C, Bovolenta M, Schmitt A, Mavilio F, Hovnanian A. Ex vivo COL7A1 correction for recessive dystrophic epidermolysis bullosa using CRISPR/Cas9 and homology-directed repair. *Mol Ther Nucleic Acids.* 2018;12:554–567.
82. Hainzl S, Peking P, Kocher T, et al. COL7A1 editing via CRISPR/Cas9 in recessive dystrophic epidermolysis bullosa. *Mol Ther.* 2017;25(11):2573–2584.
83. Jackow J, Guo Z, Hansen C, et al. CRISPR/Cas9-based targeted genome editing for correction of recessive dystrophic epidermolysis bullosa using iPS cells. *Proc Natl Acad Sci U S A.* 2019;116(52):26846–26852.
84. Itoh M, Kawagoe S, Tamai K, Nakagawa H, Asahina A, Okano HJ. Footprint-free gene mutation correction in induced pluripotent stem cell (iPSC) derived from recessive dystrophic epidermolysis bullosa (RDEB) using the CRISPR/Cas9 and piggyBac transposon system. *J Dermatol Sci.* 2020;98(3):163–172.
85. Coluccio A, Miselli F, Lombardo A, et al. Targeted gene addition in human epithelial stem cells by zinc-finger nuclease-mediated homologous recombination. *Mol Ther.* 2013;21:1695–1704.
86. Nakayama M. Homologous recombination in human iPS and ES cells for use in gene correction therapy. *Drug Discov Today.* 2010;15(5–6):198–202.
87. Osborn MJ, Newby GA, McElroy AN, et al. Base editor correction of COL7A1 in recessive dystrophic epidermolysis bullosa patient-derived fibroblasts and iPSCs. *J Invest Dermatol.* 2019;140(2):338–347.
88. Itoh M, Kiuru M, Cairo MS, Christiano AM. Generation of keratinocytes from normal and recessive dystrophic epidermolysis bullosa-induced pluripotent stem cells. *Proc Natl Acad Sci U S A.* 2011;108(21):8797–8802.
89. Itoh M, Umegaki-Arao N, Guo Z, Liu L, Higgins CA, Christiano AM. Generation of 3D skin equivalents fully reconstituted from human induced pluripotent stem cells (iPSCs). *PLoS One.* 2013;8(10): e77673.
90. Wenzel D, Bayerl J, Nystrom A, Bruckner-Tuderman L, Meixner A, Penninger JM. Genetically corrected iPSCs as cell therapy for recessive dystrophic epidermolysis bullosa. *Sci Transl Med.* 2014;6(264): 264ra165.
91. Supp DM, Hahn JM, Combs KA, et al. Collagen VII expression is required in both keratinocytes and fibroblasts for anchoring fibril formation in bilayer engineered skin substitutes. *Cell Transplant.* 2019;28(9–10):1242–1256.
92. Osborn MJ, Lees CJ, McElroy AN, et al. CRISPR/Cas9-based cellular engineering for targeted gene overexpression. *Int J Mol Sci.* 2018;19(4):946.
93. Rees HA, Liu DR. Base editing: precision chemistry on the genome and transcriptome of living cells. *Nat Rev Genet.* 2018;19(12):770–788.
94. Komor AC, Kim YB, Packer MS, Zuris JA, Liu DR. Programmable editing of a target base in genomic DNA without double-stranded DNA cleavage. *Nature.* 2016;533 (7603):420–424.
95. Gaudelli NM, Komor AC, Rees HA, et al. Programmable base editing of A*T to G*C in genomic DNA without DNA cleavage. *Nature.* 2017;551(7681):464–471.

96. Koblan LW, Doman JL, Wilson C, et al. Improving cytidine and adenine base editors by expression optimization and ancestral reconstruction. *Nat Biotechnol.* 2018;36 (9):843–846.
97. Hendel A, Bak RO, Clark JT, et al. Chemically modified guide RNAs enhance CRISPR-Cas genome editing in human primary cells. *Nat Biotechnol.* 2015;33 (9):985–989.
98. Guo Z, Higgins CA, Gillette BM, et al. Building a microphysiological skin model from induced pluripotent stem cells. *Stem Cell Res Ther.* 2013;4(Suppl. 1):S2.
99. Tsai SQ, Zheng Z, Nguyen NT, et al. GUIDE-seq enables genome-wide profiling of off-target cleavage by CRISPR-Cas nucleases. *Nat Biotechnol.* 2015;33(2):187–197.
100. Tsai SQ, Nguyen NT, Malagon-Lopez J, Topkar VV, Aryee MJ, Joung JK. CIRCLE-seq: a highly sensitive in vitro screen for genome-wide CRISPR-Cas9 nuclease off-targets. *Nat Methods.* 2017;14(6):607–614.
101. Frock RL, Hu JZ, Meyers RM, Ho YJ, Kii E, Alt FW. Genome-wide detection of DNA double-stranded breaks induced by engineered nucleases. *Nat Biotechnol.* 2015;33(2):179–186.
102. Corrigan-Curay J, O'Reilly M, Kohn DB, et al. Genome editing technologies: defining a path to clinic. *Mol Ther.* 2015;23(5):796–806.
103. Casini A, Olivieri M, Petris G, et al. A highly specific SpCas9 variant is identified by in vivo screening in yeast. *Nat Biotechnol.* 2018;36(3):265–271.
104. Chen JS, Dagdas YS, Kleinstiver BP, et al. Enhanced proofreading governs CRISPR-Cas9 targeting accuracy. *Nature.* 2017;550(7676):407–410.
105. Vakulskas CA, Dever DP, Rettig GR, et al. A high-fidelity Cas9 mutant delivered as a ribonucleoprotein complex enables efficient gene editing in human hematopoietic stem and progenitor cells. *Nat Med.* 2018;24(8):1216–1224.
106. Jinek M, Chylinski K, Fonfara I, Hauer M, Doudna JA, Charpentier E. A programmable dual-RNA-guided DNA endonuclease in adaptive bacterial immunity. *Science.* 2012;337(6096):816–821.
107. Porteus MH. A new class of medicines through DNA editing. *N Engl J Med.* 2019;380 (10):947–959.

CHAPTER FOUR

Targeted genome editing for the correction or alleviation of primary Immunodeficiencies

Christopher J. Sipe[a,b,c,d,e,†], Patricia N. Claudio Vázquez[a,b,c,d,e,†], Joseph G. Skeate[a,b,c,d], R. Scott McIvor[a,b,c,d,e], and Branden S. Moriarity[a,b,c,d,e,*]

[a]Department of Pediatrics, University of Minnesota, Minneapolis, MN, United States
[b]Masonic Cancer Center, University of Minnesota, Minneapolis, MN, United States
[c]Center for Genome Engineering, University of Minnesota, Minneapolis, MN, United States
[d]Stem Cell Institute, University of Minnesota, Minneapolis, MN, United States
[e]Department of Genetics, Cell Biology and Development, University of Minnesota, Minneapolis, MN, United States
*Corresponding author: e-mail address: mori0164@umn.edu

Contents

1. Introduction — 114
   1.1 Primary immunodeficiencies (PIDs) — 114
2. Classifications of PIDs — 117
   2.1 X-linked severe combined immunodeficiency (X-SCID) — 117
   2.2 Adenosine deaminase deficiency (ADA-SCID) — 118
   2.3 Artemis-deficient SCID (SCID-A) — 119
   2.4 Chronic granulomatous disease (CGD) — 120
   2.5 Wiskott-Aldrich syndrome (WAS) — 122
   2.6 Other PIDs — 123
3. Current PID therapies and clinical trials — 123
   3.1 Non-transplant therapies — 124
   3.2 Allogeneic HSC transplant — 125
   3.3 Gene addition therapy — 129
4. Next generation of gene-editing tools — 132
   4.1 DNA repair pathways — 133
References — 145

[†] Authors contributed equally.

Abstract

Primary immunodeficiencies (PID) are a growing list of unique disorders that result in a failure of the innate/adaptive immune systems to fully respond to disease or infection. PIDs are classified into five broad categories; B cell disorders, combined B and T cell disorders, phagocytic disorders, complement disorders, and disorders with recurrent fevers and inflammation. Many of these disorders, such as X-SCID, WAS, and CGD lead to early death in children if intervention is not implemented. At present, the predominant method of curative therapy remains an allogeneic transplant from a healthy donor, however many complications and limitations exist with his therapy such as availability of donors, graft *vs* host disease, graft rejection, and infection. More recently, gene therapy using viral based complementation vectors have successfully been implemented to functionally correct patient cells in an autologous transplant, but these methods carry significant risks, including insertional mutagenesis, and provide non-physiological gene expression. For these reasons, gene-editing reagents such as targeted nucleases, base editors (BE), and prime editors (PE) are being explored. The BE and PE tools, sometimes referred to as digital editors, are of very high interest as they provide both enhanced molecular specificity and do not rely on DNA repair pathways after DSBs to change individual base pairs or directly replace DNA sequences responsible for pathogenic phenotypes. With this in mind the purpose of this chapter is to highlight some of the most common PIDs found within the human population, discuss successes and shortcomings of previous intervention strategies, and highlight how the next generation of gene-editing tools may be deployed to directly repair the underlying genetic causes of this class of disease.

Abbreviations list

A	adenine
AAV	adeno-associated virus
ABE	adenine base editor
APOBEC	apolipoprotein B mRNA editing enzyme, catalytic polypeptide-like
alt-EJ	alternative end-joining
BE	base editor
BLESS	direct labeling of *in situ* breaks, enrichment on streptavidin and next-generation sequencing
BM	bone marrow
BMT	bone marrow transplant
bp	base pair
C	cytosine
C-NHEJ	classical/canonical nonhomologous end-joining
CAR	chimeric antigen receptor
Cas9	CRISPR associated protein 9
CBE	cytosine base editors
CGD	chronic granulomatous disease
ChIP-seq	chromatin immunoprecipitation and high-throughput sequencing
CIRCLE-seq	circularization for *in vitro* reporting of cleavage effects by sequencing
CMV	cytomegalovirus
CRISPR	clustered regularly interspaced short palindromic repeats

crRNA	CRISPR RNA
CVID	common variable immunodeficiency
Digenome-seq	digested genome sequencing
DSB	double-strand break
EBV	Epstein–Barr virus
ERT	enzyme replacement therapy
GvHD	graft-versus-host disease
gRNA	guide RNA
GUIDE-seq	genome-wide, unbiased identification of DSBs enabled by sequencing
HIV	human immunodeficiency virus
HLA	human leukocyte antigen
HDR	homology-directed repair
HR	homologous recombination
HSC	hematopoietic stem cell
HSPC	hematopoietic stem and progenitor cell
HSCT	hematopoietic stem cell transplant
I	inosine
IDLV	integrase-defective lentiviral vector
indel	insertion and/or deletion
LAD	leukocyte adhesion deficiency
LLNP	lipid-like nanoparticle
LV	lentivirus
MAC	myeloablative conditioning
MHC	major histocompatibility complex
MRD	matched related donor
MUD	match unrelated donor
NADPH	nicotinamide adenine dinucleotide phosphate
nCas9	Cas9 nickase
PAM	protospacer adjacent motif
PE	prime editor
PID	primary immunodeficiency
pmSTOP	premature STOP
rAAV	recombinant AAV
RIC	reduced-intensity conditioning
RNA	ribonucleic acid
RT	reverse transcriptase
SSA	single-strand annealing
SCD	sickle cell disease
SCID	severe combined immunodeficiency
SID	secondary immunodeficiency
sgRNA	single guide RNA
SNV	single nucleotide variants
spCas9	*Streptococcus pyogenes* Cas9
ssODNs	single-stranded oligo donors
T	thymine
TALEN	transcription activator-like effector nucleases
TF	transcription factor

tracrRNA	trans-activating CRISPR RNA
U	uracil
UGI	uracil glycosylase inhibitor
WAS	Wiskott-Aldrich syndrome
WGS	whole-genome sequencing
WT	wild type
ZFN	zinc-finger nuclease
X-MEN	X-linked immunodeficiency magnesium defect, Epstein-Barr virus, and neoplasia
X-SCID	X-linked severe combined immunodeficiency
XLP	X-linked lymphoproliferative disease

1. Introduction
1.1 Primary immunodeficiencies (PIDs)

Primary immunodeficiencies (PIDs) are a large class of genetic disorders that result in general susceptibility to infections and failure of immune surveillance by either the innate and/or adaptive immune systems. Additional complications include, but are not limited to, allergic reactions, autoinflammation, autoimmune syndromes, and the emergence of malignant disease. Inheritance patterns can be X-linked, autosomal dominant or recessive, and can exhibit complete or incomplete penetrance. As of 2019, 406 documented PIDs span 430 different genes, each of which can bear many different mutations.[1] The same committee of experts that compiled this list previously described a total of 250 different PIDs in 2014, revealing our ever-growing knowledge and discovery of this broad class of disorders.[2]

Historically PIDs, sometimes referred to as inborn errors of immunity, have been viewed as a group of extremely rare disorders affecting 1:10,000 to 1:50,000 live births.[3] Recent studies, however, suggest that the incident rate of a PID is much higher, affecting as many as 1:1000 to 1:5000 live births.[4] Furthermore, some groups have claimed that as many as ~1:500 live births may suffer from some form of a PID.[5] This wide statistical range disparity can be explained by ongoing research characterizing genes and proteins involved with PIDs, the refinement of phenotypic classifications, and advancements in genetic detection in recent years. The prevalence of PIDs is also considerably higher in populations where consanguineous inheritance is common due to familial and breeding selection.[6]

The abundant range of clinical presentations is dictated by the underlying pathologies of which gene is affected and what specific mutation is present. Original modalities of treatment such as allogeneic hematopoietic stem cell

(HSC) bone marrow (BM) transplants were utilized for some time as a one-size-fits-all method to address PIDs. However, the development of curative therapies has greatly lagged behind the discovery of new disorders. Recent advances in both genomic sequencing and enhancements in our ability to develop and implement targeted gene editing may turn the tide on this battle, and as such in this chapter we will review and discuss some of the most common forms of PIDs, how they have been mitigated historically, and ways in which they may be more effectively treated in the near future.

1.1.1 Phenotypes

In 1952, the first PID was described, X-linked agammaglobulinemia (XLA), a disorder in which B cells are not capable of maturing, and those affected lack gamma globulin.[7] Since then, there has been a categorization of PIDs into five different clustered subtypes based on their phenotypic and in some cases underlying genetic similarities. These include B cell disorders, combined B and T cell disorders, phagocytic disorders, complement disorders, and disorders with recurrent fevers and inflammation.

Classic examples of PIDs within each clustered subtype include XLA and common variable immunodeficiency (CVID) as B cell disorders, severe combined immunodeficiency (SCID), and Wiskott-Aldrich syndrome as combined B and T cell disorders, and chronic granulomatous disorder (CGD) as a phagocytic disorder.[8] Expanding on the five different PID clustered subtype categories, the *2019 Human Inborn Errors of Immunity: Update on the Classification from the International Union of Immunological Societies Expert Committee* has laid out 10 different categories for easier classification and diagnosis as is necessary for a disease group of this size.[1,2] This new categorization includes;

1. Cellular and humoral immunodeficiencies
2. Combined immunodeficiencies
3. Antibody deficiencies
4. Immune dysregulations
5. Congenital defects of phagocytosis
6. Intrinsic and innate immune defects
7. Auto-inflammation disorders
8. Complement deficiencies
9. BM failure disorders
10. Altered phenocopies of PID

Because immune cells are affected, the category of BM failure disorders such as Fanconi anemia is typically considered as PIDs and may be included as an inborn error of immunity. These PIDs arising from genetic inheritance can directly lead to increased susceptibility to secondary immunodeficiencies (SIDs) acquired later in life such as malignancies, viral hepatitis, and AIDS.[9]

1.1.2 Diagnosis
Recognizing a PID in a young child can be difficult unless the manifestation is severe such as an obvious failure to thrive or chronic infections. Physicians utilize a mixture of key co-occurring symptoms and risk factors to select individuals that may need to start on a treatment plan while laboratory verification of PID classification/diagnosis is carried out.[10] Some patients are not diagnosed until later in life after a gradual onset of symptoms has occurred, such as increased frequencies or duration of infections, failure to thrive, oral thrush, organ/skin abscesses, recurrent pneumonia, among others. Functional immunological readouts and histological analysis of blood smears are used to hone treatment regimens and necessary interventions to improve the patient's quality of life. Advanced screenings for PIDs in prenatal and newborn children can also be used to diagnose disease before the onset of symptoms. This is especially important in families that have a close relative impacted by a PID or familial history of the disease. An example of a newborn screen is the T cell receptor excision circles (TRECs) assay, which tests for T cell biomarkers.[11] Prenatal screening is also an important tool for early detection with families who have an affected individual either closely or distantly related.[12]

1.1.3 Significance
Outcomes, cases, deaths, hospital admissions, and other statistics relating to PIDs are usually reported as case studies of individual care centers across the world, making general statements about outcomes difficult. In the United States from 2003 to 2012, there were 26,794 pediatric patients hospitalized with some form of PID.[13] Overall, the mortality of children between the ages of 2 and 18 was relatively low at 1.99%. However, when considering the cohort of children ages 0–5, the mortality rate was 79.85%, a striking figure.[13] In this study, the average cost of hospital admission, which usually lasts less than a week, was $35,480. Most children require multiple hospital admissions for bacterial and other infections throughout their lives.[14] There is great difficulty accumulating more recent data post 2012 on statistical outcomes based on the wide range of different disorders associated with PIDs across several institutions and countries. Even though most subtypes of PID are rare, when combined with still unclassified disorders

they are more prevalent than previously thought. Presentations of severe PID arise from life-threatening recurrent infections and are thus detected within several months of birth. Less extreme cases on average take 12 years for a proper diagnosis, often with children's long-term mental and physical health affected.

Here, the focus will be a brief description of common PIDs, current therapeutic approaches including those that are now in or have completed clinical trials, and a discussion of the future of PID treatment with next-generation genome editing tools that provide a safer and more effective therapeutic approach compared to current strategies.

2. Classifications of PIDs
2.1 X-linked severe combined immunodeficiency (X-SCID)

Severe Combined Immunodeficiency (SCID) is a failure of the cellular and humoral immune systems with defects in T cells, B cells, NK cells, and several other immune effectors. The most common form of SCID is X-linked SCID (X-SCID) with 45–60% of all SCID cases being attributed to mutations on the X chromosome and hence inherited primarily by males.[15,16] These mutations are clustered in the Interleukin 2 Receptor Subunit Gamma (*IL2RG*) gene, which encodes the common gamma chain (γ_c), the intracellular subunit of the interleukin 2 receptor (IL-2) as well as the receptors for IL-4, IL-7, IL-9, IL-15, and IL-21. It is estimated that this form of SCID occurs in 1:50,000 to 1:100,000 live births.[17] This disease is fatal within the first 2 years of life without a transplant using either allogenic or genetically corrected autologous HSCs. For this reason, extensive effort in the development of gene therapies for PIDs has centered around SCID.

2.1.1 Symptoms

Infants with X-SCID typically present with symptoms between 3 and 6 months of age. They are characterized by a failure to thrive, lymphopenia, atrophy of the thymus, and an early onset of severe and persistent or recurring infections, many of which are caused by opportunistic organisms, accompanied by diarrhea. Opportunistic organisms like *Pneumocystis jirovecii* and *Candida albicans* are often the cause of viral and fungal infections in patients with SCID. Research has shown that between 20% and 71% of all SCID patients develop *Pneumocystis jirovecii* pneumonia.[18] *Candida albicans* causes candidiasis, a fungal infection of the skin or mucosal membranes that thrives without the control of a normally-functioning immune system in moist environments.[19] Other common symptoms are rashes, cough, congestion, fevers, and sepsis.[17]

2.1.2 Biochemistry

Over 200 pathogenic variants that lead to X-SCID have been identified.[20] They are primarily single-nucleotide changes[21] and can be found on all 8 exons of the *IL2RG* gene.[20] *IL2RG* encodes the γ_c subunit of the IL-2 receptor which, in conjunction with the α and β subunits, forms the complete IL-2 receptor. These mutations create a cascade of signaling defects that affect lymphocyte development. The γ_c subunit is necessary for proper IL-2 signaling and signals a response even at a low cytokine concentration. IL-2 has many critical roles, such as the growth, proliferation, and differentiation of naïve T cells into effector T cells and the increase of NK cell cytolytic activity. In addition to IL-2 signaling, it has been shown that γ_c is a shared receptor component for many other cytokines (IL-2, IL-4, IL-7, IL-9, IL-15, IL-21) of which the absence of signaling contributes to the lack of T and NK lymphocytes, as well as the regulation of B lymphocyte function.[22]

2.2 Adenosine deaminase deficiency (ADA-SCID)

The first PID with a specific defect characterized at the molecular level was identified as adenosine deaminase deficiency (ADA-SCID).[23] This disorder follows an autosomal recessive mode of inheritance and leads to classical SCID with depletion and/or dysfunction of T, B, and NK cells. 10–15% of all SCID cases are the result of ADA deficiency and it is sometimes referred to as T⁻ B⁻ NK⁻ SCID.[24] The absence of white blood cells (lymphocytes) leads to lymphocytopenia, essentially a complete absence of the adaptive immune system causing a severe predisposition to infections. It is estimated that the ADA form of SCID occurs in 1:200,000 to 1:500,000 live births. However, the incidence is markedly higher in specific Amish and Somalian populations where consanguineous mating is prevalent.[25] ADA-SCID occurs in 1:375,000 to 1:660,000 live births in Europe.[23]

2.2.1 Symptoms

Infants with ADA-SCID show a failure to thrive early on in life, which often results in mortality. The absence of lymphatic cells results in failed lymphatic tissue development, such as the thymus and tonsils, leading to several complications. ADA-SCID patients lack humoral and cellular immunity as there are severely reduced populations of T, B, and NK cells, and the few cells that do exist have poor function. For example, T cells that are present are unable to signal through their TCRs.[26] Besides common infections that may or may not be life-threatening, the opportunistic pathogen *Pneumocystis jirovecii* is frequently found in ADA-SCID patients, which leads to dermatitis and diarrhea.[23] Severe developmental delays that accompany the impairment of immune tissue include pulmonary, skeletal, renal, cardiovascular, hepatic, and auditory function. This is thought not only to occur because of immune cell dysregulation but also because of the effect of ADA enzyme deficiency on other systems of the body.

2.2.2 Biochemistry

There are now 70 different identified mutations that lead to ADA-SCID, all located in the ADA gene.[27] ADA is widely expressed but with the highest prevalence and activity in lymphoid cells in the gut and esophageal lining during development as well as when infections occur. Adenosine is derived from the breakdown of adenosine triphosphate (ATP) as well as from RNA, and 2′deoxyadenosine (dAdo) is derived from the breakdown of DNA. Normally, ADA converts these nucleic acids into inosine and 2′deoxyinosine, respectively, which are subsequently metabolized to uric acid. In the absence of ADA, the build-up of dAdo causes accumulation of deoxyadenosine triphosphate (dATP), which inhibits ribonucleotide reductase and, ultimately DNA synthesis during normal immune development.[28] ADA-SCID is thus categorized as a purine metabolic disorder that leads to the accumulation and buildup of toxic purine metabolites leading to improper or absent lymphocyte development.

2.3 Artemis-deficient SCID (SCID-A)

Artemis-deficient SCID (SCID-A) is an autosomal recessive form of SCID which arises from a mutation in the DNA Cross-Link Repair 1C (*DCLRE1C*) gene. This gene encodes the Artemis protein, which has endonucleolytic activity that is essential in the V(D)J (variable, diversity, joining) recombination process of T and B lymphocyte maturation. Deficiency in Artemis protein causes arrest of lymphocyte development, leading to the absence of T and B cells, but a normal number of NK cells ($T^- B^- NK^+$ SCID phenotype). This form of SCID is prevalent among Navajo and Apache Native populations, who share (Southern) Athabascan languages. In the Athabascan population, there is an incidence of 1 in 2000 live births, and the Navajo Nation has an estimated gene frequency of 2.1%.[29,30]

2.3.1 Symptoms

Although the pathology of SCID-A is different from X-SCID, it shares many of the same symptoms. Early infections such as pneumonia are prominent in patients with SCID-A, as well as lymphopenia, chronic diarrhea, oral thrush, and failure to thrive.[31] Unique to SCID-A are severe oral and genital ulcers and the impairment in DNA break repair causes sensitivity to ionizing radiation and alkylating agents.[32]

2.3.2 Biochemistry

DCLRE1C encodes Artemis, a protein that gains endonucleolytic activity on 5' and 3' overhangs and hairpins when complexed with and phosphorylated by DNA-dependent protein kinase (DNA-PK$_{cs}$). This cleaving of nucleotide chains is a key activity in the process of Non-Homologous End Joining (NHEJ), where Artemis removes damaged or mismatched nucleotides in the end processing stage.[33] The NHEJ pathway is critical in the variable (V), diversity (D), and joining (J) recombination process. The V, D, and J segments are rearranged and assembled to create immunoglobulins and T cell receptors of B and T lymphocytes, respectively.[29] During this reassembly process, a hairpin coding-end is formed, and Artemis-mediated opening of the hairpin is necessary before the segments can rejoin. In the absence of Artemis, people with SCID-A are unable to undergo DNA repair after V(D)J recombination, resulting in the absence of coding-joint formation and an early arrest of B and T cell maturation.[34]

2.4 Chronic granulomatous disease (CGD)

Chronic granulomatous disease (CGD) is a predominantly X-linked recessive disorder with a handful of autosomal recessive forms. The disease is commonly referred to as X-CGD given that 80% of cases are X-linked.[35] The disease has a range of severity depending on which specific gene is mutated, but is marked by an inability to fight bacterial and fungal infections leading to serious life-threatening complications and death. Secondary inflammatory complications often arise as well. CGD arises when mutations occur in any one of the five genes that encode the NADPH oxidase subunits. The incidence of CGD is extremely variable worldwide but occurs in 1:200,00 to 1:250,000 live births in the United States and Europe.[36] In populations that have high consanguinity, the autosomal recessive form (AR-CGD) is much more prevalent.[37]

2.4.1 Symptoms

The recurrent infections associated with CGD reach many systems of the body such as the liver, brain, bones, and kidney. However, infections are more frequent in organs with an exposed surface, such as the skin, lungs, and GI tract. The disease is usually detected a few months after birth with continued infections. Failure to thrive, colitis, dermatitis, and GI complications are also indicators.[35] Pneumonia is common, often caused by infection with *Staphylococcus aureus* or gram-negative bacteria. CGD patients frequently suffer from lymphadenitis and cutaneous abscesses also commonly associated with *S. aureus* infection.[38] Fungal infections are common and extremely persistent in CGD patients, affecting 20–40% of all cases.[39] Infected tissues accumulate defective immune cells forming granulomas, a hallmark of the disease. At sites of inflammation, macrophages cluster to form the granuloma, which then causes further complications. A classic outcome in the GI tract is a form of colitis that closely resembles and can be confused for Crohn's disease. Other complications include arthritis, gingivitis, and uveitis, an inflammation of the eye.[40]

2.4.2 Biochemistry

Before assembly, some NADPH oxidase subunits are located on the membrane of intracellular granules as well as free in the cytosol. This system is particularly important in phagocytic immune cells such as macrophages and dendritic cells for the elimination of bacteria and fungi. During phagocytosis, the NADPH oxidase subunits assemble at the membrane and release reactive oxygen species (ROS) into the phagosome. This is accomplished by the cytosolic subunits of the complex donating electrons to the apical side of the membrane which reduces oxygen, producing superoxide. This molecule is then converted into many different ROS such as hydrogen peroxide, hypochlorous acid, and reactive nitrogen species (RNS), which are highly toxic to bacteria and fungi.[35]

70% of all CGD cases are caused by mutations in the *CYBB* gene, which encodes gp91phox, a subunit found on the plasma membrane of the granulocyte. Mutations in *NCF1*, which encodes p47phox, a cytosolic subunit component, account for 20% of all cases.[41] The remaining three genes, *CYBA*, *NCF2*, and *NCF4* each represent <5% of CGD mutations. There are no founder or common mutations associated with this disease. Rather there are many different insertions, deletions, and point mutations found across all five genes.

2.5 Wiskott-Aldrich syndrome (WAS)

Wiskott-Aldrich syndrome (WAS), named after the two physicians who first characterized the disease, is an X-linked PID affecting 1:100,000 live births.[42] The disease affects both myeloid and lymphoid cell types, including T, B, and NK cells as well as neutrophils, eosinophils, macrophages, and dendritic cells which lose their ability to migrate throughout the body to sites of infection. The disease is caused by loss or dysregulation of the WAS protein (WASp), which regulates actin cytoskeleton structure, assembly, and disassembly.[43] Deletions in the WAS gene at common sites lead to the severe form of WAS with little to no protein activity. There are also more mild forms of the disease with reduced protein function. These mild manifestations include X-linked neutropenia (XLN), characterized by abnormally low levels of neutrophils, and X-linked thrombocytopenia (XLT), characterized by low levels of platelets or thrombocytes.[44]

2.5.1 Symptoms

Besides immunodeficiency and susceptibility to sinopulmonary infections, other hallmarks of WAS include eczema and thrombocytopenia. Thrombocytopenia results in excessive bleeding manifested in bruised, bloody noses, bloody stool, and poor wound healing.[45] Patients with XLN and XLT have life expectancies approaching the normal population, but patients with severe forms of WAS usually die of blood loss at a young age.[44] Pneumonia is commonplace with many occurrences caused by *Pneumocystis jirovecii*, which also infects the skin and previous eczema wounds.[44,46] Increased frequency of malignancy has also been documented with aberrant WASp function, with 13% of all WAS cases developing non-Hodgkin's lymphoma as the most prevalent form.[47]

2.5.2 Biochemistry

Much of the research and understanding of WAS as a disease arises from the importance of the WASp protein in all cells. For this reason, the protein itself has been extensively studied for its role in cellular functions through protein-protein interactions. There are over 300 different characterized mutations in the WAS gene that can lead to WASp termination or reduced activity, mostly being deletion mutations.[44] Expressed in the cytoplasm of all non-erythroid immune and hematopoietic cells, WASp interacts with the Arp2/3 complex, which polymerizes new F-actin at the leading edge of the plasma membrane. Other WAS family members that are involved in cytoskeletal rearrangements are the SCAR/WAVE and WASH proteins. The phospholipid present in the plasma membrane, phosphatidylinositol 4,5-bisphosphate or PIP2, serves to activate the WASp/Arp2/3 complex and thus regulates actin polymerization.[48]

2.6 Other PIDs

The list of uniquely classified PIDs continues to grow each year. Between 2018 and 2019 there were 64 newly discovered mutations spanning several different PIDs, some of which were newly characterized.[1] Forms of PID that receive the most attention are also those that have some type of treatment plan, those that are appropriate for a hematopoietic stem and progenitor cell (HSPC) transplant, and those that are indicated for gene therapy approaches that have moved into clinical trials.

Other PIDs include but are not limited to Leukocyte Adhesion Deficiency (LAD), common variable immunodeficiency (CVID), Autoimmune Lymphoproliferative Syndrome (ALPS), X-Linked Lymphoproliferative Disease (XLP), X-linked immunodeficiency magnesium defect, Epstein-Barr virus (EBV), and neoplasia (XMEN), and major histocompatibility complex (MHC) Class II Deficiency.

3. Current PID therapies and clinical trials

Historically, the treatment of PIDs has predominantly involved continuous antibiotic, antiviral, and antifungal administration throughout life to eliminate infections and their inherent complications. Other therapies exclusive to each disorder include enzyme replacement therapy and intravenous immunoglobulin when appropriate. These treatments are not curative and predominantly act in the short term to avoid future organ failure and other permanent damages, mainly in the respiratory tract such as bronchiectasis, auditory complications, and induction of mental disabilities. Allogeneic hematopoietic stem cell transplant (HSCT) has been effective for a wide variety of diseases that affect all parts of the immune system, as well as with tissues of the ectoderm (e.g., epidermolysis bullosa).[49] When a patient's stem cell niche is destroyed by busulfan or other pre-conditioning treatments, space is created for a healthy donor's stem cells to engraft. These stem cells then repopulate all immune and blood cells leading to healthy tissues after donor engraftment. HSCTs of this type have been applied to PIDs with success, although there are ongoing complications. General complications associated with allogeneic transplants are outlined below, as well as issues unique to patients with compromised immune systems. For this reason, a shift toward autologous *ex vivo* gene editing therapy has been pursued, whereby a patient's HSCs are isolated, sometimes expanded, gene-edited, and administered back into the patient. These corrected stem cells then fill the stem cell niche and persist, where they divide and differentiate into the effector cells of the immune system, now capable of fighting infections and participating in other necessary infections. Drawbacks of the allogeneic approach are outlined here with emphasis on the advantages of autologous *ex vivo* gene therapy.

3.1 Non-transplant therapies

When a child is first diagnosed with SCID or other forms of PID, they commonly receive a Hickman line (central line) in their arm, a tube beginning outside the skin and running directly to the heart. The administration of antibiotics, antifungals, and antivirals is frequent, and this minor installation surgery reduces patient exposure and the frequency of needle insertion. This line also serves for the administration of immunoglobulin, as SCID patients have lost the ability to produce antibodies for fighting infections. A major goal of SCID treatment is to avoid cytomegalovirus (CMV) infection, which is common and mostly harmless in the general population but deadly in SCID-affected individuals.[50] In combination with the central line, a nasogastric tube may be inserted to administer nutrients directly to the stomach such as breast milk void of CMV.

One of the few PIDs with a decent short-term therapeutic option is ADA-SCID, for which there is enzyme replacement therapy (ERT) available. Patients are infused weekly with polyethylene glycol-conjugated ADA (PEG-ADA).[51] This active form of the enzyme circulates and metabolizes the toxic purines, leading to higher levels of T, B, and NK cells. Unfortunately, the efficacy of this therapy greatly declines over time, is painful, and extremely expensive for patients undergoing lifelong treatment. This cost can reach $100,000 annually and provides an inconvenience of traveling for the procedure weekly to a hospital that has the capacity for such an intervention.[52]

Historically, CGD affected children usually die before the age of 10, succumbing to infection or a related inflammation issue.[53] Treatment regimens are always lifelong unless a successful allogeneic HSCT or an autologous gene therapy approach is achieved. Parenteral antibiotics are used to slow the spread of general infections. Sometimes before culture results have identified the infecting organism, patients are given nafcillin and ceftazidime to fight *S. aureus* and other gram-negative bacteria commonly associated with a CGD diagnosis.[35] Surgical drainage of abscesses is also a routine procedure to alleviate pressure and to remove pus/bacterial pockets in the brain, liver, and lymph nodes.[36]

In the case of WAS treatment, platelet transfusions are common and necessary to prevent bleeding caused by thrombocytopenia. Oral and topical steroids are used to prevent and abolish severe eczema in most patients.[54] In addition to strict antibiotic regimens, intravenous immunoglobulin may also be accompanied to help boost the immune system's ability to fight infections.

3.2 Allogeneic HSC transplant

The first successful bone marrow transplant (BMT) was accomplished in 1956 by Dr. E. Donnall Thomas. A child with severe leukemia was transplanted with their human leukocyte antigen (HLA)-matched twin's bone marrow.[55] Given that the genetic defects and subsequent consequences of PIDs are intrinsic to HSCs, the logical historical and predominant treatment of PID emerged as an allogeneic HLA-matched BMT. The first BMT for a patient with a PID was performed in 1968 by Dr. Robert A. Good, where an infant with SCID was transplanted with an HLA-matched sibling's BM.[56] Because PIDs are such a large class of disorders ranging from many genes and effector cell types, the progress of HSC BMTs as a general approach has been undertaken before optimizations for individual diseases could be completed. However, as SCID disease treatment options, including HSC BMTs, have been actively pursued since the first successful transplant, the rate of survival has increased from near 20% to over 90% in the last 4+ decades.[57]

3.2.1 BM transplants

Common diseases where an allotransplant is necessary or curative include hematopoietic malignancies such as leukemia and lymphoma, bone marrow failure disorders such as Fanconi anemia, and other inherited blood disorders like sickle cell disease (SCD). After an HLA-matched donor is secured, the recipient of a BMT undergoes chemotherapy to fully, or in some cases partially, ablate their hematopoietic system to create space for incoming donor HSCs. It should be noted that HSCT, with or without subsequent gene editing, does not account for extra-hematopoietic complications such as thymic stromal defects. Permanent organ damage and developmental issues associated with specific disease types, such as reduced IQ and learning disabilities, are not alleviated with this treatment. As with all types of organ transplants, additional pre-conditioning regimens are crucial for success and must be adjusted based on the organ system involved. Serotherapy is often supplemented, which consists of polyclonal and/or monoclonal antibodies to target and destroy the donor's T cells, all aimed at reducing graft-versus-host disease (GvHD).[58] The drug Alemtuzumab accomplishes this by targeting CD52, which is present on mature lymphocytes and is often a supplement to other drugs used for different conditioning procedures.[59] Myeloablative conditioning (MAC) is associated with high levels of donor cell engraftment but at the cost of high levels of toxicity. Fludarabine and busulfan are administered and greatly reduce the risk of GvHD.[59,60] A less toxic form of conditioning is reduced-intensity conditioning (RIC) which involves Alemtuzumab, fludarabine, and busulfan, but Melphalan or Treosulphan may replace busulfan. These lower doses may however reduce the engraftment rate and lead to higher levels of GvHD.[60,61]

3.2.2 HSC transplant as treatment for PIDS

Allotransplants for PIDs have evolved to the specific disorder they are aiming to treat. Manifestations and phenotypes of each PID present unique complications in general for allotransplants and are tailored according to the disease. Unfortunately, much of the data to be discussed here is limited to young children, and statistics on adults and older children are lacking.[61] Thanks to the strong cooperation between immunologists and transplant physicians, the overall survival for children with any form of PID is now over 90%.[62] This does not take into account less severe complications or if the disorder was effectively corrected. While there is much evidence that allotransplants are curative for many PIDs such as WAS, CGD, and ADA-SCID, certain PIDs would benefit greatly from an allotransplant coupled with gene therapy for survival, such as the case with X-SCID and MHC-II deficiency.[59] Several factors influence the transplant protocol and the resulting outcomes. The availability of a HLA-m, as opposed to a matched-unrelated donor (MUD), can affect the outcome. When an appropriate MRD is not available, measures can be taken to reduce the risk of GvHD, such as the depletion of T lymphocytes *ex vivo* before implantation.[59]

Different PIDs require different reagents and timing schedules. In the case of ADA-SCID, preconditioning is not recommended, and coordination between transplant and PEG-ADA enzyme replacement therapy (ERT) administration needs to be carefully timed. At the time of transplant, if an ADA-SCID patient has an infection, the chances of successful engraftment are very low.[63] A large cohort study reported that a 78% engraftment rate was achieved when ADA-SCID patients were transplanted with MRD as opposed to 56% who received MAC or RIC.[64] In contrast, in the case of WAS, conditioning is recommended, but may not necessarily be required. In a study of 194 patients where the vast majority underwent preconditioning, the 5-year survival rate was 89%.[65] In a different study of allotransplants in WAS patients, the survival of all 34 patients was 100% when two different pre-conditioning regimens were used.[66] Patients with CGD have a deformed myeloid lineage and thus an extra step must be administered to achieve myeloid donor chimerism. With CGD, a specific conditioning routine of alkylator-based RIC is recommended and was shown to produce a 2-year survival rate of 96% in CGD patients. Engraftment rates are low in general for GCD as was the case with this study.[67]

Current therapy for SCID-A includes allogeneic BMT between 2 and 4 months of age, on average. Most successful outcomes are seen when HSPCs originate from an HLA identical sibling donor, but this is only available in 15–25% of the patients in European and American Native SCID populations.[68] Less success is achieved using HLA-haploidentical donors (HID), which are normally parents, or unrelated HLA MUDs. Using either of these sources, which is most often the case, increases the risk of lower respiratory tract infections and pneumonitis, as well as GvHD.

Conditioning regimens for HSCT are not well tolerated in patients with Artemis deficiency, and lead to early mortality and/or complications. Without conditioning, however, patients develop GvHD or achieve limited T cell recovery and no B cell reconstitution, making SCID-A the most difficult form of SCID to treat by HSCT.[69]

3.2.3 Common complications
Despite the pathology of PIDs leads to the concept of HSCT as a logical curative method, there is an ever-growing list of complications both restricted to specific PIDs as well as the procedure in general. As is the case for any transplant surgery, donor availability is always an issue. Many organ system donations rely heavily on an individual's consent for donation. BMT, however, relies on the enrollment of a healthy individual to voluntarily undergo a painful extraction procedure.

The primary concern and complication of allogeneic HSCT continues to be GvHD. This occurs when the white blood cells of the donor, which may have remained in the BM compartment, recognize the recipient's cells as foreign and mount an immune response. Even if an HLA-matched donor is selected, there can still be undetected incompatibilities. Unfortunately, between ⅓ and ½ of all allogeneic HSCT recipients develop acute GvHD, which is detectable between 10 and 100 days after the initial transplant.[70] GvHD can also be triggered by pre-conditioning regimes that trigger an immune response after much cell death leading to the upregulation of MHC receptors. Especially with the chronic form of GvHD (cGvHD), the mortality rate can approach 15% of those affected.[71] The high incidence and mortality associated with GvHD are some of the main incentives for the development of autologous *ex vivo* gene editing approaches for PID.

The reciprocal of GvHD is a transplant rejection or stem cell rejection where the recipient's cells recognize the donor's cells as foreign and mount an immune response. This does not result in morbidity but instead results in host cell regeneration and persistence of the original PID. Patients may be even worse at fighting infections after having undergone a conditioning regime. Graft rejection is thought to occur in around 4% of all allotransplants but may be higher in PID patients.[72]

3.2.4 PID complications

As many PIDs arise from populations and families where there are consanguineous mating patterns, an HLA-matched sibling donor transplant can be easier to obtain than in the regular population given the high similarity of the genetic code between siblings, however these transplants are supplemented with additional complications unique to PID. The incidence of recurrent and new infections after allotransplant for PID in children is very high. In a 2020 study, non-SCID patients experienced at least one infection after transplant in 53% of cases, while for those with SCID there was an infection in 59% of cases.[73]

The treatment of certain PIDs by allotransplant is in some cases not recommended, even when this would be the only curative measure. This is the case for common variable immunodeficiency (CVID) with the depletion of antibodies, so the extreme risk of infection outweighs the therapy. The number of different PIDs also complicates the generalized approach to allotransplants for these disorders. Little to no data exists for some of the very rare PIDs where there may only be a few patients identified. Transplant physicians in these cases either do not pursue an allotransplant or have little basis to create their treatment plan.

3.3 Gene addition therapy

Recombinant viral vectors are a common way to deliver genetic material into a cell. Viruses are suitable for gene therapy due to their innate ability to infect immune cells and deliver their genetic material. Recombinant viruses have been engineered to infect cells without replicating themselves inside the host. Some of the most common types of recombinant viral vectors are retroviral/lentiviral and adeno-associated viruses (AAV). The type of viral vector used highly depends on the genetic editing to be achieved, as each type has its advantages and disadvantages.[74]

3.3.1 Retroviral and lentiviral-based corrective therapies

The effectiveness of allogeneic HSCT for PID predicts that transplantation of genetically complemented, autologous HSC will restore patient immunity without some of the most serious adverse outcomes, particularly GvHD, associated with allotransplant. Gene delivery using retroviral vectors based on Moloney murine leukemia virus was initially envisioned to treat PIDs. ADA deficiency was the first disease to be tested in clinical trials of *ex vivo* gene transfer into HSC using retroviral vectors.[75–77] During retroviral replication, the viral RNA is converted to double-stranded DNA by reverse transcriptase, and then the viral DNA is incorporated into the host cell genome by retroviral integrase. This integration step makes retroviruses extremely powerful as vectors for therapeutic gene transfer in HSC. Once integrated into the host cell genome, the vector DNA, containing the therapeutic gene, replicates along with the host cell as it proliferates and differentiates into all lymphohematopoietic lineages[78] (Fig. 1A). MoMLV-based retroviruses were developed and clinically tested first. However, it turns out that there are limitations to this vector system. These vectors are incapable of infecting non-dividing cells, which is problematic since HSCs are for the most part quiescent. Then, after the emergence of a leukemia-like syndrome in several patients being treated for X-linked SCID (see below), it was discovered that gamma-retroviruses have a preferred integration pattern that favors sequences close to transcriptional start sites, thus introducing the risk of genotoxic effects such as insertional oncogene activation.[79] Lentiviruses, a different class of retrovirus and mostly based on human immunodeficiency virus (HIV)-1, were then adopted because of their capability to infect post-mitotic cells, such as neurons, cardiac myocytes, and skeletal muscle fibers as well as HSC. Lentiviruses are also characterized by their ability to stably integrate into the host genome, favoring gene

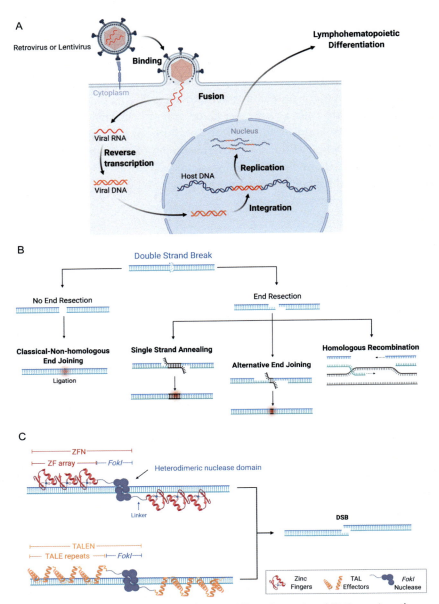

Fig. 1 Previous approaches to gene editing in PID and associated DNA repair pathways. (A) Retroviral and lentiviral replication; retroviral and lentiviral single-stranded RNA genome is inserted into the cytoplasm of a host cell, where it is reverse transcribed into DNA. This viral DNA enters the nucleus and is integrated into the host DNA, after which it proliferates and differentiates into all lymphohematopoietic lineages. (B) Double-strand break DNA repair pathways; double-strand breaks are repaired through various pathways, depending on whether end resection occurs after the break. With no end resection, they are repaired through the classical/canonical-homologous end joining pathway. With end resection, homologous recombination, single-strand annealing or alternative end-joining can occur. (C) Zinc-finger nucleases and transcription activator-like effector nucleases; ZFNs and TALENs can create DSBs at target sites. Created with BioRender.com

sequences rather than transcriptional start sites, and deliver complex genetic elements such as the ß-globin gene. Compared to other gene transfer techniques, the use of lentiviral vectors is a relatively straightforward system for vector manipulation and production.[80]

3.3.1.1 Treatments and outcomes

Gene therapy clinical trials for the treatment of X-SCID were carried out starting in the late 1990s reporting the use of first-generation retroviral vectors to achieve T and B cell immune reconstitution.[81] Unfortunately, some of the patients of these clinical trials developed T cell leukemia generated in part by vector-mediated oncogene activation. Self-inactivating retroviral and lentiviral vectors were subsequently adopted to eliminate viral enhancer elements and prevent transactivation of proto-oncogenes such as *LMO2*.[82] These engineered vectors are safer as they do not contain the viral promoter.[83] A recent clinical trial[84] aimed at treating newly diagnosed infants with X-SCID paired a low, non-myeloablative dose of busulfan as a pre-conditioning treatment with a self-inactivating lentiviral vector to transduce the IL2RG cDNA into autologous HSPCs. After a median follow-up of 16.4 months, the trial achieved immune reconstitution and efficient expression of the IL2RG transgene into T cells, B cells, NK cells, myeloid cells, and bone marrow progenitors, with only low grade acute toxic effects.[85]

The first gene therapy clinical trial was for ADA-SCID, testing retroviral transduction of patient T cells to insert an extra copy of the ADA gene.[86] ADA-SCID is an ideal candidate for gene therapy using transduced hematopoietic stem cells and has been extensively applied both in clinical trials and as an approved therapeutic product, Strimvelis™.[87] Advances in gene therapy technology provide improved methodologies for ADA-SCID. Retroviral and lentiviral vectors have been used in clinical trials of genetic therapies for ADA-SCID, using HSPCs as targets for ADA gene addition. However, the conditioning regimens provided limited engraftment and immune reconstitution.[88] As a way to make space in the bone marrow while reducing toxicity, a clinical trial reported in 2009[89] used busulfan as a nonmyeloablative pre-conditioning treatment to improve the limited clinical efficacy that had been seen thus far. Following infusion with the engineered HSPCs, the patients stably engrafted and developed myeloid and lymphoid cells after a median follow-up of 3 years post-infusion. Results from this trial showed that nonmyeloablative conditioning with busulfan is critical in achieving engraftment, restoring immune function, and protecting against severe infection.[90]

For the treatment of Artemis-deficient SCID, a lentiviral vector was generated containing the human Artemis *DCLRE1C* cDNA transcriptionally regulated by the human Artemis promoter, termed AProArt.[69] This was used to transduce human ART-SCID fibroblasts, as well as human CD34+ HSCs and murine Art$^{-/-}$ HSC. In this study, they achieved correction of radiation sensitivity in transduced human ART-SCID fibroblasts. They also successfully reconstituted T and B cell development from transduced patient-derived CD34+ ART-SCID cells and murine Art$^{-/-}$ HSC (*in vivo* and *in vitro*).[69] No integration-mediated toxicity or transformation was observed in animals or an *in vitro* insertional mutagenesis assay. This study led to an ongoing clinical trial[91] in which the patients are preconditioned with a non-myeloablative dose of busulfan and then infused with autologous Apro-Art transduced HSCs to provide a restored immune system.

3.3.1.2 Pitfalls

Clinical trials have highlighted how gene therapy by gammaretroviral and lentiviral transduction has its limitations. Both types of vectors cause insertional mutagenesis *via* different mechanisms, but at similar frequencies, and pose a risk for the upregulation of oncogenes near the vector integration site.[92] Genomic instability has also been reported in human cells after retroviral transduction disrupted normal centrosome duplication in the treatment of X-CGD.[93] Clonal dominance caused by disrupted regulation of the HMGA2 locus was also demonstrated in a patient treated for β-thalassemia using a self-inactivating lentiviral vector.[94] This study exemplifies the potential risk of genotoxicity introduced by *any* randomly integrating vector, even if self-inactivating. As such, site-directed molecular therapies have begun to make their way into the field.[95]

4. Next generation of gene-editing tools

Current methodologies of genetic therapy, although promising and in the clinic for several PIDs, have many limitations. Targeted correction of mutations, or targeted transgene integration at sites that lead to the translation of a functional protein, are much preferred over the use of randomly integrating vectors as they aim to only leave the wild type (WT) form of the protein intact. ZFNs, TALENs and CRISPR-Cas9 are derived from DNA binding proteins or naturally occurring bacterial enzymes and serve as the basis for targeted genome editing. Base editors (BEs) are capable of targeting and converting single bases without producing double-strand breaks (DSBs) and prime editors (PEs) are capable of correcting many different types of mutations also without producing DSBs. These new generations of tools and reagents have high appeal therapeutically and are considered safer because of the absence of DSBs and a severe reduction of off-target effects. This toolbox is also greatly broadened and opens the door for the therapy of disorders that was not possible before.

Viral-based approaches of *ex vivo* gene editing often involve supplementing the WT form of the mutated gene at a safe harbor locus. In this approach, the mutated sequence is still present and may still lead to translation of the deformed protein. Situations arise where this is detrimental to proper cellular function, as is the case where two forms of protein are needed for dimerization or in the case of dominant-negative mutations where the truncated or otherwise abnormal protein product blocks the proper function of the newly added protein-coding sequence. This mutated form of the protein may still bind to its effector site blocking the binding of healthy protein.[96] These pitfalls have been circumvented by targeted direct correction or alleviation of the mutation at hand by producing WT protein without the insertion of a new full gene sequence. The tools involved here rely on the creations of DSBs, and it is first important to understand how DNA is cleaved, how cells approach DNA repair, and how this natural process is manipulated to change DNA sequences.

4.1 DNA repair pathways

There are many ways by which cells repair damaged DNA. Classical/canonical nonhomologous end-joining (C-NHEJ) and homologous recombination (HR) are the two most common pathways by which double-strand breaks (DSBs) are repaired.[97] The pathway of repair in part depends on whether DNA end resection occurs at the site of the DSB, whereby several nucleotides at the 5′ end are digested to leave free 3′ ends on either side of the sequence. Without DNA end resection, the C-NHEJ pathway is favored. When DNA end resection creates the necessary single-stranded overhangs, homologous recombination can ensue. During end resection, there are two additional pathways at play that can compete with HR: single-strand annealing (SSA) and alternative end-joining (alt-EJ)[97] (Fig. 1B).

The phase of the cell cycle also determines the pathway by which a DSB is repaired. The C-NHEJ pathway is active in all phases of the cell cycle, but is dominant during the G0/G1 and G2 phases.[97] Since there is no end resection during these phases, the system ligates the blunt ends of the DSB with the help of several proteins.[97] If the end resection results in two long stretches of ssDNA, the cell commits to either HR or SSA. HR dominates in S phase, as there is a newly synthesized sister chromatid that can act as a DNA template, allowing strand invasion of the ssDNA overhang into the DNA template to copy the sequence and repair the break, and competes against C-NHEJ in the G2 phase.[97] On the other hand, SSA does not require a DNA template, as it uses large DNA sequences of homology between the two ssDNAs to repair the break. If there is only a short end recession, similar to SSA, alt-EJ repairs the break by annealing the ssDNAs, this time only requiring a short region of homology between them, termed microhomologies.[97]

These four pathways are associated with different genetic outcomes. HR, because it is a template-mediated repair, is most accurate, but can lead to loss of heterozygosity. On the other hand, C-NHEJ can be accurate, but commonly leads to small insertions and deletions.[97] Viceversa, SSA and alt-EJ are always considered mutagenic. During SSA, large deletions occur, resulting in the loss of genetic information. In the alt-EJ pathway, the micro-homologies where the sequences anneal can also be found in other places in the chromosome, and it has been shown that alt-EJ can lead to the joining of DSBs on different chromosomes, creating chromosomal translocations and mutagenic rearrangements in addition to insertions and deletions. Because of this mutagenicity, for accurate genetic engineering, it is imperative to leverage DNA repair in a direction that is less mutagenic and more amenable to precise genome editing.

4.1.1 Zinc-finger nucleases (ZFN) and transcription activator-like effector nucleases (TALENs)

The first generation of programmable gene-editing tools began with the engineering of ZFNs, specific restriction enzyme domains fused to zinc-finger DNA binding domains. ZFNs produce site-specific DSBs, which are then resolved by the cell through the aforementioned DNA repair pathways (Fig. 1C). NHEJ produces small indels leading to targeted mutagenesis, while HR requires a donor DNA template, leading to permanent genetic edits.[98] This first domain of ZFNs is composed of a series of zinc finger domains, each of which recognizes and binds to a specific 3 nucleotide DNA sequence. Between three and six zinc-finger DNA binding domains are combined, thereby comprising a recognition sequence of between 9 and 18 nucleotides. A commonly used restriction enzyme cleavage domain is FokI.[99] When two juxtaposed ZFNs bind to their targeted DNA sequences, one complimentary for the coding strand and one complimentary for the non-coding strand, the FokI domains dimerize and make a double-strand break leading to a 4-bp overhang on the 3′ strand within the target site.

The applicability of ZFNs and HR in general to human disease was demonstrated in seminal work by Urnov et al. in 2005, where ZFNs targeting exon 5 of the *IL2RG* gene were combined with a donor DNA plasmid for genetic correction in X-SCID K562 cell lines.[100] Low levels of HR were achieved with this method, spurring the use of alternative, more efficient DNA delivery vectors such as rAAV and integrase-defective lentiviral vector (IDLV). Additionally, ZFNs previously showed proof of principle in HSPC editing where now CRISPR-Cas9 was used to target exon 7 of the *CYBB* gene in tandem with single-stranded oligo donors (ssODNs) to correct a point mutation in X-CGD HSPCs. Corrected cells were transplanted into mice where they persisted for up to 5 months with an editing efficiency of 20%.[101]

An additional genome editing tool that creates DSBs in a programmable fashion is the transcription activator-like effector nuclease (TALEN) system. Similar to ZFNs, TALENs are proteins composed of DNA binding domains, each of which recognizes a single nucleotide in a DNA sequence, fused to the restriction enzyme FokI.[102] Using TALENs, any DNA binding domain sequence can be engineered and coupled with N or C terminal domains fused to a nuclease to create a DSB at a target site. Analogous to ZFNs, this system requires protein design and optimization for any given DNA target sequence.

4.1.2 CRISPR/Cas9

The 2020 Nobel Prize in Chemistry was awarded to Charpentier and Doudna for their development of the CRISPR system into a gene-editing tool.[103] Although the existence of this naturally occurring bacteria pathogen defense system had been described decades earlier, manipulation and engineering of the system into a programmable tool was first published in 2012.[104] Clustered Regularly Interspaced Short Palindromic Repeats (CRISPRs) are areas in bacterial and archaeal genomes that evolved as a form of adaptive immunity to integrate and "remember" pathogenic DNA sequences for later targeting. Foreign bacteriophage DNA is cut into short fragments by the Cas enzymes, which are then incorporated as "spacers" into the host genome at the CRISPR locus between short palindromic repeated sequences. These spacer CRISPR sequences are transcribed into RNA. Cas enzymes then cleave them into mature crRNAs. The fully formed gRNA is then bound to the Cas9 nuclease and acts to drive Cas9 to bind and cleave the bacteriophage genome in protospacer sequences complementary to the spacer portion of the gRNA.[105] For the SpCas9 enzyme to locate a protospacer DNA target site, there must be a three base pair (bp) sequence next to the guide sequence known as the protospacer adjacent motif (PAM). The SpCas9 requires a 5′-NGG-3′ PAM sequence for recognition. This toolbox has recently been expanded with the engineering of spCas9 into spCas9-NG, for which recognition of a specific DNA sequence requires only a 5′-NG-3′ PAM.[106] This system was engineered as a gene therapy tool to target specific sites in the genome for the induction of DSBs which can then either lead to the creation of indels or the insertion of new DNA sequences if an exogenous DNA template is provided (Fig. 2A).

These systems have been co-opted for genome engineering and are now widely applied to mammalian cells where RNA sequences can now be designed to target specific sites in the host genome using Cas9 nuclease.

4.1.3 CRISPR for PID

Although the exclusive use of CRISPR-Cas9 with or without a DNA donor template has yet to enter clinical trials for PID, their use as proof of principle and preclinical safety/efficacy has been shown by several groups. In the absence of donor DNA, many studies involving the correction or alleviation of disease by creating insertions and deletions (indels) at different sites are underway with some already in clinical trials. A recent example of this is the use of CRISPR-Cas9 to treat sickle cell disease by reactivating the fetal form of hemoglobin while leaving the mutated *HBB* unaltered. This was accomplished by creating indels on important transcription factor (TF) binding sites of the *BCL11A* enhancer, which participates in the silencing of the *HGB* gene. With the lack of repressive TF binding sites, the silencing of *HBG* transcription is lifted.[107] This method of therapy for SCD is currently in a clinical trial.[108]

Another example of CRISPR-Cas9 nuclease editing for PIDs is the targeting of the most common mutation leading to CGD,[101] where a single base substitution in the *CYBB* gene results in a premature STOP (pmSTOP) codon and accounts for 6% of all cases.[109] This gene encodes the gp91phox subunit of the NADPH oxidase complex. CRISPR-Cas9 and a 100 bp ssODN was employed in patient-derived CD34+ HSPCs. An editing rate

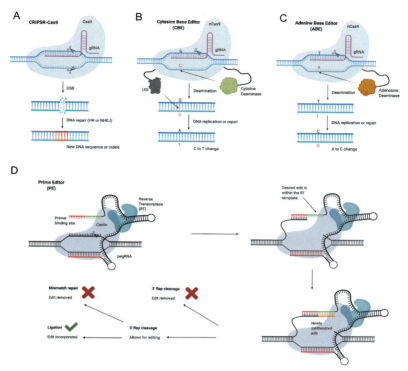

Fig. 2 Next-generation editors and their gene editing mechanisms. (A) CRISPR-Cas9; a designed sgRNA complexes with the Cas9 nuclease and binds to the complementary target site. Cas9 produces a DSB in the DNA that the cell resolves through two different DNA repair pathways. The NHEJ pathway results with indels and the HDR pathway can result in an incorporated edit if a DNA template is also provided. (B) CBE; similar mechanism to CRISPR-Cas9 but a nCas9 is used that nicks the opposite stand inducing a DNA repair pathway response. The target C in the editing window is deaminated to U, and a fused UGI prevents the cell from resolving this mismatch through the base excision repair pathway. During DNA replication or repair, the U matches with A resulting in C/G to T/A substitution. (C) ABE; similar mechanism to CBE but the target A in the editing window is deaminated to I which is then read by polymerases during DNA replication or repair as a G resulting in A/T to G/C substitution. (D) PE; a nCas9 complexes with a pegRNA containing an opposite strand complementary binding site, a primer binding site, and a RT sequence that contains the desired edit. After the target strand is nicked, the primer binding site anneals to the loose strand, and a fused RT synthesizes a new DNA strand continuing the desired edit. An equilibrium of 3′ and 5′ flaps exists where cleavage of the 3′ strand results in the removal of the edit, and cleavage of the 5′ strand results in edit incorporation after ligation. Created with BioRender.com

of 30–40% was achieved and an edit persistence of 15–20% was documented 5 months after engrafting the edited cells into immunodeficient NSG mice.[101] This led to an increase of >20% in superoxide production indicating that NADPH function was restored and thus therapeutically relevant. There were no off-target effects detected.

4.1.4 Limitations and off-target effects

For both normal cellular function and the purpose of genetic editing, repair of any DSB has the potential for error. DSBs are introduced when Cas9 is targeted by the gRNA to form sufficient base pairing with a protospacer sequence adjacent to a PAM, enabling cleavage of the targeted DNA and prompting the cell to initiate DNA repair. CRISPR-Cas9 effects can be categorized as either gRNA-dependent or gRNA-independent. gRNA-dependent off-target effects occur when there is a high degree of similarity between the gRNA sequence and some other non-targeted location in the genome. Because gRNAs, which are normally around 20 bp in length, can pair with a sequence while still allowing up to 8 bases of mismatch, DSBs can occur outside of the target site.[110] On the other hand, with the BE system, potential off-target gRNA-independent effects are more difficult to determine and is thus one of the few drawbacks of the BE system.

To ascertain off-target effects that occur during editing, several prediction methods have been developed and are categorized as either biased or unbiased. Biased tools can evaluate off-target effects in two different ways: the user can input a gRNA sequence, which is then aligned to the reference genome and will deliver to the user a list of potential off-target sites in terms of sequence homology, termed alignment models. On the other hand, in unbiased scoring models, an algorithm can choose the best gRNAs for a target and rank them based on the potential for cleavage at off-target sites. These unbiased tools focus on detecting unintended cleavage sites in living cells at the whole genomic level.[111] Unbiased methods for off-target detection include chromatin immunoprecipitation and high-throughput sequencing (ChIP-seq),[112] direct labeling of *in situ* breaks, enrichment on streptavidin and next-generation sequencing (BLESS),[113] genome-wide, unbiased identification of DSBs enabled by sequencing (GUIDE-seq[114]), circularization for *in vitro* reporting of cleavage effects by sequencing (CIRCLE-seq[115]), digested genome sequencing (Digenome-seq[116]), and whole-genome sequencing (WGS).[110,117]

Although there have been several reports demonstrating that the use of well-designed sgRNA can result in very low off-target activity, there remain concerns about Cas9 based genome editing. Several recent reports have described the potentially harmful effects of CRISPR-Cas9 induced DSBs, such as activation of the p53-mediated DNA damage response, arresting cell growth, and decreasing the efficiency of editing.[118] For HSPCs in particular, the p53 response inhibits HDR efficiency by 17-fold on average relative to transformed cell lines.[119] Inhibition of the p53 pathway has even been suggested as a way to increase the editing efficiency, but blocking the cell's check-point activity could leave it vulnerable to chromosomal rearrangements or tumorigenic effects, including those produced during the editing process.

In addition to the activation of p53, other undesirable effects of DSBs that have been reported include on-target deletions, complex rearrangements such as inversions, and endogenous/exogenous sequence insertions that can result in loss of heterozygosity, loss of function mutations, exon skipping, neoplasia, and risk of proto-oncogene activation.[82,120–122]

Overall, these concerns regarding introduction of DSB make it clear that strategies to avoid the creation of DSBs are a priority. The introduction of DSB may be beneficial in terms of gene knockout, but for gene correction or gene knock-in, there are many drawbacks. Next-generation base and prime editing tools circumvent many of the concerns about DSBs and also provide enhanced efficacy and improved safety.

4.1.5 Cas9 base editors

Cas9 base editors (BEs) represent an important milestone in the gene-editing field with rapidly expanding programmable BE variants and improved efficiency of reagents. Several components of the CRISPR-Cas9 editing system are conserved in the BE strategy. First, a nickase version of the Cas9 enzyme is used that does not produce a DSB, but instead cuts only one of the DNA strands due to an inactivating substitution in one of the Cas9 DNA cleavage domains. Second, the same type of 20 bp gRNA is used to position the enzyme to the DNA target site. BEs rely on the deaminating activity of either TadA from *E. coli*, the inducible Target-AID (PmCDA1) from sea lamprey, or the broad APOBEC[123] enzyme family.[124] APOBEC (apolipoprotein B mRNA editing enzyme, catalytic polypeptide-like) enzymes are naturally occurring deaminases that can act on either DNA or RNA. Interestingly, over 95% of all discovered human genetic defects arise from single nucleotide variants (SNV), which illustrates the importance of developing these BE technologies.[125] Of these SNVs, over 60% can be directly corrected using BEs, with 14% corrected by cytosine base editors (CBEs) and 48% corrected with adenine base editors (ABEs).[124] The utility of these technologies is broadened when considering that not all genetic disorders require direct reversion to the wild type sequence in order to ameliorate the disease. For example, an SNV that causes a pmSTOP codon results in the truncation of protein or the absence of protein leading to the phenotypic consequence of the disease. A different base in the codon besides the mutated base can be targeted to restore the pmSTOP to a different amino acid. As long as the new amino acid does not alter the protein structure and function, this strategy holds great promise for expanding the BE toolbox, however this methodology requires an additional phenotypic assessment, immunogenicity assay and/or protein structure analysis to confirm the resulting protein is working correctly and does not trigger an immune response.

4.1.6 CBE mechanism

The first engineered BE was the cytosine base editor (CBE), where cytosine (C) is deaminated and ultimately leads to a thymine transition.[126] A Cas9 nickase (nCas9) was fused *via* a peptide linker to the naturally occurring rat APOBEC1 cytidine deaminase. When paired with gRNA, the nCas9 binds the targeted DNA sequence and unwinds the DNA, producing an R-loop. The Cas9 enzyme nicks the untargeted strand of DNA, and the APOBEC1 enzyme deaminates the target C into uracil (U), which has the base-pairing properties of thymine (T). Within the protospacer sequence, the editing window extends from position 4 to 8 bases of the sgRNA pairs from the PAM. A second protein, a uracil glycosylase inhibitor (UGI), is also fused to the nCas9, preventing the cell from initiating the removal of the U through the base excision repair pathway. With the opposite strand nicked, the cell is forced to proceed with the mismatch repair pathway, and during replication and repair, the U pairs with an A and is eventually replaced with a T (Fig. 2B).

4.1.7 ABE mechanism

The next generation of BEs was adenine base editor (ABE), where adenosine is deaminated and ultimately transitioned to guanine.[124] As dsDNA deaminases do not naturally exist in nature, directed evolution was conducted, starting with the RNA deaminase TadA. Like CBE, the resulting adenine deaminase is fused *via* a peptide linker to a nCas9. The protein and gRNA localize to the target site, an R-loop is formed, edits at base positions 4–8 of the sgRNA, and the nCas9 nicks the non-target strand. The adenosine deaminase first deaminates the target A into inosine (I). The cellular polymerases read this base as a G; during DNA replication or repair the complementary C is incorporated, and then what was initially an A has been transitioned to G (Fig. 2C).

4.1.8 BE translational research

Much of the appeal of the BE process is that it does not create DSBs and thus can be considered therapeutically less genotoxic. Moreover, there is an advantage in having a targeted correction without the need for any form of donor DNA, which adds further complexity to the process. Research into identifying therapeutically relevant targets for the correction or alleviation of human disease is in its infancy, but several groups have begun this process. T cells are highly amenable to multiplex base editing, where three genes were knocked out for the improved safety and efficiency of chimeric antigen receptor (CAR) T cells for cancer immunotherapy.[127] An approach for therapy of sickle cell disease (SCD) was also investigated by using base editors. Instead of creating indels, a single base substitution on the +58 GATA box of the BCL11A enhancer was sufficient to induce HGB.[128] Neither of these strategies utilizes a knock-in approach, which will be therapeutically relevant for PIDs. Careful examination of the gene sequence for each PID must be conducted to identify potential targets such as reversion of a pSTOP codon or rescue of a disrupted splice acceptor/donor site. If BE strategies are not directly feasible for a given mutation, some other compensatory maneuver can be considered, such as targeting the erythroid specific BCL11A enhancer for therapy of SCD. Such strategies require a detailed understanding of the underlying biology of the proteins involved and their binding/interacting partners.

Base editor technology does have limitations and drawbacks, some that are inherent to all gene-editing tools and some unique to BEs. Although off-target effects are greatly reduced for BEs in comparison to CRISPR-Cas9 mediated introduction of DSB, some off-target effects persist. Such errors are not always related to the guide RNA sequence, but rather random deamination reactions at RNA and DNA bases due to the promiscuity of the deaminase enzyme used.[129] Two studies that used different levels of detection have shown non gRNA specific off-target effects in DNA, and to a larger extent, in RNA. Optimization of the BE enzyme used in these studies has since been described in part by embedding the nuclease site deep within the Cas enzyme.[130] Additionally, BEs can produce bystander edits when a second, editable nucleotide is present within the 5 bp editing window. This edit could confer an amino acid substitution that disrupts normal protein function, but the mutation might be silent. In some instances, this by-stander edit may have consequences to the resulting protein structure and function, so nearby codons must be considered carefully on a case-by-case basis. Of the 430 different genes and the multitude of mutations in these genes that have been attributed to PIDs, there are many instances in which BE could be applied to revert the sequence to wild-type. Additionally, other nucleotides within a codon containing a point mutation can be targeted to encode a different amino acid that might not affect protein function if the resulting amino acid is in some way similar to the wild type amino acid, whether that be in charge, size, ring structure, or hydrophobic interactions. Functional assays after base editing must be conducted to ensure a functioning protein is translated and that there are no harmful side effects to the cell if this method is to be employed. Limitations of BE for PIDs include situations where there are not A or C bases contained in the mutated codon, as well as PIDs resulting from chromosomal rearrangements or other larger insertions/deletions.

4.1.9 Prime editors

To date, the most advanced and cutting-edge gene-editing tools are the prime editors (PEs), often denoted as "search and replace" genome editing tools.[131] This system enables the ability to enact virtually all types of gene edits including short indels and all point mutations (transversions/transitions). Insertions of 44 bps and deletions of up to 80 bps have also been accomplished. Theoretically, this technology can correct nearly 90% of all human genetic mutations.[132]

The PE nuclease system follows the initial concept of BE in that a nCas9 localizes to a target sequence identified by a gRNA and binds to the non-edited strand. An R-loop is then exposed, but unlike BEs the nCas9 nicks the edited strand as opposed to the non-edited strand. Additionally, the PE gRNA, termed pegRNA, includes an additional RNA sequence that contains the reverse complement RNA sequence of the desired DNA sequence to be edited. This pegRNA complexes to the nCas9, which is additionally fused to a reverse transcriptase (RT) enzyme through a peptide linker. When the targeting sequence of the pegRNA binds to the unedited strand or protospacer, the other end of the pegRNA binds to the target DNA, which has been nicked and thus exposed. This end includes a sequence that anneals to the DNA followed by the sequence to be altered in the genome. The RT enzyme then extends the 3' DNA end using the novel RNA sequence in the pegRNA as a template, thereby incorporating this sequence into the genome. These edited and unedited flaps of DNA then ligate to the helix, and after DNA repair permanently install the new sequence (Fig. 2D).

Editing efficiency with PE has not yet approached levels achieved using CRISPR-Cas9 or BEs. Much effort and research have gone into the optimization of BEs, and in some cases, these can produce editing efficiencies of over 95%.[127] The PE system will likely need to undergo extensive optimization and efforts are currently underway. Additionally, the off-target effects of PE have only been evaluated in two publications, one being the original article.[131] When four different off-target sites were analyzed using 16 different pegRNAs, only one pegRNA showed activity higher than 1%. This is in stark contrast to the use of this same pegRNA with a Cas9 nuclease, which produced greater than 10% editing activity at each locus. The more recent study found that when using the next-generation sequencing tool Digenome-seq, there were only five off-target effects that had a detectable level over 0.1%.[133] This group also showed much promise for future iterations of PE conferring highly precise editing after incorporation of different Cas variants, and how the enzyme can be improved through directed evolution.[133]

PE has been applied to targets of two genetic diseases, with some success in HEK293T cells. The point mutation in the *HBB* gene that causes SCD disease, E6V, was reverted to WT with an editing efficiency of up to 52%. A 4-bp insertion in the *HEXA* gene is the most common mutation that causes Tay-Sachs disease. PE was used to create this insertion in WT cells at an editing efficiency of 31%.[131]

4.1.10 Application of BE and PE to PIDs
Many of the underlying mutations that lead to PIDs can potentially be corrected using site-directed editing technologies. Recent trends in cutting-edge gene therapy technology and reagent optimization also suggest that the future entails an *in vivo* delivery system for genetic correction or alleviation of genetic disorders. These methods would negate the need for allogeneic HSCT, as well as autologous *ex vivo* gene-edited HSC, the former having several complications and limitations and the latter requiring extensive customization, GMP qualification, and optimization. *In vivo* approaches may entail the use of lentiviral or rAAV vectors[134] or involve

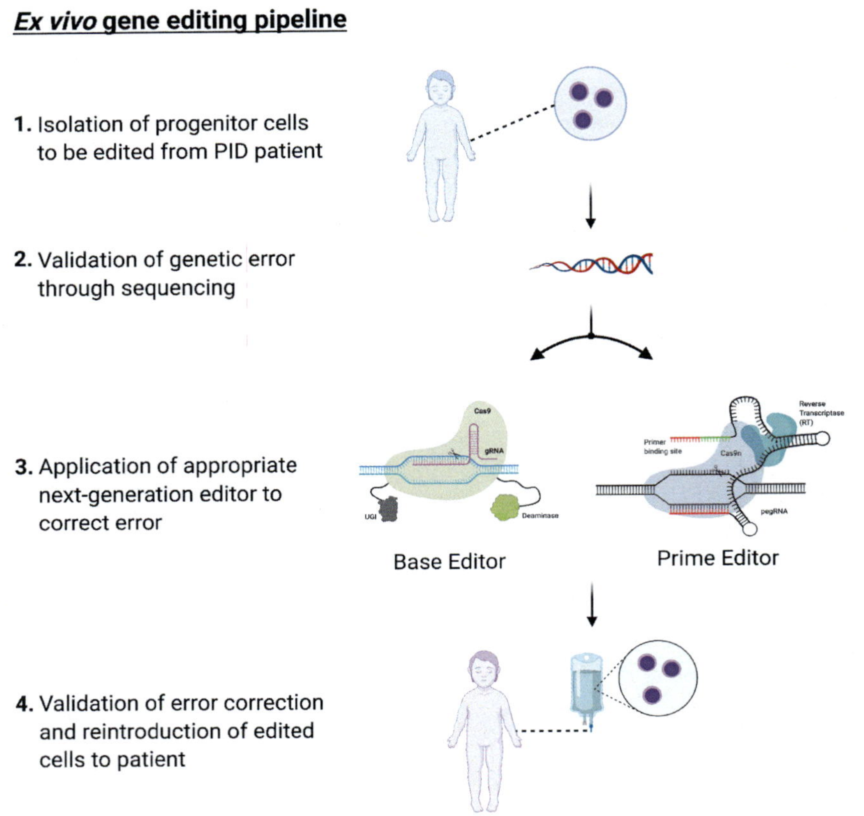

Ex vivo gene editing pipeline

1. Isolation of progenitor cells to be edited from PID patient

2. Validation of genetic error through sequencing

3. Application of appropriate next-generation editor to correct error

    Base Editor   Prime Editor

4. Validation of error correction and reintroduction of edited cells to patient

Fig. 3 Visual representation of future *ex vivo* editing pipeline for PID disease using editors. Autologous *ex vivo* gene therapy workflow for the correction or alleviation of PID using BE and PE. Created with BioRender.com

the delivery of CRISPR-Cas9 accompanied by gRNAs contained in lipid-like nanoparticles (LLNPs).[135] However, delivering gene-editing reagents *in vivo* introduces much more potential risk than delivering a cell population that has been edited *ex vivo* as there is potential for any cell in any organ system to be edited in an undesirable fashion.

Autologous *ex vivo* gene therapy approaches involving BEs and PEs (Fig. 3) for PIDs have not yet reached clinical trials. Non-clinical studies involving human HSCs engrafted into mice will be essential to show safety and efficacy. BE and PE advancements have built upon and improved previous generations of genome editing technologies. Without using a randomly integrating vector, insertional mutagenesis and potential oncogene activation brought on by viral vectors are avoided. Because BE and PE do not require a donor molecule, they are not constrained by the size of the delivery method. Most importantly, BE and PE avoid perhaps the most potentially damaging off-target gene editing effect the introduction of DSBs that can create indels leading to gene rearrangements, exon skipping, TP53 activation, and reduced cell survival.

A critical limitation of any gene therapy approach for PID is the broad spectrum of different disease categories as well as the varying mutations within each. In the case of SCD where one gene is affected, optimizations of reagents and clinical trial development are relatively simple in comparison as one system can help nearly all patients with the disease. Here, a personalized medicine approach must be created on a patient basis where a specific mutation may only affect a handful of individuals. The rapid development and evolution of BE and PE variants is of key importance to this issue as newly available reagents can be applied to patient samples *ex vivo* while different sgRNA are designed and applied with each enzyme. The delivery system of these enzymes to PID patient primary cells would most likely entail mRNA delivery for the transient nature of these RNA molecules.

To create a personalized medicine pipeline for the genetic correction or alleviation of PID using BE and PE, mutations of common occurrence must first be studied to show reproducibility across several patients. Founder mutations such as in the Artemis gene in the Athabascan-speaking Native population is an example of a mutation that should be targeted for proof of principle studies using BE and/or PE.

References

1. Bousfiha A, Jeddane L, Picard C, et al. Human inborn errors of immunity: 2019 update of the IUIS phenotypical classification. *J Clin Immunol.* 2020;40(1):66–81.
2. Al-Herz W, Bousfiha A, Casanova J-L, et al. Primary immunodeficiency diseases: an update on the classification from the international union of immunological societies expert committee for primary immunodeficiency. *Front Immunol.* 2014;5:162.
3. Bucciol G, Moens L, Bosch B, et al. Lessons learned from the study of human inborn errors of innate immunity. *J Allergy Clin Immunol.* 2019;143(2):507–527.
4. Zhang Q, Frange P, Blanche S, Casanova J-L. Pathogenesis of infections in HIV-infected individuals: insights from primary immunodeficiencies. *Curr Opin Immunol.* 2017;48:122–133.
5. McCusker C, Upton J, Warrington R. Primary immunodeficiency. *Allergy Asthma Clin Immunol.* 2018;14(Suppl 2):61.
6. Hamamy H. Consanguineous marriages: preconception consultation in primary health care settings. *J Community Genet.* 2012;3(3):185–192.
7. Bruton OC. Agammaglobulinemia. *Pediatrics.* 1952;9(6):722–728.
8. Raje N, Dinakar C. Overview of immunodeficiency disorders. *Immunol Allergy Clin North Am.* 2015;35(4):599–623.

9. Chinen J, Shearer WT. Secondary immunodeficiencies, including HIV infection. *J Allergy Clin Immunol.* 2010;125(2 Suppl 2):S195–S203.
10. Arkwright PD, Gennery AR. Ten warning signs of primary immunodeficiency: a new paradigm is needed for the 21st century. *Ann N Y Acad Sci.* 2011;1238:7–14.
11. Borte S, von Döbeln U, Hammarström L. Guidelines for newborn screening of primary immunodeficiency diseases. *Curr Opin Hematol.* 2013;20(1):48–54.
12. King JR, Hammarström L. Newborn screening for primary immunodeficiency diseases: history, current and future practice. *J Clin Immunol.* 2018;38(1):56–66.
13. Rubin Z, Pappalardo A, Schwartz A, Antoon JW. Prevalence and outcomes of primary immunodeficiency in hospitalized children in the United States. *J Allergy Clin Immunol Pract.* 2018;6(5):1705–1710.e1.
14. Kobrynski L, Powell RW, Bowen S. Prevalence and morbidity of primary immunodeficiency diseases, United States 2001–2007. *J Clin Immunol.* 2014;34(8):954–961.
15. Fischer A. Severe combined immunodeficiencies. *Immunodefic Rev.* 1992;3(2):83–100.
16. Buckley RH. Molecular defects in human severe combined immunodeficiency and approaches to immune reconstitution. *Annu Rev Immunol.* 2004;22:625–655.
17. Allenspach E, Rawlings DJ, Scharenberg AM. X-linked severe combined immunodeficiency. In: Adam MP, Ardinger HH, Pagon RA, et al., eds. *GeneReviews®*. Seattle: University of Washington; 2003.
18. Lundgren IS, Englund JA, Burroughs LM, Torgerson TR, Skoda-Smith S. Outcomes and duration of Pneumocystis jiroveci pneumonia therapy in infants with severe combined immunodeficiency. *Pediatr Infect Dis J.* 2012;31(1):95–97.
19. Nobile CJ, Johnson AD. Candida albicans biofilms and human disease. *Annu Rev Microbiol.* 2015;69:71–92.
20. Puck JM, Pepper AE, Henthorn PS, et al. Mutation analysis of IL2RG in human X-linked severe combined immunodeficiency. *Blood.* 1997;89(6):1968–1977.
21. Puck JM, Middelton L, Pepper AE. Carrier and prenatal diagnosis of X-linked severe combined immunodeficiency: mutation detection methods and utilization. *Hum Genet.* 1997;99(5):628–633.
22. Kovanen PE, Leonard WJ. Cytokines and immunodeficiency diseases: critical roles of the gamma(c)-dependent cytokines interleukins 2, 4, 7, 9, 15, and 21, and their signaling pathways. *Immunol Rev.* 2004;202:67–83.
23. Flinn AM, Gennery AR. Adenosine deaminase deficiency: a review. *Orphanet J Rare Dis.* 2018;13(1):65.
24. Gaspar HB. Bone marrow transplantation and alternatives for adenosine deaminase deficiency. *Immunol Allergy Clin North Am.* 2010;30(2):221–236.
25. Blackburn MR, Kellems RE. Adenosine deaminase deficiency: metabolic basis of immune deficiency and pulmonary inflammation. *Adv Immunol.* 2005;86:1–41.
26. Whitmore KV, Gaspar HB. Adenosine deaminase deficiency—more than just an immunodeficiency. *Front Immunol.* 2016;7:314.
27. Hellani A, Almassri N, Abu-Amero KK. A novel mutation in the ADA gene causing severe combined immunodeficiency in an Arab patient: a case report. *J Med Case Reports.* 2009;3:6799.
28. Hirschhorn R. Overview of biochemical abnormalities and molecular genetics of adenosine deaminase deficiency. *Pediatr Res.* 1993;33(1 Suppl):S35–S41.
29. Li L, Moshous D, Zhou Y, et al. A founder mutation in Artemis, an SNM1-like protein, causes SCID in Athabascan-speaking Native Americans. *J Immunol.* 2002;168 (12):6323–6329.
30. Jones JF, Ritenbaugh CK, Spence MA, Hayward A. Severe combined immunodeficiency among the Navajo. I. Characterization of phenotypes, epidemiology, and population genetics. *Hum Biol.* 1991;63(5):669–682.

31. O'Marcaigh AS, DeSantes K, Hu D, et al. Bone marrow transplantation for T-B- severe combined immunodeficiency disease in Athabascan-speaking native Americans. *Bone Marrow Transplant.* 2001;27(7):703–709.
32. Nicolas N, Moshous D, Cavazzana-Calvo M, et al. A human severe combined immunodeficiency (SCID) condition with increased sensitivity to ionizing radiations and impaired V(D)J rearrangements defines a new DNA recombination/repair deficiency. *J Exp Med.* 1998;188(4):627–634.
33. Ma Y, Pannicke U, Schwarz K, Lieber MR. Hairpin opening and overhang processing by an Artemis/DNA-dependent protein kinase complex in nonhomologous end joining and V(D)J recombination. *Cell.* 2002;108(6):781–794.
34. Moshous D, Li L, Chasseval R, et al. A new gene involved in DNA double-strand break repair and V(D)J recombination is located on human chromosome 10p. *Hum Mol Genet.* 2000;9(4):583–588.
35. Roos D. Chronic granulomatous disease. *Br Med Bull.* 2016;118(1):50–63.
36. van den Berg JM, van Koppen E, Ahlin A, et al. Chronic granulomatous disease: the European experience. *PLoS One.* 2009;4(4):e5234.
37. Wolach B, Gavrieli R, de Boer M, et al. Chronic granulomatous disease: clinical, functional, molecular, and genetic studies. The Israeli experience with 84 patients. *Am J Hematol.* 2017;92(1):28–36.
38. Bennett N, Maglione PJ, Wright BL, Zerbe C. Infectious complications in patients with chronic granulomatous disease. *J Pediatric Infect Dis Soc.* 2018;7 (suppl_1):S12–S17.
39. Marciano BE, Spalding C, Fitzgerald A, et al. Common severe infections in chronic granulomatous disease. *Clin Infect Dis.* 2015;60(8):1176–1183.
40. Manzi S, Urbach AH, McCune AB, et al. Systemic lupus erythematosus in a boy with chronic granulomatous disease: case report and review of the literature. *Arthritis Rheum.* 1991;34(1):101–105.
41. Arnold DE, Heimall JR. A review of chronic granulomatous disease. *Adv Ther.* 2017;34 (12):2543–2557.
42. Stray-Pedersen A, Abrahamsen TG, Frøland SS. Primary immunodeficiency diseases in Norway. *J Clin Immunol.* 2000;20(6):477–485.
43. Moulding DA, Record J, Malinova D, Thrasher AJ. Actin cytoskeletal defects in immunodeficiency. *Immunol Rev.* 2013;256(1):282–299.
44. Buchbinder D, Nugent DJ, Fillipovich AH. Wiskott-Aldrich syndrome: diagnosis, current management, and emerging treatments. *Appl Clin Genet.* 2014;7:55–66.
45. Modell V, Gee B, Lewis DB, et al. Global study of primary immunodeficiency diseases (PI)--diagnosis, treatment, and economic impact: an updated report from the Jeffrey Modell Foundation. *Immunol Res.* 2011;51(1):61–70.
46. Weintraub HD, Wilson WJ. Pneumocystis carinii pneumonia in wiskott-aldrich syndrome. *Am J Dis Child.* 1964;108:198–200.
47. Sullivan KE, Mullen CA, Blaese RM, Winkelstein JA. A multiinstitutional survey of the Wiskott-Aldrich syndrome. *J Pediatr.* 1994;125(6 Pt 1):876–885.
48. Pollitt AY, Insall RH. WASP and SCAR/WAVE proteins: the drivers of actin assembly. *J Cell Sci.* 2009;122(Pt 15):2575–2578.
49. Wagner JE, Ishida-Yamamoto A, McGrath JA, et al. Bone marrow transplantation for recessive dystrophic epidermolysis bullosa. *N Engl J Med.* 2010;363(7):629–639.
50. Vora SB, Englund JA. Cytomegalovirus in immunocompromised children. *Curr Opin Infect Dis.* 2015;28(4):323–329.
51. Chan B, Wara D, Bastian J, et al. Long-term efficacy of enzyme replacement therapy for adenosine deaminase (ADA)-deficient severe combined immunodeficiency (SCID). *Clin Immunol.* 2005;117(2):133–143.

52. Kohn DB, Gaspar HB. How we manage adenosine deaminase-deficient severe combined immune deficiency (ADA SCID). *J Clin Immunol.* 2017;37(4):351–356.
53. Thomas DC. How the phagocyte NADPH oxidase regulates innate immunity. *Free Radic Biol Med.* 2018;125:44–52.
54. Rivers E, Worth A, Thrasher AJ, Burns SO. How I manage patients with Wiskott Aldrich syndrome. *Br J Haematol.* 2019;185(4):647–655.
55. Thomas ED, Epstein RB. Bone marrow transplantation in acute leukemia. *Cancer Res.* 1965;25(9):1521–1524.
56. Buckley RH. Transplantation of hematopoietic stem cells in human severe combined immunodeficiency: longterm outcomes. *Immunol Res.* 2011;49(1–3):25–43.
57. de la Morena MT, Nelson Jr RP. Recent advances in transplantation for primary immune deficiency diseases: a comprehensive review. *Clin Rev Allergy Immunol.* 2014; 46(2):131–144.
58. Willemsen L, Jol-van der Zijde CM, Admiraal R, et al. Impact of serotherapy on immune reconstitution and survival outcomes after stem cell transplantations in children: thymoglobulin versus alemtuzumab. *Biol Blood Marrow Transplant.* 2015;21(3): 473–482.
59. Castagnoli R, Delmonte OM, Calzoni E, Notarangelo LD. Hematopoietic stem cell transplantation in primary immunodeficiency diseases: current status and future perspectives. *Front Pediatr.* 2019;7:295.
60. Byrne JL, Stainer C, Cull G, et al. The effect of the serotherapy regimen used and the marrow cell dose received on rejection, graft-versus-host disease and outcome following unrelated donor bone marrow transplantation for leukaemia. *Bone Marrow Transplant.* 2000;25(4):411–417.
61. Gennery AR, Albert MH, Slatter MA, Lankester A. Hematopoietic stem cell transplantation for primary immunodeficiencies. *Front Pediatr.* 2019;7:445.
62. Laberko A, Gennery AR. Clinical considerations in the hematopoietic stem cell transplant management of primary immunodeficiencies. *Expert Rev Clin Immunol.* 2018; 14(4):297–306.
63. Pai S-Y, Logan BR, Griffith LM, et al. Transplantation outcomes for severe combined immunodeficiency, 2000–2009. *N Engl J Med.* 2014;371(5):434–446.
64. Kohn DB, Hershfield MS, Puck JM, et al. Consensus approach for the management of severe combined immune deficiency caused by adenosine deaminase deficiency. *J Allergy Clin Immunol.* 2019;143(3):852–863.
65. Moratto D, Giliani S, Bonfim C, et al. Long-term outcome and lineage-specific chimerism in 194 patients with Wiskott-Aldrich syndrome treated by hematopoietic cell transplantation in the period 1980–2009: an international collaborative study. *Blood.* 2011;118(6):1675–1684.
66. Elfeky RA, Furtado-Silva JM, Chiesa R, et al. One hundred percent survival after transplantation of 34 patients with Wiskott-Aldrich syndrome over 20 years. *J Allergy Clin Immunol.* 2018;142(5):1654–1656.e7.
67. Güngör T, Teira P, Slatter M, et al. Reduced-intensity conditioning and HLA-matched haemopoietic stem-cell transplantation in patients with chronic granulomatous disease: a prospective multicentre study. *Lancet.* 2014;383(9915):436–448.
68. Grunebaum E, Roifman CM. Bone marrow transplantation using HLA-matched unrelated donors for patients suffering from severe combined immunodeficiency. *Immunol Allergy Clin North Am.* 2010;30(1):63–73.
69. Punwani D, Kawahara M, Yu J, et al. Lentivirus mediated correction of artemis-deficient severe combined immunodeficiency. *Hum Gene Ther.* 2017;28(1):112–124.
70. Ramachandran V, Kolli SS, Strowd LC. Review of graft-versus-host disease. *Dermatol Clin.* 2019;37(4):569–582.
71. Jagasia M, Arora M, Flowers MED, et al. Risk factors for acute GVHD and survival after hematopoietic cell transplantation. *Blood.* 2012;119(1):296–307.

72. Olsson RF, Logan BR, Chaudhury S, et al. Primary graft failure after myeloablative allogeneic hematopoietic cell transplantation for hematologic malignancies. *Leukemia.* 2015;29(8):1754–1762.
73. Zając-Spychała O, Zaucha-Prażmo A, Zawitkowska J, et al. Infectious complications after hematopoietic stem cell transplantation for primary immunodeficiency in children: a multicenter nationwide study. *Pediatr Allergy Immunol.* 2020;31(5):537–543.
74. Manfredsson FP, Mandel RJ. Development of gene therapy for neurological disorders. *Discov Med.* 2010;9(46):204–211.
75. Hoogerbrugge PM, Vossen JM, v Beusechem VW, Valerio D. Treatment of patients with severe combined immunodeficiency due to adenosine deaminase (ADA) deficiency by autologous transplantation of genetically modified bone marrow cells. *Hum Gene Ther.* 1992;3(5):553–558.
76. Kohn DB, Weinberg KI, Nolta JA, et al. Engraftment of gene-modified umbilical cord blood cells in neonates with adenosine deaminase deficiency. *Nat Med.* 1995;1(10):1017–1023.
77. Bordignon C, Mavilio F, Ferrari G, et al. Transfer of the ADA gene into bone marrow cells and peripheral blood lymphocytes for the treatment of patients affected by ADA-deficient SCID. *Hum Gene Ther.* 1993;4(4):513–520.
78. Hindmarsh P, Leis J. Retroviral DNA integration. *Microbiol Mol Biol Rev.* 1999;63(4):836–843. table of contents.
79. Lukashev AN, Zamyatnin Jr AA. Viral vectors for gene therapy: current state and clinical perspectives. *Biochemistry.* 2016;81(7):700–708.
80. Sakuma T, Barry MA, Ikeda Y. Lentiviral vectors: basic to translational. *Biochem J.* 2012;443(3):603–618.
81. Hacein-Bey-Abina S, Hauer J, Lim A, et al. Efficacy of gene therapy for X-linked severe combined immunodeficiency. *N Engl J Med.* 2010;363(4):355–364.
82. Zhou S, Mody D, DeRavin SS, et al. A self-inactivating lentiviral vector for SCID-X1 gene therapy that does not activate LMO2 expression in human T cells. *Blood.* 2010;116(6):900–908.
83. Zufferey R, Dull T, Mandel RJ, et al. Self-inactivating lentivirus vector for safe and efficient in vivo gene delivery. *J Virol.* 1998;72(12):9873–9880.
84. *Gene Transfer for X-Linked Severe Combined Immunodeficiency in Newly Diagnosed Infants—Full Text View—ClinicalTrials.Gov*; 2020. Accessed December 29 https://clinicaltrials.gov/ct2/show/NCT01512888.
85. Mamcarz E, Zhou S, Lockey T, et al. Lentiviral gene therapy combined with low-dose busulfan in infants with SCID-X1. *N Engl J Med.* 2019;380(16):1525–1534.
86. Blaese RM, Culver KW, Miller AD, et al. T lymphocyte-directed gene therapy for ADA- SCID: initial trial results after 4 years. *Science.* 1995;270(5235):475–480.
87. Aiuti A, Roncarolo MG, Naldini L. Gene therapy for ADA-SCID, the first marketing approval of an ex vivo gene therapy in Europe: paving the road for the next generation of advanced therapy medicinal products. *EMBO Mol Med.* 2017;9(6):737–740.
88. Cicalese MP, Ferrua F, Castagnaro L, et al. Gene therapy for adenosine deaminase deficiency: a comprehensive evaluation of short- and medium-term safety. *Mol Ther.* 2018;26(3):917–931.
89. *ADA Gene Transfer Into Hematopoietic Stem/Progenitor Cells for the Treatment of ADA-SCID—Full Text View—ClinicalTrials.Gov*; 2020. Accessed December 29 https://clinicaltrials.gov/ct2/show/NCT00598481.
90. Aiuti A, Cattaneo F, Galimberti S, et al. Gene therapy for immunodeficiency due to adenosine deaminase deficiency. *N Engl J Med.* 2009;360(5):447–458.
91. *Autologous Gene Therapy for Artemis-Deficient SCID*; 2020. Accessed December 29 https://clinicaltrials.gov/ct2/show/NCT03538899.
92. Bokhoven M, Stephen SL, Knight S, et al. Insertional gene activation by lentiviral and gammaretroviral vectors. *J Virol.* 2009;83(1):283–294.

93. Stein S, Ott MG, Schultze-Strasser S, et al. Genomic instability and myelodysplasia with monosomy 7 consequent to EVI1 activation after gene therapy for chronic granulomatous disease. *Nat Med.* 2010;16(2):198–204.
94. Cavazzana-Calvo M, Payen E, Negre O, et al. Transfusion independence and HMGA2 activation after gene therapy of human β-thalassaemia. *Nature.* 2010;467(7313): 318–322.
95. Lino CA, Harper JC, Carney JP, Timlin JA. Delivering CRISPR: a review of the challenges and approaches. *Drug Deliv.* 2018;25(1):1234–1257.
96. Zhou R, Yang G, Shi Y. Dominant negative effect of the loss-of-function γ-secretase mutants on the wild-type enzyme through heterooligomerization. *Proc Natl Acad Sci U S A.* 2017;114(48):12731–12736.
97. Ceccaldi R, Rondinelli B, D'Andrea AD. Repair pathway choices and consequences at the double-strand break. *Trends Cell Biol.* 2016;26(1):52–64.
98. Carroll D. Genome engineering with zinc-finger nucleases. *Genetics.* 2011;188(4): 773–782.
99. Wah DA, Bitinaite J, Schildkraut I, Aggarwal AK. Structure of FokI has implications for DNA cleavage. *Proc Natl Acad Sci U S A.* 1998;95(18):10564–10569.
100. Urnov FD, Miller JC, Lee Y-L, et al. Highly efficient endogenous human gene correction using designed zinc-finger nucleases. *Nature.* 2005;435(7042):646–651.
101. De Ravin SS, Li L, Wu X, et al. CRISPR-Cas9 gene repair of hematopoietic stem cells from patients with X-linked chronic granulomatous disease. *Sci Transl Med.* 2017; 9(372):1–10. https://doi.org/10.1126/scitranslmed.aah3480.
102. Joung JK, Sander JD. TALENs: a widely applicable technology for targeted genome editing. *Nat Rev Mol Cell Biol.* 2013;14(1):49–55.
103. *The Nobel Prize in Chemistry*; 2020. Accessed December 21, 2020 https://www.nobelprize.org/prizes/chemistry/2020/press-release/.
104. Jinek M, Chylinski K, Fonfara I, Hauer M, Doudna JA, Charpentier E. A programmable dual-RNA-guided DNA endonuclease in adaptive bacterial immunity. *Science.* 2012;337(6096):816–821.
105. Gupta RM, Musunuru K. Expanding the genetic editing tool kit: ZFNs, TALENs, and CRISPR-Cas9. *J Clin Invest.* 2014;124(10):4154–4161.
106. Nishimasu H, Shi X, Ishiguro S, et al. Engineered CRISPR-Cas9 nuclease with expanded targeting space. *Science.* 2018;361(6408):1259–1262.
107. Wu Y, Zeng J, Roscoe BP, et al. Highly efficient therapeutic gene editing of human hematopoietic stem cells. *Nat Med.* 2019;25(5):776–783.
108. *A Safety and Efficacy Study Evaluating CTX001 in Subjects With Severe Sickle Cell Disease*; 2020. Accessed December 29 https://clinicaltrials.gov/ct2/show/NCT03745287.
109. Kuhns DB, Alvord WG, Heller T, et al. Residual NADPH oxidase and survival in chronic granulomatous disease. *N Engl J Med.* 2010;363(27):2600–2610.
110. Li J, Hong S, Chen W, Zuo E, Yang H. Advances in detecting and reducing off-target effects generated by CRISPR-mediated genome editing. *J Genet Genomics.* 2019;46 (11):513–521.
111. Naeem M, Majeed S, Hoque MZ, Ahmad I. Latest developed strategies to minimize the off-target effects in CRISPR-cas-mediated genome editing. *Cell.* 2020;9(7): 1–23. https://doi.org/10.3390/cells9071608.
112. Kuscu C, Arslan S, Singh R, Thorpe J, Adli M. Genome-wide analysis reveals characteristics of off-target sites bound by the Cas9 endonuclease. *Nat Biotechnol.* 2014; 32(7):677–683.
113. Crosetto N, Mitra A, Silva MJ, et al. Nucleotide-resolution DNA double-strand break mapping by next-generation sequencing. *Nat Methods.* 2013;10(5):361–365.
114. Tsai SQ, Zheng Z, Nguyen NT, et al. GUIDE-seq enables genome-wide profiling of off-target cleavage by CRISPR-Cas nucleases. *Nat Biotechnol.* 2015;33(2):187–197.

115. Tsai SQ, Nguyen NT, Malagon-Lopez J, Topkar VV, Aryee MJ, Joung JK. CIRCLE-seq: a highly sensitive in vitro screen for genome-wide CRISPR-Cas9 nuclease off-targets. *Nat Methods*. 2017;14(6):607–614.
116. Kim D, Bae S, Park J, et al. Digenome-seq: genome-wide profiling of CRISPR-Cas9 off-target effects in human cells. *Nat Methods*. 2015;12(3):237–243. 1 p following 243.
117. Wu X, Kriz AJ, Sharp PA. Target specificity of the CRISPR-Cas9 system. *Quant Biol*. 2014;2(2):59–70.
118. Haapaniemi E, Botla S, Persson J, Schmierer B, Taipale J. CRISPR-Cas9 genome editing induces a p53-mediated DNA damage response. *Nat Med*. 2018;24(7): 927–930.
119. Ihry RJ, Worringer KA, Salick MR, et al. p53 inhibits CRISPR-Cas9 engineering in human pluripotent stem cells. *Nat Med*. 2018;24(7):939–946.
120. Kosicki M, Tomberg K, Bradley A. Repair of double-strand breaks induced by CRISPR-Cas9 leads to large deletions and complex rearrangements. *Nat Biotechnol*. 2018;36(8):765–771.
121. Chen D, Tang J-X, Li B, Hou L, Wang X, Kang L. CRISPR/Cas9-mediated genome editing induces exon skipping by complete or stochastic altering splicing in the migratory locust. *BMC Biotechnol*. 2018;18(1):60.
122. Kapahnke M, Banning A, Tikkanen R. Random splicing of several exons caused by a single base change in the target exon of CRISPR/Cas9 mediated gene knockout. *Cell*. 2016;5(4):1–12. https://doi.org/10.3390/cells5040045.
123. Nishida K, Arazoe T, Yachie N, et al. Targeted nucleotide editing using hybrid prokaryotic and vertebrate adaptive immune systems. *Science*. 2016;353(6305):1248. https://doi.org/10.1126/science.aaf8729.
124. Gaudelli NM, Komor AC, Rees HA, et al. Programmable base editing of A·T to G·C in genomic DNA without DNA cleavage. *Nature*. 2017;551(7681):464–471.
125. Landrum MJ, Lee JM, Benson M, et al. ClinVar: public archive of interpretations of clinically relevant variants. *Nucleic Acids Res*. 2016;44(D1):D862–D868.
126. Komor AC, Kim YB, Packer MS, Zuris JA, Liu DR. Programmable editing of a target base in genomic DNA without double-stranded DNA cleavage. *Nature*. 2016; 533(7603):420–424.
127. Webber BR, Lonetree C-L, Kluesner MG, et al. Highly efficient multiplex human T cell engineering without double-strand breaks using Cas9 base editors. *Nat Commun*. 2019;10(1):5222.
128. Zeng J, Wu Y, Ren C, et al. Therapeutic base editing of human hematopoietic stem cells. *Nat Med*. 2020;26(4):535–541.
129. Park S, Beal PA. Off-target editing by CRISPR-guided DNA base editors. *Biochemistry*. 2019;58(36):3727–3734.
130. Liu Y, Zhou C, Huang S, et al. A Cas-embedding strategy for minimizing off-target effects of DNA base editors. *Nat Commun*. 2020;11(1):6073.
131. Anzalone AV, Randolph PB, Davis JR, et al. Search-and-replace genome editing without double-strand breaks or donor DNA. *Nature*. 2019;576(7785):149–157.
132. Matsoukas IG. Prime editing: genome editing for rare genetic diseases without double-strand breaks or donor DNA. *Front Genet*. 2020;11:528.
133. Kim DY, Moon SB, Ko J-H, Kim Y-S, Kim D. Unbiased investigation of specificities of prime editing systems in human cells. *Nucleic Acids Res*. 2020;48(18): 10576–10589.
134. Breuer CB, Hanlon KS, Natasan J-S, et al. In vivo engineering of lymphocytes after systemic exosome-associated AAV delivery. *Sci Rep*. 2020;10(1):4544.
135. Wei T, Cheng Q, Min Y-L, Olson EN, Siegwart DJ. Systemic nanoparticle delivery of CRISPR-Cas9 ribonucleoproteins for effective tissue specific genome editing. *Nat Commun*. 2020;11(1):3232.

CHAPTER FIVE

Genome editing approaches to β-hemoglobinopathies

Mégane Brusson* and Annarita Miccio*

Université de Paris, Imagine Institute, Laboratory of Chromatin and Gene Regulation During Development, INSERM UMR 1163, Paris, France
*Corresponding authors: e-mail address: megane.brusson@institutimagine.org; annarita.miccio@institutimagine.org

Contents

1. Introduction	155
2. Genome editing tools	157
2.1 Nuclease-mediated genome editing	157
2.2 Base and prime editing	161
3. Correcting β-globin gene mutations to treat β-hemoglobinopathies	162
3.1 Correction of the SCD mutation	162
3.2 Correction of β-thalassemia mutations	166
4. De-repression of the endogenous fetal γ-globin genes	169
4.1 Mimicking HPFH mutations in *HBG* promoters	169
4.2 Targeting γ-globin transcriptional repressors	174
4.3 Clinical trials for β-hemoglobinopathies based on genome editing approaches	176
Author contributions	177
Acknowledgments	177
Conflicts of interest	177
References	177

Abstract

β-hemoglobinopathies are the most common monogenic disorders worldwide and are caused by mutations in the β-globin locus altering the production of adult hemoglobin (HbA). Transplantation of autologous hematopoietic stem cells (HSCs) corrected by lentiviral vector-mediated addition of a functional β-like globin raised new hopes to treat sickle cell disease and β-thalassemia patients; however, the low expression of the therapeutic gene per vector copy is often not sufficient to fully correct the patients with a severe clinical phenotype.

Recent advances in the genome editing field brought new possibilities to cure β-hemoglobinopathies by allowing the direct modification of specific endogenous loci. Double-strand breaks (DSBs)-inducing nucleases (i.e., ZFNs, TALENs and CRISPR-Cas9) or DSB-free tools (i.e., base and prime editing) have been used to directly correct the disease-causing mutations, restoring HbA expression, or to reactivate the expression of the fetal hemoglobin (HbF), which is known to alleviate clinical symptoms of β-hemoglobinopathy patients.

Here, we describe the different genome editing tools, their application to develop therapeutic approaches to β-hemoglobinopathies and ongoing clinical trials using genome editing strategies.

Abbreviations list

ABEs	adenine BEs
AsCas12a	*Acidaminococcus* sp. Cas12a
BEs	base editors
BSs	binding sites
Cas9n	Cas9 nickase
CBEs	cytosine BEs
CRISPR-Cas9	clustered regularly interspaced short palindromic repeat (CRISPR)-associated Cas9 nucleases
dCas9	dead Cas9
DSBS	double-strand breaks
gRNA	guide RNA
hA3A	human APOBEC3A deaminase
HbA	adult hemoglobin
HBB	β-globin gene
HbE	hemoglobin E
HbF	fetal hemoglobin
HBG	γ-globin gene
HbS	sickle hemoglobin
HDR	homology-directed repair
hiPSCs	human induced pluripotent stem cells
HPFH	hereditary persistence of fetal hemoglobin
HSCs	hematopoietic stem cells
HSCT	hematopoietic stem cell transplantation
HSPCs	hematopoietic stem/progenitor cells
InDels	insertions or deletions
LbCas12a	*L. bacterium* Cas12a
LV	lentiviral vector
NHEJ	non-homologous end-joining
NuRD	nucleosome remodeling and deacetylase
PAM	protospacer adjacent motif
PE	prime editor
pegRNA	prime editing guide RNA
PNA	peptide nucleic acid
rAAV6	recombinant adeno-associated virus serotype 6
RBC	red blood cell
RNPs	ribonucleoproteins
RT	reverse transcriptase
SaCas9	*Staphylococcus aureus* Cas9
SCD	sickle cell disease

SpCas9	*Streptococcus pyogenes* Cas9
ssODN	single-stranded oligonucleotide
TadA	tRNA adenine deaminase
TALENs	transcriptional activator-like effector nucleases
TDT	transfusion-dependent β-thalassemia
UGI	uracil glycosylase inhibitor
ZFNs	zinc finger nucleases

1. Introduction

Hemoglobin (Hb) is a tetramer composed of two α-like and two β-like globin chains. Its main function is to deliver oxygen to the tissues. Genes encoding the α-like and β-like globin chains are clustered on chromosome 16 and 11, respectively (Fig. 1). The locus control region (LCR) in the β-globin locus contains potent enhancers that sequentially activates β-like globin genes during development (embryonic *HBE* ε-globin gene, fetal *HBG1* and *HBG2* γ-globin genes, adult *HBB* β-globin and *HBD* δ-globin genes). Fetal Hb (HbF) is composed of two α- and two β-like γ-globin chains ($α_2γ_2$). Adult hemoglobin mainly contains two α- and two β-globin chains (HbA, $α_2β_2$) (Fig. 1).

β-hemoglobinopathies are the most common monogenic disorders worldwide and affect hundreds of thousands of newborns each year.[1] These blood disorders are due to mutations in the β-globin locus that cause defective β-globin production and affect the formation of HbA.

β-thalassemias are caused by more than 300 mutations in the *HBB* locus,[2] leading to reduced ($β^+$) or abolished ($β^0$) β-globin expression. This alters the balance between α- and β-globin chains and causes the

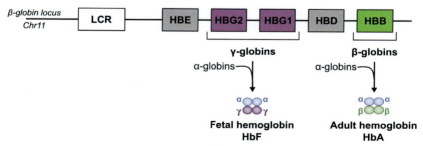

Fig. 1 The β-globin locus. β-like globin genes are sequentially expressed during development. The LCR is responsible for boosting β-like globin gene expression. *HBE*, embryonic ε-globin gene; *HBG1* and *HBG2*, fetal γ-globin genes; *HBB* and *HBD*, adult β- and δ-globin genes; LCR, locus control region.

formation of α-precipitates in erythroid precursors, ineffective erythropoiesis, dysregulation of iron homeostasis and anemia.[3] Red blood cell (RBC) transfusion is the main treatment to alleviate anemia in β-thalassemia patients, but it is a lifelong treatment, which contributes to iron overload and organ damage (e.g., to the heart and the liver).[4]

In sickle cell disease (SCD), a point mutation in the *HBB* gene results in the substitution of a valine for a glutamic acid (E6V) and gives rise to the formation of the abnormal sickle β^S-globin composing the sickle hemoglobin (HbS, $α_2β^S_2$). Upon deoxygenation, the HbS tetramers polymerize in long fibers, resulting in the generation of sickle-shaped RBCs, which cause vaso-occlusive crises, multi-organ damage and anemia.[5] RBC exchange transfusion is used as a supportive treatment for severe SCD patients to lower the number of sickle RBCs, but is not a definitive treatment and the frequent transfusions impair the patients' quality of life.[6]

Allogenic hematopoietic stem cell transplantation (HSCT) is the only curative treatment for both β-thalassemia and SCD patients; however, the lack of compatible donors and the HSCT-associated immunological complications (e.g., graft versus host disease) limit the use of allogenic HSCT only to a fraction of the patients.[7–12]

Transplantation of autologous HSCs corrected by lentiviral vector (LV)-mediated addition of a functional β-globin represents a therapeutic option for patients lacking a suitable donor.[13–15] However, the low expression of the LV-derived therapeutic β-globin is beneficial mostly for β-thalassemic patients with residual β-globin production (β$^+$) and is not sufficient to completely correct the phenotype of severe β-thalassemia and SCD patients.[16–18] Despite the new hopes raised by gene addition approaches, further improvements are still required to cure all the patients. Firstly, a higher expression level of the therapeutic β-like globin per cell is required for treating the patients with the most severe clinical phenotypes; however, this is difficult to achieve with classical gene addition strategies as the entire LCR, responsible for high β-like globin expression, cannot fit within a LV. Secondly, the persistence of high mutant β^S-globin expression in LV-transduced SCD cells prevents the full correction of the sickle RBC phenotype. Another limitation of LV-based gene therapy is the high cost of LV manufacturing that could prevent its application to a large number of patients.[19] Finally, despite the safe integration profile of LVs, the preferential integration of LV in genes could lead to gene inactivation and aberrantly spliced cellular transcripts.[20–22] Recent advances in the genome editing field brought new possibilities to treat blood disorders, such as β-hemoglobinopathies, by allowing the modification of specific endogenous loci. Genome editing-based therapies (usually based on non-viral delivery methods) might be less expensive than LV-based treatment allowing a broader use of gene therapy for β-hemoglobinopathies.

In this chapter, we describe the different genome editing tools that are currently used to develop gene therapy approaches, such as protein- or RNA-guided nucleases and the recently developed base and prime editing technologies. Then, we discuss the therapeutic approaches developed to correct the β-hemoglobinopathy-causing mutations or to reactivate the expression of fetal hemoglobin (HbF, $α_2γ_2$), which ameliorates the clinical phenotype of patients with β-thalassemia and SCD. Finally, we describe the ongoing clinical trials based on genome editing approaches for treating β-hemoglobinopathies.

2. Genome editing tools
2.1 Nuclease-mediated genome editing

The development of site-specific nucleases opened new perspectives to treat genetic diseases, such as β-hemoglobinopathies, by directly correcting disease-causing mutations or by modulating the expression of disease-modifier genes.

The first generation of site-specific nucleases are protein-based editing tools. The zinc finger nucleases (ZFNs) and the transcriptional activator-like effector nucleases (TALENs) harbor a DNA-binding domain that recognizes specific sequences in the genome (Fig. 2). The second generation of nucleases, the clustered regularly interspaced short palindromic repeat (CRISPR)-associated Cas9 nucleases, are guided by an engineered RNA (guide RNA, gRNA) allowing the recognition of a specific DNA sequence.

These nucleases induce double-strand breaks (DSBs) at the target locus, which are repaired by the endogenous DNA repair machinery to maintain genome integrity. The two major cellular DNA repair pathways are non-homologous end-joining (NHEJ) or homology-directed repair (HDR).[23]

NHEJ is the most active DNA repair pathway in human cells and repairs DSBs without using a repair template. NHEJ-mediated DNA repair can introduce small insertions or deletions (InDels) at the cleavage site. InDels can disrupt the target gene by generating frameshift mutations and premature stop codons in the coding sequence. *Cis*-regulatory regions,

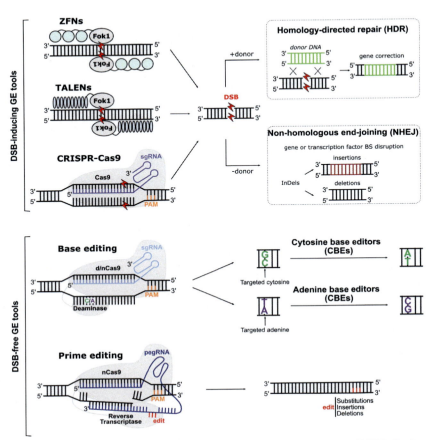

Fig. 2 Genome editing (GE) tools. DSBs generated by ZFNs, TALENs or CRISPR-Cas9 are repaired by homologous directed repair (HDR) in the presence of a DNA donor allowing gene correction or by non-homologous end joining (NHEJ) introducing InDels to disrupt genes or transcription factor BSs. Base and prime editors are DSB-free GE tools containing a dead or a nickase Cas9 (d/nCas9). Base editing allows precise base conversion (C-to-T for CBEs and A-to-G for ABEs) and prime editing can generate all the different base substitutions, insertions and deletions.

such as binding sites for transcriptional activators or repressors, can also be inactivated via NHEJ-mediated InDel generation to reduce or increase gene expression, respectively.[24] This pathway can also be exploited to restore correct gene expression, for example, by disrupting aberrant splice sites[25,26] (Fig. 2).

HDR is a high-fidelity DNA repair pathway mainly active in S/G2 phases of the cell cycle and requires the presence of a DNA template to precisely correct the DNA lesion. The repair template is usually the homologous DNA sequence in the sister chromatid. An exogenous donor template carrying the correct gene can be used by the HDR pathway to repair a mutant gene.[27] The donor template containing homologous arms (sequences homologous to the regions flanking the DSB) can be delivered to the cells as single or double-stranded oligonucleotides or by a viral vector. The HDR pathway can be exploited to correct genetic defects causing hematopoietic disorders; however, its poor efficiency in quiescent HSCs, the target cell population in gene therapy, represent an obstacle for the development of therapeutic solutions based on HDR[28] (Fig. 2).

NHEJ- and HDR-mediated genome editing can be exploited to develop therapeutic solutions for β-hemoglobinopathies; however, DSBs can lead to large chromosomal rearrangements, p53 pathway activation and apoptosis, raising safety concerns when developing nuclease-based therapies.[29,30] To avoid these drawbacks and offer a safer therapeutic solution to genetic disorders including β-hemoglobinopathies, novel genome editing tools were recently developed. In particular, base and prime editing are two new DSB-free genome editing tools that use catalytically inactive nucleases (see paragraph "Base and prime editing").[31,32]

2.1.1 ZFNs and TALENs

ZFNs and TALENs are engineered fusion proteins that recognize and cleave specific DNA sequences.

ZFNs are composed of zinc finger DNA-binding domains (three to six, each of which recognizes a 3-base pair (bp) DNA sequence) fused to the endonuclease domain of the *FokI* restriction enzyme from the *Flavobacterium okeanokoites*.[33] As *FokI*-dependent DNA cleavage requires dimerization of the cleavage domains, two ZFNs are required for binding DNA (on both strands and in opposite direction) to align *FokI* domains and allow DNA cleavage. After DNA binding, the dimerized cleavage domains generate DSBs, which can be repaired by NHEJ or HDR (Fig. 2).

TALENs are composed by TALE DNA-binding domains fused to *FokI*. Multiple engineered TALE domains need to be combined and each of them binds independently to a single nucleotide, facilitating the design of TALENs compared to ZFNs.[34] As for ZFNs, TALEN-mediated DSBs require the dimerization of the *FokI* domains (which is possible by antiparallel DNA binding of two TALENs) and are repaired by NHEJ or HDR (Fig. 2).

ZFNs and TALENs have been shown to edit numerous loci in multiple cell types. TALENs are often associated with a lower cytotoxicity compared to ZFNs, but their larger size could impair their delivery in clinically-relevant cells for gene therapy applications.[35] Both ZFNs and TALENs can generate off-targets, which could be deleterious for the target cells. Finally, their complex design is a potential limitation to their development for genome editing approaches.

2.1.2 CRISPR-Cas9 nuclease

Less than 10 years ago, Charpentier and Doudna exploited the CRISPR-Cas9 system from *Streptococcus pyogenes* (SpCas9), which is involved in the immune adaptive response of bacteria and archaea against viruses and plasmids, to develop a new genome editing technology.[36,37] In 2020, they received the Nobel Prize in Chemistry for their discovery that brought novel genome editing-based therapeutic solutions for genetic disorders including β-hemoglobinopathies.

Contrary to ZFNs and TALENs, the SpCas9 nuclease relies on a guide RNA (gRNA) to target and cleave a specific DNA locus. The CRISPR-Cas9 activity requires the binding of the gRNA to the target sequence (protospacer) and the recognition of a 3-nucleotide DNA sequence downstream of the protospacer (named protospacer adjacent motif or PAM) by the Cas9 nuclease. This RNA/DNA binding unwinds DNA and allows Cas9-mediated cleavage of the targeted sequence, 3 to 4 nucleotides upstream the PAM. DSBs are then repaired by endogenous cellular machinery by NHEJ or HDR (Fig. 2).

The major advantages of the CRISPR-Cas system compared to ZFNs and TALENs are: (1) the simple design and cloning, (2) the high genome editing efficiency, (3) the low cytotoxicity and the transient expression when delivered as ribonucleoproteins (RNPs) in comparison to mRNA delivery (typically used for ZFNs and TALENs) and (4) the possibility of multiplexing allowing the simultaneous targeting of different loci.[29] Altogether, these advantages make CRISPR-Cas technology the genome editing tool of choice to develop therapeutic approaches.

To enlarge the targeting scope of CRISPR-Cas9, which is limited to DNA loci containing a 5′-NGG-3′ PAM (specific of the SpCas9), many laboratories have characterized other Cas9 orthologs or engineered Cas9 variants with different PAM compatibility. By the way of example, the SaCas9 from *Staphylococcus aureus* targets a 5′-NNGRRT-3′ PAM, while the Cas12a from *L. bacterium* (LbCas12a) or *Acidaminococcus* sp.

(AsCas12a) recognize a T-rich PAM (5′-TTTN-3′) located upstream the protospacer.[38,39] Examples of engineered Cas orthologs are SpCas9-VRQR, SpCas9-VRER and SaCas9-KHH recognizing NGA, NGCG and NNNRRT PAMs, respectively. To minimize off-target activity "high fidelity" Cas9 nucleases have been developed: HypaCas9,[40] evoCas9,[41] Sniper-Cas9[42] and HiFi Cas9.[43] However, among these SpCas9 variants, the HiFi Cas9 is the only one able to reduce off-targets while maintaining a high on-target genome editing efficiency, as demonstrated by Vakulska et al. in human CD34$^+$ hematopoietic stem/progenitor cells (HSPCs).[43]

2.2 Base and prime editing

Recently, Liu and coworkers engineered the CRISPR-Cas9 tool to develop base and prime editing technologies (Fig. 2). The major advantage of these tools is their ability to modify DNA without generating DSBs, improving the safety profile of genome editing-based therapies. To this aim, they introduced amino acid substitutions in the Cas9 nuclease to generate a catalytically inactive or dead Cas9 (dCas9, D10A and H840A), suppressing its nuclease activity, or a Cas9 nickase (Cas9n, D10A or H840A), generating only single strand DNA breaks.

More precisely, base editors (BEs) consist of dCas9 or Cas9n fused to a deaminase, which is driven to the target locus by a gRNA and induces pinpoint mutations without introducing DSBs. Two types of BEs were developed: cytosine BEs or CBEs, which convert a C•G to a T•A bp,[44] and adenine BEs or ABEs allowing A•T to G•C bp conversions[45] (Fig. 2).

In the first-generation of CBEs (BE1 and 2), a rat deaminase (rAPOBEC1) was selected as the best performing cytosine deaminase and fused to the dCas9 (BE1).[44] The addition of an uracil glycosylase inhibitor (UGI) to the dCas9 C-terminus further improved base conversion (BE2) by inhibiting base excision repair.[44] The replacement of the dCas9 by a nCas9 in BE3 favored the repair of the nicked and non-edited strand using the edited strand as a DNA template.[44] In BE4, a second UGI was added to BE3 to improve base conversion efficiency and product purity (i.e., to reduce C > non-T edits).[46]

Contrary to CBEs, the development of ABEs was more complex due to the absence of a natural DNA adenine deaminase. To create ABEs, Gaudelli et al. exploited a tRNA adenine deaminase from *E. coli* (TadA) that works natively as a homodimer. In particular, they fused Cas9n to the wild-type

TadA and an evolved TadA* that efficiently deaminates adenines in the DNA, and created ABE variants enabling A-to-G base conversion.[45] Among all ABE variants, the ABE7.10 showed the highest editing efficiency.[45] Notably, ABEs insert point mutations with a higher efficiency and reduced InDel formation compared to Cas9 nuclease-mediated HDR approaches.[45]

BEs have been extensively engineered to improve base conversion and correct a higher number of point mutations causing genetic disorders. As for CRISPR-Cas9 system, the use of Cas9 orthologs or engineered variants led to the development of BEs displaying an extended targeting scope.[47] Modifications of the deaminase reduced bystander edits (edits of neighboring non-target bases), limiting unwanted base conversion, or modified the activity or editing window (i.e., targeted bases located in a specific DNA stretch of the protospacer) to enlarge the BE targeting scope.[48]

Prime editing is a more versatile genome editing tool as it allows insertions, deletions and all base conversions without generating DSBs.[31] The prime editor (PE) consists of a Cas9n fused to an engineered reverse transcriptase (RT) guided to the target region by a prime editing guide RNA (pegRNA), which also carries a template containing the desired edits that is eventually inserted into the genomic DNA after reverse transcription (Fig. 2). Compared to base editing, prime editing produces less bystander edits, does not require a close PAM and is not restricted to a particular editing window, enabling to target a wider variety of loci.[31] Moreover, prime editing generates less off-targets compared to Cas9 nuclease.[31]

3. Correcting β-globin gene mutations to treat β-hemoglobinopathies

Correcting the disease-causing mutations is a promising therapeutic solution for β-hemoglobinopathies. In fact, *HBB* gene correction permits high-level expression of the endogenous wild-type gene.

3.1 Correction of the SCD mutation

SCD is due to a single point mutation in the *HBB* gene, facilitating the development of a unique genome editing strategy applicable to all patients. Different approaches have been developed and most of them exploit the HDR-mediated DNA repair after DNA cleavage to correct the mutant β^S-globin gene. More recently, the development of the base and prime

Editing strategies for β-hemoglobinopathies 163

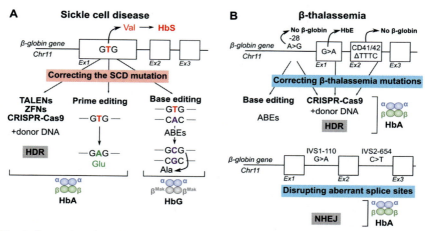

Fig. 3 Correcting disease-causing mutations by GE. (A) DSB-inducing nucleases, base and prime editors have been used to correct the SCD point mutation. TALEN-, ZFN- or CRISPR-Cas9-HDR strategies and prime editing allow the production of the wild type HbA. ABEs can be used to convert the mutated GTG codon in a GCG triplet, generating the non-pathogenic HbG Makassar. (B) GE strategies for correcting β-thalassemia mutations are based on the use of DSB-inducing nucleases or ABEs. (top panel) CRISPR-Cas9 combined to a donor template can be used to correct mutations in HBB gene and promoter via HDR. The -28A > G mutation in HBB promoter can be also corrected using ABEs. (bottom panel) DSBs following CRISPR-Cas9 cleavage induce InDels enabling the disruption of aberrant splice sites generated by the IVS1-110G > A and IVS2-654C > T mutations.

editing systems offered new promising therapeutic approaches to correct the most frequent disease-causing point mutation (Fig. 3A).

The first approach of genome editing to correct the SCD mutation was based on ZFNs to target a GFP reporter gene carrying nucleotides surrounding the sickle HBB mutation. The co-delivery of engineered ZFNs and a plasmid DNA template allowed gene correction and showed the first evidences of efficacy of a genome editing strategy based on nuclease-mediated HDR in mammalian cells.[49] Based on this approach, two different studies showed that ZFNs co-delivered with a DNA template into patient-derived human induced pluripotent stem cells (hiPSCs) enabled correction of the SCD mutation in the endogenous locus, advancing the development of genome editing-based therapies for SCD patients.[50,51] However, due to the low efficiency of gene correction in hiPSCs, these strategies required the addition of a drug resistant gene in the donor DNA template to select the corrected clones. Removal of this gene using the Cre recombinase permitted the expression of the corrected β-globin. However, in these studies, a loxP motif was still present in the *HBB* gene and this could alter β-globin expression.[50,51]

TALENs have also been used to correct the SCD mutation by HDR. Different studies reported the use of engineered TALENs in combination with a DNA template in the human erythroleukemic K562 cell line[52] and in hiPSCs.[53,54] Even if HDR following TALEN-induced DNA cleavage seems more efficient than ZFN-based strategies,[52] this approach in hiPSCs still required the addition of a selection gene cassette in the donor DNA template to isolate corrected clones. This cassette is then removed by a secondary excision step leaving a scar in the genome, which could potentially affect β-globin expression.[53] Sun and Zhao developed a safer, seamless approach enabling the complete removal of the selection cassette.[54] Importantly, corrected hiPSC clones do not present off-targets, maintain their pluripotency capacity and expressed β-globin once differentiated into erythroid cells.[53,54]

hiPSCs could be easily generated from patient-derived fibroblasts or peripheral blood mononuclear cells and their use was essential for the development and characterization of new genome editing approaches for SCD. However, correcting HSPCs is preferred to demonstrate the feasibility of genome editing approaches for SCD, as they are the most clinically-relevant cell source for autologous transplantation. Hoban et al. used for the first time ZFNs targeting the SCD mutation that were delivered as mRNA together with a donor DNA template (delivered as a single-stranded oligonucleotide [ssODN] or provided by an integrative deficient LV) into human HSPCs.[55] SCD CD34$^+$ HSPCs showed around 20% of corrected alleles enabling the production of HbA in HSPC-derived erythroid cells.[55] Similarly, Dewitt et al. used CRISPR-Cas9 and a ssODN to correct the SCD mutation in HSPCs and obtained around 12% of HDR-based gene correction.[56] In these two studies, although the modified HSPCs were able to engraft in immunodeficient mice and produce all the cell lineages, the long-term repopulating HSCs displayed a lower gene correction efficiency compared to input cells (0.85%[55] and 2%[56]). This reduced efficacy in the target cell population of gene therapy approaches for hematopoietic disorders still represents an obstacle to the success of HDR-based genome editing strategies.[55,56]

To improve gene correction after transplantation, Dever et al. combined the CRISPR-Cas9 cleavage of the *HBB* gene with donor delivery by recombinant adeno-associated virus serotype 6 (rAAV6) to correct the SCD mutation in patient HSPCs.[57] Interestingly, they obtained a high frequency of corrected alleles ranging from 30% to 70% *in vitro* in CD34+ cells, enabling the production of wild-type β-globin mRNA in HSPC-derived erythrocytes.[57] However, they also observed an important decrease in the proportion of corrected cells *in vivo* in immunodeficient mice transplanted with edited HSPCs.[57] Comparison of ZFN and CRISPR-Cas9 efficiency in HSPCs revealed that both nucleases generate a similar level of DSBs and that the reduced gene correction frequency *in vivo* is not due to defects in DNA cleavage.[58] Moreover, different donor templates were compared in terms of efficiency of gene correction. While the donor template delivery by rAAV6 generated a higher proportion of corrected cells *in vitro*, the frequency of ssODN-corrected cells after xenotransplantation was similar or even higher than the percentage of rAAV6-modified cells. These results suggest that ssODNs are either more efficient in HSCs or rAAV6 transduction impairs HSC viability and/or engraftment.[58,59] Park et al. ameliorated the ssODN design, thus increasing the efficiency of CRISPR-Cas9-based SCD gene correction to 24% in patients' HSPCs.[60] However, less than 10% of corrected CD34+ cells persisted in mice 14–16 weeks after transplantation confirming the poor rate of HDR in quiescent HSCs and/or the sub-standard long-term engraftment capability of the modified cells.[57,59,60]

Notably, in these nuclease-based gene correction approaches for SCD, failed HDR-mediated gene correction can lead to *HBB* gene inactivation by NHEJ, thus potentially generating β-thalassemia-like mutations.[61] Amelioration of the donor template design and delivery, as well as optimization of culture conditions to preserve HSC stemness, will likely benefit this approach and move it forward to the clinic.[62]

The recent development of base and prime editing technologies brought new possibilities of correcting the SCD mutation. Compared to the previously described approaches, base and prime editing do not generate DSBs, limiting the risk of *HBB* gene inactivation and circumvent the need of HDR-mediated gene correction, which is poorly efficient in HSCs[28] (Fig. 2).

While no BE can revert the SCD mutation and convert the mutated GTG codon (Valine) in a wild-type GAG codon (Glutamic acid), BEs can be used to convert the mutated base and generate a GCG codon translated into an Alanine, thus reproducing the non-pathogenic HbG Makassar allele.[63,64] Miller et al. evolved SpCas9 to improve the targeting scope of ABEs. These new ABE variants enabled for the first time to target the SCD mutation thanks to their compatibility with the previously inaccessible NRHH (R=A or G, H=A or C or T) PAMs.[47] Different ABE variants and gRNAs targeting a protospacer associated with CATG or CACC PAMs were tested in HEK293T cells homozygous for the SCD mutation. These novel ABE variants showed the highest base conversion activity when using a gRNA followed by the CACC PAM and the best performing ABE-NRCH variant generated up to 40% of A•T-to-G•C base conversion.[47]

Prime editing was also used to correct the SCD mutation, by reverting the mutated GTG codon into the wild-type GAG triplet. In HEK293T cells, Anzalone et al. obtained a frequency of gene correction of up to 50% using prime editing, while maintaining a low level of InDels. Importantly, both these parameters (i.e., high gene correction and low InDels) were more favorable in prime-edited cells compared to cells corrected by Cas9 nuclease-mediated HDR.[31]

The results of these two strategies are promising for the treatment of SCD; however, they need to be tested in clinically-relevant cells (i.e., patient-derived HSPCs) to be translated to the clinical realm.

3.2 Correction of β-thalassemia mutations

Contrary to SCD, caused by a single point mutation, for which the same genome editing strategy can be applied to all the patients, more than 300 mutations have been described in β-thalassemia (including point mutations and deletions). This could hamper the development of a unique genome editing-based therapeutic approach for all the β-thalassemia patients. However, 90% of β-thalassemia patients harbor only about 40 of these mutations,[65] highlighting the possibility to develop genome editing-based therapies for the vast majority of patients (Fig. 3).

Several groups developed HDR-based strategies to correct β-thalassemia mutations in the *HBB* gene. Most of these approaches are based on the use of CRISPR-Cas9 and different donor templates, such as plasmids[66] or ssODNs.[67-69]

The most common β-thalassemia mutation in Southeast Asia and Southern China,[70] the CD41/42ΔTTTC located in *HBB* exon 2, was efficiently corrected in patient-derived hiPSCs using HDR-based approaches. In particular, two scarless strategies employed ssODNs harboring the corrective gene fragment and a selection gene (puromycin resistance gene or GFP reporter gene) introduced in the CRISPR-Cas9 expressing plasmid to enrich for corrected hiPSC clones.[67,68]

The G > A point mutation in codon 26 of *HBB*, resulting in the production of the abnormal Hemoglobin E (HbE), was targeted via an HDR-based approach using CRISPR-Cas9 and a ssODN donor; however, less than 3% of hiPSC clones showed efficient gene correction due to poor transfection rate and HDR efficiency.[69]

Universal approaches were also developed to propose a therapeutic solution to all the β-thalassemia patients harboring mutations in introns or exons. In fact, the use of TALENs[71] or CRISPR-Cas9[72] together with a donor template carrying the entire *HBB* coding sequence allowed the integration of a wild-type *HBB* gene in its endogenous locus in hiPSCs. Corrected hiPSC clones expressed wild-type β-globin after erythroid differentiation.[71,72]

Notably, a nuclease-free approach based on HDR was developed to correct the aberrant splice site generated by the IVS2-654C > T mutation (the most common β-thalassemia-causing variant in Southeast Asia[73,74]) using a peptide nucleic acid (PNA) targeting the *HBB* gene and a donor DNA that replaces the mutant sequence.[75,76] In vivo nanoparticle-based delivery of PNAs and donor DNA led to the correction of the β-thalassemic phenotype in mice harboring the IVS2-654C > T mutation. HDR-mediated editing of the *HBB* gene, using PNA and a donor DNA containing the IVS2-654C > T mutation, was also achieved in human HSCs from healthy donors with a frequency of around 3%.[75,76] Therefore, increasing the proportion of edited alleles in human HSCs is necessary to achieve a therapeutic benefit.

Overall, the efficacy of these HDR-based strategies for correcting β-thalassemia mutations needs to be evaluated in patient HSCs for a potential clinical translation.

Interestingly, NHEJ-based approaches have been developed to target the most frequent point mutations occurring in *HBB* introns, which are known to generate aberrant splice sites affecting *HBB* mRNA splicing and leading to abnormal β-globin production. In particular, the NHEJ-mediated InDel formation induced by nucleases allows to disrupt the aberrant splice sites.[25,26,77] This is a promising strategy, as it exploits the NHEJ mechanism that is highly active in HSCs compared to HDR.[28]

Different groups used TALENs or CRISPR-Cas9 to disrupt the aberrant splice site generated by the IVS2-654C > T mutation.[25,26,77] In the absence of an NGG PAM (recognized by the classical SpCas9) to target the IVS2-654 mutation, Xu et al. used a CRISPR-Cas system based on LbCas12a (which is compatible with a TTTV PAM) to target this mutation[26] in patient HSPCs by RNP electroporation. After differentiation toward the erythroid lineage, the edited cells presented more than 75% of InDels, which was associated with a higher production of wild-type β-globin mRNA and an increased proportion of HbA compared to non-edited cells.[26]

The IVS1-110G > A mutation, which is one of the most common mutations in the Mediterranean and Middle East areas,[65] was also targeted using CRISPR-Cas9.[26] After RNP electroporation of patients' HSPCs and erythroid differentiation, Xu et al. observed an InDel frequency of up to 73%, which restored *HBB* mRNA splicing and increased wild-type β-globin chain expression.[26] Notably, 1 bp-insertions at the IVS1-110 site were sufficient to restore normal mRNA splicing.[26]

Finally, the -28A > G mutation in the *HBB* promoter, which is highly prevalent in Southeast Asia[2], was corrected using the CRISPR-Cas9 nuclease technology and more recently with base editing. In a first study, the -28A > G mutation was efficiently corrected by HDR using the CRISPR-Cas9 nuclease in hiPSCs from SCD patients restoring the expression of the wild-type *HBB* gene in the erythroid progeny of edited clones.[66] More recently, the -28A > G mutation was efficiently reverted using BE3 in patient's fibroblasts; however, the simultaneous bystander edit at position -25 generated a new β-thalassemic mutation that can impair β-globin production.[78] Hence, Gehrke et al. developed novel CBEs displaying a narrowed editing window.[79] The eA3A(N57Q)-BE3 variant (containing the human APOBEC3A (hA3A) deaminase harboring the N57Q mutation) outperformed all the other BE3 enzymes in efficiency of base conversion, while limiting bystander edits. RNP treatment of erythroid precursors from β-thalassemia patients led to an increased *HBB* expression; however, this system generated off-targets at four out of six tested sites.[79] Using the same strategy, Zeng et al. demonstrated the successful correction of the -28A > G mutation in heterozygous ($\beta^0/\beta^{-28A>G}$) patient-derived HSPCs upon RNP electroporation, thus restoring *HBB* expression in HSPC-derived erythroid cells.[80] It is important to note that despite the 68% of corrected edits, around 10% of bystander edits were still observed at position -25.[80] However, this DSB-free strategy has been proved efficient in clinically-relevant cells and provided the proof of principle for the applicability of base editing to treat β-thalassemia.

4. De-repression of the endogenous fetal γ-globin genes

After birth, HbF expression is switched off and RBCs mainly express adult Hb. In patients affected by β-hemoglobinopathies, the Hb switching marks the appearance of disease manifestations. A benign condition causing hereditary persistence of fetal hemoglobin (HPFH) after birth alleviates the β-hemoglobinopathy clinical phenotypes.[81] Therefore, HbF reactivation represents a potent strategy to treat β-hemoglobinopathies. Importantly, de-repressing HbF expression is a universal approach that can be used to treat all the patients (including, for example, patients affected with β-thalassemia caused by large deletions encompassing the *HBB* gene that cannot be corrected using the previously described approaches). Furthermore, γ-globin up-regulation is often associated with adult β-globin down-regulation, which can be beneficial for SCD patients to reduce HbS levels.

The strategies aimed at reactivating HbF expression can be divided in two categories: (1) approaches mimicking HPFH mutations in the promoters of the two γ-globin (*HBG1* and *HBG2*) genes to modify binding sites (BSs) for transcription factors (e.g., γ-globin activator or repressor BSs) (Fig. 4) and (2) approaches downregulating the expression of γ-globin repressors (e.g., BCL11A) (Fig. 5).

4.1 Mimicking HPFH mutations in *HBG* promoters

When co-inherited with β-globin gene mutations, HPFH mutations maintain HbF expression in adult RBCs and attenuate clinical complications of patients with β-globin disorders.[81] HPFH point mutations or small deletions disrupt the BSs for γ-globin repressors (e.g., BCL11A and LRF) or create activator (e.g., KLF1, TAL1 and GATA1) BSs. HPFH mutations are located in three main clusters in the *HBG* promoters (at position -115, -175 and -200 upstream of *HBG1* and *HBG2* transcription start sites) (Fig. 4).

Nuclease-induced DSBs is a promising approach to disrupt repressor BSs. In fact, nuclease-induced DSBs are repaired by the NHEJ pathway, which generates InDels that can inactivate γ-globin repressor BSs. Similarly, base editing can introduce HPFH point mutations that disrupt repressor BSs. Interestingly, base editing can also be exploited to introduce HPFH mutations that create BSs for γ-globin activators in the *HBG* promoters.

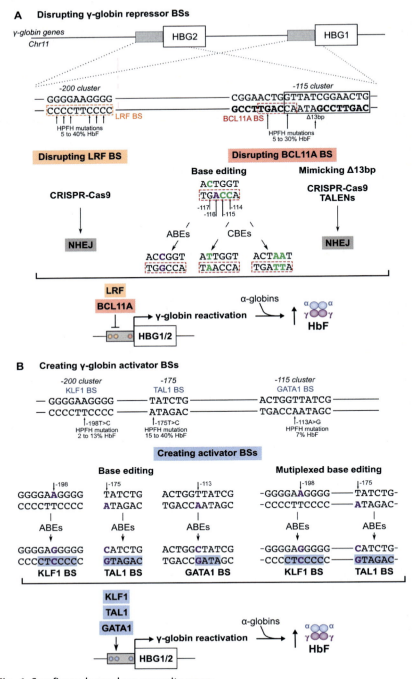

Fig. 4 See figure legend on opposite page.

4.1.1 Disrupting γ-globin repressor BSs in the HBG promoters

The sequence surrounding the -115 position of the *HBG* promoters contains a BS for the BCL11A γ-globin repressor that is disrupted by several HPFH mutations (-117G>A, -114C>T/A/G or a 13-bp deletion [Δ13bp]).[82] The -117G>A and the Δ13bp mutations are associated with HbF levels of up to 30% when present individually in patients carrying β-globin mutations (Fig. 4A).

By using a CRISPR-Cas9 strategy targeting the -115 region, Traxler et al. efficiently edited HSPCs from SCD patients and reproduced the Δ13bp mutation in around half of the edited alleles.[83] The Δ13bp was generated thanks to the 8-nt tandem repeats flanking the Cas9 cleavage site, which favor the generation of the 13-bp deletion through the microhomology-mediated end joining repair pathway.[84] After erythroid differentiation of edited HSPCs, increased HbF expression was observed in 90% of the cells, leading to a reduced RBC sickling *in vitro*.[83] A similar strategy was tested by Lux et al. using TALENs in healthy donor HSPCs.[85] A high InDel frequency was obtained in both *HBG1* and *HBG2* promoters resulting in a nearly 5-fold increase of γ-globin expression in erythroid cells.[85] Importantly, xenotransplantation experiments showed that CRISPR-Cas9- or TALEN-edited HSPCs were able to engraft and reconstitute all lineages in primary and secondary recipient mice. Despite a decreased editing rate between input and engrafted cells, HbF reactivation persisted in HSC-derived erythroid cells.[85,86] Lastly, autologous HSPCs edited at the -115 region of the *HBG* promoters by CRISPR-Cas9 engrafted in non-human primates, resulting in 6–18% of F-cells (cells expressing HbF) in peripheral blood.[87] Further evaluations are required to confirm if this increase in γ-globin expression can be sufficient to correct the β-hemoglobinopathy clinical phenotype. Interestingly, an approach targeting *in vivo* HSCs has been proposed using a non-integrating adenoviral vector expressing the CRISPR-Cas9 system to disrupt the BCL11A

Fig. 4 Mimicking HPFH mutations in *HBG* promoters to re-activate HbF expression. HPFH mutations can be reproduced in *HBG* promoters to (A) disrupt γ-globin repressor binding sites (BSs) or (B) generate γ-globin activator BSs in order to de-repress the γ-globin genes. LRF BS (in orange, -200 cluster) was disrupted using CRISPR-Cas9 to generate DSBs and introduce InDels via NHEJ. BCL11A BS (in red, -115 cluster) was precisely modified using ABEs or CBEs or disrupted by reproducing the Δ13bp HPFH mutation using CRISPR-Cas9 or ZFNs. Homologous sequences (in bold) favor MMEJ. Contrary to repressor BSs, DNA motifs for transcriptional activators (e.g., KLF1, TAL1 and GATA1, in blue) can be generated only with base editing.

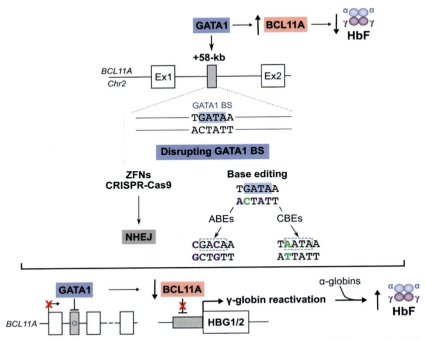

Fig. 5 Disrupting a *BCL11A* erythroid-specific enhancer to induce HbF expression. The transcriptional activator GATA1 is thought to bind to a *BCL11A* intronic enhancer (located 58 kb downstream of the *BCL11A* TSS) and activate its expression. BCL11A, in turn, represses HbF expression. The disruption of the GATA1 BS in the *BCL11A* enhancer by nucleases (via NHEJ) or by ABEs and CBEs efficiently reactivate γ-globin expression.

BS in the *HBG* promoters.[88] The results obtained in a mouse model harboring the human *HBB* locus are encouraging and might lead to the development of an *in vivo* gene therapy approach that would bypass the transplantation procedure, thus reducing the potential HSCT-related toxicities and expanding the use of gene therapy for β-hemoglobinopathies.

Disrupting the BCL11A BS by nuclease-induced InDels is a promising therapeutic solution for patients with β-hemoglobinopathies as it causes a potent HbF upregulation and does not impair HSC engraftment and multi-lineage differentiation; however, nuclease-mediated targeting of the BCL11A BS generates large deletions between the *HBG1* and *HBG2* promoters, leading to the loss of *HBG2* gene, which could affect the total HbF expression.[83,88] As an alternative solution, the DSBs-free base editing technology was used to install HPFH or HPFH-like mutations (e.g., −116A > G,

−117G>A, −115C>T or −114C>T) in the *HBG* promoters of erythroid cell lines or primary cells.[89–91] In primary HSPCs derived from β-thalassemic patients, an editing efficiency of 20% in the BCL11A BS was sufficient to re-activate HbF levels in HSPC-derived erythroid cells, while less than 2% of InDels and no variation in the *HBG* gene copy number were observed.[90]

Finally, many HPFH mutations mapping to the −200 region of the *HBG* promoters impair the recruitment of the HbF repressor LRF (ZBTB7A) and induce HbF levels accounting for 5–40% of the total hemoglobin in β-hemoglobinopathy patients.[82] Based on these findings, Weber et al. developed a CRISPR-Cas9 nuclease strategy to target the −200 region and disrupt the LRF BS.[92] Editing the LRF binding site reactivated γ-globin expression resulting in 80% of F-cells enabling the correction of the RBC sickling phenotype. In addition, the −200-edited HSPCs successfully engrafted in immunodeficient mice confirming that the LRF BS is a potent target for genome editing treatment of β-hemoglobinopathies.[92]

4.1.2 Creating γ-globin activator BSs in the HBG promoters

Contrary to the CRISPR-Cas9 nuclease system, BEs can allow the creation of activator BSs in the *HBG* promoters (Fig. 4B).

The −198T>C HPFH mutation creates a BS for the erythroid transcription factor KLF1.[93] This T>C substitution is responsible for driving HbF levels representing 5–20% of the total Hb types.[94,95] The −198T>C HPFH mutation was first introduced in the *HBG* promoters of HEK293T cells using ABE7.10.[45] Then, the RNA delivery of the novel and highly efficient ABE8s variants allowed to install this mutation in HSPCs derived from healthy donors, achieving >40% of γ-globin chain expression after erythroid differentiation compared to 10% in control cells.[96]

TAL1 and GATA1 binding sites were also introduced in the *HBG* promoters of HEK293T cells by installing HPFH mutations (−175T>C for TAL1; −113A>G for GATA1) using an optimized version of ABE7.10 (ABEmax).[89] Further analysis are required to evaluate if the introduction of these mutations in clinically-relevant cells can reproduce the HbF levels observed in patients harboring the −175T>C (up to 40% HbF)[81,97,98] or the −113A>G mutation (7% HbF).[99]

Finally, base editing allowed the development of a multiplex strategy to simultaneously generate the −198 and −175 HPFH mutations with the aim of further increasing HbF levels.[100] Indeed, a multiplex approach using nucleases to target two regions in the *HBG* promoters is not optimal, as it can lead to the deletion of the intervening sequence, likely affecting γ-globin expression. Dual base editing of around 20% of *HBG* promoters was achieved in HEK293T, validating the feasibility of this strategy; however, analyses in primary cells are necessary to compare the extent of HbF reactivation upon single or dual editing of the *HBG* promoters.[100]

In conclusion, reproducing HPFH or HPFH-like mutations using nucleases or base editing is a promising therapeutic approach to β-hemoglobinopathies, as it can ameliorate the clinical phenotype of all the patients regardless of the specific mutations and does not require HDR, which is poorly active in quiescent HSCs.[28]

4.2 Targeting γ-globin transcriptional repressors

4.2.1 Downregulating BCL11A by disrupting its erythroid-specific enhancer

The γ-globin repressor BCL11A is another interesting target to develop genome editing therapies for β-hemoglobinopathies by reactivating HbF. This approach aims at disrupting a *BCL11A* erythroid-specific enhancer to limit BCL11A silencing to the erythroid compartment and avoid side effects in HSCs (i.e., reduced engraftment capability) and other non-erythroid lineages (e.g., B-cells).[101–105] The +58-kb *BCL11A* intronic enhancer contains a BS for GATA1, which is believed to sustain BCL11A-lineage restricted expression[102] (Fig. 5).

The GATA motif was first targeted using ZFNs in HSPCs, leading to γ-globin reactivation in differentiated RBCs.[106] Xenotransplantation experiments in immunodeficient mice revealed that editing of the *BCL11A* enhancer does not affect the long-term engraftment of HSCs, demonstrating that this region is a safe target to cure β-hemoglobinopathies.[106] Similarly, a Cas9-RNP complex was used to target the GATA1 BS in primary cells derived from SCD and β-thalassemia patients.[107] BCL11A downregulation led to γ-globin expression in RBCs differentiated from edited HSPCs, resulting in the correction of both SCD and β-thalassemia phenotypes. CRISPR-Cas9-treated HSPCs were able to engraft in immunodeficient mice and high and therapeutically relevant editing efficiency persisted after primary and secondary xenotransplantation.[107] Importantly, a recent study revealed that targeting the *BCL11A* enhancer does not affect the cellular transcriptome.[108] Finally, Demirci et al. used a non-human primate model to evaluate engraftment, differentiation and therapeutic effect of HSPCs edited at the *BCL11A* enhancer using Cas9-RNP transfection. HSPCs edited at the *BCL11A* enhancer or at a control locus engrafted similarly. Edited cells persisted in peripheral blood and bone marrow for almost 2 years and *BCL11A* editing led to upregulation of HbF expression.[109] These results confirmed the potential of this strategy to cure β-hemoglobinopathies.

Base editing strategies have also been developed to disrupt the GATA1 motif in the *BCL11A* enhancer by generating precise base substitutions. The use of the novel ABE8e variant enabled to simultaneously install two A•T to G•C edits in the GATA1 motif with an efficiency of 50% in HEK293T cells.[100] eA3A(N57Q)-BE3 was also used to convert a C•G to a T•A bp in the GATA1 BS in patient HSPCs enabling more than 90% of base conversion after RNP transfection.[80] This single base conversion upregulated HbF, and corrected the sickling phenotype in SCD HSPC-derived RBCs and the α-/non-α-globin imbalance in β-thalassemic RBCs. Importantly, comparable multilineage engraftment was observed for unedited and edited cells in primary and secondary xenotransplantation experiments. Finally, base editing persisted in both primary and secondary bone marrow recipient mice demonstrating that long-term repopulating HSCs were edited and confirming the potential of this approach to correct patient HSCs.[80]

Altogether, these results confirmed that *BCL11A* enhancer is a potent and safe therapeutic target to treat β-hemoglobinopathies.

4.2.2 Down-regulation of transcription factors and co-factors involved in HbF expression

KLF1 haploinsufficiency is associated with high HbF levels in adult life.[110] Therefore, *KLF1* gene disruption was evaluated as a therapeutic approach to de-repress HbF. In K562 cells, a CRISPR-Cas9 approach targeting the *KLF1* coding sequence generated an InDels frequency of around 24% resulting in a eightfold increased HbF expression.[111] A high HbF upregulation was observed also in the erythroid progeny of *KLF1*-edited HSPCs. However, KLF1 down-regulation affected important genes involved in cell cycle, apoptosis and immune pathways.[108] Hence, *KLF1* is probably not the ideal therapeutic target to develop a clinical treatment for β-hemoglobinopathies. However, novel strategies can be developed to achieve a fine modulation of KLF1 expression, which could lead to HbF reactivation without affecting erythroid cell development.

SOX6 is a γ-globin repressor that cooperates with BCL11A to repress HbF expression.[112] Editing of *SOX6* gene via ZFNs or CRISPR-Cas9 in K562 cells increased γ-globin expression and HbF production.[113,114] SOX6 is important for mouse erythropoiesis[115] and while in human

SOX6 overexpression accelerates erythroid maturation[116] its down-regulation seems not to affect erythropoiesis.[117] Thus, *SOX6* is an interesting target to treat β-hemoglobinopathies, but further investigations are required before translating this strategy to the clinics.

Knock-down of the components of the Nucleosome remodeling and deacetylase (NuRD) complex also leads to HbF reactivation in primary erythroblasts.[118–122] Therefore, genome editing approaches aimed at down-regulating members of the NuRD complex might be envisioned; however, further studies are required to evaluate the safety of these strategies.

4.3 Clinical trials for β-hemoglobinopathies based on genome editing approaches

Several ongoing clinical trials aim at evaluating the safety and efficacy of genome editing-based treatments for β-hemoglobinopathies.

Two clinical trials are currently been conducted by the Shanghai Bioray Laboratory in China to evaluate safety and efficacy of CRISPR-Cas9 strategies aimed either at restoring β-globin expression in β-thalassemic patients harboring the IVS-2-654C>T mutation (NCT04205435) or at de-repressing γ-globin in transfusion-dependent β-thalassemia (TDT) patients (NCT04211480); however, no results are yet available.

Disrupting the *BCL11A* erythroid-specific enhancer is the more advanced therapeutic strategy.[106,107,109] Sangamo Therapeutics developed ZFNs to edit the *BCL11A* enhancer in HSPCs derived from TDT and SCD patients (ST-400 and BIVV003 therapies, respectively).[102,106,123] More than 50% of edited cells showed bi-allelic editing resulting in a high γ-globin expression at mRNA and protein levels. Importantly, *BCL11A* enhancer editing was maintained after xenotransplantation in immunodeficient mice.[124] Two ongoing phase 1/2 clinical trials sponsored by Sangamo Therapeutics and Sanofi aim at evaluating safety and tolerability of ST-400 (NCT03432364) and BIVV003 (NCT03653247) drug products in TDT and SCD, respectively. However, the first results in TDT patients showed a relatively low genome editing efficiency associated with poor HbF expression.[125]

BCL11A enhancer disruption by CRISPR-Cas9 is also currently evaluated in two phase 1/2 clinical trials for TDT in Europe (NCT03655678) and SCD patients in USA (NCT03745287) by CRISPR Therapeutics and Vertex Pharmaceuticals (CTX001 therapy).[126] Preliminary results in two patients—one with TDT and one with SCD—revealed a high editing efficiency (between 69% and 83%) in the infused drug product, which was maintained in the bone marrow and in peripheral blood cells more than a year after gene therapy, demonstrating the engraftment potential of CRISPR-Cas9-edited HSCs.[126] HbF levels and the proportion of F-cells increased rapidly after infusion of edited cells improving Hb concentration and suppressing transfusion requirement in both patients. Importantly, the SCD patient did not experience any vaso-occlusive episode in more than 16 months post-treatment.[126] Altogether, these early clinical data demonstrated the potential of genome editing approaches to cure β-hemoglobinopathies.

Author contributions

M.B.: writing—original draft preparation; M.B. and A.M.: writing—review and editing.

Acknowledgments

This work was supported by State funding from the Agence Nationale de la Recherche (under "Investissements d'avenir" program, grant number ANR-10-IAHU-01 and ANR-PRC HEMOLEN), and by the European Research Council, grant number 865797 DITSB.

Conflicts of interest

The Authors have no conflict of interest.

References

1. Weatherall D. The inherited disorders of haemoglobin: an increasingly neglected global health burden. *Indian J Med Res*. 2011;134:493–497.
2. Thein SL. The molecular basis of β-thalassemia. *Cold Spring Harb Perspect Med*. 2013; 3(5):a011700.
3. Weatherall DJ. Phenotype-genotype relationships in monogenic disease: lessons from the thalassaemias. *Nat Rev Genet*. 2001;2(4):245–255.
4. Cappellini MD, Porter JB, Viprakasit V, Taher AT. A paradigm shift on beta-thalassaemia treatment: how will we manage this old disease with new therapies? *Blood Rev*. 2018;32(4):300–311.
5. Piel FB, Steinberg MH, Rees DC. Sickle cell disease. *N Engl J Med*. 2017;376 (16):1561–1573.
6. Kato GJ, Piel FB, Reid CD, et al. Sickle cell disease. *Nat Rev Dis Primers*. 2018;4:18010.
7. Baronciani D, Angelucci E, Potschger U, et al. Hemopoietic stem cell transplantation in thalassemia: a report from the European Society for Blood and Bone Marrow Transplantation Hemoglobinopathy Registry, 2000–2010. *Bone Marrow Transplant*. 2016;51(4):536–541.
8. Hsieh MM, Fitzhugh CD, Weitzel RP, et al. Nonmyeloablative HLA-matched sibling allogeneic hematopoietic stem cell transplantation for severe sickle cell phenotype. *JAMA*. 2014;312(1):48–56.
9. Hsieh MM, Kang EM, Fitzhugh CD, et al. Allogeneic hematopoietic stem-cell transplantation for sickle cell disease. *N Engl J Med*. 2009;361(24):2309–2317.

10. Leonard A, Tisdale JF. Stem cell transplantation in sickle cell disease: therapeutic potential and challenges faced. *Expert Rev Hematol*. 2018;11(7):547–565.
11. Locatelli F, Merli P, Strocchio L. Transplantation for thalassemia major: alternative donors. *Curr Opin Hematol*. 2016;23(6):515–523.
12. Locatelli F, Crotta A, Ruggeri A, et al. Analysis of risk factors influencing outcomes after cord blood transplantation in children with juvenile myelomonocytic leukemia: a EUROCORD, EBMT, EWOG-MDS, CIBMTR study. *Blood*. 2013;122(12): 2135–2141.
13. Cavazzana M, Bushman FD, Miccio A, André-Schmutz I, Six E. Gene therapy targeting haematopoietic stem cells for inherited diseases: progress and challenges. *Nat Rev Drug Discov*. 2019;18(6):447–462.
14. Miccio A, Cesari R, Lotti F, et al. In vivo selection of genetically modified erythroblastic progenitors leads to long-term correction of β-thalassemia. *Proc Natl Acad Sci U S A*. 2008;105(30):10547–10552.
15. Weber L, Poletti V, Magrin E, et al. An optimized lentiviral vector efficiently corrects the human sickle cell disease phenotype. *Mol Ther Methods Clin Dev*. 2018;10: 268–280.
16. Magrin E, Miccio A, Cavazzana M. Lentiviral and genome-editing strategies for the treatment of β-hemoglobinopathies. *Blood*. 2019;134(15):1203–1213.
17. Marktel S, Scaramuzza S, Cicalese MP, et al. Intrabone hematopoietic stem cell gene therapy for adult and pediatric patients affected by transfusion-dependent ß-thalassemia. *Nat Med*. 2019;25(2):234–241.
18. Thompson AA, Walters MC, Kwiatkowski J, et al. Gene therapy in patients with transfusion-dependent β-thalassemia. *N Engl J Med*. 2018;378(16):1479–1493.
19. Coquerelle S, Ghardallou M, Rais S, et al. Innovative curative treatment of beta thalassemia: cost-efficacy analysis of gene therapy versus allogenic hematopoietic stem-cell transplantation. *Hum Gene Ther*. 2019;30(6):753–761.
20. Moiani A, Paleari Y, Sartori D, et al. Lentiviral vector integration in the human genome induces alternative splicing and generates aberrant transcripts. *J Clin Invest*. 2012;122 (5):1653–1666.
21. Cavazzana-Calvo M, Payen E, Negre O, et al. Transfusion independence and HMGA2 activation after gene therapy of human β-thalassaemia. *Nature*. 2010;467 (7313):318–322.
22. Fraietta JA, Nobles CL, Sammons MA, et al. Disruption of TET2 promotes the therapeutic efficacy of CD19-targeted T cells. *Nature*. 2018;558(7709):307–312.
23. Rothkamm K, Krüger I, Thompson LH, Löbrich M. Pathways of DNA double-strand break repair during the mammalian cell cycle. *Mol Cell Biol*. 2003;23(16):5706–5715.
24. Uddin F, Rudin CM, Sen T. CRISPR gene therapy: applications, limitations, and implications for the future. *Front Oncol*. 2020;10:1387.
25. Fang Y, Cheng Y, Lu D, et al. Treatment of β654-thalassaemia by TALENs in a mouse model. *Cell Prolif*. 2018;51(6):e12491.
26. Xu S, Luk K, Yao Q, et al. Editing aberrant splice sites efficiently restores β-globin expression in β-thalassemia. *Blood*. 2019;133(21):2255–2262.
27. Shrivastav M, De Haro LP, Nickoloff JA. Regulation of DNA double-strand break repair pathway choice. *Cell Res*. 2008;18(1):134–147.
28. Antony JS, Latifi N, Haque AKMA, et al. Gene correction of HBB mutations in CD34 + hematopoietic stem cells using Cas9 mRNA and ssODN donors. *Mol Cell Pediatr*. 2018;5(1):9.
29. Cromer MK, Vaidyanathan S, Ryan DE, et al. Global transcriptional response to CRISPR/Cas9-AAV6-based genome editing in CD34+ hematopoietic stem and progenitor cells. *Mol Ther*. 2018;26(10):2431–2442.

30. Kosicki M, Tomberg K, Bradley A. Repair of double-strand breaks induced by CRISPR-Cas9 leads to large deletions and complex rearrangements. *Nat Biotechnol.* 2018;36(8):765–771.
31. Anzalone AV, Randolph PB, Davis JR, et al. Search-and-replace genome editing without double-strand breaks or donor DNA. *Nature.* 2019;576(7785):149–157.
32. Rees HA, Liu DR. Base editing: precision chemistry on the genome and transcriptome of living cells. *Nat Rev Genet.* 2018;19(12):770–788.
33. Klug A. The discovery of zinc fingers and their applications in gene regulation and genome manipulation. *Annu Rev Biochem.* 2010;79:213–231.
34. Boch J, Bonas U. Xanthomonas AvrBs3 family-type III effectors: discovery and function. *Annu Rev Phytopathol.* 2010;48:419–436.
35. Lee J, Chung J-H, Kim HM, Kim D-W, Kim H. Designed nucleases for targeted genome editing. *Plant Biotechnol J.* 2016;14(2):448–462.
36. Doudna JA. The promise and challenge of therapeutic genome editing. *Nature.* 2020;578(7794):229–236.
37. Jinek M, Chylinski K, Fonfara I, Hauer M, Doudna JA, Charpentier E. A programmable dual-RNA-guided DNA endonuclease in adaptive bacterial immunity. *Science.* 2012;337(6096):816–821.
38. Nishimasu H, Cong L, Yan WX, et al. Crystal structure of Staphylococcus aureus Cas9. *Cell.* 2015;162(5):1113–1126.
39. Zetsche B, Gootenberg JS, Abudayyeh OO, et al. Cpf1 is a single RNA-guided endonuclease of a class 2 CRISPR-Cas system. *Cell.* 2015;163(3):759–771.
40. Chen JS, Dagdas YS, Kleinstiver BP, et al. Enhanced proofreading governs CRISPR-Cas9 targeting accuracy. *Nature.* 2017;550(7676):407.
41. Casini A, Olivieri M, Petris G, et al. A highly specific SpCas9 variant is identified by in vivo screening in yeast. *Nat Biotechnol.* 2018;36(3):265–271.
42. Lee JK, Jeong E, Lee J, et al. Directed evolution of CRISPR-Cas9 to increase its specificity. *Nat Commun.* 2018;9(1):3048.
43. Vakulskas CA, Dever DP, Rettig GR, et al. A high-fidelity Cas9 mutant delivered as a ribonucleoprotein complex enables efficient gene editing in human hematopoietic stem and progenitor cells. *Nat Med.* 2018;24(8):1216–1224.
44. Komor AC, Kim YB, Packer MS, Zuris JA, Liu DR. Programmable editing of a target base in genomic DNA without double-stranded DNA cleavage. *Nature.* 2016;533 (7603):420–424.
45. Gaudelli NM, Komor AC, Rees HA, et al. Programmable base editing of A•T to G•C in genomic DNA without DNA cleavage. *Nature.* 2017;551(7681):464–471.
46. Komor AC, Zhao KT, Packer MS, et al. Improved base excision repair inhibition and bacteriophage Mu Gam protein yields C:G-to-T:A base editors with higher efficiency and product purity. *Sci Adv.* 2017;3(8):eaao4774.
47. Miller SM, Wang T, Randolph PB, et al. Continuous evolution of SpCas9 variants compatible with non-G PAMs. *Nat Biotechnol.* 2020;38(4):471–481.
48. Antoniou P, Miccio A, Brusson M. Base and prime editing technologies for blood disorders. *Front Genome Ed.* 2021;3:618406. https://doi.org/10.3389/fgeed.2021.618406.
49. Porteus MH. Mammalian gene targeting with designed zinc finger nucleases. *Mol Ther.* 2006;13(2):438–446.
50. Sebastiano V, Maeder ML, Angstman JF, et al. In situ genetic correction of the sickle cell anemia mutation in human induced pluripotent stem cells using engineered zinc finger nucleases. *Stem Cells.* 2011;29(11):1717–1726.
51. Zou J, Mali P, Huang X, Dowey SN, Cheng L. Site-specific gene correction of a point mutation in human iPS cells derived from an adult patient with sickle cell disease. *Blood.* 2011;118(17):4599–4608.

52. Voit RA, Hendel A, Pruett-Miller SM, Porteus MH. Nuclease-mediated gene editing by homologous recombination of the human globin locus. *Nucleic Acids Res.* 2014;42(2):1365–1378.
53. Ramalingam S, Annaluru N, Kandavelou K, Chandrasegaran S. TALEN-mediated generation and genetic correction of disease-specific human induced pluripotent stem cells. *Curr Gene Ther.* 2014;14(6):461–472.
54. Sun N, Zhao H. Seamless correction of the sickle cell disease mutation of the HBB gene in human induced pluripotent stem cells using TALENs. *Biotechnol Bioeng.* 2014;111(5):1048–1053.
55. Hoban MD, Cost GJ, Mendel MC, et al. Correction of the sickle cell disease mutation in human hematopoietic stem/progenitor cells. *Blood.* 2015;125(17):2597–2604.
56. DeWitt MA, Magis W, Bray NL, et al. Selection-free genome editing of the sickle mutation in human adult hematopoietic stem/progenitor cells. *Sci Transl Med.* 2016;8(360):360ra134.
57. Dever DP, Bak RO, Reinisch A, et al. CRISPR/Cas9 β-globin gene targeting in human haematopoietic stem cells. *Nature.* 2016;539(7629):384–389.
58. Romero Z, Lomova A, Said S, et al. Editing the sickle cell disease mutation in human hematopoietic stem cells: comparison of endonucleases and homologous donor templates. *Mol Ther.* 2019;27(8):1389–1406.
59. Pattabhi S, Lotti SN, Berger MP, et al. In vivo outcome of homology-directed repair at the HBB gene in HSC using alternative donor template delivery methods. *Mol Ther Nucleic Acids.* 2019;17:277–288.
60. Park SH, Lee CM, Dever DP, et al. Highly efficient editing of the β-globin gene in patient-derived hematopoietic stem and progenitor cells to treat sickle cell disease. *Nucleic Acids Res.* 2019;47(15):7955–7972.
61. Magis W, DeWitt MA, Wyman SK, et al. High-level correction of the sickle mutation amplified in vivo during erythroid differentiation. *bioRxiv.* 2018. Published online October 3.
62. Ferrari S, Jacob A, Beretta S, et al. Efficient gene editing of human long-term hematopoietic stem cells validated by clonal tracking. *Nat Biotechnol.* 2020;38(11):1298–1308.
63. Mohamad AS, Hamzah R, Selvaratnam V, Yegapan S, Sathar J. Human hemoglobin G-Makassar variant masquerading as sickle cell anemia. *Hematol Rep.* 2018;10(3):7210.
64. Viprakasit V, Wiriyasateinkul A, Sattayasevana B, Miles KL, Laosombat V. Hb G-Makassar [beta6(A3)Glu—>Ala; codon 6 (GAG—>GCG)]: molecular characterization, clinical, and hematological effects. *Hemoglobin.* 2002;26(3):245–253.
65. Kountouris P, Lederer CW, Fanis P, Feleki X, Old J, Kleanthous M. IthaGenes: an interactive database for haemoglobin variations and epidemiology. *PLoS One.* 2014;9(7):e103020.
66. Xie F, Ye L, Chang JC, et al. Seamless gene correction of β-thalassemia mutations in patient-specific iPSCs using CRISPR/Cas9 and piggyBac. *Genome Res.* 2014;24(9):1526–1533.
67. Liu Y, Yang Y, Kang X, et al. One-step biallelic and scarless correction of a β-thalassemia mutation in patient-specific iPSCs without drug selection. *Mol Ther Nucleic Acids.* 2017;6:57–67.
68. Niu X, He W, Song B, et al. Combining single strand oligodeoxynucleotides and CRISPR/Cas9 to correct gene mutations in β-thalassemia-induced pluripotent stem cells. *J Biol Chem.* 2016;291(32):16576–16585.
69. Wattanapanitch M, Damkham N, Potirat P, et al. One-step genetic correction of hemoglobin E/beta-thalassemia patient-derived iPSCs by the CRISPR/Cas9 system. *Stem Cell Res Ther.* 2018;9(1):46.

70. Xu XM, Zhou YQ, Luo GX, et al. The prevalence and spectrum of alpha and beta thalassemia in Guangdong Province: implications for the future health burden and population screening. *J Clin Pathol*. 2004;57(5):517–522.
71. Ma N, Liao B, Zhang H, et al. Transcription activator-like effector nuclease (TALEN)-mediated gene correction in integration-free β-thalassemia induced pluripotent stem cells. *J Biol Chem*. 2013;288(48):34671–34679.
72. Cai L, Bai H, Mahairaki V, et al. A universal approach to correct various HBB gene mutations in human stem cells for gene therapy of Beta-thalassemia and sickle cell disease. *Stem Cells Transl Med*. 2018;7(1):87–97.
73. Cheng TC, Orkin SH, Antonarakis SE, et al. Beta-thalassemia in Chinese: use of in vivo RNA analysis and oligonucleotide hybridization in systematic characterization of molecular defects. *Proc Natl Acad Sci U S A*. 1984;81(9):2821–2825.
74. Kazazian HH, Boehm CD. Molecular basis and prenatal diagnosis of beta-thalassemia. *Blood*. 1988;72(4):1107–1116.
75. Bahal R, Ali McNeer N, Quijano E, et al. In vivo correction of anaemia in β-thalassemic mice by γPNA-mediated gene editing with nanoparticle delivery. *Nat Commun*. 2016;7:13304.
76. Ricciardi AS, Bahal R, Farrelly JS, et al. In utero nanoparticle delivery for site-specific genome editing. *Nat Commun*. 2018;9(1):2481.
77. Xu P, Tong Y, Liu X, et al. Both TALENs and CRISPR/Cas9 directly target the HBB IVS2-654 (C > T) mutation in β-thalassemia-derived iPSCs. *Sci Rep*. 2015;5:12065.
78. Liang P, Ding C, Sun H, et al. Correction of β-thalassemia mutant by base editor in human embryos. *Protein Cell*. 2017;8(11):811–822.
79. Gehrke JM, Cervantes O, Clement MK, et al. An APOBEC-Cas9 base editor with minimized bystander and off-target activities. *Nat Biotechnol*. 2018;36(10):977–982.
80. Zeng J, Wu Y, Ren C, et al. Therapeutic base editing of human hematopoietic stem cells. *Nat Med*. 2020;26(4):535–541.
81. Forget BG. Molecular basis of hereditary persistence of fetal hemoglobin. *Ann N Y Acad Sci*. 1998;850:38–44.
82. Martyn GE, Wienert B, Yang L, et al. Natural regulatory mutations elevate the fetal globin gene via disruption of BCL11A or ZBTB7A binding. *Nat Genet*. 2018;50(4):498–503.
83. Traxler EA, Yao Y, Wang Y-D, et al. A genome-editing strategy to treat β-hemoglobinopathies that recapitulates a mutation associated with a benign genetic condition. *Nat Med*. 2016;22(9):987–990.
84. Truong LN, Li Y, Shi LZ, et al. Microhomology-mediated end joining and homologous recombination share the initial end resection step to repair DNA double-strand breaks in mammalian cells. *Proc Natl Acad Sci U S A*. 2013;110(19):7720–7725.
85. Lux CT, Pattabhi S, Berger M, et al. TALEN-mediated gene editing of HBG in human hematopoietic stem cells leads to therapeutic fetal hemoglobin induction. *Mol Ther Methods Clin Dev*. 2019;12:175–183.
86. Métais J-Y, Doerfler PA, Mayuranathan T, et al. Genome editing of HBG1 and HBG2 to induce fetal hemoglobin. *Blood Adv*. 2019;3(21):3379–3392.
87. Humbert O, Radtke S, Samuelson C, et al. Therapeutically relevant engraftment of a CRISPR-Cas9-edited HSC-enriched population with HbF reactivation in non-human primates. *Sci Transl Med*. 2019;11(503):eaaw3768. https://doi.org/10.1126/scitranslmed.aaw3768.
88. Li C, Psatha N, Sova P, et al. Reactivation of γ-globin in adult β-YAC mice after ex vivo and in vivo hematopoietic stem cell genome editing. *Blood*. 2018;131(26):2915–2928.

89. Koblan LW, Doman JL, Wilson C, et al. Improving cytidine and adenine base editors by expression optimization and ancestral reconstruction. *Nat Biotechnol.* 2018;36(9): 843–846.
90. Wang L, Li L, Ma Y, et al. Reactivation of γ-globin expression through Cas9 or base editor to treat β-hemoglobinopathies. *Cell Res.* 2020;30(3):276–278.
91. Zhang X, Chen L, Zhu B, et al. Increasing the efficiency and targeting range of cytidine base editors through fusion of a single-stranded DNA-binding protein domain. *Nat Cell Biol.* 2020;22(6):740–750.
92. Weber L, Frati G, Felix T, et al. Editing a γ-globin repressor binding site restores fetal hemoglobin synthesis and corrects the sickle cell disease phenotype. *Sci Adv.* 2020;6(7) eaay9392. https://doi.org/10.1126/sciadv.aay9392.
93. Wienert B, Martyn GE, Kurita R, Nakamura Y, Quinlan KGR, Crossley M. KLF1 drives the expression of fetal hemoglobin in British HPFH. *Blood.* 2017;130 (6):803–807.
94. Weatherall DJ, Cartner R, Clegg JB, Wood WG, Macrae IA, Mackenzie A. A form of hereditary persistence of fetal haemoglobin characterized by uneven cellular distribution of haemoglobin F and the production of haemoglobins A and A2 in homozygotes. *Br J Haematol.* 1975;29(2):205–220.
95. Wood WG, MacRae IA, Darbre PD, Clegg JB, Weatherall DJ. The British type of non-deletion HPFH: characterization of developmental changes in vivo and erythroid growth in vitro. *Br J Haematol.* 1982;50(3):401–414.
96. Gaudelli NM, Lam DK, Rees HA, et al. Directed evolution of adenine base editors with increased activity and therapeutic application. *Nat Biotechnol.* 2020;38(7): 892–900.
97. Martin DI, Tsai SF, Orkin SH. Increased gamma-globin expression in a nondeletion HPFH mediated by an erythroid-specific DNA-binding factor. *Nature.* 1989;338(6214): 435–438.
98. Wienert B, Funnell APW, Norton LJ, et al. Editing the genome to introduce a beneficial naturally occurring mutation associated with increased fetal globin. *Nat Commun.* 2015;6:7085.
99. Martyn GE, Wienert B, Kurita R, Nakamura Y, Quinlan KGR, Crossley M. A natural regulatory mutation in the proximal promoter elevates fetal globin expression by creating a de novo GATA1 site. *Blood.* 2019;133(8):852–856.
100. Richter MF, Zhao KT, Eton E, et al. Phage-assisted evolution of an adenine base editor with improved Cas domain compatibility and activity. *Nat Biotechnol.* 2020;38 (7):883–891.
101. Bauer DE, Peng C, Smith EC, Orkin SH. Identification of BCL11A structure-function domains for fetal hemoglobin silencing. *Blood.* 2013;122(21):435.
102. Canver MC, Smith EC, Sher F, et al. BCL11A enhancer dissection by Cas9-mediated in situ saturating mutagenesis. *Nature.* 2015;527(7577):192–197.
103. Smith EC, Luc S, Croney DM, et al. Strict in vivo specificity of the Bcl11a erythroid enhancer. *Blood.* 2016;128(19):2338–2342.
104. Brendel C, Guda S, Renella R, et al. Lineage-specific BCL11A knockdown circumvents toxicities and reverses sickle phenotype. *J Clin Invest.* 2016;126(10):3868–3878.
105. Tsang JCH, Yu Y, Burke S, et al. Single-cell transcriptomic reconstruction reveals cell cycle and multi-lineage differentiation defects in Bcl11a-deficient hematopoietic stem cells. *Genome Biol.* 2015;16:178.
106. Chang K-H, Smith SE, Sullivan T, et al. Long-term engraftment and fetal globin induction upon BCL11A gene editing in bone-marrow-derived CD34 + hematopoietic stem and progenitor cells. *Mol Ther Methods Clin Dev.* 2017;4:137–148.
107. Wu Y, Zeng J, Roscoe BP, et al. Highly efficient therapeutic gene editing of human hematopoietic stem cells. *Nat Med.* 2019;25(5):776–783.

108. Lamsfus-Calle A, Daniel-Moreno A, Antony JS, et al. Comparative targeting analysis of KLF1, BCL11A, and HBG1/2 in CD34+ HSPCs by CRISPR/Cas9 for the induction of fetal hemoglobin. *Sci Rep.* 2020;10(1):10133.
109. Demirci S, Zeng J, Wu Y, et al. BCL11A enhancer-edited hematopoietic stem cells persist in rhesus monkeys without toxicity. *J Clin Invest.* 2020;130(12):6677–6687.
110. Borg J, Papadopoulos P, Georgitsi M, et al. Haploinsufficiency for the erythroid transcription factor KLF1 causes hereditary persistence of fetal hemoglobin. *Nat Genet.* 2010;42(9):801–805.
111. Shariati L, Khanahmad H, Salehi M, et al. Genetic disruption of the KLF1 gene to over-express the γ-globin gene using the CRISPR/Cas9 system. *J Gene Med.* 2016;18(10):294–301.
112. Xu J, Sankaran VG, Ni M, et al. Transcriptional silencing of {gamma}-globin by BCL11A involves long-range interactions and cooperation with SOX6. *Genes Dev.* 2010;24(8):783–798.
113. Modares Sadeghi M, Shariati L, Hejazi Z, Shahbazi M, Tabatabaiefar MA, Khanahmad H. Inducing indel mutation in the SOX6 gene by zinc finger nuclease for gamma reactivation: an approach towards gene therapy of beta thalassemia. *J Cell Biochem.* 2018;119(3):2512–2519.
114. Shariati L, Rohani F, Heidari Hafshejani N, et al. Disruption of SOX6 gene using CRISPR/Cas9 technology for gamma-globin reactivation: an approach towards gene therapy of β-thalassemia. *J Cell Biochem.* 2018;119(11):9357–9363.
115. Dumitriu B, Patrick MR, Petschek JP, et al. Sox6 cell-autonomously stimulates erythroid cell survival, proliferation, and terminal maturation and is thereby an important enhancer of definitive erythropoiesis during mouse development. *Blood.* 2006;108(4):1198–1207.
116. Cantù C, Ierardi R, Alborelli I, et al. Sox6 enhances erythroid differentiation in human erythroid progenitors. *Blood.* 2011;117(13):3669–3679.
117. Li J, Lai Y, Luo J, et al. SOX6 downregulation induces γ-globin in human β-thalassemia major erythroid cells. *Biomed Res Int.* 2017;2017:9496058.
118. Amaya M, Desai M, Gnanapragasam MN, et al. Mi2β-mediated silencing of the fetal γ-globin gene in adult erythroid cells. *Blood.* 2013;121(17):3493–3501.
119. Lan X, Ren R, Feng R, et al. ZNF410 uniquely activates the NuRD component CHD4 to silence fetal hemoglobin expression. *Mol Cell.* 2021;81(2):239–254.e8. https://doi.org/10.1016/j.molcel.2020.11.006.
120. Sher F, Hossain M, Seruggia D, et al. Rational targeting of a NuRD subcomplex guided by comprehensive in situ mutagenesis. *Nat Genet.* 2019;51(7):1149–1159.
121. Xu J, Bauer DE, Kerenyi MA, et al. Corepressor-dependent silencing of fetal hemoglobin expression by BCL11A. *Proc Natl Acad Sci U S A.* 2013;110(16):6518–6523.
122. Yu X, Azzo A, Bilinovich SM, et al. Disruption of the MBD2-NuRD complex but not MBD3-NuRD induces high level HbF expression in human adult erythroid cells. *Haematologica.* 2019;104(12):2361–2371.
123. Psatha N, Reik A, Phelps S, et al. Disruption of the BCL11A erythroid enhancer reactivates fetal hemoglobin in erythroid cells of patients with β-thalassemia major. *Mol Ther Methods Clin Dev.* 2018;10:313–326.
124. Holmes MC, Reik A, Rebar EJ, et al. A potential therapy for beta-thalassemia (ST-400) and sickle cell disease (BIVV003). *Blood.* 2017;130(Suppl. 1):2066.
125. Smith AR, Schiller GJ, Vercellotti GM, et al. Preliminary results of a phase 1/2 clinical study of zinc finger nuclease-mediated editing of BCL11A in autologous hematopoietic stem cells for transfusion-dependent beta thalassemia. *Blood.* 2019;134(Suppl. 1):3544.
126. Frangoul H, Altshuler D, Cappellini MD, et al. CRISPR-Cas9 gene editing for sickle cell disease and β-thalassemia. *N Engl J Med.* 2021;384(3):252–260. https://doi.org/10.1056/NEJMoa2031054.

CHAPTER SIX

Rewriting *CFTR* to cure cystic fibrosis

Giulia Maule[a,b,†], Marjolein Ensinck[c,†], Mattijs Bulcaen[c], and Marianne S. Carlon[c,*]

[a]Department CIBIO, University of Trento, Trento, Italy
[b]Institute of Biophysics, National Research Council, Trento, Italy
[c]Molecular Virology and Gene Therapy, Department of Pharmaceutical and Pharmacological Sciences, KU Leuven, Flanders, Belgium
*Corresponding author: e-mail address: marianne.carlon@kuleuven.be

Contents

1. Cystic fibrosis—Where do we stand today	187
1.1 State of the art treatments—CFTR modulators	188
1.2 New CF therapies in the pipeline	190
1.3 Gene therapy for CF	191
1.4 Why gene addition/editing for CF?	191
2. Gene editing—Towards rewriting *CFTR*	192
2.1 Programmable nucleases	192
2.2 Base editors	199
2.3 Prime editing	202
3. Modulation of *CFTR* expression levels as an alternative next-generation approach	203
3.1 Transcriptional activation of *CFTR*	203
3.2 Long-noncoding RNA regulation	203
3.3 MicroRNA downregulation	204
3.4 Nonsense mediated mRNA decay inhibition	205
4. From Petri-dish to patient	206
4.1 *Ex vivo* or *in vivo* gene editing	206
4.2 Choice of delivery agent for efficient and safe gene editing	209
5. Conclusion	211
Acknowledgments	212
References	212

Abstract

Cystic fibrosis (CF) is an autosomal recessive monogenic disease caused by mutations in the Cystic Fibrosis Transmembrane conductance Regulator (*CFTR*) gene. Although F508del is the most frequent mutation, there are in total 360 confirmed disease-causing

[†] Shared first authors

CFTR mutations, impairing CFTR production, function and stability. Currently, the only causal treatments available are CFTR correctors and potentiators that directly target the mutant protein. While these pharmacological advances and better symptomatic care have improved life expectancy of people with CF, none of these treatments provides a cure. The discovery and development of programmable nucleases, in particular CRISPR nucleases and derived systems, rekindled the field of CF gene therapy, offering the possibility of a permanent correction of the *CFTR* gene. In this review we will discuss different strategies to restore CFTR function via gene editing correction of *CFTR* mutations or enhanced *CFTR* expression, and address how best to deliver these treatments to target cells.

Abbreviations

AAV	adeno-associated viral vector
AdV	adenoviral vector
ABE	adenine base editor
ASO	antisense oligonucleotide
Cas	CRISPR associated
Cas9n	Cas9 nickase
CBE	cytosine base editor
CGBE	C-to-G base editor
CF	cystic fibrosis
CFBE41o	cystic fibrosis bronchial epithelial cell line, homozygous for F508del mutation
CFF	cystic fibrosis foundation
CFTR	cystic fibrosis transmembrane conductance regulator
CFTR2	clinical and functional translation of CFTR
CRISPR	clustered regularly interspaced palindromic repeats
DDR	DNA damage repair
DSB	double strand break
EJC	exon junction complex
HDR	homology directed repair
HPV	human papilloma virus
GBE	C-to-G base editor
GMP	good manufacturing practice
iPSCs	induced pluripotent stem cells
lncRNA	long noncoding RNA
miRNA	microRNA
MRE	miRNA recognition element
ncRNA	noncoding RNA
NHEJ	nonhomologous end joining
NMD	nonsense mediated mRNA decay
PAM	protospacer adjacent motif
PBS	primer binding site
PCBP1	poly(rC)-binding protein 1
PE	prime editor
pegRNA	prime editing guide RNA
PNA	peptide nucleic acid

PTC	premature termination codon
PwCF	people with cystic fibrosis
RNP	ribonucleoprotein
RT	reverse transcriptase
scRNAseq	single-cell RNA sequencing
sgRNA	single guide RNA
SMA	spinal muscular atrophy
SpCas9	Cas9 from *Streptococcus pyogenes*
TALE	transcription activator like effector
TALEN	transcription activator-like effector nuclease
TGA	targeted gene activation
TRID	translational readthrough inducing drugs
TSB	target site blocker
UABC	upper airway basal cell
UGI	uracil DNA glycosylase inhibitor
UNG	uracil DNA glycosylase
UTR	untranslated region
VLP	virus like particles
WT	wild type
ZFN	zinc finger nucleases

1. Cystic fibrosis—Where do we stand today

Cystic fibrosis (CF) is perhaps the archetype of an autosomal recessive disorder. With over 85,000 people with CF (PwCF) worldwide it is the most common monogenic disease, particularly frequent in people of Caucasian descent with a prevalence that ranges between 1:1400 (Ireland) and 1:3500 (United States).[1] Many organs, including the pancreas, liver and intestines, are affected in CF, however lung disease remains the major cause of morbidity and mortality. CF is caused by mutations in the Cystic Fibrosis Transmembrane conductance Regulator (*CFTR*) gene,[2–4] that regulates fluid transport across epithelia as a cAMP-regulated anion channel. Over 2000 mutations in *CFTR* have been described (www.genet.sickkids.on.ca), of which currently 360 confirmed disease-causing (www.CFTR2.org; "CFTR2"; update July 2020). *CFTR* mutations disturb one or more aspects of CFTR biogenesis and are grouped according to how the mutation disturbs the production (class I, II, V, VII), function (class III, IV) or stability (class VI) of the CFTR protein.[5,6] Almost 70% of CF alleles are accounted for by a single three-nucleotide deletion which results in loss of phenylalanine at position 508, F508del (c.1521-1523delCTT). By comparison, all other mutations are rare: only four other mutations reach a frequency of >1% (Table 1).

Table 1 List of main *CFTR* mutations.

Legacy name	cDNA name	Allele frequency in CFTR2 (%)	Approved for CFTR modulator treatment
F508del	c.1521_1523delCTT	69.74	Orkambi™, Symdeko™, Trikafta™
G542X	c.1624G>T	2.54	—
G551D	c.1652G>A	2.10	Kalydeco™, Symdeko™, Trikafta™
N1303K	c.3909C>G	1.58	—
W1282X	c.3846G>A	1.22	—
R553X	c.1657C>T	0.93	—
3849+10kbC->T	c.3718-2477C>T	0.82	Kalydeco™, Symdeko™
2789+5G->A	c.2657+5G>A	0.72	Kalydeco™, Symdeko™
R1162X	c.3484C>T	0.46	—
3272-26A->G	c.3140-26A>G	0.33	Kalydeco™, Symdeko™
1811+1.6kbA>G	c.1680-886A>G	0.057	—
R785X	c.2353C>T	0.023	—

CFTR mutations cited in the chapter with relative allele frequency and approved modulator therapy.

1.1 State of the art treatments—CFTR modulators

Currently, the only FDA/EMA approved treatments targeting the underlying protein defect in CF are CFTR modulators (Fig. 1). These small molecules have been developed to improve mutant CFTR folding and trafficking to the plasma membrane ("correctors") or to potentiate channel function at the plasma membrane ("potentiators"). Four CFTR modulators are available as mono- or combination therapy: potentiator ivacaftor (VX-770), first-generation correctors lumacaftor (VX-809) and its derivative tezacaftor (VX-661), and second-generation corrector elexacaftor (VX-445). Kalydeco™ (ivacaftor monotherapy) was initially approved for the gating-mutation G551D (c.1652G>A),[7] but its approval has since been expanded to other gating and residual function mutations, covering ~8% of PwCF.[8] For F508del, a combination of correction and potentiation is necessary in order to overcome the multiple defects and restore anion transport. In 2015, a first combination treatment of lumacaftor and ivacaftor (Orkambi™) was FDA approved for F508del homozygous PwCF.[9] Lumacaftor was later replaced

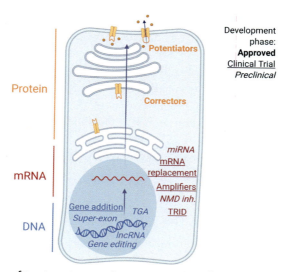

Fig. 1 Overview of treatment strategies to restore CFTR function in CF. Different strategies are being explored in order to restore anion transport through the CFTR channel. These strategies are directed either at the CFTR protein, mRNA, or DNA. To date, only therapies targeting the CFTR protein ("CFTR modulators") have received market approval. These treatments however, need to be taken on a daily base. Several mRNA-based strategies are currently being evaluated in clinical trials or in pre-clinical development. Amplifiers stabilize *CFTR* mRNA, resulting in more immature protein. For PwCF carrying PTC-mutations, translational readthrough inducing drugs (TRIDs) are being developed as well as treatments that inhibit nonsense mediated mRNA decay (NMD). Supplying the cells with a correct *CFTR* mRNA allows for a mutation agnostic approach with an increased treatment interval. Alternatively, TGA can potentially increase CFTR expression for mutations with some residual function. Alternatively, this can be achieved by modulation of specific miRNAs and lncRNAs. However, none of these strategies target the underlying *CFTR* gene defect. Several strategies aim at restoring the mutant *CFTR* DNA sequence, either by integration of a (partial) *CFTR* cDNA or by gene editing the specific mutation site. These strategies hold promise to provide a long-term functional rescue (depending on the longevity of the corrected target cell) and in the case of gene editing, a highly mutation-specific correction. All of these strategies are still in preclinical development, with the exception of *CFTR* cDNA delivery.

with tezacaftor, which has fewer side effects, in Symdeko/Symkevi™.[10] Although both therapies improved lung function by 2.6–4%point,[11,12] the increase was moderate compared to the 10.6%point improvement seen with Kalydeco™ in people with gating mutations.[13] Therefore, second-generation correctors with a different mechanism of action were developed to form a triple combination.[14] In October 2019 the FDA approved Trikafta™ (combination of elexacaftor, tezacaftor and ivacaftor; Kaftrio™ in Europe/UK) for PwCF with at least one F508del allele,[15] after phase III trials had shown a 13.8%point (against placebo) improvement in lung function in PwCF heterozygous for F508del with a minimal-function mutation on the other allele and 10.0% point (against tezacaftor-ivacaftor) for homozygous F508del PwCF.[16,17] Combined, there is now highly effective CFTR modulator treatment available for 90% of PwCF. However, there are no approved causal treatments for the remaining 10% of PwCF who carry two rare, minimal-function alleles. Of note, of the five most common *CFTR* mutations (F508del, G542X (c.1624G>T), G551D, N1303K (c.3909C>G), W1282X (c.3846G>A), only F508del and G551D have been approved for CFTR modulator treatment to date (Table 1).

1.2 New CF therapies in the pipeline

Although multiple CFTR modulator treatments have become available in the past 10 years, many more are currently in clinical development.[18] These include novel correctors and potentiators as well as a new class of CFTR modulators which increases the amount of immature protein ("amplifiers") by selectively enhancing CFTR translation and mRNA stability through interaction with poly(rC)-binding protein 1 (PCBP1) and *CFTR* mRNA.[19] This immature protein can, if necessary, be corrected and potentiated with other CFTR modulators in another triple combination (NCT03500263[20]).

All CFTR modulators have in common that they require the expression of a full length CFTR (immature) protein or mRNA. In their absence, other strategies are needed. For those mutations that result in premature termination codons (PTCs), accounting for ~5% of disease-causing CF alleles, translational readthrough inducing drugs (TRIDs) are being developed that incorporate near-cognate tRNAs at the position of the PTC. Aminoglycoside analogues like Ataluren (Translarna™) have already been market approved for other genetic diseases like Duchenne Muscular Dystrophy.[21] While Ataluren failed to significantly improve lung function in nonsense-mutation CF,[22] a novel TRID, ELX-02 is currently in Phase II trials for PwCF with at least one G542X allele (NCT04135495[23]). Alternatively, a new copy of the *CFTR* mRNA can be delivered to the cell, which allows translation of the wild type (WT) CFTR protein: MRT5005 is an inhaled mRNA therapeutic currently in Phase I/II (NCT03375047[24]). As a new copy of the *CFTR* mRNA is introduced, the treatment is mutation agnostic. This is a particularly interesting approach for disorders like CF as, despite their monogenic nature, >2000 mutations have been reported. CFTR modulators and mRNA therapies do not cure the underlying gene defect, as they aim to restore the CFTR protein and mRNA, respectively, and therefore provide transient treatments.

1.3 Gene therapy for CF

In contrast to CFTR modulators and mRNA therapies, gene therapy aims to restore the gene defect at the DNA level, thus providing a long-term therapy. Traditionally, gene therapy efforts were focused on delivering a copy of the *CFTR* cDNA, providing cells with the means to produce WT CFTR protein. While mutation agnostic, clinical trials investigating different gene addition strategies for CF in the past have not yet resulted in substantial improvements in lung function (reviewed in Refs. 25–27). However, there has been a renewed interest into CF gene therapy research after successes in other genetic diseases have led to market approval of gene therapeutic agents, such as for retinal dystrophy,[28] spinal muscular atrophy (SMA) type 1,[29] β-thalassemia[30] and several forms of cancer (reviewed in Ref. 31). In recent years, besides the gene addition efforts, proof-of-concept studies have shown repair of the mutated *CFTR* gene by gene editing approaches in relevant CF models, which will be discussed in detail in Section 2 of this chapter.

1.4 Why gene addition/editing for CF?

Although highly effective CFTR modulators are approved for the majority of PwCF, they do not cure CF and have to be taken twice-daily. They improve lung function and quality of life, but are added onto the current treatment regimen, increasing the already high treatment burden.[32] Furthermore, adverse effects have been reported. A French study investigating Orkambi™ in a real-world setting, showed 18.2% of participants discontinued treatment, mainly due to respiratory side effects.[33] Also, ocular and hepatic side effects have been documented.[34] Besides adverse effects, treatment responses have been reported variable in PwCF, including for the highly effective Trikafta™.[16,17] Finally, although Kalydeco™ initially boosts lung function significantly, CF lung disease continues to worsen, although at a slower pace.[35] This underscores that CFTR modulators do not cure the disease. Importantly, there remains a group of PwCF in need of novel therapies, i.e. those who do not tolerate CFTR modulator treatment, for whom the treatment is ineffective, and finally those that are not eligible. The latter consists of PwCF carrying nonsense mutations, drug-refractory missense mutations, frameshifts and large indels. Providing a therapy and preferably even a cure for all PwCF, is the current priority of the CF community, as exemplified by the recent "Path to a Cure" mission announced by the Cystic Fibrosis Foundation (CFF).[36]

2. Gene editing—Towards rewriting *CFTR*

The gene editing toolbox has expanded rapidly over the last decade, providing different technologies that are constantly evolving. In particular, implementation of CRISPR (clustered regularly interspaced short palindromic repeats) systems opened up a new era for treating genetic diseases, by enabling targeted genome modifications. The possibility to "rewrite" and correct mutations at their endogenous loci offers new therapeutic strategies for monogenic disorders like CF, preserving endogenous gene expression and regulation, in contrast to gene addition. A limitation however is that gene editing is highly mutation specific and that one therapy is unlikely to be repurposed to other mutations. Considering that the mutations causing CF are a diverse set of base conversions and both small and larger indels, different mutations call for different strategies (Table 2; Fig. 2). In the following part, we will discuss the different tools for gene editing and their potential to repair *CFTR* mutations and rescue CFTR function.

2.1 Programmable nucleases

2.1.1 ZFN, TALEN and Cas nucleases

Technologies for precise and programmable gene editing in eukaryotic cells have been available since 2001. As a first tool for specific and targeted DNA cleavage, zinc finger nucleases (ZFN) were shown to enhance homologous recombination in *Xenopus laevis* oocytes.[37] These programmable nucleases were built by fusing multiple zinc fingers (DNA-recognizing protein motifs) to the nuclease domain of Type-IIS *Fok*I nuclease. Target specificity is determined by triplets of DNA base pairs that are recognized by each of the zinc finger motifs. DNA cleavage activity is activated by proximity when the two ZFN are recruited to the target sequence, allowing the dimerization of the FokI nuclease domain required to create a double strand break (DSB). A similar technology, known as TALENs (transcription activator-like effector nucleases) was first described in 2010.[38] TALENs use the DNA-recognizing ability of transcription activator-like effectors (TALEs) to guide the FokI nuclease domains to the site of interest. Although each application requires custom protein design, ZFNs and TALENs have proven

Table 2 Methods for the correction of *CFTR* mutations in CF models.

Correction method	*CFTR* mutation	Model	References
CFTR cDNA delivery	All	Primary cells, CF animal model, Clinical trial	176,177,194,196,197
HDR	F508del	Cell lines, primary cells, iPSCs	67,68,70,72,73,198
NHEJ	Intronic splicing mutations	Minigenes, primary cells	78,79
Super-exon	All mutations downstream of targeted integration	Cell lines, primary cells	77,82
Base editors	Point mutations (R553X, R785X, R1162X, W1282X)	Cell lines, primary cells	107,108
Prime editor	F508del	Primary cells	113
Peptide nucleic acids	F508del	Cell lines, CF mouse model	85
NMD prevention	W1282X	Human bronchial epithelial cells	141
dCas transcriptional activator	F508del	Human nasal epithelial cells	116
Long noncoding RNA	F508del	Human nasal epithelial cells	116
Micro-RNA	F508del	Human bronchial epithelial cells	127

List of gene therapy strategies applied to correct the underlying defect causing CF. For each correction method, the mutations for which it was used and the CF models were reported. HDR, homology-directed repair; NHEJ, nonhomologous end joining; NMD, nonsense mediated mRNA decay.

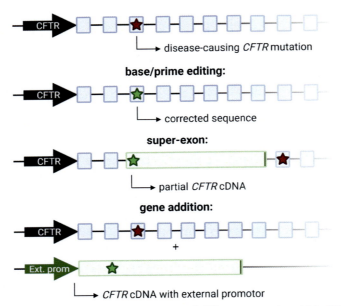

Fig. 2 Strategies to rewrite *CFTR*. Overview of different ways in which *CFTR* can be rewritten as part of a CF cure. Three different strategies can be distinguished. Precise edits at the site of the mutation can be made with base or prime editors, resulting in the corrected sequence. This is a highly mutation-specific strategy. Alternatively, multiple mutations can be covered by the targeted integration of a super-exon into the *CFTR* locus, a partial *CFTR* cDNA that includes the sequence downstream of the integration site. Finally, in a gene addition approach, a *CFTR* cDNA is integrated into the genome, together with an external promoter, which allows the transcription of a correct *CFTR* mRNA and subsequent translation of CFTR protein, and can be used irrespective of *CFTR* genotype but at the cost of losing endogenous regulation.

useful in a plethora of fundamental and translational research fields to better understand the function of genes or develop therapeutic gene editing strategies for human diseases, respectively. For a more general overview of these technologies and their applications, we refer to.[39,40]

In 2012, the gene editing field took a shift in paradigm with the discovery of CRISPR-associated (Cas) nucleases from a prokaryotic defense system, known as CRISPR, able to be harnessed for targeted DSB generation.[41,42] The Cas9 nuclease from *Streptococcus pyogenes* (SpCas9) was the first Cas to be characterized and engineered to perform targeted genome editing in eukaryotic cells. Via expression of a single guide RNA (sgRNA), SpCas9 can be retargeted toward any locus of interest. The sgRNA contains a 5′ spacer sequence that defines the target DNA sequence (protospacer), and a 3′ scaffold that allows structural recognition by SpCas9. In order to generate a DSB, SpCas9 requires an NGG nucleotide motif (known as the protospacer adjacent motif or PAM) directly downstream of the protospacer. Upon recognition of a PAM sequence, the base pairing between spacer and protospacer will temporarily denature the dsDNA. A complete match will activate the nuclease activity of SpCas9, resulting in cleavage of both DNA strands. Because CRISPR-Cas9 does not require custom protein design, but merely customization of a sgRNA, it quickly became the technology that enabled worldwide application of gene editing. After SpCas9, many more CRISPR systems from different species were closely investigated for use in gene editing (reviewed in Refs. 43 and 44). The discovery of alternative Cas enzymes, like Cas12a (Cpf1),[45] recognizing different PAM sequences, notably expanded the number of CRISPR-Cas targetable genomic loci. Besides, several Cas nucleases have also been engineered to "relax" their PAM requirement.[46–50] Moreover, the identification of smaller proteins, such as *Campylobacter jejuni* Cas9[51] or CasΦ from bacteriophages,[52] increases the number of suitable vector systems for its delivery, in particular for adeno-associated viral vectors (AAV) which are limited in their packaging capacity (<5 kb).

2.1.2 Just cutting won't solve the problem, how do we paste?

The plethora of applications that quickly arose from Cas9 DSB generation relies on the mechanisms of mammalian DNA damage repair (DDR). Two pathways are predominantly responsible for coping with a DSB: homology-directed repair (HDR) and nonhomologous end joining (NHEJ). In mammalian cells NHEJ is the most frequent response to a DSB and functions through the re-ligation of the generated 5′ and 3′ ends.[53] This system however will allow some base pairs to become inserted or deleted, generating indels which can produce frameshifts and can consequently knock-out proteins. While this mechanism is useful to study gene function, it cannot be used to correct most disease-causing loss-of-function mutations. Alternatively, the HDR pathway can be exploited. This route of DNA repair employs a donor DNA molecule as a template to repair the DSB. For the restoration of disease-causing mutations, a donor-template that contains the correct sequence is provided. Correction via the HDR pathway will result in the desired editing.[54] HDR evolved as a protective mechanism during cell division, since DNA is partly denatured during S-phase to allow replication. As a result, it is only active in actively dividing cells,[55] which rather limits its use in many adult cell types.

2.1.3 Limitations to CRISPR therapy: Off-target effects

The main concern for therapeutic applications using CRISPR nucleases is off-target activity. Due to their ability to tolerate mismatches between sgRNA and the protospacer sequence, Cas proteins can produce unwanted gene edits at other sites, causing unpredictable alterations in cellular metabolism and viability. DSBs generally are genotoxic and trigger multiple protective pathways, including apoptosis,[56] and multiple DSBs could result in large deletions, insertions or translocations, all of which can drive oncogenesis.[57] For this reason, several high-fidelity variants have been engineered, demonstrating high efficiency and specificity.[50,58–60] In addition, strategies to minimize the duration of nuclease expression have been associated with reduced off-targets.[61–63] Finally, off-target screening methodologies now allow assessing Cas off-target activity,[64] that, together with in silico prediction and statistical or data-based models, offer a good evaluation of putative Cas induced cleavages.[65]

2.1.4 Application of nucleases to correct CFTR mutations

Programmable nucleases have been thoroughly investigated for rewriting the mutated *CFTR* gene. The "cut-and-paste"-like approach with DSB introduction and HDR repair, has been applied in several preclinical, but translational, CF cell models, including human intestinal organoids and induced pluripotent stem cells (iPSCs). All of these initial studies focused on the most common mutation, F508del. ZFNs were the first nucleases designed to target *F508del-CFTR* in cell lines[66] and in CF patient-derived tracheal epithelial cells and iPSCs.[67,68] A first TALEN-mediated repair of this mutation in iPSCs was reported in 2014.[69] In 2016, another study confirmed that edited iPSCs maintained their differentiation potential and allowed restoring chloride secretion.[70] A recent TALEN-based study reported that a puromycin selection-based strategy was able to correct patient-derived iPSCs with ~10% efficacy.[71] Although CRISPR-Cas nucleases were only added to the gene editing toolbox in 2012, their first application to correct F508del in patient-derived intestinal organoids was already published in 2013 by Schwank et al.[72] Functional rescue of the CFTR protein was assessed by CFTR-dependent organoid swelling after stimulation with the cAMP agonist forskolin in selected edited organoids. Although these ZFN, TALEN and CRISPR-Cas studies provide a valuable proof-of-concept, they reported low editing efficiencies and required enrichment of corrected cells through selection. In 2015, Firth et al. developed a strategy that allowed for selection, yet also delivered a marker-free restoration of F508del.[73] Here, the mutation was corrected via a repair template that contained a selection cassette with an antibiotic resistance and a thymidine kinase gene. Subsequently, the selection cassette could be removed via PiggyBac transposition, and cells that had not excised the cassette could be removed via ganciclovir selection.[73] Although this approach enables footprint-free correction of cells, it employs two steps of selection, which is incompatible with *in vivo* gene therapy. In 2016, Hollywood et al. were able to achieve selection-free correction via optimization of the repair template in CF tracheal epithelial cells, however with a modest editing efficiency of 1.9%.[74] More recently, Ruan et al. reported correction efficiencies >20% in iPSCs via an optimized ribonucleoprotein (RNP) electroporation technique.[75] Even more recent work on correcting *F508del-CFTR* features two studies in which high correction levels were achieved in basal cells, i.e. airway progenitor cells. The first study employed electroporation of Cas9-sgRNA RNPs and AAV serotype 6 (AAV6) transduction of a repair template to correct homozygous F508del with an efficiency of 42%.[76] Alternatively, Suzuki et al. used electroporated ZFN mRNA and AAV6-donor template to restore F508del with an efficiency of 31%.[77]

In addition to HDR-facilitated correction, a small subset of mutations is eligible for repair by nuclease-induced NHEJ. While NHEJ cannot restore mutations in the coding sequence of *CFTR*, it can be used to correct intronic splicing mutations. These generate novel cryptic donor or acceptor splice sites that are recognized by the splicing machinery and lead to the inclusion of extra nucleotides in the mature mRNA. Targeted excision or disruption of these sites restores the *CFTR* mRNA sequence. In 2017, Sanz et al. combined SpCas9 with two sgRNAs to excise the cryptic intronic splicing mutations 1811+1.6kbA>G (c.1680-886A>G), 3237-26A>G (c.3140-26A>G) and 3849+10kbC>T (c.3718-2477C>T) using a minigene model in cell lines.[78] Maule et al. demonstrated correction of 3272-26A>G and 3849+10kbC>T using a single mutation-specific sgRNA and Cas12a nuclease.[79] This allowed a highly allele-specific correction of mutant *CFTR* in patient-derived intestinal organoids and airway cells.[79] The advantage of this approach is to use the more frequent NHEJ pathway as well as abolishing the need to co-deliver a donor template. Currently, a similar strategy is being clinically evaluated to treat Leber congenital amaurosis by gene editing (EDIT-101, NCT03872479[80]).

2.1.5 Why stop at correcting only a single mutation?—The use of super-exons

All previously mentioned examples are highly mutation-specific and require tailored sgRNAs (and donor templates). The targeted integration of a super-exon is an alternative approach that offers the possibility of correcting different mutations with a single strategy (reviewed in Ref. 81). A super-exon codes for a portion of the *CFTR* cDNA that gets integrated into the *CFTR* locus and remains under control of the CFTR promoter, thereby maintaining endogenous regulation of protein expression, which is a potential advantage over classical gene addition. Super-exons can correct all *CFTR* mutations located downstream of the super-exon integration site, therefore choosing the optimal location is key in order to maximize the fraction of PwCF that could benefit from this treatment. A thorough understanding of regulatory elements inside the *CFTR* gene, however, is indispensable, as these sequences should not be destroyed by super-exon integration. Also, not all delivery vehicles, e.g. AAV, allow the delivery of large super-exons spanning the entire *CFTR* cDNA and additional homology arms.[81]

The feasibility of super-exon correction of CF was initially demonstrated by Bednarski et al. in a bronchial epithelial cell line (CFBE41o-), homozygous for the F508del mutation.[82] A super-exon, encoding CFTR exons 11–27, was integrated into exon 11 (containing the F508del mutation) of the endogenous *CFTR* locus exploiting ZFN cleavage and HDR repair. Despite the low integration efficiency, selected edited clones showed restored mRNA levels and ion transport across the cell membrane indicating functional CFTR. Although only F508del correction was assessed in this study, this super-exon strategy can potentially rescue all other mutations located in exons 11–27 of the *CFTR* gene, covering 226 out of 360 disease-causing mutations.[82] Recently, Suzuki et al. proposed a similar strategy by integrating a super-exon *CFTR* cDNA containing exons 9–27 into intron 8 by means of ZFNs, expanding the number of mutations with an additional 19 mutations.[77] Electroporation of ZFN combined with AAV6 transduction to deliver the repair template, showed 56.5% integration efficiency and the restoration of therapeutically relevant levels of CFTR function (~40% of WT-CFTR channel activity) in patient-derived airway basal cells. This approach achieved site-specific targeted integration with high efficiency and no need for antibiotic or clonal selection, thereby obtaining a pool of corrected cells suitable for *ex vivo* cell therapy.

2.1.6 Nuclease-independent gene editing
Alternatively, triplex-forming peptide nucleic acids (PNA)-mediated gene editing does not require expression of an exogenous nuclease. PNA molecules mimic DNA, however purine and pyrimidine bases are attached to a

polyamide backbone which is resistant to nuclease and protease degradation, thus increasing its stability. PNA can base pair with target DNA forming a clamp by generating a PNA/DNA/PNA triplex that will displace the other DNA strand and form a D-loop. This will initiate DNA repair mechanisms depending on the nucleotide excision repair factor xeroderma pigmentosum complementation group A protein. Simultaneous co-delivery of a donor DNA template will promote site-directed recombination.[83] For an overview of the use of PNA for site-specific gene editing, we refer to Ref.[84] PNAs have been used to correct F508del in CF models.[85] Specifically, nanoparticle-based delivery of the PNA molecule and a short donor template (50 nucleotides) led to editing efficiencies of 9.2% in CFBE41o- cells and 5.7% and 1.2% in nasal and pulmonary epithelium respectively, of F508del mice. Moreover, off-targets were very low in these models, underscoring the potential of this alternative gene editing strategy for CF.

2.2 Base editors

Only a few years after the initial papers describing CRISPR-Cas9, a new gene editing technology was introduced: base editors. The first of such enzymes, the cytosine base editor (CBE) relied on a very elegant yet simple architecture: a Cas9 nickase (Cas9n) fused to the rAPOBEC deaminase.[86] While the Cas9 part of the complex defines the target site via its sgRNA, the enzyme will deaminate cytosines at the target site after R-loop formation by Cas9. Deaminated cytosine renders uracil which is restored to thymine by the DDR pathways.[87] To shift the DDR machinery towards a T-A base pair rather than C-G base pair (which would reverse the envisioned edit), the Cas9n places a nick in the targeted and nondeaminated strand. To further improve the efficacy of base editing, Komor et al. fused an uracil glycosylase inhibitor (UGI) to the C-terminal of CBE,[86,88] that by inhibition of base excision repair increased base editing efficiency. The CBE was rapidly followed by the adenine base editor (ABE), which was published in 2017,[89] constituting of a Cas9 nickase fused to an adenine deaminase. Targeted A-to-G conversion by an ABE can theoretically restore 107 *CFTR* mutations, whereas CBE can only correct 35 (Fig. 3). Two recent publications reported on a derivative of CBE that performs targeted C-to-G transversions. This conversion cannot be achieved by deamination alone. Indeed, the C-to-G base editor (CGBE or GBE) arose from the early observation that C-to-U transversions are often repaired to guanine when the uracil DNA glycosylase (UNG) pathway is not inhibited by UGI of the CBE.[90,91] The mechanism remains incompletely

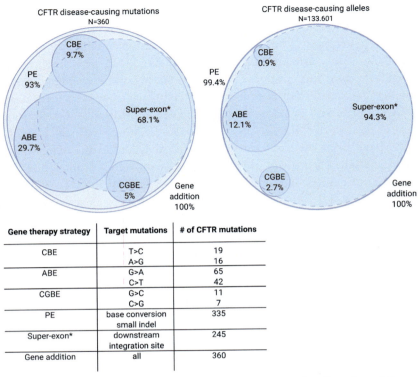

Fig. 3 Targeting *CFTR* with different gene therapeutic strategies. Overview of the percentage of disease-causing mutations (left, n = 360) and alleles (right, n = 133.601) that can be targeted by different gene therapy strategies (based on CFTR2 database, July 2020 update). Cytosine base editors (CBE), adenine base editors (ABE) and C-to-G base editors (CGBE) can make C > T, A > G and C > G edits, respectively. Prime editing allows all base conversions as well as small indels. Super-exons contain a part of the *CFTR* cDNA and can be used to correct all mutations downstream of the integration site (*[77]). Finally, gene addition is mutation agnostic and does not rely on the CF genotype.

understood but both Kurt et al. and Zhao et al. realized its potential and exploited it by removing the UGIs from a state-of-the-art CBE (BE4max).[90,91] To enhance excision of the generated uracil, the UNG coding sequence was added N-terminally to nCas9.

2.2.1 Advantages and drawbacks of base editors

Base editors allow to correct the genetic code without inducing DSBs, thereby preventing possible adverse genotoxic events. Furthermore, cells do not need to be actively dividing in order to achieve base editing.

Nevertheless, gRNA design can be more difficult compared to SpCas9-nucleases since edits can only take place within the editing window (usually 4–5 nucleotides).[43] Also, bystander edits can be generated, possibly causing undesired protein sequence alterations, if the editing window contains multiple bases amenable to editing.[86,89] Incorporating Cas proteins with different PAM requirements, as well as altering the editing window range, expands the utility to base edit more targets.[46,47,92–95] In silico tools such as BE-HIVE are of great help to predict which base preferentially gets deaminated, allowing better guided decisions in sgRNA design.[96] While base editors in general cause only limited DSBs and resulting indels, unexpected SpCas9-independent off-target editing was reported by Jin et al. and Zuo et al.[97,98] Moreover, deamination of RNA molecules was identified as base editor specific off-target, in particular for ABE.[99–101] Optimization of Cas-enzymes and deaminases however is continuously being pursued by many groups, thereby improving specificity, efficiency and reduced off-target effects.[102–105]

2.2.2 Base conversion for CF treatment

Both CBE and ABE have been validated in multiple cell and animal models[43,106] leading to the first therapeutic applications by pioneering companies such as BEAM Therapeutics. Geurts et al. recently published on the rescue of four *CFTR* nonsense mutations using ABE in patient-derived intestinal organoids.[107] Plasmids encoding canonical or xCas9 ABE were electroporated into R553X (c.1657C>T), R785X (c.2353C>T), R1162X (c.3484C>T) or W1282X organoids. Efficiencies after selection ranged from 1.43% to 8.8% depending on the mutation and the ABE variant used. Correction of the W1282X mutation was also reported by another group, where mRNA delivery of ABE lead to base conversion up to 26.4%, rescuing 10% of CFTR expression.[108] Besides the four mutations tested, many more *CFTR* mutations are in theory amenable to base editing (Fig. 3). Most of these mutations however are very rare. The most common mutation, F508del, cannot be rescued by any base editor because it consists of a three-nucleotide deletion. While only 18 out of 360 CF-causing mutations can potentially be restored by CGBE as opposed to 35 for CBE, CGBE covers 2.7% of all CF-causing alleles in the CFTR2 database compared to only 0.9% for CBE, since the fourth most common mutation, N1303K (~1.7% of disease-causing alleles in CFTR2), is amenable to CGBE (Fig. 3).

2.3 Prime editing

The latest CRISPR tool is the prime editor (PE), able to re-write genetic information at specific genomic loci.[109] A Cas9 nickase is fused to an engineered reverse transcriptase (RT) and uses a particular gRNA, the prime editing guide RNA (pegRNA). The pegRNA contains both a spacer and a scaffold, as well as a 3′ extension that provides an RT template and a primer binding site (PBS), needed to insert the desired edit into the genome. The Cas9n generates a nick in the PAM strand, forming a 3′-flap that anneals with the PBS. Starting from the PBS, the RT template is reverse transcribed, which results in a branched intermediate with a 3′ ssDNA flap containing the edit, and a 5′ flap, which contains the original sequence. If the 5′ flap is excised and the 3′ flap ligated, a heteroduplex DNA helix is formed that will be repaired by cellular mechanisms. The PE efficiency has been improved by mutations in the RT enzyme (PE2), and the addition of another gRNA to generate a nick in the unedited strand (PE3).[109] PE greatly expands the modifications that can be inserted into the genome without the need for DSBs. Not only can all base conversions be performed by PE, but also small insertions or deletions. The design of pegRNAs however is less straightforward compared to sgRNAs, since the length of both PBS and RT template can vary and to date there is no consensus on the optimal lengths of these parts. The possibility to design RT templates of variable length though, results in an increased flexibility of the system, making PAM requirements less restrictive. Recently, computational models have been developed to help in optimizing pegRNA design.[110,111] Due to the novelty of the PE, to date there are only a few publications applying this technology. As such, off-targets have not been studied yet to the same extent as for Cas9 or base editors, although researchers continue to define the possibilities and limitations of PE. Nevertheless, off-target analyses to date revealed a significant reduction in off-target cleavage by PE compared to Cas9. Also, possible side-effects of RT have been evaluated, demonstrating no difference in cell viability and minimal perturbation of the transcriptome.[109,112] Geurts et al. very recently (currently still a pre-print) reported prime editing to repair *CFTR* mutations.[113] Electroporation of PE3 encoding plasmids in patient-derived organoids carrying the F508del or R785X mutation, resulted in genetic correction and recovery of CFTR channel function. Despite the low efficiency compared to HDR, forskolin-responsive organoids showed a heterozygous correction of F508del. In some cases, undesired indels were observed in the second F508del allele, but nevertheless, organoids derived from edited clones demonstrated functional responses, underscoring the feasibility of prime editing in primary cells to correct a *CFTR* mutation.

3. Modulation of *CFTR* expression levels as an alternative next-generation approach

Another potential therapeutic approach involves increasing *CFTR* mRNA levels, which is of particular interest for mutations with residual CFTR function. Alternatively, this can be applied to increase mutant protein expression for subsequent correction by existing CFTR modulators. Altered expression of noncoding RNAs (ncRNAs) by post-transcriptional regulation of gene expression is implicated in the pathogenesis of many diseases. Therefore, ncRNAs have emerged as possible therapeutic targets to treat genetic disorders.[114] In the following paragraphs, we will focus on transcriptional activation of *CFTR*, modulation of long noncoding RNAs (lncRNAs) and microRNAs (miRNAs), and enhancement of mRNA stability for nonsense mutations that get rapidly degraded by nonsense mediated mRNA decay (NMD).

3.1 Transcriptional activation of *CFTR*

Transcriptional upregulation of the endogenous *CFTR* gene provides another potential strategy to treat CF. Many disease-causing mutations, including F508del, display some residual function. Therefore, increasing the amount of *CFTR* mRNA could result in enhanced channel activity due to an increased number of CFTR proteins with residual function. ZFNs, TALENs and CRISPR-Cas-based systems have been developed for targeted gene activation (TGA). For a more elaborated discussion on the tools for targeted gene modulation, we refer to.[115] Briefly, expression modulating enzymes were fused to inactive nucleases that guide the complex to the genomic target site, allowing to modulate gene expression, without compromising the underlying genetic code. For CF, only one study has attempted to increase CFTR function by gene activation. Villamizar et al. achieved a 3.6-fold increase in mRNA expression by targeting the dead Cas9-tripartite gene activator to the *CFTR* promoter in F508del human nasal epithelial cells, which correlated with an increase in CFTR maturation and function.[116]

3.2 Long-noncoding RNA regulation

LncRNAs are RNA molecules over 200 nucleotides long that are not translated into proteins but are involved in many regulatory pathways. To date, the role of only one lncRNA in *CFTR* transcriptional regulation has been characterized.[117] Saayman et al. observed that overexpression of the lncRNA BGas resulted in a strong downregulation of CFTR in human airway epithelial 1HAEo- and CFPAC cells. In line with this, downregulating BGas increased CFTR expression. This lncRNA is expressed from intron 11 of the *CFTR* gene in the antisense orientation relative to *CFTR*, and consists of two exons with a total length of 3944 nucleotides. Studies on the mechanism of action revealed that BGas acts as an *in cis* element by recruiting chromatin associated proteins like HMGA1, HMGB1 and WIBG to the previously identified regulatory region of *CFTR* intron 11.[118] This causes distortions in the chromatin structure and subsequent transcription interruption. A possible therapeutic application by targeting BGas with antisense oligonucleotides (ASOs) was later shown by the same group.[116] The ASO efficiently repressed BGas, thereby enhancing *CFTR* transcription and CFTR expression up to fourfold, after exosome delivery to F508del human nasal cells. Recently, a comparative analysis of lncRNA in PwCF and healthy individuals led to the identification of 636 and 1974 lncRNAs differentially expressed in CF airway epithelium and CF lung parenchyma, respectively.[119] These newly identified lncRNAs may be involved in biological processes relevant to CF pathogenesis, so the identification of their mechanism of action in CF may reveal new therapeutic targets.

3.3 MicroRNA downregulation

MiRNAs are short RNA molecules (20–25 nucleotides) that are involved in post-transcriptional regulation of gene expression. In contrast to lncRNAs, the role of several miRNAs has been investigated in CF. In particular, altered expression of miRNAs involved in innate immune responses and inflammation pathways, such as miR-126, miR-93 and miR-17, has been reported in PwCF. Moreover, expression of specific miRNAs has been associated with impairment of macrophage activity (miR-181b) or lung fibrosis (miR-155). For a complete overview of miRNAs in CF we refer to.[120] Among the CF-related miRNAs, several have been shown to directly target either *WT* or *F508del-CFTR*. In particular, miR-101-3p, miR-145-5p, miR-223-3p, miR-494-3p and miR-509-3p have been identified as negative regulators of *CFTR* expression by binding specific elements present in the 3' untranslated region (UTR) of the *CFTR* mRNA resulting in its degradation or translational repression.[121–127] Targeting CFTR regulatory miRNAs with PNAs resulted in increased expression of WT and F508del-CFTR protein in different cell lines.[128–130] Because miRNAs usually have several target genes, silencing of specific miRNAs might alter the expression of many genes. Moreover, it has been shown that PNA inhibition of miR-145-5p, which increased *CFTR* expression, resulted also in the inhibition of other miRNAs. These off-target effects limit its potential as a therapeutic strategy and should be carefully evaluated.

Alternatively, target site blockers (TSBs) have been designed to prevent *CFTR* inhibition by miR-145-5p and miR-223-3p.[127] TSBs are locked nucleic acid ASOs that base pair with miRNA recognition elements (MREs) on the target RNA, competing with the miRNA. De Santi et al. demonstrated that, after transfection of TSBs specifically recognizing the binding sites of miR-145-5p or miR-223-3p, an increase in CFTR expression and function was measured in two bronchial epithelial cell lines homozygous for F508del.[127] All in all, miRNA modulation could represent a promising strategy to treat CF.

3.4 Nonsense mediated mRNA decay inhibition

According to the Human Gene Mutation Database (http://www.hgmd.cf.ac.uk), ~11% of all disease-causing mutations are nonsense mutations. In CF, nonsense mutations account for ~5% of disease-causing alleles in CFTR2 and are associated with severe disease. Not only do nonsense mutations give rise to C-terminally truncated proteins which mostly are not functional, the PTC-containing mRNA is often also rapidly degraded by NMD. NMD is a cellular quality control mechanism for mRNA which protects cells from expressing truncated proteins that might be harmful to the cell or could clog the endoplasmic reticulum. It also allows cells to respond quickly to changing environments by modulating mRNA levels. It is often triggered when a long 3'UTR is present or exon junction complexes (EJC) are formed downstream of the PTC (reviewed in Ref. 131). The level of NMD is highly variable between samples and the number of transcripts available correlates with the response to TRIDs in cell models.[132] Indeed, if the PTC-carrying *CFTR* mRNA is efficiently degraded by NMD machinery, insufficient mRNA levels are available for translational readthrough by TRIDs. Recognizing this, is important to overcome NMD in order to rescue nonsense mutations, and in that regard, several strategies have been proposed to inhibit this process. Inhibition, by small molecules or ASOs, of NMD-factor Serine/Threonine-protein kinase SMG1 that activates RNase H1 mediated mRNA degradation, prevented NMD and thereby increased W1282X-CFTR expression in human bronchial and primary nasal epithelial cells.[133,134] In combination with TRIDs and CFTR modulators, this restored CFTR function in primary nasal epithelial cells at ~50% of that observed in non-CF samples. Because of its physiological cellular functions, NMD inhibition for specific PTC-containing transcripts might be more desirable than general NMD inhibition. ASOs can also be used to physically block protein binding sites on the pre-mRNA without degrading it, thereby preventing specific aberrant mRNA processing steps that for instance result in altered splicing. Splice altering ASOs have been market approved, for example Nusinersen (Spinranza™) which was the first approved causal therapy for SMA.[135,136] In CF, splice altering ASOs have been used to restore splicing of 2789+5G>A (c.2657+5G>A)[137] and 3849+10kbC>T[138] in mini-gene systems and for the latter, more recently also in primary human bronchial epithelial cells.[139] By the same principle, NMD could also be prevented with ASOs that block deposition of EJC on the mRNA.[140] Recently, Erwood et al. described an allele-specific approach to inhibit NMD by removing the *CFTR* sequence downstream of the W1282X mutation.[141] Using CRISPR-Cas9 with two gRNAs targeting exon 23 (W1282X-allele specific) and exon 27, a 24 kb deletion was generated, which led to a 2.4-fold higher expression of the truncated *CFTR* mRNA and W1282X-CFTR protein in human bronchial epithelial cells, which subsequently was rescued by CFTR modulators. This approach is likely best suited for mutations generating PTCs near the C-terminus of proteins, since deletion of the gene downstream of the PTC prevents a combination approach with TRIDs.

4. From Petri-dish to patient

As the toolbox with strategies to edit the *CFTR* gene or alter its expression is rapidly expanding, it is important to keep in mind which challenges need to be overcome before any of these strategies can be of benefit to PwCF. In the final part of this chapter we will compare *ex vivo* and *in vivo* approaches and discuss how best to deliver different gene editing components to cells.

4.1 Ex vivo or in vivo gene editing

To date, most of the clinical trials involving CRISPR have focused on editing *ex vivo*, followed by re-introduction of modified cells into the patient (for a complete overview of clinical trials involving gene editing, we refer to Refs. 142 and 143). For diseases such as specific cancers and blood disorders, e.g. sickle cell disease or β-thalassemia, the *ex vivo* approach is appealing as blood cells are good targets for *ex vivo* gene editing. These approaches have proven highly effective in preclinical studies,[144,145] setting the stage for clinical testing (NCT03745287, NCT04443907[146]). For other diseases however, there is a rationale for direct *in vivo* editing, because the target organ is attractive, e.g. the eye being a small immune-privileged site, or identification, isolation and expansion of target cells *ex vivo* is challenging. In that light, the first *in vivo* use of CRISPR in patients is on its way for the inherited blindness disorder, Leber congenital amaurosis (NCT03872479[147]).

For CF, it remains elusive whether *in vivo* or *ex vivo* gene editing will result in a clinically meaningful improvement of lung disease, or other affected organs. Much can be learnt however from the preclinical and clinical studies conducted in the field of "classical" gene addition gene therapy, i.e. the *in vivo* delivery of a normal copy of the *CFTR* cDNA. While safety and tolerability of different vector systems have been proven, none have led to a persistent clinical benefit. However, these efforts allowed uncovering some of the major bottlenecks in the development of an airway-targeted gene therapy product for CF. Correction of sufficient numbers of target cells is a prerequisite, which asks for robust and sensitive read-outs of CFTR expression. Recently, single-cell RNA sequencing (scRNAseq) of primary human lung samples, allowed identification of CFTR expressing cells in a sensitive and unbiased manner.[148,149] These data suggest that ionocytes are a major source of CFTR expression and activity in the airway epithelium despite representing only 1–2% of epithelial cells. This contrasts with older studies, where ciliated cells were identified as important CFTR expressers.[150] Additionally, the longevity of gene expression would ideally be lifelong to provide a cure. Clinical trials with viral and nonviral vectors delivering a *CFTR* cDNA all resulted in transient increases in CFTR expression, function or lung function, suggestive of a loss of transgene expression (reviewed in Refs. 26 and 27). This underscores the likely need for repeated administrations, where innate immune responses and pre-existing or acquired neutralizing antibodies against viral vector components can diminish gene therapy efficacy.[151–153]

Besides immune responses to vector components, the immunogenicity of Cas9 and sgRNA poses a challenge towards its clinical translation (reviewed in Ref. 154). Recent studies have demonstrated an innate and adaptive cellular immune response to Cas9 in mouse models[155,156] and the presence of anti-Cas9 antibodies and cytotoxic T-cells in human plasma.[157,158] Cytosolic nucleic acids, including sgRNA or DNA/RNA encoding Cas9, can elicit an innate immune response. Transfection of in vitro-transcribed sgRNAs in both human and mouse cells induced an interferon response and subsequent cell death.[159,160] Interestingly, the timing of gene delivery largely impacts the potential induction of a deleterious immune response. We have shown that fetal and neonatal AAV5 delivery to mouse lungs allowed re-administration by avoiding the induction of anti-capsid neutralizing antibodies, by taking advantage of an immature immune system.[161] Similarly, systemic or intramuscular delivery of AAV-SaCas9 to neonatal mice did not induce reactive T-cell responses or antibody development against Cas9.[162] An early intervention thus holds promise for avoiding unwanted immune responses, but also responds to the clinical need for PwCF to receive disease modifying treatments as soon as possible, before the onset of irreversible organ damage (reviewed in Ref. 163). A final point of consideration, is the possible induction of an adaptive immune response to CFTR, in the case of PwCF with two class I mutations, who do not express full-length CFTR, or even with only minor differences between mutant and WT CFTR, as for example in F508del. This was investigated by the Wilson lab experimentally after adenoviral vector mediated (AdV-hCFTR) delivery to murine (m)CFTR-KO mice and via MHC-binding prediction programs for F508del homozygous PwCF.[164,165] Indeed, their mouse data suggest that class I mutations are more at risk of adaptive immune responses to CFTR, based on the activation of T-cells in mCFTR-KO mice (absent in heterozygous and WT mice) to an epitope that is conserved between mouse and human CFTR.[165] Also the prediction algorithm identified a few high-scoring MHC-Class I binders, suggesting a possible host immune response to CFTR, even when the difference between therapeutic and host CFTR is a single amino acid.[164] Accompanying experimental evidence is however needed to further support this in vitro prediction of immunogenicity, in the specific context where F508del is expressed in the absence of any residual CFTR from the other CF allele.

From what we have learnt in the past on *in vivo CFTR* cDNA delivery, better informed decisions can be made for future gene therapies for CF, in particular regarding efficiency and longevity of gene expression. For *ex vivo* gene therapy, no clinical trials have been performed for CF to date[18] and the body of preclinical work is limited. Seminal work however was recently provided by Vaidyanathan and colleagues, setting the first steps toward transplanting *ex vivo* gene-corrected autologous cells into PwCF.[76] Specifically, they showed that upper airway basal cells (UABCs) edited with Cas9 and a donor DNA template restoring the F508del mutation could graft on an

FDA-approved biodegradable scaffold, without loss of differentiation capacity or CFTR function. Follow-up experiments in CF animal models will be key to address the success rate of engraftment in an inflamed lung lined with viscous mucus. Indeed, successful engraftment of stem cells in the lung represents a longstanding therapeutic challenge. Rosen et al. showed that overcoming two major barriers facilitate this process: (1) prior removal of the host stem cell niche by a preconditioning regimen, and (2) the use of embryonic lung-derived progenitor cells.[166] This proof-of-concept study in mice provides new insights in how to increase lung engraftment, but further optimization is necessary prior to clinical use. For a comprehensive review on the successes and challenges of cell therapy for CF, we refer to.[167–169]

4.2 Choice of delivery agent for efficient and safe gene editing

The delivery of CRISPR machinery brings challenges, including a complex set of parts that must be introduced simultaneously to achieve efficient editing. Timing is equally important. The longer Cas9 lingers in the nucleus, the higher the number of possible off-target cuts.[170,171] The consensus is therefore to keep treatment transient. CRISPR-Cas components (the "cargo") can furthermore be delivered in different forms: (1) DNA plasmid encoding both the Cas9 protein and the sgRNA, (2) mRNA for Cas9 translation together with a separate sgRNA, and (3) Cas9 protein complexed with a sgRNA (RNP). A repair template, if needed, can be added as ss/dsDNA. The delivery vehicle used will often determine which cargo to choose, and whether the system is suited for in vitro and/or *in vivo* applications.[172] Each vehicle has its pros and cons, which will be discussed in the following section.

From a detailed overview of delivery methods used in clinical trials for *in vivo* or *ex vivo* editing,[143] it is clear that a variety of delivery vehicles have been used. For *ex vivo* delivery these include adenoviral vectors (AdV) and electroporation of mRNA or plasmids, to treat HIV, various cancers and blood disorders. The limited *in vivo* gene editing studies conducted so far, include AAV as delivery vehicle to target hepatocytes or retinal cells, and plasmid containing polymer gels for vaginal delivery for human papilloma virus (HPV)-related malignancies.[143] Results from these early-phase clinical trials have been promising so far. However, given the small number of participants in these trials, the safety of gene editing therapies has not yet been fully elucidated.[173] Therefore, preclinical studies are of importance to study in a well-controlled environment not only on-target efficacy, but also potential deleterious off-target edits, which vary between gene editors and delivery systems.

4.2.1 Viral vectors
In general, viral vectors are more efficient than nonviral vectors in achieving robust gene expression levels in a high number of target cells, depending on target organ and delivery route. Their tropism can be tailored by distinct viral glycoproteins or protein capsids, naturally occurring or engineered, to transduce target cells.[174–177] However, different viral vectors suffer from varying degrees of innate and adaptive immunity, ranging from acute inflammatory to pre-existing or acquired neutralizing antibodies and cytotoxic T-cell responses[152,161,178,179] (for an extensive review, we refer to Ref. 180). These immune responses can reduce viral vector transduction efficiencies or induce rapid clearance of transduced cells, as was the case for first-generation AdV.[181,182] A better understanding of the mechanisms by which viral vectors induce immune responses, has shed a light on how to improve viral vector design (e.g. gutless AdV,[182,183] engineered AAV capsids[175]) or transiently suppress the immune system to overcome some of these immunological hurdles encountered in early preclinical and clinical research.[180] Most viral vectors confer medium to long-term gene expression, which is undesirable for gene editing purposes. Nevertheless, promising proof-of-concept studies have been reported using viral-vector mediated gene editing in vitro and in animal models. Examples include LV-mediated Cas12a delivery to correct intronic splicing mutations in CFTR,[79] an AAV-CRISPR exon skipping strategy for Duchenne muscular dystrophy[162] and the use of a dual–AAV system to deliver the cytosine base editor into the inner ear of the animals with hearing loss.[184]

4.2.2 Virus-like particles
In that light, virus like particles (VLPs) provide an attractive alternative as they combine the efficiency of a viral vector to enter the host cell and nucleus, with a limited lifespan of expression in target cells. VLPs are multiprotein structures that mimic the organization and conformation of native viruses but lack the viral genome. As such, they serve as delivery vehicles of RNA/protein cargo. Different types of viruses have been mimicked by VLPs (reviewed in Ref. 185). Several VLPs have been tailored to deliver CRISPR-Cas cargo in the form of RNPs to cells in vitro, *ex vivo* and *in vivo* in lab animals, termed nanoMEDIC, Nanoblades, VesiCas, or

Cas9P LV.[61,186–188] Higher on- to off-target editing ratios were achieved for all compared to plasmid-based CRISPR-Cas delivery, mainly because of the transient expression of Cas9 and sgRNA.[61,186,187] This was specifically assessed for VesiCas delivery of CRISPR-Cas9 RNPs where SpCas9 protein expression was minimal 24 h post-delivery, compared to peak levels at 24–48 h for plasmid transfection.[61] In conclusion, VLPs hold promise for the delivery of gene editing cargo because of their efficient cell entry and high on-target editing. The limited timespan of nuclease expression strongly reduces off-targets and importantly also minimizes the window of immune activation for both CRISPR-Cas components and viral (glyco-)proteins.

4.2.3 Nonviral vectors

Finally, synthetic vectors provide a third strategy for the delivery of nucleic acids or RNPs and can be divided into lipid, polymeric and inorganic particles (reviewed in Ref. 189). They are generally less immunogenic than viral vectors and not restricted in packaging capacity. To confer cell specificity, synthetic vectors can be chemically functionalized with ligands. Disadvantages however include possible bio-incompatibility and toxicity, and problems with therapeutic cargo release. Indeed, achieving endosomal escape to avoid lysosomal degradation, presents one of the biggest challenges for nonviral vectors, although continuous efforts are being made to promote this release.[190,191] In CF, nonviral vectors have been preclinically used for transient delivery of *CFTR* mRNA[192] or for Sleeping Beauty transposase and *CFTR* cDNA plasmids to confer long-term gene replacement by *CFTR* cDNA integration.[193] The UK CF Gene Therapy Consortium showed that monthly doses of CFTR plasmid-cationic lipid complexes led to a modest stabilizing effect in lung function in PwCF.[194] Nonviral RNP delivery by means of synthetic vectors has, to our knowledge, not yet been applied for *CFTR* gene editing, but the promise of such a strategy is exemplified in other disease areas, such as Duchenne Muscular Dystrophy.[195]

5. Conclusion

Despite recent advances in the pharmacological treatment of CF, with highly effective CFTR modulator therapy available for ∼90% of PwCF, the need for a CF cure remains as current treatments do not cover all PwCF and consist of life-long therapies with the accompanying side-effects. Development of new genome editing tools provide for the first time the possibility to permanently correct the *CFTR* gene. Here we discussed programmable nuclease-based gene editing, as well as the CRISPR-derivatives, base and prime editing, and preclinical work that has already applied these techniques in cells and animal models. CF is caused by at least 360 *CFTR* mutations of all kinds and different mutations call for different gene editing strategies (summarized in Fig. 3 and Table 2). Besides, we provided an overview of approaches aiming to enhance *CFTR* expression, e.g. by TGA or modulating of ncRNA, which is of particular interest for mutations with some residual CFTR function. Finally, we discussed an important hurdle for the translation of any of these therapies toward PwCF: delivery. Both *in vivo* and *ex vivo* approaches have been proposed for CF gene therapy. The choice of the delivery vehicle—viral vectors, virus like particles or non-viral synthetic vectors—each with its pros and cons, will need to be weighed for the specific gene editing strategy and target organ envisioned, together with their potential for good manufacturing practice (GMP) production and upscaling capacity for further use in PwCF. In conclusion, it is likely that multiple gene editing strategies will need to be developed in order to provide a treatment for all PwCF. The ongoing preclinical work on correction of *CFTR* mutations as well pioneering work on delivery of gene editing therapies into patients in other fields, will provide valuable insights into the opportunities and challenges of each strategy. This information will form the basis for extensively optimizing and testing CF-specific gene editing treatments before moving to clinical trials in the future.

Acknowledgments

Images were created with BioRender.com. M.E. is supported by an FWO-SB (Flemish Research Foundation) doctoral fellowship 1S29917N, M.S.C. by a senior post-doctoral FWO scholarship 12Z5920N. Cystic fibrosis research is funded by the Belgian CF patient Association and Fund Alphonse Jean Forton from the King Baudouin Foundation (2020-J1810150-E015) and by the Italian CF Patient Society FFC.

References

1. Corriveau S, Sykes J, Stephenson AL. Cystic fibrosis survival: the changing epidemiology. *Curr Opin Pulm Med*. Nov 2018;24(6):574–578. https://doi.org/10.1097/MCP.0000000000000520.
2. Kerem B, Rommens JM, Buchanan JA, et al. Identification of the cystic fibrosis gene: genetic analysis. *Science*. 1989;245(4922):1073–1080. https://doi.org/10.1126/science.2570460.
3. Riordan JR, Rommens JM, Kerem B, et al. Identification of the cystic fibrosis gene: cloning and characterization of complementary DNA. *Science*. 1989;245(4922):1066–1073. https://doi.org/10.1126/science.2475911.

4. Rommens JM, Iannuzzi MC, Kerem B, et al. Identification of the cystic fibrosis gene: chromosome walking and jumping. *Science.* 1989;245(4922):1059–1065. https://doi.org/10.1126/science.2772657.
5. De Boeck K, Amaral MD. Progress in therapies for cystic fibrosis. *Lancet Respir Med.* 2016;4(8):662–674. https://doi.org/10.1016/S2213-2600(16)00023-0.
6. Veit G, Avramescu RG, Chiang AN, et al. From CFTR biology toward combinatorial pharmacotherapy: expanded classification of cystic fibrosis mutations. *Mol Biol Cell.* 2016;27(3):424–433. https://doi.org/10.1091/mbc.E14-04-0935.
7. *Press Announcements—FDA approves Kalydeco to treat rare form of cystic fibrosis*; 2012. Accessed 2020/10/14 https://wayback.archive-it.org/7993/20170112023947/http://www.fda.gov/NewsEvents/Newsroom/PressAnnouncements/ucm289633.htm.
8. Vertex. *Kalydeco Label.* https://www.accessdata.fda.gov/drugsatfda_docs/label/2019/203188s029,207925s008lbl.pdf.
9. *Press Announcements > FDA approves new treatment for cystic fibrosis*; 2015. Accessed 2020/10/14 https://wayback.archive-it.org/7993/20170112023800/http://www.fda.gov/NewsEvents/Newsroom/PressAnnouncements/ucm453565.htm.
10. FDA. *Drug Trials Snapshots: SYMDEKO | FDA.* Accessed 2020/10/14 https://www.fda.gov/drugs/drug-approvals-and-databases/drug-trials-snapshots-symdeko.
11. Wainwright CE, Elborn JS, Ramsey BW, et al. Lumacaftor-Ivacaftor in patients with cystic fibrosis homozygous for Phe508del CFTR. *N Engl J Med.* 2015;373(3):220–231. https://doi.org/10.1056/NEJMoa1409547.
12. Taylor-Cousar JL, Munck A, McKone EF, et al. Tezacaftor-Ivacaftor in patients with cystic fibrosis homozygous for Phe508del. *N Engl J Med.* 2017;377(21):2013–2023. https://doi.org/10.1056/NEJMoa1709846.
13. Ramsey BW, Davies J, McElvaney NG, et al. A CFTR potentiator in patients with cystic fibrosis and the G551D mutation. *N Engl J Med.* 2011;365(18):1663–1672. https://doi.org/10.1056/NEJMoa1105185.
14. Veit G, Roldan A, Hancock MA, et al. Allosteric folding correction of F508del and rare CFTR mutants by elexacaftor-tezacaftor-ivacaftor (Trikafta) combination. *JCI Insight.* 2020;5(18):e139983. https://doi.org/10.1172/jci.insight.139983.
15. *FDA approves new breakthrough therapy for cystic fibrosis*; 2019. Accessed 2020/10/14 https://www.fda.gov/news-events/press-announcements/fda-approves-new-breakthrough-therapy-cystic-fibrosis.
16. Heijerman HGM, McKone EF, Downey DG, et al. Efficacy and safety of the elexacaftor plus tezacaftor plus ivacaftor combination regimen in people with cystic fibrosis homozygous for the F508del mutation: a double-blind, randomised, phase 3 trial. *Lancet.* 2019;394(10212):1940–1948. https://doi.org/10.1016/S0140-6736(19)32597-8.
17. Middleton PG, Mall MA, Drevinek P, et al. Elexacaftor-tezacaftor-ivacaftor for cystic fibrosis with a single Phe508del allele. *N Engl J Med.* 2019;381(19):1809–1819. https://doi.org/10.1056/NEJMoa1908639.
18. CFF. *Drug Development Pipeline | CFF Clinical Trials Tool.* Accessed 2020/10/26 https://www.cff.org/Trials/pipeline.
19. Dukovski D, Villella A, Bastos C, et al. Amplifiers co-translationally enhance CFTR biosynthesis via PCBP1-mediated regulation of CFTR mRNA. *J Cyst Fibros.* 2020; 19(5):733–741. https://doi.org/10.1016/j.jcf.2020.02.006.
20. Miller JP, Drew L, Green O, et al. CFTR amplifiers: a new class of CFTR modulator that complements the substrate limitations of other CF therapeutic modalities. *Am J Respir Crit Care Med.* 2016;193: A5574.
21. McDonald CM, Campbell C, Torricelli RE, et al. Ataluren in patients with nonsense mutation Duchenne muscular dystrophy (ACT DMD): a multicentre, randomised, double-blind, placebo-controlled, phase 3 trial. *Lancet.* 2017;390(10101):1489–1498. https://doi.org/10.1016/S0140-6736(17)31611-2.

22. Kerem E, Konstan MW, De Boeck K, et al. Ataluren for the treatment of nonsense-mutation cystic fibrosis: a randomised, double-blind, placebo-controlled phase 3 trial. *Lancet Respir Med.* 2014;2(7):539–547. https://doi.org/10.1016/S2213-2600(14)70100-6.
23. Kerem E. ELX-02: an investigational read-through agent for the treatment of nonsense mutation-related genetic disease. *Expert Opin Investig Drugs.* 2020;29(12):1347–1354. https://doi.org/10.1080/13543784.2020.1828862.
24. *Translate Bio Announces Interim Results from Phase 1/2 Clinical Trial of MRT5005 in Patients with Cystic Fibrosis*; 2019. Accessed 2020/10/15 https://investors.translate.bio/news-releases/news-release-details/translate-bio-announces-interim-results-phase-12-clinical-trial.
25. Cooney AL, McCray Jr PB, Sinn PL. Cystic fibrosis gene therapy: looking back, looking forward. *Genes (Basel).* 2018;9(11): 538. https://doi.org/10.3390/genes9110538.
26. Sondhi D, Stiles KM, De BP, Crystal RG. Genetic modification of the lung directed toward treatment of human disease. *Hum Gene Ther.* 2017;28(1):3–84. https://doi.org/10.1089/hum.2016.152.
27. Alton EW, Boyd AC, Davies JC, et al. Genetic medicines for CF: hype versus reality. *Pediatr Pulmonol.* 2016;51(S44):S5–S17. https://doi.org/10.1002/ppul.23543.
28. Russell S, Bennett J, Wellman JA, et al. Efficacy and safety of voretigene neparvovec (AAV2-hRPE65v2) in patients with RPE65-mediated inherited retinal dystrophy: a randomised, controlled, open-label, phase 3 trial. *Lancet.* 2017;390(10097):849–860. https://doi.org/10.1016/S0140-6736(17)31868-8.
29. Mendell JR, Al-Zaidy S, Shell R, et al. Single-dose gene-replacement therapy for spinal muscular atrophy. *N Engl J Med.* 2017;377(18):1713–1722. https://doi.org/10.1056/NEJMoa1706198.
30. Harrison C. First gene therapy for beta-thalassemia approved. *Nat Biotechnol.* 2019;37 (10):1102–1103. https://doi.org/10.1038/d41587-019-00026-3.
31. Beyar-Katz O, Gill S. Advances in chimeric antigen receptor T cells. *Curr Opin Hematol.* 2020;27(6):368–377. https://doi.org/10.1097/MOH.0000000000000614.
32. Gifford AH, Mayer-Hamblett N, Pearson K, Nichols DP. Answering the call to address cystic fibrosis treatment burden in the era of highly effective CFTR modulator therapy. *J Cyst Fibros.* 2020;19(5):762–767. https://doi.org/10.1016/j.jcf.2019.11.007.
33. Burgel PR, Munck A, Durieu I, et al. Real-life safety and effectiveness of lumacaftor-ivacaftor in patients with cystic fibrosis. *Am J Respir Crit Care Med.* 2020;201(2): 188–197. https://doi.org/10.1164/rccm.201906-1227OC.
34. Talamo Guevara M, McColley SA. The safety of lumacaftor and ivacaftor for the treatment of cystic fibrosis. *Expert Opin Drug Saf.* 2017;16(11):1305–1311. https://doi.org/10.1080/14740338.2017.1372419.
35. Volkova N, Moy K, Evans J, et al. Disease progression in patients with cystic fibrosis treated with ivacaftor: data from national US and UK registries. *J Cyst Fibros.* 2020;19 (1):68–79. https://doi.org/10.1016/j.jcf.2019.05.015.
36. CFF. *Path to a Cure: Many Routes, One Mission | CF Foundation.* Accessed 2020/10/17 https://www.cff.org/Research/About-Our-Research/Path-to-a-Cure-Many-Routes-One-Mission/.
37. Bibikova M, Carroll D, Segal DJ, et al. Stimulation of homologous recombination through targeted cleavage by chimeric nucleases. *Mol Cell Biol.* 2001;21(1):289–297. https://doi.org/10.1128/MCB.21.1.289-297.2001.
38. Christian M, Cermak T, Doyle EL, et al. Targeting DNA double-strand breaks with TAL effector nucleases. *Genetics.* 2010;186(2):757–761. https://doi.org/10.1534/genetics.110.120717.
39. Urnov FD, Rebar EJ, Holmes MC, Zhang HS, Gregory PD. Genome editing with engineered zinc finger nucleases. *Nat Rev Genet.* 2010;11(9):636–646. https://doi.org/10.1038/nrg2842.

40. Joung JK, Sander JD. TALENs: a widely applicable technology for targeted genome editing. *Nat Rev Mol Cell Biol.* 2013;14(1):49–55. https://doi.org/10.1038/nrm3486.
41. Jinek M, Chylinski K, Fonfara I, Hauer M, Doudna JA, Charpentier E. A programmable dual-RNA-guided DNA endonuclease in adaptive bacterial immunity. *Science.* 2012;337(6096):816–821. https://doi.org/10.1126/science.1225829.
42. Gasiunas G, Barrangou R, Horvath P, Siksnys V. Cas9-crRNA ribonucleoprotein complex mediates specific DNA cleavage for adaptive immunity in bacteria. *Proc Natl Acad Sci U S A.* 2012;109(39):E2579–E2586. https://doi.org/10.1073/pnas.1208507109.
43. Anzalone AV, Koblan LW, Liu DR. Genome editing with CRISPR-Cas nucleases, base editors, transposases and prime editors. *Nat Biotechnol.* 2020;38(7):824–844. https://doi.org/10.1038/s41587-020-0561-9.
44. Makarova KS, Wolf YI, Iranzo J, et al. Evolutionary classification of CRISPR–Cas systems: a burst of class 2 and derived variants. *Nat Rev Microbiol.* 2020;18(2):67–83. https://doi.org/10.1038/s41579-019-0299-x.
45. Zetsche B, Jonathan O, et al. Cpf1 is a single rna-guided endonuclease of a class 2 CRISPR-Cas system. *Cell.* 2015;163(3):759–771. https://doi.org/10.1016/j.cell.2015.09.038.
46. Miller SM, Wang T, Randolph PB, et al. Continuous evolution of SpCas9 variants compatible with non-G PAMs. *Nat Biotechnol.* 2020;38(4):471–481. https://doi.org/10.1038/s41587-020-0412-8.
47. Walton RT, Christie KA, Whittaker MN, Kleinstiver BP. Unconstrained genome targeting with near-PAMless engineered CRISPR-Cas9 variants. *Science.* 2020;368(6488):290–296. https://doi.org/10.1126/science.aba8853.
48. Gao L, Cox DBT, Yan WX, et al. Engineered Cpf1 variants with altered PAM specificities. *Nat Biotechnol.* 2017;35(8):789–792. https://doi.org/10.1038/nbt.3900.
49. Kleinstiver BP, Prew MS, Tsai SQ, et al. Broadening the targeting range of Staphylococcus aureus CRISPR-Cas9 by modifying PAM recognition. *Nat Biotechnol.* 2015;33(12):1293–1298. https://doi.org/10.1038/nbt.3404.
50. Kleinstiver BP, Sousa AA, Walton RT, et al. Engineered CRISPR–Cas12a variants with increased activities and improved targeting ranges for gene, epigenetic and base editing. *Nat Biotechnol.* 2019;37(3):276–282. https://doi.org/10.1038/s41587-018-0011-0.
51. Kim E, Koo T, Park SW, et al. In vivo genome editing with a small Cas9 orthologue derived from Campylobacter jejuni. *Nat Commun.* 2017;8(1): 14500. https://doi.org/10.1038/ncomms14500.
52. Pausch P, Al-Shayeb B, Bisom-Rapp E, et al. CRISPR-CasPhi from huge phages is a hypercompact genome editor. *Science.* 2020;369(6501):333–337. https://doi.org/10.1126/science.abb1400.
53. Guirouilh-Barbat J, Huck S, Bertrand P, et al. Impact of the KU80 pathway on NHEJ-induced genome rearrangements in mammalian cells. *Mol Cell.* 2004;14(5):611–623. https://doi.org/10.1016/j.molcel.2004.05.008.
54. Lukacsovich T, Yang D, Waldman AS. Repair of a specific double-strand break generated within a mammalian chromosome by yeast endonuclease I-SceI. *Nucleic Acids Res.* 1994;22(25):5649–5657. https://doi.org/10.1093/nar/22.25.5649.
55. Chapman JR, Taylor MR, Boulton SJ. Playing the end game: DNA double-strand break repair pathway choice. *Mol Cell.* 2012;47(4):497–510. https://doi.org/10.1016/j.molcel.2012.07.029.
56. Haapaniemi E, Botla S, Persson J, Schmierer B, Taipale J. CRISPR-Cas9 genome editing induces a p53-mediated DNA damage response. *Nat Med.* 2018;24(7):927–930. https://doi.org/10.1038/s41591-018-0049-z.
57. Kosicki M, Tomberg K, Bradley A. Repair of double-strand breaks induced by CRISPR-Cas9 leads to large deletions and complex rearrangements. *Nat Biotechnol.* 2018;36(8):765–771. https://doi.org/10.1038/nbt.4192.

58. Kleinstiver BP, Pattanayak V, Prew MS, et al. High-fidelity CRISPR–Cas9 nucleases with no detectable genome-wide off-target effects. *Nature*. 2016;529(7587):490–495. https://doi.org/10.1038/nature16526.
59. Slaymaker IM, Gao L, Zetsche B, Scott DA, Yan WX, Zhang F. Rationally engineered Cas9 nucleases with improved specificity. *Science*. 2016;351(6268):84–88. https://doi.org/10.1126/science.aad5227.
60. Casini A, Olivieri M, Petris G, et al. A highly specific SpCas9 variant is identified by in vivo screening in yeast. *Nat Biotechnol*. 2018;36(3):265–271. https://doi.org/10.1038/nbt.4066.
61. Montagna C, Petris G, Casini A, et al. VSV-G-enveloped vesicles for traceless delivery of CRISPR-Cas9. *Mol Ther–Nucleic Acids*. 2018;12:453–462. https://doi.org/10.1016/j.omtn.2018.05.010.
62. Petris G, Casini A, Montagna C, et al. Hit and go CAS9 delivered through a lentiviral based self-limiting circuit. *Nat Commun*. 2017;8(1): 15334. https://doi.org/10.1038/ncomms15334.
63. Kim S, Kim D, Cho SW, Kim J, Kim JS. Highly efficient RNA-guided genome editing in human cells via delivery of purified Cas9 ribonucleoproteins. *Genome Res*. 2014; 24(6):1012–1019. https://doi.org/10.1101/gr.171322.113.
64. Kim D, Luk K, Wolfe SA, Kim JS. Evaluating and enhancing target specificity of gene-editing nucleases and deaminases. *Annu Rev Biochem*. 2019;88:191–220. https://doi.org/10.1146/annurev-biochem-013118-111730.
65. Broeders M, Herrero-Hernandez P, Ernst MPT, van der Ploeg AT, Pijnappel W. Sharpening the molecular scissors: advances in gene-editing technology. *iScience*. 2020;23(1): 100789. https://doi.org/10.1016/j.isci.2019.100789.
66. Maeder ML, Thibodeau-Beganny S, Osiak A, et al. Rapid "open-source" engineering of customized zinc-finger nucleases for highly efficient gene modification. *Mol Cell*. 2008;31(2):294–301. https://doi.org/10.1016/j.molcel.2008.06.016.
67. Crane AM, Kramer P, Bui JH, et al. Targeted correction and restored function of the CFTR gene in cystic fibrosis induced pluripotent stem cells. *Stem Cell Rep*. 2015;4(4): 569–577. https://doi.org/10.1016/j.stemcr.2015.02.005.
68. Lee CM, Flynn R, Hollywood JA, Scallan MF, Harrison PT. Correction of the DeltaF508 mutation in the cystic fibrosis transmembrane conductance regulator gene by zinc-finger nuclease homology-directed repair. *Biores Open Access*. 2012;1(3): 99–108. https://doi.org/10.1089/biores.2012.0218.
69. Sargent RG, Suzuki S, Gruenert DC. Nuclease-mediated double-strand break (DSB) enhancement of small fragment homologous recombination (SFHR) gene modification in human-induced pluripotent stem cells (hiPSCs). *Methods Mol Biol*. 2014;1114: 279–290. https://doi.org/10.1007/978-1-62703-761-7_18.
70. Suzuki S, Sargent RG, Illek B, et al. TALENs facilitate single-step seamless SDF correction of F508del CFTR in airway epithelial submucosal gland cell-derived CF-iPSCs. *Mol Ther Nucleic Acids*. 2016;5:e273. https://doi.org/10.1038/mtna.2015.43.
71. Fleischer A, Vallejo-Díez S, Martín-Fernández JM, et al. iPSC-derived intestinal organoids from cystic fibrosis patients acquire CFTR activity upon TALEN-mediated repair of the p.F508del mutation. *Mol Ther*. 2020;17:858–870. https://doi.org/10.1016/j.omtm.2020.04.005.
72. Schwank G, Koo BK, Sasselli V, et al. Functional repair of CFTR by CRISPR/Cas9 in intestinal stem cell organoids of cystic fibrosis patients. *Cell Stem Cell*. 2013;13(6): 653–658. https://doi.org/10.1016/j.stem.2013.11.002.
73. Firth AL, Menon T, Parker GS, et al. Functional gene correction for cystic fibrosis in lung epithelial cells generated from patient iPSCs. *Cell Rep*. 2015;12(9):1385–1390. https://doi.org/10.1016/j.celrep.2015.07.062.

74. Hollywood JA, Lee CM, Scallan MF, Harrison PT. Analysis of gene repair tracts from Cas9/gRNA double-stranded breaks in the human CFTR gene. *Sci Rep*. 2016;6: 32230. https://doi.org/10.1038/srep32230.
75. Ruan J, Hirai H, Yang D, et al. Efficient gene editing at major CFTR mutation Loci. *Mol Ther Nucleic Acids*. 2019;16:73–81. https://doi.org/10.1016/j.omtn.2019.02.006.
76. Vaidyanathan S, Salahudeen AA, Sellers ZM, et al. High-efficiency, selection-free gene repair in airway stem cells from cystic fibrosis patients rescues CFTR function in differentiated epithelia. *Cell Stem Cell*. 2020;26(2):161–171.e4. https://doi.org/10.1016/j.stem.2019.11.002.
77. Suzuki S, Crane AM, Anirudhan V, et al. Highly efficient gene editing of cystic fibrosis patient-derived airway basal cells results in functional CFTR correction. *Mol Ther*. 2020;28(7):1684–1695. https://doi.org/10.1016/j.ymthe.2020.04.021.
78. Sanz DJ, Hollywood JA, Scallan MF, Harrison PT. Cas9/gRNA targeted excision of cystic fibrosis-causing deep-intronic splicing mutations restores normal splicing of CFTR mRNA. *PLoS One*. 2017;12(9): e0184009. https://doi.org/10.1371/journal.pone.0184009.
79. Maule G, Casini A, Montagna C, et al. Allele specific repair of splicing mutations in cystic fibrosis through AsCas12a genome editing. *Nat Commun*. 2019;10(1):3556. https://doi.org/10.1038/s41467-019-11454-9.
80. Maeder ML, Stefanidakis M, Wilson CJ, et al. Development of a gene-editing approach to restore vision loss in Leber congenital amaurosis type 10. *Nat Med*. 2019;25(2): 229–233. https://doi.org/10.1038/s41591-018-0327-9.
81. Mention K, Santos L, Harrison PT. Gene and base editing as a therapeutic option for cystic fibrosis-learning from other diseases. *Genes (Basel)*. 2019;10(5):387. https://doi.org/10.3390/genes10050387.
82. Bednarski C, Tomczak K, Vom Hovel B, Weber WM, Cathomen T. Targeted integration of a super-exon into the CFTR locus leads to functional correction of a cystic fibrosis cell line model. *PLoS One*. 2016;11(8): e0161072. https://doi.org/10.1371/journal.pone.0161072.
83. Rogers FA, Vasquez KM, Egholm M, Glazer PM. Site-directed recombination via bifunctional PNA-DNA conjugates. *Proc Natl Acad Sci U S A*. 2002;99(26):16695–16700. https://doi.org/10.1073/pnas.262556899.
84. Ricciardi AS, Quijano E, Putman R, Saltzman WM, Glazer PM. Peptide nucleic acids as a tool for site-specific gene editing. *Molecules*. 2018;23(3):632. https://doi.org/10.3390/molecules23030632.
85. McNeer NA, Anandalingam K, Fields RJ, et al. Nanoparticles that deliver triplex-forming peptide nucleic acid molecules correct F508del CFTR in airway epithelium. *Nat Commun*. 2015;6:6952. https://doi.org/10.1038/ncomms7952.
86. Komor AC, Kim YB, Packer MS, Zuris JA, Liu DR. Programmable editing of a target base in genomic DNA without double-stranded DNA cleavage. *Nature*. 2016;533 (7603):420–424. https://doi.org/10.1038/nature17946.
87. Krokan HE, Drablos F, Slupphaug G. Uracil in DNA—occurrence, consequences and repair. *Oncogene*. 2002;21(58):8935–8948. https://doi.org/10.1038/sj.onc.1205996.
88. Kunz C, Saito Y, Schar P. DNA Repair in mammalian cells: mismatched repair: variations on a theme. *Cell Mol Life Sci*. 2009;66(6):1021–1038. https://doi.org/10.1007/s00018-009-8739-9.
89. Gaudelli NM, Komor AC, Rees HA, et al. Programmable base editing of A*T to G*C in genomic DNA without DNA cleavage. *Nature*. 2017;551(7681):464–471. https://doi.org/10.1038/nature24644.
90. Kurt IC, Zhou R, Iyer S, et al. CRISPR C-to-G base editors for inducing targeted DNA transversions in human cells. *Nat Biotechnol*. 2021;39:41–46. https://doi.org/10.1038/s41587-020-0609-x.

91. Zhao D, Li J, Li S, et al. Glycosylase base editors enable C-to-A and C-to-G base changes. *Nat Biotechnol.* 2021;39:35–40. https://doi.org/10.1038/s41587-020-0592-2.
92. Kim YB, Komor AC, Levy JM, Packer MS, Zhao KT, Liu DR. Increasing the genome-targeting scope and precision of base editing with engineered Cas9-cytidine deaminase fusions. *Nat Biotechnol.* 2017;35(4):371–376. https://doi.org/10.1038/nbt.3803.
93. Huang TP, Zhao KT, Miller SM, et al. Circularly permuted and PAM-modified Cas9 variants broaden the targeting scope of base editors. *Nat Biotechnol.* 2019;37(6):626–631. https://doi.org/10.1038/s41587-019-0134-y.
94. Wang X, Ding C, Yu W, et al. Cas12a base editors induce efficient and specific editing with low DNA damage response. *Cell Rep.* 2020;31(9): 107723. https://doi.org/10.1016/j.celrep.2020.107723.
95. Thuronyi BW, Koblan LW, Levy JM, et al. Continuous evolution of base editors with expanded target compatibility and improved activity. *Nat Biotechnol.* 2019;37(9):1070–1079. https://doi.org/10.1038/s41587-019-0193-0.
96. Arbab M, Shen MW, Mok B, et al. Determinants of base editing outcomes from target library analysis and machine learning. *Cell.* 2020;182(2):463–480.e30. https://doi.org/10.1016/j.cell.2020.05.037.
97. Zuo E, Sun Y, Wei W, et al. Cytosine base editor generates substantial off-target single-nucleotide variants in mouse embryos. *Science.* 2019;364(6437):289–292. https://doi.org/10.1126/science.aav9973.
98. Jin S, Zong Y, Gao Q, et al. Cytosine, but not adenine, base editors induce genome-wide off-target mutations in rice. *Science.* 2019;364(6437):292–295. https://doi.org/10.1126/science.aaw7166.
99. Grunewald J, Zhou R, Garcia SP, et al. Transcriptome-wide off-target RNA editing induced by CRISPR-guided DNA base editors. *Nature.* 2019;569(7756):433–437. https://doi.org/10.1038/s41586-019-1161-z.
100. Rees HA, Wilson C, Doman JL, Liu DR. Analysis and minimization of cellular RNA editing by DNA adenine base editors. *Sci Adv.* 2019;5(5):eaax5717. https://doi.org/10.1126/sciadv.aax5717.
101. Grunewald J, Zhou R, Iyer S, et al. CRISPR DNA base editors with reduced RNA off-target and self-editing activities. *Nat Biotechnol.* 2019;37(9):1041–1048. https://doi.org/10.1038/s41587-019-0236-6.
102. Gehrke JM, Cervantes O, Clement MK, et al. An APOBEC3A-Cas9 base editor with minimized bystander and off-target activities. *Nat Biotechnol.* 2018;36(10):977–982. https://doi.org/10.1038/nbt.4199.
103. Koblan LW, Doman JL, Wilson C, et al. Improving cytidine and adenine base editors by expression optimization and ancestral reconstruction. *Nat Biotechnol.* 2018;36(9):843–846. https://doi.org/10.1038/nbt.4172.
104. Richter MF, Zhao KT, Eton E, et al. Phage-assisted evolution of an adenine base editor with improved Cas domain compatibility and activity. *Nat Biotechnol.* 2020;38(7):883–891. https://doi.org/10.1038/s41587-020-0453-z.
105. Doman JL, Raguram A, Newby GA, Liu DR. Evaluation and minimization of Cas9-independent off-target DNA editing by cytosine base editors. *Nat Biotechnol.* 2020;38(5):620–628. https://doi.org/10.1038/s41587-020-0414-6.
106. Levy JM, Yeh WH, Pendse N, et al. Cytosine and adenine base editing of the brain, liver, retina, heart and skeletal muscle of mice via adeno-associated viruses. *Nat Biomed Eng.* 2020;4(1):97–110. https://doi.org/10.1038/s41551-019-0501-5.
107. Geurts MH, de Poel E, Amatngalim GD, et al. CRISPR-based adenine editors correct nonsense mutations in a cystic fibrosis organoid biobank. *Cell Stem Cell.* 2020;26(4):503–510.e7. https://doi.org/10.1016/j.stem.2020.01.019.

108. Jiang T, Henderson JM, Coote K, et al. Chemical modifications of adenine base editor mRNA and guide RNA expand its application scope. *Nat Commun.* 2020;11:1979. https://doi.org/10.1038/s41467-020-15892-8.
109. Anzalone AV, Randolph PB, Davis JR, et al. Search-and-replace genome editing without double-strand breaks or donor DNA. *Nature.* 2019;576(7785):149–157. https://doi.org/10.1038/s41586-019-1711-4.
110. Hsu JY, Anzalone AV, Grünewald J, et al. PrimeDesign software for rapid and simplified design of prime editing guide RNAs. *bioRxiv.* 2020. https://doi.org/10.1101/2020.05.04.077750. 2020.05.04.077750.
111. Kim HK, Yu G, Park J, et al. Predicting the efficiency of prime editing guide RNAs in human cells. *Nat Biotechnol.* 2020. https://doi.org/10.1038/s41587-020-0677-y.
112. Schene IF, Joore IP, Oka R, et al. Prime editing for functional repair in patient-derived disease models. *Nat Commun.* 2020;11:5352. https://doi.org/10.1038/s41467-020-19136-7.
113. Geurts MH, de Poel E, Pleguezuelos-Manzano C, et al. Evaluating CRISPR-based prime editing for cancer modeling and CFTR repair in intestinal organoids. *bioRxiv.* 2020. https://doi.org/10.1101/2020.10.05.325837. 2020.10.05.325837.
114. Cohen-Cymberknoh M, Kerem E, Ferkol T, Elizur A. Airway inflammation in cystic fibrosis: molecular mechanisms and clinical implications. *Thorax.* 2013;68(12):1157–1162. https://doi.org/10.1136/thoraxjnl-2013-203204.
115. Thakore PI, Black JB, Hilton IB, Gersbach CA. Editing the epigenome: technologies for programmable transcription and epigenetic modulation. *Nat Methods.* 2016;13(2):127–137. https://doi.org/10.1038/nmeth.3733.
116. Villamizar O, Waters SA, Scott T, et al. Targeted activation of cystic fibrosis transmembrane conductance regulator. *Mol Ther.* 2019;27(10):1737–1748. https://doi.org/10.1016/j.ymthe.2019.07.002.
117. Saayman SM, Ackley A, Burdach J, et al. Long non-coding rna bgas regulates the cystic fibrosis transmembrane conductance regulator. *Mol Ther.* 2016;24(8):1351–1357. https://doi.org/10.1038/mt.2016.112.
118. Ott CJ, Blackledge NP, Kerschner JL, et al. Intronic enhancers coordinate epithelial-specific looping of the active CFTR locus. *Proc Natl Acad Sci U S A.* 2009;106(47):19934–19939. https://doi.org/10.1073/pnas.0900946106.
119. Kumar P, Sen C, Peters K, Frizzell RA, Biswas R. Comparative analyses of long non-coding RNA profiles in vivo in cystic fibrosis lung airway and parenchyma tissues. *Respir Res.* 2019;20(1):284. https://doi.org/10.1186/s12931-019-1259-8.
120. Glasgow AMA, De Santi C, Greene CM. Non-coding RNA in cystic fibrosis. *Biochem Soc Trans.* 2018;46(3):619–630. https://doi.org/10.1042/BST20170469.
121. Ramachandran S, Karp PH, Osterhaus SR, et al. Post-transcriptional regulation of cystic fibrosis transmembrane conductance regulator expression and function by microRNAs. *Am J Respir Cell Mol Biol.* 2013;49(4):544–551. https://doi.org/10.1165/rcmb.2012-0430OC.
122. Oglesby IK, Chotirmall SH, McElvaney NG, Greene CM. Regulation of cystic fibrosis transmembrane conductance regulator by microRNA-145, -223, and -494 is altered in DeltaF508 cystic fibrosis airway epithelium. *J Immunol.* 2013;190(7):3354–3362. https://doi.org/10.4049/jimmunol.1202960.
123. Gillen AE, Gosalia N, Leir SH, Harris A. MicroRNA regulation of expression of the cystic fibrosis transmembrane conductance regulator gene. *Biochem J.* 2011;438(1):25–32. https://doi.org/10.1042/BJ20110672.
124. Hassan F, Nuovo GJ, Crawford M, et al. MiR-101 and miR-144 regulate the expression of the CFTR chloride channel in the lung. *PLoS One.* 2012;7(11): e50837. https://doi.org/10.1371/journal.pone.0050837.

125. Megiorni F, Cialfi S, Dominici C, Quattrucci S, Pizzuti A. Synergistic post-transcriptional regulation of the cystic fibrosis transmembrane conductance regulator (CFTR) by miR-101 and miR-494 specific binding. *PLoS One*. 2011;6(10): e26601. https://doi.org/10.1371/journal.pone.0026601.
126. Viart V, Bergougnoux A, Bonini J, et al. Transcription factors and miRNAs that regulate fetal to adult CFTR expression change are new targets for cystic fibrosis. *Eur Respir J*. 2015;45(1):116–128. https://doi.org/10.1183/09031936.00113214.
127. De Santi C, Fernandez Fernandez E, Gaul R, et al. Precise targeting of miRNA sites restores CFTR activity in CF bronchial epithelial cells. *Mol Ther*. 2020;28(4): 1190–1199. https://doi.org/10.1016/j.ymthe.2020.02.001.
128. Amato F, Tomaiuolo R, Nici F, et al. Exploitation of a very small peptide nucleic acid as a new inhibitor of miR-509-3p involved in the regulation of cystic fibrosis disease-gene expression. *Biomed Res Int*. 2014;2014. https://doi.org/10.1155/2014/610718, 610718.
129. Fabbri E, Tamanini A, Jakova T, et al. A peptide nucleic acid against MicroRNA miR-145-5p enhances the expression of the cystic fibrosis transmembrane conductance regulator (CFTR) in Calu-3 cells. *Molecules*. 2018;23(1):71. https://doi.org/10.3390/molecules23010071.
130. Lutful Kabir F, Ambalavanan N, Liu G, et al. MicroRNA-145 antagonism reverses TGF-beta inhibition of F508del CFTR correction in airway epithelia. *Am J Respir Crit Care Med*. 2018;197(5):632–643. https://doi.org/10.1164/rccm.201704-0732OC.
131. Kurosaki T, Maquat LE. Nonsense-mediated mRNA decay in humans at a glance. *J Cell Sci*. 2016;129(3):461–467. https://doi.org/10.1242/jcs.181008.
132. Linde L, Boelz S, Nissim-Rafinia M, et al. Nonsense-mediated mRNA decay affects nonsense transcript levels and governs response of cystic fibrosis patients to gentamicin. *J Clin Invest*. 2007;117(3):683–692. https://doi.org/10.1172/JCI28523.
133. Keenan MM, Huang L, Jordan NJ, et al. Nonsense-mediated RNA decay pathway inhibition restores expression and function of W1282X CFTR. *Am J Respir Cell Mol Biol*. 2019;61(3):290–300. https://doi.org/10.1165/rcmb.2018-0316OC.
134. Laselva O, Eckford PD, Bartlett C, et al. Functional rescue of c.3846G>A (W1282X) in patient-derived nasal cultures achieved by inhibition of nonsense mediated decay and protein modulators with complementary mechanisms of action. *J Cyst Fibros*. 2020; 19(5):717–727. https://doi.org/10.1016/j.jcf.2019.12.001.
135. Hua Y, Sahashi K, Hung G, et al. Antisense correction of SMN2 splicing in the CNS rescues necrosis in a type III SMA mouse model. *Genes Dev*. 2010;24(15):1634–1644. https://doi.org/10.1101/gad.1941310.
136. Aartsma-Rus A. FDA approval of nusinersen for spinal muscular atrophy makes 2016 the year of splice modulating oligonucleotides. *Nucleic Acid Ther*. 2017;27(2):67–69. https://doi.org/10.1089/nat.2017.0665.
137. Igreja S, Clarke LA, Botelho HM, Marques L, Amaral MD. Correction of a cystic fibrosis splicing mutation by antisense oligonucleotides. *Hum Mutat*. 2016;37(2):209–215. https://doi.org/10.1002/humu.22931.
138. Friedman KJ, Kole J, Cohn JA, Knowles MR, Silverman LM, Kole R. Correction of aberrant splicing of the cystic fibrosis transmembrane conductance regulator (CFTR) gene by antisense oligonucleotides. *J Biol Chem*. 1999;274(51):36193–36199. https://doi.org/10.1074/jbc.274.51.36193.
139. Michaels WE, Bridges RJ, Hastings ML. Antisense oligonucleotide-mediated correction of CFTR splicing improves chloride secretion in cystic fibrosis patient-derived bronchial epithelial cells. *Nucleic Acids Res*. 2020;48(13):7454–7467. https://doi.org/10.1093/nar/gkaa490.

140. Nomakuchi TT, Rigo F, Aznarez I, Krainer AR. Antisense oligonucleotide-directed inhibition of nonsense-mediated mRNA decay. *Nat Biotechnol.* 2016;34(2):164–166. https://doi.org/10.1038/nbt.3427.
141. Erwood S, Laselva O, Bily TMI, et al. Allele-specific prevention of nonsense-mediated decay in cystic fibrosis using homology-independent genome editing. *Mol Ther Methods Clin Dev.* 2020;17:1118–1128. https://doi.org/10.1016/j.omtm.2020.05.002.
142. Hampton T. With first CRISPR trials, gene editing moves toward the clinic. *JAMA.* 2020;323(16):1537–1539. https://doi.org/10.1001/jama.2020.3438.
143. Hirakawa MP, Krishnakumar R, Timlin JA, Carney JP, Butler KS. Gene editing and CRISPR in the clinic: current and future perspectives. *Biosci Rep.* 2020;40(4): BSR20200127. https://doi.org/10.1042/BSR20200127.
144. Dever DP, Bak RO, Reinisch A, et al. CRISPR/Cas9 beta-globin gene targeting in human haematopoietic stem cells. *Nature.* 2016;539(7629):384–389. https://doi.org/10.1038/nature20134.
145. Ye L, Wang J, Tan Y, et al. Genome editing using CRISPR-Cas9 to create the HPFH genotype in HSPCs: An approach for treating sickle cell disease and beta-thalassemia. *Proc Natl Acad Sci U S A.* 2016;113(38):10661–10665. https://doi.org/10.1073/pnas.1612075113.
146. Corbacioglu S, Cappellini MD, Chapin J, et al. *Initial safety and efficacy results with a single dose of autologous CRISPR-Cas9 modified CD34+ hematopoietic stem and progenitor cells in transfusion-dependent B-thalassemia and sickle cell disease*; 2020:S280.
147. *Allergan And Editas Medicine Announce Dosing Of First Patient In Landmark Phase 1/2 Clinical Trial Of CRISPR Medicine AGN-151587 (EDIT-101) For The Treatment Of LCA10.* Editas Medicine; 2020. March 4, 2020.
148. Plasschaert LW, Zilionis R, Choo-Wing R, et al. A single-cell atlas of the airway epithelium reveals the CFTR-rich pulmonary ionocyte. *Nature.* 2018;560(7718): 377–381. https://doi.org/10.1038/s41586-018-0394-6.
149. Montoro DT, Haber AL, Biton M, et al. A revised airway epithelial hierarchy includes CFTR-expressing ionocytes. *Nature.* 2018;560(7718):319–324. https://doi.org/10.1038/s41586-018-0393-7.
150. Kreda SM, Mall M, Mengos A, et al. Characterization of wild-type and deltaF508 cystic fibrosis transmembrane regulator in human respiratory epithelia. *Mol Biol Cell.* 2005;16 (5):2154–2167. https://doi.org/10.1091/mbc.e04-11-1010.
151. Hurlbut GD, Ziegler RJ, Nietupski JB, et al. Preexisting immunity and low expression in primates highlight translational challenges for liver-directed AAV8-mediated gene therapy. *Mol Ther.* 2010;18(11):1983–1994. https://doi.org/10.1038/mt.2010.175.
152. Vidovic D, Gijsbers R, Quiles Jimenez A, et al. Noninvasive imaging reveals stable transgene expression in mouse airways after delivery of a nonintegrating recombinant adeno-associated viral vector. *Hum Gene Ther.* 2016;27(1):60–71. https://doi.org/10.1089/hum.2015.109.
153. Bottermann M, Foss S, van Tienen LM, et al. TRIM21 mediates antibody inhibition of adenovirus-based gene delivery and vaccination. *Proc Natl Acad Sci U S A.* 2018;115 (41):10440–10445. https://doi.org/10.1073/pnas.1806314115.
154. Mehta A, Merkel OM. Immunogenicity of Cas9 protein. *J Pharm Sci.* 2020;109 (1):62–67. https://doi.org/10.1016/j.xphs.2019.10.003.
155. Chew WL, Tabebordbar M, Cheng JK, et al. A multifunctional AAV-CRISPR-Cas9 and its host response. *Nat Methods.* 2016;13(10):868–874. https://doi.org/10.1038/nmeth.3993.
156. Wang D, Mou H, Li S, et al. Adenovirus-mediated somatic genome editing of Pten by CRISPR/Cas9 in mouse liver in spite of Cas9-specific immune responses. *Hum Gene Ther.* 2015;26(7):432–442. https://doi.org/10.1089/hum.2015.087.

157. Charlesworth CT, Deshpande PS, Dever DP, et al. Identification of preexisting adaptive immunity to Cas9 proteins in humans. *Nat Med.* 2019;25(2):249–254. https://doi.org/10.1038/s41591-018-0326-x.
158. Simhadri VL, McGill J, McMahon S, Wang J, Jiang H, Sauna ZE. Prevalence of pre-existing antibodies to CRISPR-associated nuclease Cas9 in the USA population. *Mol Ther Methods Clin Dev.* 2018;10:105–112. https://doi.org/10.1016/j.omtm.2018.06.006.
159. Kim S, Koo T, Jee HG, et al. CRISPR RNAs trigger innate immune responses in human cells. *Genome Res.* 2018;28(3):367–373. https://doi.org/10.1101/gr.231936.117.
160. Wienert B, Shin J, Zelin E, Pestal K, Corn JE. In vitro-transcribed guide RNAs trigger an innate immune response via the RIG-I pathway. *PLoS Biol.* 2018;16(7): e2005840. https://doi.org/10.1371/journal.pbio.2005840.
161. Carlon MS, Vidović D, Dooley J, et al. Immunological ignorance allows long-term gene expression after perinatal recombinant adeno-associated virus-mediated gene transfer to murine airways. *Hum Gene Ther.* 2014;25(6):517–528. https://doi.org/10.1089/hum.2013.196.
162. Nelson CE, Wu Y, Gemberling MP, et al. Long-term evaluation of AAV-CRISPR genome editing for Duchenne muscular dystrophy. *Nat Med.* 2019;25(3):427–432. https://doi.org/10.1038/s41591-019-0344-3.
163. Carlon MS, Vidovic D, Birket S. Roadmap for an early gene therapy for cystic fibrosis airway disease. *Prenat Diagn.* 2017;37(12):1181–1190. https://doi.org/10.1002/pd.5164.
164. Figueredo J, Limberis MP, Wilson JM. Prediction of cellular immune responses against CFTR in patients with cystic fibrosis after gene therapy. *Am J Respir Cell Mol Biol.* 2007;36(5):529–533. https://doi.org/10.1165/rcmb.2006-0313CB.
165. Limberis MP, Figueredo J, Calcedo R, Wilson JM. Activation of CFTR-specific T Cells in cystic fibrosis mice following gene transfer. *Mol Ther.* 2007;15(9): 1694–1700. https://doi.org/10.1038/sj.mt.6300210.
166. Rosen C, Shezen E, Aronovich A, et al. Preconditioning allows engraftment of mouse and human embryonic lung cells, enabling lung repair in mice. *Nat Med.* 2015;21(8): 869–879. https://doi.org/10.1038/nm.3889.
167. Berical A, Lee RE, Randell SH, Hawkins F. Challenges facing airway epithelial cell-based therapy for cystic fibrosis. *Front Pharmacol.* 2019;10:74. https://doi.org/10.3389/fphar.2019.00074.
168. Koh KD, Erle DJ. Steps toward cell therapy for cystic fibrosis. *Am J Respir Cell Mol Biol.* 2020;63(3):275–276. https://doi.org/10.1165/rcmb.2020-0235ED.
169. Duchesneau P, Waddell TK, Karoubi G. Cell-based therapeutic approaches for cystic fibrosis. *Int J Mol Sci.* 2020;21(15):5219. https://doi.org/10.3390/ijms21155219.
170. Fu Y, Foden JA, Khayter C, et al. High-frequency off-target mutagenesis induced by CRISPR-Cas nucleases in human cells. *Nat Biotechnol.* 2013;31(9):822–826. https://doi.org/10.1038/nbt.2623.
171. Lin Y, Cradick TJ, Brown MT, et al. CRISPR/Cas9 systems have off-target activity with insertions or deletions between target DNA and guide RNA sequences. *Nucleic Acids Res.* 2014;42(11):7473–7485. https://doi.org/10.1093/nar/gku402.
172. Lino CA, Harper JC, Carney JP, Timlin JA. Delivering CRISPR: a review of the challenges and approaches. *Drug Deliv.* 2018;25(1):1234–1257. https://doi.org/10.1080/10717544.2018.1474964.
173. Han HA, Pang JKS, Soh BS. Mitigating off-target effects in CRISPR/Cas9-mediated in vivo gene editing. *J Mol Med (Berl).* 2020;98(5):615–632. https://doi.org/10.1007/s00109-020-01893-z.

174. Limberis MP, Vandenberghe LH, Zhang L, Pickles RJ, Wilson JM. Transduction efficiencies of novel AAV vectors in mouse airway epithelium in vivo and human ciliated airway epithelium in vitro. *Mol Ther*. 2009;17(2):294–301. https://doi.org/10.1038/mt.2008.261.
175. Zinn E, Pacouret S, Khaychuk V, et al. In silico reconstruction of the viral evolutionary lineage yields a potent gene therapy vector. *Cell Rep*. 2015;12(6):1056–1068. https://doi.org/10.1016/j.celrep.2015.07.019.
176. Cooney AL, Abou Alaiwa MH, Shah VS, et al. Lentiviral-mediated phenotypic correction of cystic fibrosis pigs. *JCI Insight*. 2016;1(14):e88730. https://doi.org/10.1172/jci.insight.88730.
177. Vidović D, Carlon MS, Da Cunha MF, et al. rAAV-CFTRΔR rescues the cystic fibrosis phenotype in human intestinal organoids and cystic fibrosis mice. *Am J Respir Crit Care Med*. 2016;193(3):288–298. https://doi.org/10.1164/rccm.201505-0914oc.
178. Boutin S, Monteilhet V, Veron P, et al. Prevalence of serum IgG and neutralizing factors against adeno-associated virus (AAV) types 1, 2, 5, 6, 8, and 9 in the healthy population: implications for gene therapy using AAV vectors. *Hum Gene Ther*. 2010;21(6):704–712. https://doi.org/10.1089/hum.2009.182.
179. Mingozzi F, Meulenberg JJ, Hui DJ, et al. AAV-1-mediated gene transfer to skeletal muscle in humans results in dose-dependent activation of capsid-specific T cells. *Blood*. 2009;114(10):2077–2086. https://doi.org/10.1182/blood-2008-07-167510.
180. Shirley JL, de Jong YP, Terhorst C, Herzog RW. Immune responses to viral gene therapy vectors. *Mol Ther*. 2020;28(3):709–722. https://doi.org/10.1016/j.ymthe.2020.01.001.
181. Crystal RG. Adenovirus: the first effective in vivo gene delivery vector. *Hum Gene Ther*. 2014;25(1):3–11. https://doi.org/10.1089/hum.2013.2527.
182. Seiler MP, Cerullo V, Lee B. Immune response to helper dependent adenoviral mediated liver gene therapy: challenges and prospects. *Curr Gene Ther*. 2007;7(5):297–305. https://doi.org/10.2174/156652307782151452.
183. Lee D, Liu J, Junn HJ, Lee EJ, Jeong KS, Seol DW. No more helper adenovirus: production of gutless adenovirus (GLAd) free of adenovirus and replication-competent adenovirus (RCA) contaminants. *Exp Mol Med*. 2019;51(10):1–18. https://doi.org/10.1038/s12276-019-0334-z.
184. Yeh WH, Shubina-Oleinik O, Levy JM, et al. In vivo base editing restores sensory transduction and transiently improves auditory function in a mouse model of recessive deafness. *Sci Transl Med*. 2020;12(546):eaay9101. https://doi.org/10.1126/scitranslmed.aay9101.
185. Roldão A, Silva AC, MCM M. Viruses and virus-like particles in biotechnology: fundamentals and applications. In: *Comprehensive Biotechnology*. vol. 1. 3rd ed; 2017:633–656. https://doi.org/10.1016/B978-0-12-809633-8.09046-4.
186. Gee P, Lung MSY, Okuzaki Y, et al. Extracellular nanovesicles for packaging of CRISPR-Cas9 protein and sgRNA to induce therapeutic exon skipping. *Nat Commun*. 2020;11(1):1334. https://doi.org/10.1038/s41467-020-14957-y.
187. Mangeot PE, Risson V, Fusil F, et al. Genome editing in primary cells and in vivo using viral-derived Nanoblades loaded with Cas9-sgRNA ribonucleoproteins. *Nat Commun*. 2019;10(1):45. https://doi.org/10.1038/s41467-018-07845-z.
188. Choi JG, Dang Y, Abraham S, et al. Lentivirus pre-packed with Cas9 protein for safer gene editing. *Gene Ther*. 2016;23(7):627–633. https://doi.org/10.1038/gt.2016.27.
189. Wilbie D, Walther J, Mastrobattista E. Delivery aspects of CRISPR/Cas for in vivo genome editing. *Acc Chem Res*. 2019;52(6):1555–1564. https://doi.org/10.1021/acs.accounts.9b00106.

190. Chang J, Chen X, Glass Z, et al. Integrating combinatorial lipid nanoparticle and chemically modified protein for intracellular delivery and genome editing. *Acc Chem Res.* 2019;52(3):665–675. https://doi.org/10.1021/acs.accounts.8b00493.
191. Patel S, Ashwanikumar N, Robinson E, et al. Boosting intracellular delivery of lipid nanoparticle-encapsulated mRNA. *Nano Lett.* 2017;17(9):5711–5718. https://doi.org/10.1021/acs.nanolett.7b02664.
192. Sainz-Ramos M, Villate-Beitia I, Gallego I, et al. Non-viral mediated gene therapy in human cystic fibrosis airway epithelial cells recovers chloride channel functionality. *Int J Pharm.* 2020;588:119757. https://doi.org/10.1016/j.ijpharm.2020.119757.
193. Guan S, Munder A, Hedtfeld S, et al. Self-assembled peptide-poloxamine nanoparticles enable in vitro and in vivo genome restoration for cystic fibrosis. *Nat Nanotechnol.* 2019;14(3):287–297. https://doi.org/10.1038/s41565-018-0358-x.
194. Alton E, Armstrong DK, Ashby D, et al. Repeated nebulisation of non-viral CFTR gene therapy in patients with cystic fibrosis: a randomised, double-blind, placebo-controlled, phase 2b trial. *Lancet Respir Med.* 2015;3(9):684–691. https://doi.org/10.1016/S2213-2600(15)00245-3.
195. Wei T, Cheng Q, Min YL, Olson EN, Siegwart DJ. Systemic nanoparticle delivery of CRISPR-Cas9 ribonucleoproteins for effective tissue specific genome editing. *Nat Commun.* 2020;11(1):3232. https://doi.org/10.1038/s41467-020-17029-3.
196. Steines B, Dickey DD, Bergen J, et al. CFTR gene transfer with AAV improves early cystic fibrosis pig phenotypes. *JCI Insight.* 2016;1(14):e88728. https://doi.org/10.1172/jci.insight.88728.
197. Alton EWFW, Beekman JM, Boyd AC, et al. Preparation for a first-in-man lentivirus trial in patients with cystic fibrosis. *Thorax.* 2017;72(2):137–147. https://doi.org/10.1136/thoraxjnl-2016-208406.
198. Merkert S, Bednarski C, Göhring G, Cathomen T, Martin U. Generation of a gene-corrected isogenic control iPSC line from cystic fibrosis patient-specific iPSCs homozygous for p.Phe508del mutation mediated by TALENs and ssODN. *Stem Cell Res.* 2017;23:95–97. https://doi.org/10.1016/j.scr.2017.07.010.

CHAPTER SEVEN

Gene editing and modulation for Duchenne muscular dystrophy

Anthony A. Stephenson[a] and Kevin M. Flanigan[a,b,c,]*

[a]Center for Gene Therapy, Abigail Wexner Research Institute, Nationwide Children's Hospital, Columbus, OH, United States
[b]Department of Pediatrics, College of Medicine, The Ohio State University, Columbus, OH, United States
[c]Department of Neurology, College of Medicine, The Ohio State University, Columbus, OH, United States
*Corresponding author: e-mail address: kevin.flanigan@nationwidechildrens.org

Contents

1. Introduction to Duchenne muscular dystrophy	226
1.1 Overview	226
1.2 The DMD gene	228
1.3 Dystrophin structure and function	229
1.4 Dystrophinopathies	230
2. First-generation gene therapies for DMD	232
2.1 Overview	232
2.2 Gene replacement: Miniaturized dystrophins	233
2.3 Exon skipping with antisense oligonucleotides (AONs)	234
2.4 Vectorized exon-skipping with modified U7 small nuclear RNAs	235
3. Second-generation approaches to DMD	237
3.1 Overview	237
3.2 Multiplex gene editing for exon deletion	239
3.3 Single-cut gene editing for exon skipping and reframing	241
3.4 Homology-directed gene repair for DMD mutation correction	242
4. Challenges for current and future DMD therapies	244
4.1 Overview	244
4.2 Immunity issues	244
4.3 Durability of adeno-associated virus episomes	245
4.4 Safety	246
5. Conclusions	247
References	248

Abstract

Duchenne muscular dystrophy (DMD) is a progressive muscle disease caused by loss of dystrophin protein, encoded by the *DMD* gene. DMD manifests early in childhood as difficulty walking, progresses to loss of ambulation by the teens, and leads to death in early adulthood. Adeno-associated virus-vectorized gene therapies to restore dystrophin protein expression using gene replacement or antisense oligonucleotide-mediated pre-mRNA splicing modulation have emerged, making great strides in uncovering barriers to gene therapies for DMD and other genetic diseases. While this first-generation of DMD therapies are being evaluated in ongoing clinical trials, uncertainties regarding durability and therapeutic efficacy prompted the development of new experimental therapies for DMD that take advantage of somatic cell gene editing. These experimental therapies continue to advance toward clinic trials, but questions remain unanswered regarding safety and translatable efficacy. Here we review the advancements toward treatment of DMD using gene editing and modulation therapies, with an emphasis on those nearest to clinical applications.

Abbreviations

AAV	adeno-associated virus
ABD	actin-binding domain
AONs	antisense oligonucleotides
BMD	Becker muscular dystrophy
DEx50	dog model of DMD harboring an exon 50 deletion mutation
DMD	Duchenne muscular dystrophy
ECEs	extrachromosomal elements
EF	EF-hand motif
GRDM	golden retriever DMD model
gRNA	guide RNA
H	hinge
HDR	homology directed repair
HITI	homology-independent targeted integration
indels	small insertions or deletions
PMO	phosphorodiamidate morpholino oligomer
Sm	Smith antigen
smOPT	consensus Sm binding sequence
snRNAs	small nuclear RNAs
SR	spectrin-like repeat
ssODNs	single-stranded oligodeoxynucleotides
vg/kg	vector genomes per kilogram subject weight

1. Introduction to Duchenne muscular dystrophy

1.1 Overview

Duchenne muscular dystrophy (DMD) is a fatal muscle-wasting disease that almost exclusively affects males (one in approximately 5000 male births) and is caused by mutations in the *DMD* gene on the X chromosome.[1–3] *DMD* encodes a muscle-specific structural protein which contains N- and C-terminal actin- and dystroglycan-binding domains, respectively (Fig. 1A).[4] By acting as a physical link between the cytoskeleton and the membrane-associated dystroglycan complex (Fig. 1A), dystrophin provides

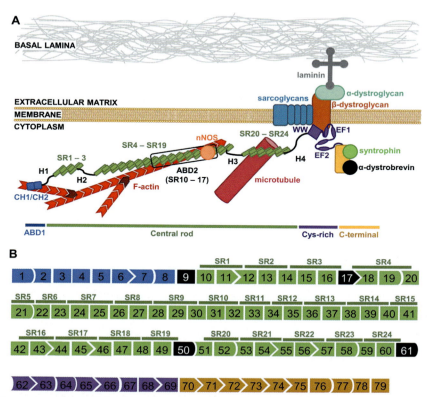

Fig. 1 Structure and function of dystrophin (A) Schematic of the role of dystrophin in muscle cells. ABD1 and ABD2; actin-binding domains 1 and 2, C-terminal; C-terminal domain, CH1 and CH2; Calponin-homology domains 1 and 2, Cys-rich; cysteine-rich domain, EF1 and EF2; EF-hand motifs 1 and 2; H1–H4; hinges 1–4, SR1–SR24; spectrin-like repeats 1–24, WW; WW domain. Helices of the 24 SR helical bundles are depicted as intercalated short and long green cylinders that pack together to form long rod-like filaments. Each SR is made up of a one short helix and one long helix. The short and long helices of one SR are packed into a triple-helical coiled-coil with the long helix of an adjacent SR. (B) Diagram of *DMD* gene exon reading frame phasing. Exons are colored as in Fig. 1A based on the approximate locations of the protein domains. Shapes of the exons indicate 0 (flat), +1 (round), or +2 (pointed) reading frames at the start and end of the exon coding sequences (i.e., exon phasing). Notably, repeat phasing of the alternating short and long helices of the SRs is asynchronous with the *DMD* exon phasing. Approximate locations of SRs are marked above the exons.

shock absorption and anchors the cytoskeleton to the extracellular matrix to protect the membrane integrity of muscle cells from the shearing forces of muscle contractions.[4] In the absence of dystrophin, skeletal muscle and cardiac tissue undergo chronic cycles of degeneration leading to fibrosis and fatty replacement, resulting in a decline in muscle mass, strength, cardiac performance, and ventilation over time.[1–3] DMD manifests in affected boys aged 3–5 years and is typically fatal before the end of the third decade of life.[1–3] There is currently no cure for DMD, but several gene replacement therapies are in trials, as are other experimental therapies aimed at restoring dystrophin expression via gene reprogramming.

1.2 The DMD gene

Spanning a region of more than 2 million bases on the p arm of the X chromosome, *DMD* is the largest known human gene.[5,6] *DMD* comprises at least 7 promoters that drive expression of various tissue-specific transcripts from the 79 exons that splice together to generate an open reading frame of more than 11,000 bases for the longest transcripts.[5,7–10] Expression of the longest *DMD* transcripts are driven by three different promoters that each encode 427 kDa proteins that differ only by the coding sequence of exon 1 and respective tissue-specific expression in the cerebral cortex (Dp427c), muscle (Dp427m), and Purkinje (Dp427p) cells.[8–10] Although the roles of Dp427m in muscle physiology are relatively well established, those of the Dp427c and Dp427p isoforms are less understood[11,12]. Four additional promoters reside upstream of exon 30 (Dp260), exon 45 (Dp140), exon 56 (Dp116), and exon 63 (Dp71), respectively, driving expression of shorter *DMD* transcripts encoding unique N-terminal amino acids and generally resulting in N-truncated dystrophin isoforms.[13–18] Similar to the role of Dp427m in cardiac and skeletal muscles, dystrophin isoforms in the brain and peripheral nervous system are generally thought to support membrane structures. Dp260 is found in the outer plexiform layer of the retina where it localizes to post-synaptic membranes.[13,19,20] Dp116 is abundant in myelin sheaths of Schwann cells in the peripheral nervous system.[17] Dp140 encodes a long 5′ untranslated region and initiates translation in exon 51, generating a protein that localizes to post-synaptic membranes in the cerebral cortex.[18] Dp71 is exceptional as it is ubiquitously expressed and has roles in many membrane-associated structures.[14,15] The promoter responsible for Dp71 also generates a mature transcript called Dp40, which includes intron 70 and results in an alternative early stop codon and polyadenylation signal.[15,16] Dp40 expression remains high in the cerebrum throughout human develop and also localizes to membranes of synaptic vesicles.[16,21,22] In addition to these various transcripts, a highly functional 412 kDa isoform can be expressed in muscle cells of individuals with truncating mutations in the 5' end of the gene via utilization of a highly-conserved exon 5 internal ribosome entry site, the native function of which is yet undescribed.[23] Importantly, little more than localization is known about the molecular biology of the non-muscle dystrophin isoforms. This should be a focal point of future research in the field of neurology as clinical data demonstrate a high instance of behavioral and cognitive deficits in boys with *DMD* mutations,[16,24–26] highlighting the potential importance of dystrophin brain isoforms as a target for gene therapies. However, the focus of this chapter is on Dp427m, which is the most abundant dystrophin isoform in heart and skeletal muscles, as current therapies are focused on restoring expression of full-length or a functional mutant dystrophin in these tissues.

1.3 Dystrophin structure and function

The Dp427m transcript comprises 79 exons and encodes a 427 kDa protein that provides structural support to muscle cells by linking the intracellular actin filaments to β-dystroglycan and other proteins at the sarcolemma (Fig. 1A).[27] Dystrophin generally adopts an extended structure with four main functional domains; an N-terminal actin-binding domain (ABD1), a central rod domain, a cysteine-rich domain, and a carboxyl-terminus (Fig. 1A). ABD1 contains two calponin homology domains (CH1 and CH2, Fig. 1A) that impart binding to F actin, directly linking dystrophin to the intracellular actin cytoskeleton.[27] The central rod domain is made up of 24 spectrin-like repeats (SR1–24, Fig. 1) interspaced and bookended by four hinges (H1–H4, Fig. 1A). The spectrin-like repeats adopt alpha-helices that pack together into triple-helical bundles and form long filaments.[28–30] The rod domain is thought to act as a spring-like flexible linker to absorb shearing forces of muscle contractions.[28–30] The central rod domain also directly mediates interactions with the cytoskeleton. SR10–SR17 harbor a cation-rich actin-binding motif (ABD2) without calponin homology which mediates binding to acidic actin filaments through electrostatic interactions and provides lateral binding to the cytoskeleton (Fig. 1A).[31] SR20–SR23 bind to microtubules which supports proper organization of the microtubule network in muscle cells (Fig. 1A).[32,33] Moreover, SR16–SR17 recruit neuronal nitrous oxide synthetase to the sarcolemma to facilitate nitrous oxide secretion by skeletal muscles which induces vasodilation during sustained physical activity (Fig. 1A).[34] The cysteine-rich domain contains two EF-hand motifs (EF1 and EF2, Fig. 1A) along with a WW domain that mediate binding to the transmembrane protein β-dystroglycan at the sarcolemma

(Fig. 1A).[35] The dystrophin C-terminal domain provides a scaffold for assembly of other sarcolemmal proteins such as α-dystrobrevin and syntrophin (Fig. 1A).[36,37]

The identification of patients with large rod domain deletion mutations yet relatively mild symptoms suggested the critical importance of the ABD1 and the cysteine-rich domain in dystrophin function.[38] Subsequently, extensive studies in cell and animal models have focused on determining the minimally functional domain structure required. These studies have led to the development of synthetically miniaturized dystrophin isoforms, or microdystrophins, designed to be compatible with the 5 kb limit for packaging within adeno-associated virus (AAV) genomes (Fig. 2). These synthetic isoforms typically comprise deletions within the central rod domain and C-terminal domain; three of which have been shown to improve muscle integrity and function in dystrophic animal models and are currently in clinical trials (Fig. 2).[39–41]

1.4 Dystrophinopathies

DMD is categorized as the most severe within a spectrum of disorders caused by mutations in the *DMD* gene called dystrophinopathies. DMD is marked by progressive muscle degeneration and weakness that, without corticosteroid therapy, progresses to a loss of ambulation by 12 years of age.[1–3] Once in a wheelchair, patients experience steep declines in cardiac and respiratory functions, which ultimately lead to fatality in early adulthood.[1–3] Becker muscular dystrophy (BMD) is the least severe, historically characterized by retention of ambulation beyond 12 years. Large phenotypic variability is observed among BMD patients with some losing ambulation in their teens

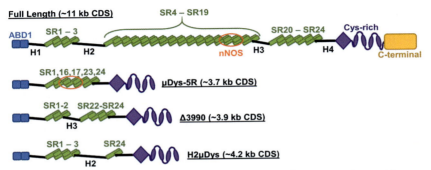

Fig. 2 Domain structures of full-length dystrophin and three microdystrophins currently in clinical trials.

and others retaining the ability to walk into the eighth decade of life.[1-3] The severity of dystrophinopathies generally tracks with the level of dystrophin protein expression in muscle fibers.[2,42-46] DMD is usually caused by mutations that result in premature translation termination, such as a nonsense mutation or by a shift in the reading frame, that ablate dystrophin expression. BMD is typically associated with retention of an open reading frame and expression of mutant dystrophin proteins with diminished accumulation, stability, and/or function. The standard clinical therapy for boys with DMD consists of daily or intermittent (e.g., weekends only) dosing of corticosteroids, which has been shown to prolong ambulation by up to 3 years and also delay the need for ventilatory support.[1-3,47] After loss of ambulation, therapeutic attention shifts to cardiac function and, with increasing age, palliative care.

The most common pathogenic *DMD* mutation class, affecting nearly two-thirds of patients, are large deletions of one or more exons, likely resulting from unequal crossing-over events between regions within the large *DMD* introns and generally involving segments hundreds of kilobases in length.[45,48] Large exonic duplications account for around 10% of patients, small insertions or deletions (indels) another 10%, and about 15% of patients have nonsense mutations.[45,48] The remaining approximately 5% patients have mutations in exons or introns that affect proper splicing resulting in exclusion of an exon or inclusion of intronic sequences in mature transcripts (i.e., pseudoexons).[45,48] Importantly, variable residual protein expression may be observed in patients carrying truncating mutations, likely attributed to low-levels of natural exon skipping during splicing that reframe the coding sequence, which may contribute partly to the minor phenotypic variability among DMD patients.[42] Phenotypic variability among BMD patients is greater and multifactorial, affected by mutant transcript splicing efficiency and/or stability or functionality/stability of in-frame mutant dystrophin protein isoforms.[43,44] The rate of disease progression is difficult to infer from genotype alone but can be more accurately inferred by evaluation of muscle histopathology, dystrophin protein expression, ambulation, and cardiac function.[1,2,49]

The correlation between clinical severity and the level of dystrophin protein expression must be approached with caution because of the wide variation in methods of quantifying dystrophin expression used in studies over many years and many different labs, but general conclusions may be made. An early study suggested that patients with severe DMD or cardiomyopathies typically have dystrophin protein levels lower than 10% of normal.[50] Other studies suggest that levels of 20% of normal may provide significant protection from pathology while 30–60% may be enough to avoid muscular dystrophy altogether.[42–44] Correlative studies are further confounded by the variability of the mutated dystrophin expressed in BMD patient; a level as low as 15% of a nearly normal 412 kDa isoform was compatible with minimal symptoms and ambulation into the eighth decade, and levels as low as 3% may significantly ameliorate the phenotype.[51–53] In contrast, studies in mdx mice have suggested as little as 15% of near-full-length dystrophin is enough to delay heart failure and restore normal skeletal muscle physiology.[46,54] This discrepancy may be due to differences between mouse and human physiology or the challenges of accurately measuring dystrophin protein levels by western blotting. More studies are therefore needed to fully elucidate the relationship between dystrophin protein levels and dystrophic phenotypes.

2. First-generation gene therapies for DMD
2.1 Overview

A highly compelling therapeutic solution to genetic diseases caused by loss-of-function mutations is virus-mediated gene replacement therapy, wherein a functional copy, or a cDNA, of the affected gene is inserted into the cells of a patient's affected tissue with a non-pathogenic virus like adeno-associated virus (AAV). Gene replacement therapy has had success in several vectorized therapies for genetic diseases such as congenital blindness and spinal muscular atrophy.[55,56] However, the long coding sequence of *DMD* remains incompatible with the small packaging limit of the current viral vectors for gene therapy—less than 5 kb for AAV genomes—necessitating internal deletion mutants of dystrophin for clinical translation (Fig. 2).[57] An alternative approach is to use antisense oligonucleotides (AONs) complementary to the *DMD* primary transcript at intron/exon junctions to promote exon skipping and re-establish an open reading frame in the context of truncating mutations (Fig. 3). Both approaches have been vectorized with AAV and comprise the first-generation of DMD gene therapies. Importantly, despite their small packaging limit, AAVs are non-pathogenic, essentially non-integrating, and exhibit long-term stability in slowly dividing or non-dividing cells persisting as extrachromosomal elements (ECEs). Moreover, several AAV serotypes exhibit strong muscle- and heart-specific tropism and high transduction efficiency with systemic delivery.[58–60] Thus, AAV will likely remain the vector of choice for DMD gene therapies for the immediate future.

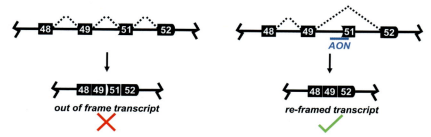

Fig. 3 AON-mediated exon 51 skipping to reframe the DMD gene in the context of an exon 50 deletion mutation. (Left side) Splicing (dotted lines) of exon 49 to exon 51 results in a frameshift in the dystrophin coding sequence and loss of dystrophin expression (red X). (Right side) Blocking exon 51 splicing with an AON (blue) during pre-mRNA maturation (arrow) leads to splicing of exon 49 to exon 52 restoring the open reading frame.

2.2 Gene replacement: Miniaturized dystrophins

Although it is not the main focus of this chapter, the progression of vectorized DMD gene replacement therapy with engineered miniaturized dystrophin isoforms provides context for the design of recent cutting-edge gene modulation and editing for DMD. The quest for a miniaturized dystrophin gene for AAV-mediated gene replacement therapy was inspired by two patients harboring deletion of exons 17–48 (Δ17–48 mini-dystrophin, Fig. 2), corresponding to SR4–SR19 and H3, who were described as ambulant at age 61 and a body-builder at age 25, respectively.[38] Although this 6.2 kb Δ17–48 mini-dystrophin is still too large for AAV, it gave clinical evidence that very large internal deletion mutant isoforms of dystrophin can provide significant protection to skeletal and cardiac muscles throughout life. Thus, researchers have put in exhaustive efforts to determine the minimal domain structure for dystrophin function with a coding sequence size limitation of around 4 kb (i.e., microdystrophin). The first microdystrophin was described in the late 1990s and lacked H2, H3, and all but one SR, but did not ameliorate the DMD phenotype in dystrophic mice.[61–63] Subsequent studies of SR structures and molecular biology of dystrophin found that a larger rod domain was required, with four or five SRs, but that the C-terminal domain was dispensable for the role of dystrophin in muscle integrity.[39,41] Moreover, the "phasing" of the SRs with regard to alternating short and long helices (Fig. 1A), which are not synchronous with DMD exon phasing (Fig. 1B), was found to be critical for dystrophin function in patients.[64–66] As summarized elsewhere, microdystrophin studies have culminated in the discovery of more than 30 different isoforms capable of supporting muscle integrity in vivo.[63] Three isoforms, named μDys-5R (3.7 kb CDS), Δ3990 (3.9 kb CDS), and H2μDys (4.2 kb CDS) (Fig. 2), have advanced to ongoing clinical trials (ClinicalTrials.gov NCT04626674, NCT03368742, NCT04281485, NCT03769116, NCT03362502, and NCT03375164).[39–41] Although very promising, these microdystrophins are synthetic dystrophin isoforms that have never been found to naturally occur in humans; their long-term benefits are thus unknown, and research continues to focus on developing therapies that can result in more predictable benefits based on expression of dystrophin isoforms equivalent to those found in patients with in-frame mutations and mild BMD.

2.3 Exon skipping with antisense oligonucleotides (AONs)

Although microdystrophin gene therapy shows promise, the first FDA-approved genetic therapies were directed toward expression of near-full-length dystrophin using AONs to induce exon skipping and restore an open reading frame. Unlike DNA AONs, which mediated transcript down-regulation through RNase H-dependent degradation of hybrid DNA-RNA duplexes, RNA (or modified RNA) AONs can mediated exon skipping during splicing via hybridizing to splice recognition or regulatory elements of targeted exons in primary transcripts, blocking their recognition by the splicing machinery, and leading to exclusion in mature transcripts (Fig. 3). Therapies developed with this technology can theoretically result in expression of Becker-like dystrophin isoforms in patients with mutations compatible with restoration of an open reading frame by skipping of a single exon (estimated ~70% of patients).[45,67] Three such drugs, utilizing phosphorodiamidate morpholino oligomer (PMO) chemistries, have been approved by the FDA for use in DMD. Eteplirsen targets exon 51, skipping of which would therapeutically reframe the *DMD* transcript in about 13% of patients, whereas golodirsen and viltolarsen both target exon 53, potentially therapeutic for around 8–10% of patients.[67] Despite evidence of clinical benefits, these drugs have been shown to induce only very modest dystrophin expression in clinical trials.[68–70] For example, by western blotting, treatment with eteplirsen for 48 weeks resulted in an increase of mean dystrophin levels from 0.16% of normal to 0.44%, a significant but relatively small increase.[71] Golodirsen treatment over the same period resulted in a similar small, but significant, increase in mean dystrophin from 0.095 to 1.019% of normal levels.[72] Viltolarsen utilizes a slightly shorter target sequence than golodirsen, and was reported to result in a higher degree of protein expression, with an increase in the mean level in the high dose group of 0.6–5.7% of normal at 25 weeks post-treatment.[73,74] Notably, these targeted exons reside within the central rod domain and thus phasing of spectrin repeats that remain after exon skipping depends upon the patient's underlying mutation (Fig. 1B). This may be predicted to result in variable outcomes, although in contrast to microdystrophin therapies one may reasonably expect that the existence of BMD patients carrying the equivalent mutations may help to predict the maximum limit of function that might result from exon skipping.[75]

2.4 Vectorized exon-skipping with modified U7 small nuclear RNAs

In addition to the relative limited degree of protein expression, AON therapies currently involve a burdensome treatment regimen of weekly systemic injections throughout life. These barriers prompted research into AAV-vectorized AONs, the most successful example of which utilizes modified U7 small nuclear RNAs (snRNAs). U7 snRNA is a 62 nucleotide transcript that normally functions during histone pre-mRNA processing via antisense binding to a downstream element on histone primary transcripts which recruits cleavage factors involved in maturation of the pre-mRNA 3′ end.[76,77] The recruitment of these cleavage factors is mediated via a Smith antigen (Sm) protein binding sequence in the middle of the U7 snRNA (5′-AAUUUGUCUAG-3′) which differs significantly from the consensus Sm binding sequence (smOPT, 5′-AAUUUUUGGAG-3′) of other snRNAs such as the core spliceosomal snRNAs U1, U2, U4, and U5.[78,79] The Sm binding sequence in spliceosomal snRNAs recruits the seven standard Sm proteins B/B′, D3, D1, D2, E, F, and G. However, the U7 Sm sequence recruits Lsm 10 and Lsm 11 in place of Sm D1 and Sm D2, which mediate U7 snRNA-specific interactions with other proteins bound to the histone pre-mRNA.[78–80] Mutation of the U7 snRNA Sm binding sequence into smOPT sequence (smOPT U7 snRNA) ablates its function in histone pre-mRNA processing, increases its stability, and enables localization to regions of active splicing in the nucleus.[79,81,82] Further modification of the snRNA 5′ end, which normally harbors an eight nucleotide antisense sequence corresponding to the histone pre-mRNA downstream element, can be used to target the smOPT U7 snRNA (Fig. 4) to bind other transcripts.[82] These modified smOPT U7 snRNAs can be used to target intronic or exonic splice recognition or regulatory elements of primary transcripts to

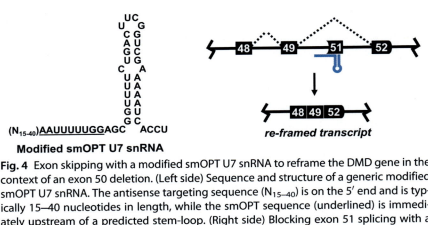

Fig. 4 Exon skipping with a modified smOPT U7 snRNA to reframe the DMD gene in the context of an exon 50 deletion. (Left side) Sequence and structure of a generic modified smOPT U7 snRNA. The antisense targeting sequence (N_{15-40}) is on the 5′ end and is typically 15–40 nucleotides in length, while the smOPT sequence (underlined) is immediately upstream of a predicted stem-loop. (Right side) Blocking exon 51 splicing with a modified smOPT U7 snRNA (blue) during pre-mRNA maturation (arrow) leads to splicing of exon 49 to exon 52 restoring the open reading frame.

block binding of native splice machinery and induce exon skipping during pre-mRNA splicing (Fig. 4).[82–84]

The earliest proof-of-concept study used retrovirus to stably express smOPT U7 constructs targeting the splice junctions of exon 51 in myoblasts from a DMD patient with a deletion of exons 48–50, resulting in efficient (30–40%) exclusion of exon 51 in mRNA.[83] Subsequently, AAV-vectorized smOPT U7 snRNA was shown to induce skipping of *DMD* exon 23 mdx mice, with greater than 60% skipping of exon 23, along with expression of dystrophin at levels of 50–80% of normal in the tibialis anterior muscles 4 weeks after intramuscular injection.[85] This high efficiency enables multiplexing of modified smOPT U7 snRNAs for multiple exon skipping and has been demonstrated for a DMD mutation not amenable to single-exon skipping. The golden retriever DMD model (GRDM) carries a single-point mutation within the splice acceptor site of intron 7, causing either exclusion of exon 7 from or inclusion of intron 6 into the mature transcript.[86] This mutation cannot be corrected by single exon skipping, but multiplexing of modified smOPT U7 snRNAs to skip both exons 6 and 8 can lead to production of an in-frame transcript by splicing of exon 5 to exon 9.[87] Local forelimb injections of AAV serotype 1 encoding two U7 snRNA cassettes targeting exons 6 and 8 led to significant expression (up to ∼25% of normal) of dystrophin throughout the muscles (up to ∼80% of fibers) in injected limbs, albeit the levels of dystrophin restoration were approximately two- to threefold lower than those demonstrated with smOPT U7 snRNA-mediated single exon skipping in mice.[85,87]

These results suggest that in vivo efficiency of multiplexed exon skipping may rapidly drop-off with increasing numbers of targeted exons. In the Dup2 mouse model, which carries an exon 2 duplication (the most common exon duplication in DMD patients), delivery of an AAV serotype 9 vector carrying four copies of modified smOPT U7 snRNAs targeting exon 2 splice acceptor and donor site results in efficient exon skipping, restoring two therapeutic transcripts: wild-type, and an exon 2-deleted transcript that is translated via an exon 5 internal ribosome entry site into a highly functional dystrophin isoform.[23] Vectorized exon skipping is in fact close behind gene replacement therapy in clinical development, and a small open-label Phase I/IIa clinical trial of AAV-vectorized modified smOPT U7 snRNAs was initiated in 2020 in boys with exon 2 duplications (ClinicalTrials.gov NCT04240314).

3. Second-generation approaches to DMD
3.1 Overview

RNA splice modulation approaches like exon skipping are transient therapies in that they require constant expression of the modified U7 snRNAs from AAV genomes in muscle cells. While it is widely accepted that AAV genomes persist in cells as ECEs, the long-term durability of AAV ECEs in muscle cells is not well-established. In the case of U7 snRNAs, evidence from 5 years of follow-up after local forelimb injection in GRDM dogs demonstrated that the AAV genomes were rapidly depleted within the first year after treatment and virtually absent by the third year despite a lack of apparent immune response.[87] The authors attributed this loss of AAV genomes to a lack of wild-type muscle integrity bestowed by the resultant Becker-like dystrophin lacking exons 6–8. It remains unclear whether the persistence of AAV genomes in muscle cells is robust enough for life-long benefits to be achieved by transient therapies like gene replacement and exon skipping. Furthermore, re-dosing of AAV therapies is expected to be significantly hindered by immunity against AAV capsids.[88] Thus, the second generation of DMD gene therapies in development are focused on permanent modification of the *DMD* gene using gene editing (Fig. 5).

The concept of gene editing developed over several decades in the mid-20th century alongside the discovery and dissection of mechanisms of DNA repair. As reviewed elsewhere, several technologies have emerged which mediate gene editing by inducing a DNA damage repair response at a specific genomic location via endonucleolytic cleavage of the DNA backbone including zinc-finger nucleases, meganucleases, and transcription-

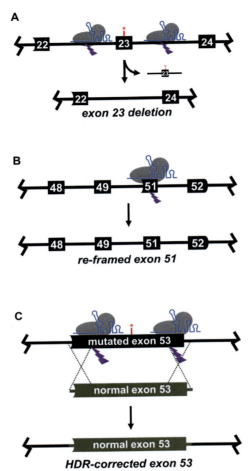

Fig. 5 Gene editing strategies for DMD. (A) Multiplex editing for exon 23 deletion in the context of the mdx mouse antisense mutation (red asterisk). Cas9 (gray) is guided to genomic DNA target sites in intron 22 and intron 23 by two different gRNAs (blue). Simultaneous cleavage of the target sites by Cas9 (purple bolt) followed by DNA repair (arrow) leads to removal of exon 23 and restoration of dystrophin expression. (B) Single-cut gene editing for exon 50 skipping in the context of an exon 50 deletion mutation. Cas9 is guided to a genomic target site within exon 51 by a gRNAs (blue). Cleavage of the target site followed by DNA repair (arrow) leads to indels that reframe exon 51 and restore an open reading frame. (C) Homology-directed gene repair for *DMD* mutation correction in the context of a nonsense mutation in exon 53 (red asterisk). Cas9 is guided to genomic DNA target sites in exon 53 by two different gRNAs and a homologous donor DNA encoding normal exon 53 is provided in trans (green). Simultaneous cleavage of the target sites by Cas9 activates DNA repair. Homology between the donor DNA and the genomic DNA (dotted-line X's) leads to HDR (arrow) and drives knock-in of normal exon 53 in place of mutated exon 53.

activator-like effector nucleases.[89] These systems share a mechanism of protein-mediated targeted sequence recognition, requiring significant protein engineering efforts to program a new target. CRISPR-Cas9 was realized as a gene editing technology in 2012 and greatly simplified the gene editing workflow, enabling laboratories across the world to successfully integrate gene editing into their research.[90–93] Briefly, CRISPR-Cas9 gene editing is primarily mediated by the DNA repair mechanisms of cells after cleavage of a DNA sequence by the Cas9 enzyme, targeted to a specific sequence with homology to a small, easily-reprogrammed guide RNA (gRNA).[90,92] For gene knock-out, researchers typically target the coding sequence of a gene, resulting in frameshifting indel mutations that ablate expression.[91–93] However, two gRNAs can be used together (i.e., multiplexed) to result in more elegant and specific genetic modifications.[92]

Inspired by the success of exon skipping, vectorized gene editing for DMD has primarily focused on permanently installing exon deletion or exon-skipping mutations at the genomic DNA level. Two distinct methods have been described for gene editing for *DMD* exon skipping which rely on either multiplexed genomic DNA cleavage to mediate exon deletion (Fig. 5A), or single-cut gene editing to generate indels that either ablate exon recognition or reframe an exon (Fig. 5B). While exon deletion or skipping can theoretically result in full-length dystrophin expression in patients with duplications, for other classes of mutations (including, most commonly, deletions) these would result in Becker-like dystrophin isoforms; instead, DNA knock-in would be required to restore full-length dystrophin expression in these patients. The canonical method of knock-in gene editing takes advantage of the homology directed repair pathway (HDR) in cells (Fig. 5C).[92] Few strategies harnessing HDR have been tested in DMD models and, as discussed in more detail below, the low efficiency of HDR in muscle tissue remains a barrier to development of knock-in gene editing for DMD.

3.2 Multiplex gene editing for exon deletion

Successful CRISPR-Cas9 editing of the *DMD* locus in human cells was first demonstrated in vitro in patient-derived immortalized myoblasts in which a deletion of exons 45–55 (del44–55) was induced, corresponding to a patient mutation typically associated with a very mild BMD phenotype.[94,95] Evidence for the feasibility of in vivo editing was subsequently demonstrated by the simultaneous publication by three independent groups showing that AAV-vectorized multiplex gene editing could delete exons (Fig. 5A) to correct *DMD* mutations in live mice.[96–98] Two of the three groups evaluated both local and systemic injections of AAVs encoding Cas9 and gRNAs targeting sequences adjacent to exon 23 in mdx mice to delete exon 23 and restore near full-length dystrophin expression.[96,97] It is difficult to directly compare their results as different methods of systemic delivery and dystrophin protein quantitation were used, but based on limited western blot analyses, dystrophin protein restoration reached more than 10% of normal levels in skeletal and cardiac muscles following systemic AAV delivery. Moreover, all three studies showed significant improvements in other hallmarks of DMD, including measures of muscle physiology (strength and force), histopathology, and serum CK levels.[96–98] Each group reported exon deletion efficiency based on analysis of transcripts, but only one group assessed the efficiency of genomic level editing and its correlation to dystrophin restoration; they showed that even low levels (less than 3%) of genomic DNA editing could restore significant levels of dystrophin protein (5–10% of normal) in approximately 60% of fibers throughout injected muscles, which lends confidence to further evaluation of systemic gene editing as a therapy for DMD.[98]

Although these results are promising, the direct applicability of these vectors to humans is limited, as the mdx mouse (which carries a nonsense mutation in exon 23) is not genetically representative of DMD patients, in whom mutations are clustered in hot spots across the gene. Around 30% have mutations in the central rod domain hotspot, between exons 45–55, and exon 23 mutations in boys with DMD are extremely rare; thus a therapy to correct a deletion of exon 23 is not widely applicable.[45,48] A mouse model more relevant to human gene corrective therapies contains the entire human *DMD* locus on an mdx mutation background, in which a deletion of human *DMD* exon 45 was introduced; correction of the humanized locus results in expression of dystrophin, providing a powerful model for clinical development.[99] Electroporation-mediated delivery of plasmids encoding CRISPR-Cas9 with gRNAs to simultaneously target intron 44 and intron 55 to delete more than 340 kb of genomic DNA, resulting in the del45–55 genotype that would be predicted to be therapeutic for nearly one-third of DMD patients.[45,48,99] Although vectorized multiplex gene editing is promising, there are caveats to this approach. First, the use of multiple gRNAs raises the risks of editing at unintended sites (i.e., off-target effects). Second, there is a propensity for large rearrangements, translocations, and other complex aberrant outcomes with large genomic

DNA deletions.[100] Finally, the lack of conservation of human intronic sequences targeted by gRNAs with intronic sequence in other animal models besides the humanized mouse may impair the translatability of gene editing therapies into the clinic.

3.3 Single-cut gene editing for exon skipping and reframing

The single-cut strategy (Fig. 5B) for *DMD* exon deletion and/or reframing overcomes some of the limitations of multiplex gene editing. By targeting within an exon coding sequence, indels can be generated that either (i) result in inactivating mutations of splice elements or (ii) reframe the exon (Fig. 5B).[101,102] In the context of exon deletion mutations, such editing can result in a reframed *DMD* gene and expression of near-full-length Becker-like dystrophin isoforms. Single-cut genome editing limits the likelihood of off-target DNA cleavage, large rearrangements, and increases translatability due to the greater conservation between dystrophin coding sequences of humans and research animals. In the seminal paper using this approach, performed in a mouse model with a deletion of *Dmd* exon 51, intramuscular injection of two AAV serotype 9 vectors encoding Cas9 and an exon 51-targeting gRNA resulted in dystrophin expression in virtually all fibers of injected muscles, and the levels of near-full-length dystrophin expression mirrored those of healthy control animals.[102] Importantly, they found that although a fraction of editing outcomes resulted in exon skipping, the primary therapeutic outcome was a single nucleotide insertion at the Cas9 cut site within exon 51 that reframed exon 51 putting it in phase with exons 49 and 52 as in Fig. 5B.[102] This approach has shown excellent efficiency in reframing the *DMD* gene for a handful of different mutations in mice and human cells including multiexon deletions or duplications.[103–107] It has also been applied to deep intronic mutations that result in activation of cryptic or creation of novel splice definition signals that result in out-of-frame pseudoexons; correction of these result in expression of full length dystrophin.[103]

Among the various gene editing approaches for DMD, this approach is the furthest along in pre-clinical proof-of-concept studies and expected to be among the first of a few AAV-vectorized systemic in vivo gene editing therapies for genetic diseases. In support of clinical applicability, single-cut editing was tested in a dog model of DMD harboring a spontaneous exon 50 deletion mutation (DEx50).[108] Four DEx50 dogs were tested; two received intramuscular injections and two received system injections of the same two AAV serotype 9 vectors encoding Cas9 and a gRNA to target exon 51 as was used in the mouse studies.[108] For each group of two dogs, one received a total dose of 4×10^{13} and the other 2×10^{14} vector genomes per kilogram weight (vg/kg). At the highest dose of 2×10^{14} vg/kg, restoration of near-full-length dystrophin to levels similar as those in wild-type dogs were detected in heart (92%), diaphragm (58%), biceps (64%), and triceps (25%) 8 weeks after systemic injection.[108] Gene editing at the Cas9 target site in exon 51 mirrored that of the mouse studies, with the primary therapeutic outcome being a single adenosine base insertion that reframed exon 51.[102,108] Interestingly, therapeutic gene editing efficiency at the genomic level was less than 5% in most tissues, reinforcing the notion that low levels of editing can result in therapeutic levels of dystrophin expression.[108] Pathology was greatly ameliorated by the gene editing treatment, evidenced by a major reduction in necrotic fibers and fibrosis compared to untreated DEx50 dogs.[108] Toxicity was virtually absent based on hematology counts, serum electrolytes, as well as liver and kidney function blood tests, suggesting that gene editing to reframe or skip exon 51 is potentially safe and poised for clinical trials.[108]

3.4 Homology-directed gene repair for DMD mutation correction

In contrast to gene editing approaches directed toward restoring an open reading frame to express a Becker-like dystrophin protein, an alternate approach focuses on restoring full-length dystrophin expression. As the most common DMD mutations are large exonic deletions, only DNA knock-in gene editing can achieve full-length expression in most patients. The canonical HDR-mediated method of knock-in gene editing with CRISPR-Cas9 begins with a targeted double-stranded break at the desired knock-in location (Fig. 5C).[92] Providing an excess of donor DNA in the form of a single-stranded oligodeoxynucleotides (ssODNs) or a double-stranded DNA molecules, flanked by homologous segments corresponding to the regions near the cut site (tens to thousands of bases in length) leads to donor-mediated repair and installation of precise changes as well as small and large knock-ins.[92] As reviewed elsewhere, HDR was one of the first approaches to be tested for DMD gene correction after the inception of CRISPR/Cas9 gene editing.[109] Primarily, these studies have not addressed systemic somatic cell gene editing, but rather have used HDR to correct DMD mutations in either (i) mouse germline cells or (ii) patient-derived stem cells followed by transplantation into dystrophic mouse muscles.[101,110–112] Importantly, embryonic DMD correction is not translatable to the clinic as germline gene editing is considered unethical and may be more dangerous than somatic cell gene editing by installing heritable defects in treated individuals.[113,114] Although stem cell transplantation remains an active area of research as a potential DMD therapy,[115] it is not likely with current technologies that local injections of modified muscle stem cells can achieve the levels of body-wide dystrophin expression required for meaningful therapeutic benefits for DMD, especially in the heart and diaphragm.

Only one study has explored vectorized postnatal somatic cell HDR in mice, using intramuscular injections of AAV serotype 6 to deliver Cas9 and two gRNAs flanking a nonsense mutation within exon 53 (the mdx^{4cv} mouse allele) as well as a donor cassette carrying the wild-type exon 53 sequence.[116] Following concomitant cleavage of the two exon 53 target sites by Cas9 in transduced cells, the donor molecule can be used during HDR to repair exon 53 (Fig. 5C).[116] Although only a single Cas9 cut site is required to induce DNA repair and HDR, cleavage on either side of the exon 53 mutation enabled an in-frame deletion to result from NHEJ in the absence of HDR repair, maximizing the therapeutic potential of non-HDR editing outcomes.[116] Efficiency was substantially lower than that observed for multiplex and single-target exon deletion or reframing in mice. Only 0.18% of total genomes in injected muscles exhibited HDR-mediated editing and led to normal exon 53 in only 0.8% of total dystrophin transcripts 4 weeks post-injection.[116] As expected based on single-target exon deletion studies, some near-full-length dystrophin protein restoration occurred due to serendipitous deletion or reframing of exon 53.[116] However, dystrophin protein levels were too low (less than 5%) for meaningful benefits in muscle function.[116] This result is consistent with the observation that HDR is generally restricted to mitotic cells and virtually inactive in post-mitotic tissues such as skeletal muscle.[117,118] Precise modifications to elicit specific changes or replace missing elements via DNA knock-in remains a significant challenge for future somatic gene editing therapies. However, an alternative method of DNA knock-in, termed homology-independent targeted integrated (HITI), was recently described and takes advantage of NHEJ for DNA knock-in instead of HDR. HITI has not yet been established as an in vivo gene editing knock-in strategy to correct DMD gene mutations, but has demonstrated robust knock-in efficiency in mouse muscle tissue when vectorized with AAV and used systemically in neonatal mice.[119]

4. Challenges for current and future DMD therapies

4.1 Overview

Despite robust performance during pre-clinical development in cells and animals, several aspects of the current experimental therapies limit their clinical applicability as well as make long-term safety and efficacy uncertain. The primary barrier is patient immunity against transgenes, rescued dystrophin isoforms, and vector capsids.[120–122] However, several strategies to circumvent or suppress immunity provide confidence to continued development of these therapies.[123] Additionally, it is unclear how long AAV episomes remain in muscle and heart tissues which raises the question of if and when re-treatment will be required.[87,124] Gene editing could circumvent the issue of AAV durability by correction of the genetic cause of DMD. However, researchers and regulatory agencies remain sensibly cautious in moving gene editing technologies into clinical trials and continue to explore routes to reduce the risks unintended genetic consequences. The next-generation of gene editing and modulation therapies will be aimed at overcoming these challenges and exploring new technologies to restore dystrophin expression in greater numbers of patients.

4.2 Immunity issues

In common with all AAV-based therapies, environmentally-acquired antibodies to AAV capsid proteins—present in as much as 30–50% of the general population—have circulating neutralizing antibodies against AAVs, which may limit efficacy of AAV therapy.[122,125] Gene therapy itself induces significant antibody responses that may preclude any repeat dosing. This problem is not unique to DMD gene therapies and many strategies are being explored to overcome this barrier such as modification of transgenes or vectors to evade immune response as well as temporary suppression or depletion of a patient's immune system.[123] An additional concern is the development of immune responses to the newly expressed gene product. In the case of DMD, T-cells primed for dystrophin epitopes have been identified; the role they play in disease pathogenesis is unclear, but their presence suggests the possibility that expression of transgene-encoded epitopes may encounter pre-existing immunity.[126] This was demonstrated by the first AAV-vectorized miniaturized dystrophin gene replacement clinical trial in which transgene expression was virtually absent despite evidence of significant transduction, a finding attributed in part to T cell mediated responses to dystrophin peptides.[120,121] This highlights a significant challenge to current gene editing and modulation therapies in which novel epitopes may be generated via exon skipping or reframing. Moreover, replacement of missing or aberrant exons via DNA knock-in may result in immune response to natural dystrophin epitopes not previously present in the patient.

CRISPR-Cas9 also presents specific immunity-related challenges for clinical development. Cas9 is a bacterial protein and the isoforms used in gene editing are usually from pathogenic bacteria like *S. pyogenes*, *S. aureus*, and *C. jejuni*. As reviewed elsewhere, pre-existing immunity in the form of antibodies and/or reactive T-cells against Cas9 may be present in up to approximately 80% of adults based on some reports.[127–129] However, others have estimated the prevalence of pre-existing Cas9 immunity to be 10% or less in adult populations.[127,130,131] Moreover, it is unclear what the prevalence of pre-existing Cas9 immunity is within pediatric populations (including DMD patients) or how pre-existing Cas9 immunity affects the safety and efficacy of gene editing therapies. Future research should aim to answer these questions and better illuminate the potential challenges of immunity against Cas9.

4.3 Durability of adeno-associated virus episomes

In general, the durability of expression from AAV-associated episomes in non-dividing tissues is presumed to be significant; for example, a long-term follow up study with AAV for treatment of severe hemophilia B demonstrated persistence of AAV episomes in muscle tissue 10 years after local administration.[124] However, a long-term DMD animal study suggested that AAV genome depletion occurs rapidly within 1 year after administration, at least in some muscles, and may be virtually absent after 3 years, suggesting that degeneration of dystrophic muscles may contribute to rapid AAV genome depletion.[87] For transient gene replacement and modulation, this presents the potential need for re-administration of the vectorized therapy, which may not be possible due to the immune responses following initial treatment. However, for gene editing therapies, loss or silencing of episomes may be beneficial as long-term expression of Cas9 and gRNAs could lead to accumulation of off-target, pathogenic mutations and thus depletion of AAV genomes after administration could be beneficial for safety. However, muscle fibers are repaired throughout life by asymmetric expansion of satellite cells from their typical quiescent state to create new myoblasts that fuse to form new myofibers. Limited research is available in the literature for the efficiency of AAV-mediated CRISPR-Cas9 gene editing in muscle progenitor cells, but recent studies showed that AAV serotype 9 can transduce satellite cells in mouse skeletal muscles and that satellite cells are amenable to AAV-vectorized Cas9 gene editing.[132,133] It was found that approximately 60% of satellite cells in mdx mouse tibialis anterior muscles were transduced following local injections of AAV serotype 9 carrying Cre recombinase. Local injection of AAV serotype 9 vectors encoding Cas9 and gRNAs to delete exon 23 resulted in editing of 0.01–1% of satellite cell genomes which was approximately 10-fold lower than that measured in the treated whole skeletal muscle tissues.[132,133] These results reinforce confidence in the durability of gene editing in skeletal muscles and highlight a potential avenue of optimization for future development of gene editing therapeutics for DMD.

4.4 Safety

Many systemic AAV-based therapies are now approved or in clinical trials and have generally shown a reasonable safety profile.[134] However, studies in large animals have suggested high-dose toxicity, especially in the liver.[135] Pre-existing liver dysfunction may play an important role in AAV toxicity as has been suggested recently following the tragic deaths of three pediatric participants in the high dose cohort of a small (24 participants) clinical trial of AAV gene replacement therapy for X-linked myotubular myopathy (ClinicalTrials.gov NCT03199469), a muscle disease that is typically fatal early in childhood and correlated with liver disease.[136] In addition to these AAV-associated risks, gene editing presents additional safety risks, particularly regarding the potential for unintended editing outcomes like off-target gene editing.[137] While animal studies with Cas9 for DMD have not revealed significant editing at predicted off-target sites by next-generation sequencing, it is prudent to realize that the off-target editing landscape in humans is likely different from that of research animals due to their genetic divergence.[96–98,108] Importantly, integration of AAV genomes at Cas9 target sites was recently reported as an unintended on-target gene editing outcome during AAV-mediated CRISPR-Cas9 gene editing in mice.[138,139] In fact, AAV integration occurred at similar, usually higher, levels (up to ~5%) as the intended deletion (up to ~2.5%) in heart and skeletal muscles of mdx mice analyzed 8-weeks after systemic injection of AAVs encoding CRISPR-Cas9 for multiplex gene editing to delete exon 23.[139] Aside from potentially reducing the frequency of intended editing outcomes, the impact of AAV integration at Cas9 cut sites in the *DMD* gene is unknown. However, this should be monitored carefully in future studies as a potential source of genotoxicity. Intriguingly, AAV integration likely occurs through NHEJ between the linear AAV genomes and the free DNA ends at the genomic DNA Cas9 cut site suggesting that NHEJ-mediated DNA knock-in (e.g., HITI) may be an effective method of gene correction for DMD. Altogether, the first use of AAV-vectorized CRISPR-Cas9 for DMD will need to devise a proper route to ensuring safety regarding unintended gene editing outcomes. A major milestone for therapeutic gene editing occurred in November of 2020 when the first participant was injected with nanoparticle-vectorized CRISPR/Cas9 for systemic gene editing in a clinical trial for hereditary transthyretin amyloidosis (ClinicalTrials.gov NCT04601051). The gene editing research community eagerly awaits results from this first-in-human trial, which is sure to inform future gene editing therapies for DMD.

5. Conclusions

In only a few decades since the discovery of the *DMD* gene product near the end of the 1980s, great strides have been made in understanding the molecular basis of DMD pathology.[140] Clinical observations of genotype-phenotype correlations in dystrophinopathy patients and characterization of the relationship between dystrophin structure and function have been the major driving forces for development of DMD gene editing and modulation therapies. The first-generation of these therapies include now approved AON exon-skipping therapies as well as therapies in clinical trials, such as gene replacement with microdystrophins and transient exon skipping with AONs or AAV-vectorized modified smOPT U7 snRNAs. Several questions regarding many of these first-generation therapies remain unanswered, including their clinical benefits and life-long durability. The second generation of therapies for DMD, based on CRISPR-Cas9 gene editing, may overcome some of the limitations of microdystrophin and transient exon skipping but are still in the pre-clinical development stage. For these therapeutic approaches salient challenges remain, including a lack of significant efficacy, unknown long-term durability, and potential safety issues (primarily involving immunity to AAV or foreign transgenes). Ongoing clinical trials of gene replacement and splice alteration in DMD and of gene editing for other genetic diseases, will illuminate potential barriers and inform future efforts to overcome these barriers via next-generation DMD therapies. Advances and refinements in gene editing technologies, such as alternative routes for DNA knock-in and methods to circumvent patient immunity against Cas9, can be anticipated. And just as efforts at developing therapies for DMD have provided insights and uncovered barriers to gene splicing and replacement therapies for a wide variety of disorders, development of therapies for DMD will likely to continue to be a model for cutting-edge advancements in gene editing and modulation for other genetic diseases for years to come.

References

1. Flanigan KM. Duchenne and Becker muscular dystrophies. *Neurol Clin.* 2014;32 (3):671–688 [viii].
2. Koeks Z, Bladen CL, Salgado D, et al. Clinical outcomes in Duchenne muscular dystrophy: a study of 5345 patients from the TREAT-NMD DMD global database. *J Neuromuscul Dis.* 2017;4(4):293–306.
3. O'Brien KF, Kunkel LM. Dystrophin and muscular dystrophy: past, present, and future. *Mol Genet Metab.* 2001;74(1–2):75–88.
4. Ahn AH, Kunkel LM. The structural and functional diversity of dystrophin. *Nat Genet.* 1993;3(4):283–291.
5. Koenig M, Hoffman EP, Bertelson CJ, Monaco AP, Feener C, Kunkel LM. Complete cloning of the Duchenne muscular dystrophy (DMD) cDNA and preliminary genomic organization of the DMD gene in normal and affected individuals. *Cell.* 1987;50 (3):509–517.
6. Davies KE, Pearson PL, Harper PS, et al. Linkage analysis of two cloned DNA sequences flanking the Duchenne muscular dystrophy locus on the short arm of the human X chromosome. *Nucleic Acids Res.* 1983;11(8):2303–2312.
7. Boyd Y, Buckle V, Holt S, Munro E, Hunter D, Craig I. Muscular dystrophy in girls with X; autosome translocations. *J Med Genet.* 1986;23(6):484–490.
8. Gorecki DC, Monaco AP, Derry JM, Walker AP, Barnard EA, Barnard PJ. Expression of four alternative dystrophin transcripts in brain regions regulated by different promoters. *Hum Mol Genet.* 1992;1(7):505–510.
9. Holder E, Maeda M, Bies RD. Expression and regulation of the dystrophin Purkinje promoter in human skeletal muscle, heart, and brain. *Hum Genet.* 1996;97(2):232–239.
10. Boyce FM, Beggs AH, Feener C, Kunkel LM. Dystrophin is transcribed in brain from a distant upstream promoter. *Proc Natl Acad Sci USA.* 1991;88(4):1276–1280.
11. Kueh SL, Head SI, Morley JW. GABA(A) receptor expression and inhibitory post-synaptic currents in cerebellar Purkinje cells in dystrophin-deficient mdx mice. *Clin Exp Pharmacol Physiol.* 2008;35(2):207–210.
12. Snow WM, Anderson JE, Fry M. Regional and genotypic differences in intrinsic electrophysiological properties of cerebellar Purkinje neurons from wild-type and dystrophin-deficient mdx mice. *Neurobiol Learn Mem.* 2014;107:19–31.
13. Dsouza VN, Man NT, Morris GE, Karges W, Pillers DAM, Ray PN. A novel dystrophin isoform is required for normal retinal electrophysiology. *Hum Mol Genet.* 1995;4 (5):837–842.
14. Lederfein D, Levy Z, Augier N, et al. A 71-kilodalton protein is a major product of the Duchenne muscular-dystrophy gene in brain and other nonmuscle tissues. *Proc Natl Acad Sci USA.* 1992;89(12):5346–5350.
15. Austin RC, Howard PL, Dsouza VN, Klamut HJ, Ray PN. Cloning and characterization of alternatively spliced isoforms of Dp71. *Hum Mol Genet.* 1995;4(9):1475–1483.

16. Doorenweerd N, Mahfouz A, van Putten M, et al. Timing and localization of human dystrophin isoform expression provide insights into the cognitive phenotype of Duchenne muscular dystrophy. *Sci Rep*. 2017;7;12575.
17. Byers TJ, Lidov HG, Kunkel LM. An alternative dystrophin transcript specific to peripheral nerve. *Nat Genet*. 1993;4(1):77–81.
18. Lidov HGW, Selig S, Kunkel LM. Dp140—a novel 140-Kda Cns transcript from the Dystrophin locus. *Hum Mol Genet*. 1995;4(3):329–335.
19. Pillers DA, Bulman DE, Weleber RG, et al. Dystrophin expression in the human retina is required for normal function as defined by electroretinography. *Nat Genet*. 1993;4(1):82–86.
20. Drenckhahn D, Holbach M, Ness W, Schmitz F, Anderson LV. Dystrophin and the dystrophin-associated glycoprotein, beta-dystroglycan, co-localize in photoreceptor synaptic complexes of the human retina. *Neuroscience*. 1996;73(2):605–612.
21. Waite A, Brown SC, Blake DJ. The dystrophin-glycoprotein complex in brain development and disease. *Trends Neurosci*. 2012;35(8):487–496.
22. Tozawa T, Itoh K, Yaoi T, et al. The shortest isoform of dystrophin (Dp40) interacts with a group of presynaptic proteins to form a presumptive novel complex in the mouse brain. *Mol Neurobiol*. 2012;45(2):287–297.
23. Wein N, Vulin A, Falzarano MS, et al. Translation from a DMD exon 5 IRES results in a functional dystrophin isoform that attenuates dystrophinopathy in humans and mice. *Nat Med*. 2014;20(9):992–1000.
24. Chamova T, Guergueltcheva V, Raycheva M, et al. Association between loss of dp140 and cognitive impairment in duchenne and Becker dystrophies. *Balkan J Med Genet*. 2013;16(1):21–30.
25. Felisari G, Martinelli Boneschi F, Bardoni A, et al. Loss of Dp140 dystrophin isoform and intellectual impairment in Duchenne dystrophy. *Neurology*. 2000;55(4):559–564.
26. Bardoni A, Felisari G, Sironi M, et al. Loss of Dp140 regulatory sequences is associated with cognitive impairment in dystrophinopathies. *Neuromuscul Disord*. 2000;10(3):194–199.
27. Ervasti JM, Campbell KP. A role for the dystrophin-glycoprotein complex as a transmembrane linker between laminin and actin. *J Cell Biol*. 1993;122(4):809–823.
28. Delalande O, Molza AE, Dos Santos MR, et al. Dystrophin's central domain forms a complex filament that becomes disorganized by in-frame deletions. *J Biol Chem*. 2018;293(18):6637–6646.
29. Koenig M, Monaco AP, Kunkel LM. The complete sequence of dystrophin predicts a rod-shaped cytoskeletal protein. *Cell*. 1988;53(2):219–228.
30. Koenig M, Kunkel LM. Detailed analysis of the repeat domain of dystrophin reveals four potential hinge segments that may confer flexibility. *J Biol Chem*. 1990;265(8):4560–4566.
31. Amann KJ, Renley BA, Ervasti JM. A cluster of basic repeats in the dystrophin rod domain binds F-actin through an electrostatic interaction. *J Biol Chem*. 1998;273(43):28419–28423.
32. Prins KW, Humston JL, Mehta A, Tate V, Ralston E, Ervasti JM. Dystrophin is a microtubule-associated protein. *J Cell Biol*. 2009;186(3):363–369.
33. Belanto JJ, Mader TL, Eckhoff MD, et al. Microtubule binding distinguishes dystrophin from utrophin. *Proc Natl Acad Sci U S A*. 2014;111(15):5723–5728.
34. Lai Y, Thomas GD, Yue Y, et al. Dystrophins carrying spectrin-like repeats 16 and 17 anchor nNOS to the sarcolemma and enhance exercise performance in a mouse model of muscular dystrophy. *J Clin Invest*. 2009;119(3):624–635.
35. Rentschler S, Linn H, Deininger K, Bedford MT, Espanel X, Sudol M. The WW domain of dystrophin requires EF-hands region to interact with beta-dystroglycan. *Biol Chem*. 1999;380(4):431–442.

36. Sadoulet-Puccio HM, Rajala M, Kunkel LM. Dystrobrevin and dystrophin: an interaction through coiled-coil motifs. *Proc Natl Acad Sci U S A*. 1997;94(23):12413–12418.
37. Suzuki A, Yoshida M, Ozawa E. Mammalian alpha 1- and beta 1-syntrophin bind to the alternative splice-prone region of the dystrophin COOH terminus. *J Cell Biol*. 1995;128(3):373–381.
38. England SB, Nicholson LV, Johnson MA, et al. Very mild muscular dystrophy associated with the deletion of 46% of dystrophin. *Nature*. 1990;343(6254):180–182.
39. Harper SQ, Hauser MA, DelloRusso C, et al. Modular flexibility of dystrophin: implications for gene therapy of Duchenne muscular dystrophy. *Nat Med*. 2002;8(3):253–261.
40. Hakim CH, Wasala NB, Pan X, et al. A five-repeat micro-Dystrophin gene ameliorated dystrophic phenotype in the severe DBA/2J-mdx model of Duchenne muscular dystrophy. *Mol Ther Methods Clin Dev*. 2017;6:216–230.
41. Wang B, Li J, Xiao X. Adeno-associated virus vector carrying human minidystrophin genes effectively ameliorates muscular dystrophy in mdx mouse model. *Proc Natl Acad Sci U S A*. 2000;97(25):13714–13719.
42. Neri M, Torelli S, Brown S, et al. Dystrophin levels as low as 30% are sufficient to avoid muscular dystrophy in the human. *Neuromuscul Disord*. 2007;17(11–12):913–918.
43. Anthony K, Cirak S, Torelli S, et al. Dystrophin quantification and clinical correlations in Becker muscular dystrophy: Implications for clinical trials. *Brain*. 2011;134(Pt. 12):3547–3559.
44. Beggs AH, Hoffman EP, Snyder JR, et al. Exploring the molecular basis for variability among patients with Becker muscular dystrophy: dystrophin gene and protein studies. *Am J Hum Genet*. 1991;49(1):54–67.
45. Flanigan KM, Dunn DM, von Niederhausern A, et al. Mutational spectrum of DMD mutations in dystrophinopathy patients: application of modern diagnostic techniques to a large cohort. *Hum Mutat*. 2009;30(12):1657–1666.
46. Godfrey C, Muses S, McClorey G, et al. How much dystrophin is enough: the physiological consequences of different levels of dystrophin in the mdx mouse. *Hum Mol Genet*. 2015;24(15):4225–4237.
47. Guglieri M, Bushby K, McDermott MP, et al. Developing standardized corticosteroid treatment for Duchenne muscular dystrophy. *Contemp Clin Trials*. 2017;58:34–39.
48. Bladen CL, Salgado D, Monges S, et al. The TREAT-NMD DMD global database: analysis of more than 7,000 Duchenne muscular dystrophy mutations. *Hum Mutat*. 2015;36(4):395–402.
49. Dent KM, Dunn DM, von Niederhausern AC, et al. Improved molecular diagnosis of dystrophinopathies in an unselected clinical cohort. *Am J Med Genet A*. 2005;134(3):295–298.
50. Hoffman EP, Fischbeck KH, Brown RH, et al. Characterization of dystrophin in muscle-biopsy specimens from patients with Duchenne's or Becker's muscular dystrophy. *N Engl J Med*. 1988;318(21):1363–1368.
51. Gurvich OL, Maiti B, Weiss RB, Aggarwal G, Howard MT, Flanigan KM. DMD exon 1 truncating point mutations: amelioration of phenotype by alternative translation initiation in exon 6. *Hum Mutat*. 2009;30(4):633–640.
52. Flanigan KM, Dunn DM, von Niederhausern A, et al. DMD Trp3X nonsense mutation associated with a founder effect in North American families with mild Becker muscular dystrophy. *Neuromuscul Disord*. 2009;19(11):743–748.
53. Waldrop MA, Gumienny F, El Husayni S, Frank DE, Weiss RB, Flanigan KM. Low-level dystrophin expression attenuating the dystrophinopathy phenotype. *Neuromuscul Disord*. 2018;28(2):116–121.
54. van Putten M, van der Pijl EM, Hulsker M, et al. Low dystrophin levels in heart can delay heart failure in mdx mice. *J Mol Cell Cardiol*. 2014;69:17–23.

55. Pierce EA, Bennett J. The status of RPE65 gene therapy trials: safety and efficacy. *Cold Spring Harb Perspect Med.* 2015;5(9):a017285.
56. Meyer K, Ferraiuolo L, Schmelzer L, et al. Improving single injection CSF delivery of AAV9-mediated gene therapy for SMA: a dose-response study in mice and nonhuman primates. *Mol Ther.* 2015;23(3):477–487.
57. Wu Z, Yang H, Colosi P. Effect of genome size on AAV vector packaging. *Mol Ther.* 2010;18(1):80–86.
58. Daya S, Berns KI. Gene therapy using adeno-associated virus vectors. *Clin Microbiol Rev.* 2008;21(4):583–593.
59. Pacak CA, Mah CS, Thattaliyath BD, et al. Recombinant adeno-associated virus serotype 9 leads to preferential cardiac transduction in vivo. *Circ Res.* 2006;99(4):e3–e9.
60. Wang D, Zhong L, Nahid MA, Gao G. The potential of adeno-associated viral vectors for gene delivery to muscle tissue. *Expert Opin Drug Deliv.* 2014;11(3):345–364.
61. Yuasa K, Ishii A, Miyagoe Y, Takeda S. Introduction of rod-deleted dystrophin cDNA, delta DysM3, into mdx skeletal muscle using adenovirus vector [in Japanese]. *Nihon Rinsho.* 1997;55(12):3148–3153.
62. Takeda S. Development of new therapy on muscular dystrophy [in Japanese]. *Rinsho Shinkeigaku.* 2001;41(12):1154–1156.
63. Duan D, Systemic AAV. Micro-dystrophin gene therapy for Duchenne muscular dystrophy. *Mol Ther.* 2018;26(10):2337–2356.
64. Kaspar RW, Allen HD, Ray WC, et al. Analysis of dystrophin deletion mutations predicts age of cardiomyopathy onset in Becker muscular dystrophy. *Circ Cardiovasc Genet.* 2009;2(6):544–551.
65. Ruszczak C, Mirza A, Menhart N. Differential stabilities of alternative exon-skipped rod motifs of dystrophin. *Biochim Biophys Acta.* 2009;1794(6):921–928.
66. Gao QQ, McNally EM. The Dystrophin complex: structure, function, and implications for therapy. *Compr Physiol.* 2015;5(3):1223–1239.
67. Aartsma-Rus A, Fokkema I, Verschuuren J, et al. Theoretic applicability of antisense-mediated exon skipping for Duchenne muscular dystrophy mutations. *Hum Mutat.* 2009;30(3):293–299.
68. Alfano LN, Charleston JS, Connolly AM, et al. Long-term treatment with eteplirsen in nonambulatory patients with Duchenne muscular dystrophy. *Medicine (Baltimore).* 2019;98(26):e15858.
69. Mendell JR, Goemans N, Lowes LP, et al. Longitudinal effect of eteplirsen versus historical control on ambulation in Duchenne muscular dystrophy. *Ann Neurol.* 2016;79 (2):257–271.
70. Mendell JR, Rodino-Klapac LR, Sahenk Z, et al. Eteplirsen for the treatment of Duchenne muscular dystrophy. *Ann Neurol.* 2013;74(5):637–647.
71. EXONDYS 51. *Package Insert.* Sarepta Therapeutics, Inc.; 2020.
72. Frank DE, Schnell FJ, Akana C, et al. Increased dystrophin production with golodirsen in patients with Duchenne muscular dystrophy. *Neurology.* 2020;94(21):e2270–e2282.
73. Clemens PR, Rao VK, Connolly AM, et al. Safety, tolerability, and efficacy of viltolarsen in boys with Duchenne muscular dystrophy amenable to exon 53 skipping: a phase 2 randomized clinical trial. *JAMA Neurol.* 2020;77(8):982–991.
74. Dzierlega K, Yokota T. Optimization of antisense-mediated exon skipping for Duchenne muscular dystrophy. *Gene Ther.* 2020;27(9):407–416.
75. Waldrop MA, Yaou RB, Lucas KK, et al. Clinical phenotypes of DMD exon 51 skip equivalent deletions: a systematic review. *J Neuromuscul Dis.* 2020;7(3):217–229.
76. Birchmeier C, Schumperli D, Sconzo G, Birnstiel ML. 3′ editing of mRNAs: sequence requirements and involvement of a 60-nucleotide RNA in maturation of histone mRNA precursors. *Proc Natl Acad Sci U S A.* 1984;81(4):1057–1061.

77. Smith HO, Tabiti K, Schaffner G, Soldati D, Albrecht U, Birnstiel ML. Two-step affinity purification of U7 small nuclear ribonucleoprotein particles using complementary biotinylated 2'-O-methyl oligoribonucleotides. *Proc Natl Acad Sci U S A*. 1991;88(21):9784–9788.
78. Pillai RS, Grimmler M, Meister G, et al. Unique Sm core structure of U7 snRNPs: assembly by a specialized SMN complex and the role of a new component, Lsm11, in histone RNA processing. *Genes Dev*. 2003;17(18):2321–2333.
79. Stefanovic B, Hackl W, Luhrmann R, Schumperli D. Assembly, nuclear import and function of U7 Snrnps studied by microinjection of synthetic U7 Rna into xenopus oocytes. *Nucleic Acids Res*. 1995;23(16):3141–3151.
80. Schumperli D, Pillai RS. The special Sm core structure of the U7 snRNP: far-reaching significance of a small nuclear ribonucleoprotein. *Cell Mol Life Sci*. 2004;61(19–20):2560–2570.
81. Grimm C, Stefanovic B, Schumperli D. The low abundance of U7 Snrna is partly determined by its Sm binding-site. *EMBO J*. 1993;12(3):1229–1238.
82. Gorman L, Suter D, Emerick V, Schumperli D, Kole R. Stable alteration of pre-mRNA splicing patterns by modified U7 small nuclear RNAs. *Proc Natl Acad Sci USA*. 1998;95(9):4929–4934.
83. De Angelis FG, Sthandier O, Berarducci B, et al. Chimeric snRNA molecules carrying antisense sequences against the splice junctions of exon 51 of the dystrophin pre-mRNA induce exon skipping and restoration of a dystrophin synthesis in Delta 48-50 DMD cells. *Proc Natl Acad Sci USA*. 2002;99(14):9456–9461.
84. Brun C, Suter D, Pauli C, et al. U7 snRNAs induce correction of mutated dystrophin pre-mRNA by exon skipping. *Cell Mol Life Sci*. 2003;60(3):557–566.
85. Goyenvalle A, Vulin A, Fougerousse F, et al. Rescue of dystrophic muscle through U7 snRNA-mediated exon skipping. *Science*. 2004;306(5702):1796–1799.
86. Sharp NJ, Kornegay JN, Van Camp SD, et al. An error in dystrophin mRNA processing in golden retriever muscular dystrophy, an animal homologue of Duchenne muscular dystrophy. *Genomics*. 1992;13(1):115–121.
87. Vulin A, Barthelemy I, Goyenvalle A, et al. Muscle function recovery in golden retriever muscular dystrophy after AAV1-U7 exon skipping. *Mol Ther*. 2012;20(11):2120–2133.
88. Tse LV, Moller-Tank S, Asokan A. Strategies to circumvent humoral immunity to adeno-associated viral vectors. *Expert Opin Biol Ther*. 2015;15(6):845–855.
89. Carroll D. Genome engineering with targetable nucleases. *Annu Rev Biochem*. 2014;83:409–439.
90. Jinek M, Chylinski K, Fonfara I, Hauer M, Doudna JA, Charpentier E. A programmable dual-RNA-guided DNA endonuclease in adaptive bacterial immunity. *Science*. 2012;337(6096):816–821.
91. Jinek M, East A, Cheng A, Lin S, Ma E, Doudna J. RNA-programmed genome editing in human cells. *elife*. 2013;2:e00471.
92. Mali P, Yang L, Esvelt KM, et al. RNA-guided human genome engineering via Cas9. *Science*. 2013;339(6121):823–826.
93. Ran FA, Hsu PD, Wright J, Agarwala V, Scott DA, Zhang F. Genome engineering using the CRISPR-Cas9 system. *Nat Protoc*. 2013;8(11):2281–2308.
94. Ousterout DG, Kabadi AM, Thakore PI, Majoros WH, Reddy TE, Gersbach CA. Multiplex CRISPR/Cas9-based genome editing for correction of dystrophin mutations that cause Duchenne muscular dystrophy. *Nat Commun*. 2015;6:6244.
95. Taglia A, Petillo R, D'Ambrosio P, et al. Clinical features of patients with dystrophinopathy sharing the 45-55 exon deletion of DMD gene. *Acta Myol*. 2015;34(1):9–13.

96. Tabebordbar M, Zhu K, Cheng JKW, et al. In vivo gene editing in dystrophic mouse muscle and muscle stem cells. *Science*. 2016;351(6271):407–411.
97. Long C, Amoasii L, Mireault AA, et al. Postnatal genome editing partially restores dystrophin expression in a mouse model of muscular dystrophy. *Science*. 2016;351(6271):400–403.
98. Nelson CE, Hakim CH, Ousterout DG, et al. In vivo genome editing improves muscle function in a mouse model of Duchenne muscular dystrophy. *Science*. 2016;351(6271):403–407.
99. Young CS, Mokhonova E, Quinonez M, Pyle AD, Spencer MJ. Creation of a novel humanized dystrophic mouse model of Duchenne muscular dystrophy and application of a CRISPR/Cas9 gene editing therapy. *J Neuromuscul Dis*. 2017;4(2):139–145.
100. Borsenberger V, Croux C, Daboussi F, Neuveglise C, Bordes F. Developing methods to circumvent the conundrum of chromosomal rearrangements occurring in multiplex gene edition. *ACS Synth Biol*. 2020;9(9):2562–2575.
101. Li HL, Fujimoto N, Sasakawa N, et al. Precise correction of the dystrophin gene in duchenne muscular dystrophy patient induced pluripotent stem cells by TALEN and CRISPR-Cas9. *Stem Cell Rep*. 2015;4(1):143–154.
102. Amoasii L, Long C, Li H, et al. Single-cut genome editing restores dystrophin expression in a new mouse model of muscular dystrophy. *Sci Transl Med*. 2017;9(418):eaan8081.
103. Long C, Li H, Tiburcy M, et al. Correction of diverse muscular dystrophy mutations in human engineered heart muscle by single-site genome editing. *Sci Adv*. 2018;4(1):eaap9004.
104. Min YL, Li H, Rodriguez-Caycedo C, et al. CRISPR-Cas9 corrects Duchenne muscular dystrophy exon 44 deletion mutations in mice and human cells. *Sci Adv*. 2019;5(3):eaav4324.
105. Amoasii L, Li H, Zhang Y, et al. In vivo non-invasive monitoring of dystrophin correction in a new Duchenne muscular dystrophy reporter mouse. *Nat Commun*. 2019;10(1):4537.
106. Zhang Y, Li H, Min YL, et al. Enhanced CRISPR-Cas9 correction of Duchenne muscular dystrophy in mice by a self-complementary AAV delivery system. *Sci Adv*. 2020;6(8):eaay6812.
107. Min YL, Chemello F, Li H, et al. Correction of three prominent mutations in mouse and human models of Duchenne muscular dystrophy by single-cut genome editing. *Mol Ther*. 2020;28(9):2044–2055.
108. Amoasii L, Hildyard JCW, Li H, et al. Gene editing restores dystrophin expression in a canine model of Duchenne muscular dystrophy. *Science*. 2018;362(6410):86–91.
109. Lim KRQ, Yoon C, Yokota T. Applications of CRISPR/Cas9 for the treatment of Duchenne muscular dystrophy. *J Pers Med*. 2018;8(4):38.
110. Long C, McAnally JR, Shelton JM, Mireault AA, Bassel-Duby R, Olson EN. Prevention of muscular dystrophy in mice by CRISPR/Cas9-mediated editing of germline DNA. *Science*. 2014;345(6201):1184–1188.
111. Zhang Y, Long C, Li H, et al. CRISPR-Cpf1 correction of muscular dystrophy mutations in human cardiomyocytes and mice. *Sci Adv*. 2017:3(4), e1602814.
112. Zhu P, Wu F, Mosenson J, Zhang H, He TC, Wu WS. CRISPR/Cas9-mediated genome editing corrects Dystrophin mutation in skeletal muscle stem cells in a mouse model of muscle dystrophy. *Mol Ther–Nucleic Acids*. 2017;7:31–41.
113. Wolinetz CD, Collins FS. NIH supports call for moratorium on clinical uses of germline gene editing. *Nature*. 2019;567(7747):175.
114. Cwik B. Responsible translational pathways for germline gene editing? *Curr Stem Cell Rep*. 2020;6:126–133.

115. Judson RN, Rossi FMV. Towards stem cell therapies for skeletal muscle repair. *NPJ Regen Med.* 2020;5:10.
116. Bengtsson NE, Hall JK, Odom GL, et al. Muscle-specific CRISPR/Cas9 dystrophin gene editing ameliorates pathophysiology in a mouse model for Duchenne muscular dystrophy. *Nat Commun.* 2017;8:14454.
117. Heyer WD, Ehmsen KT, Liu J. Regulation of homologous recombination in eukaryotes. *Annu Rev Genet.* 2010;44:113–139.
118. van Gent DC, van der Burg M. Non-homologous end-joining, a sticky affair. *Oncogene.* 2007;26(56):7731–7740.
119. Suzuki K, Tsunekawa Y, Hernandez-Benitez R, et al. In vivo genome editing via CRISPR/Cas9 mediated homology-independent targeted integration. *Nature.* 2016;540(7631):144–149.
120. Mendell JR, Rodino-Klapac L, Sahenk Z, et al. Gene therapy for muscular dystrophy: Lessons learned and path forward. *Neurosci Lett.* 2012;527(2):90–99.
121. Mendell JR, Campbell K, Rodino-Klapac L, et al. Dystrophin immunity in Duchenne's muscular dystrophy. *N Engl J Med.* 2010;363(15):1429–1437.
122. Halbert CL, Miller AD, McNamara S, et al. Prevalence of neutralizing antibodies against adeno-associated virus (AAV) types 2, 5, and 6 in cystic fibrosis and normal populations: Implications for gene therapy using AAV vectors. *Hum Gene Ther.* 2006;17(4):440–447.
123. Sack BK, Herzog RW. Evading the immune response upon in vivo gene therapy with viral vectors. *Curr Opin Mol Ther.* 2009;11(5):493–503.
124. Buchlis G, Podsakoff GM, Radu A, et al. Factor IX expression in skeletal muscle of a severe hemophilia B patient 10 years after AAV-mediated gene transfer. *Blood.* 2012;119(13):3038–3041.
125. Zygmunt DA, Crowe KE, Flanigan KM, Martin PT. Comparison of serum rAAV serotype-specific antibodies in patients with Duchenne muscular dystrophy, Becker muscular dystrophy, inclusion body myositis, or GNE myopathy. *Hum Gene Ther.* 2017;28(9):737–746.
126. Flanigan KM, Campbell K, Viollet L, et al. Anti-dystrophin T cell responses in Duchenne muscular dystrophy: prevalence and a glucocorticoid treatment effect. *Hum Gene Ther.* 2013;24(9):797–806.
127. Gough V, Gersbach CA. Immunity to Cas9 as an obstacle to persistent genome editing. *Mol Ther.* 2020;28(6):1389–1391.
128. Charlesworth CT, Deshpande PS, Dever DP, et al. Identification of preexisting adaptive immunity to Cas9 proteins in humans. *Nat Med.* 2019;25(2):249–254.
129. Wagner DL, Amini L, Wendering DJ, et al. High prevalence of streptococcus pyogenes Cas9-reactive T cells within the adult human population. *Nat Med.* 2019;25(2):242–248.
130. Ferdosi SR, Ewaisha R, Moghadam F, et al. Multifunctional CRISPR-Cas9 with engineered immunosilenced human T cell epitopes. *Nat Commun.* 2019;10(1):1842.
131. Simhadri VL, McGill J, McMahon S, Wang J, Jiang H, Sauna ZE. Prevalence of pre-existing antibodies to CRISPR-associated nuclease Cas9 in the USA population. *Mol Ther Methods Clin Dev.* 2018;10:105–112.
132. Kwon JB, Ettyreddy AR, Vankara A, et al. In vivo gene editing of muscle stem cells with Adeno-associated viral vectors in a mouse model of Duchenne muscular dystrophy. *Mol Ther Methods Clin Dev.* 2020;19:320–329.
133. Nance ME, Shi R, Hakim CH, et al. AAV9 edits muscle stem cells in Normal and dystrophic adult mice. *Mol Ther.* 2019;27(9):1568–1585.
134. Mendell JR, Al-Zaidy S, Shell R, et al. Single-dose gene-replacement therapy for spinal muscular atrophy. *N Engl J Med.* 2017;377(18):1713–1722.

135. Hinderer C, Katz N, Buza EL, et al. Severe toxicity in nonhuman primates and piglets following high-dose intravenous administration of an adeno-associated virus vector expressing human SMN. *Hum Gene Ther.* 2018;29(3):285–298.
136. Dowling JJ, Lawlor MW, Das S. X-linked myotubular myopathy. In: Adam MP, Ardinger HH, Pagon RA, et al., eds. *GeneReviews.* Seattle, WA: University of Washington. 1993.
137. Zhang XH, Tee LY, Wang XG, Huang QS, Yang SH. Off-target effects in CRISPR/Cas9-mediated genome engineering. *Mol Ther–Nucleic Acids.* 2015;4:e264.
138. Hanlon KS, Kleinstiver BP, Garcia SP, et al. High levels of AAV vector integration into CRISPR-induced DNA breaks. *Nat Commun.* 2019;10(1):4439.
139. Nelson CE, Wu Y, Gemberling MP, et al. Long-term evaluation of AAV-CRISPR genome editing for Duchenne muscular dystrophy. *Nat Med.* 2019;25(3):427–432.
140. Hoffman EP. The discovery of dystrophin, the protein product of the Duchenne muscular dystrophy gene. *FEBS J.* 2020;287(18):3879–3887.

CHAPTER EIGHT

Genome editing in the human liver: Progress and translational considerations

Samantha L. Ginn[a], Sharntie Christina[a], and Ian E. Alexander[a,b],*

[a]Gene Therapy Research Unit, Children's Medical Research Institute, Faculty of Medicine and Health, The University of Sydney and Sydney Children's Hospitals Network, Westmead, NSW, Australia
[b]Discipline of Child and Adolescent Health, Sydney Medical School, Faculty of Medicine and Health, The University of Sydney, Westmead, NSW, Australia
*Corresponding author: e-mail address: ian.alexander@health.nsw.gov.au

Contents

1. Introduction — 258
2. The liver: A high value therapeutic target — 259
3. Liver biology and implications for genome editing — 263
   3.1 Metabolic zonation, blood flow and fenestration — 263
   3.2 Liver growth and implications for the pediatric population — 265
4. Genome editing technologies — 266
5. Disease specific challenges of genome editing in the liver — 270
   5.1 Cell autonomous and non-cell autonomous liver diseases — 271
   5.2 Hemophilia as an exemplar of a non-cell autonomous disease target — 271
   5.3 Ornithine transcarbamylase deficiency as an exemplar of a cell autonomous disease target — 272
6. Strategies to achieve genome editing outcomes in the liver — 275
   6.1 Locus-specific gene disruption — 275
   6.2 Selective expansion of genome engineered cells — 276
   6.3 Targeted insertion of therapeutic transgenes into the liver — 277
7. Translational considerations — 279
   7.1 Cellular responses to Cas proteins — 279
   7.2 Introducing unwanted mutations into the genome — 280
   7.3 Logistic and commercial constraints — 281
8. Concluding remarks — 281
References — 282

Abstract

Liver-targeted genome editing offers the prospect of life-long therapeutic benefit following a single treatment and is set to rapidly supplant conventional gene addition approaches. Combining progress in liver-targeted gene delivery with genome editing technology, makes this not only feasible but realistically achievable in the near term.

However, important challenges remain to be addressed. These include achieving therapeutic levels of editing, particularly *in vivo*, avoidance of off-target effects on the genome and the potential impact of pre-existing immunity to bacteria-derived nucleases, when used to improve editing rates. In this chapter, we outline the unique features of the liver that make it an attractive target for genome editing, the impact of liver biology on therapeutic efficacy, and disease specific challenges, including whether the approach targets a cell autonomous or non-cell autonomous disease. We also discuss strategies that have been used successfully to achieve genome editing outcomes in the liver and address translational considerations as genome editing technology moves into the clinic.

Abbreviations

AAV	adeno-associated virus
ASS	argininosuccinate synthetase
bp	base pair
cDNA	complementary DNA
CRISPR	clustered regularly interspaced short palindromic repeats
DNA	deoxyribonucleic acid
DSB	double-strand break
HDR	homology-directed repair
HITI	homology-independent targeted integration
HR	homologous recombination
Indel	insertion and deletion
MMEJ	microhomology-mediated end joining
NHEJ	non-homologous end joining
nm	nanometer
OTC	ornithine transcarbamylase
PAM	protospacer adjacent motif
PCSK9	*proprotein convertase subtilisin/kexin 9*
PITCh	Precise Integration into Target Chromosome
sgRNA	single guide RNA
TALEN	transcription activator-like effector nucleases
ZRN	zinc-finger nucleases

1. Introduction

Building conceptually on the use of synthetic RNA sequences to crack the genetic code, Nirenberg foresaw the possibility of reprogramming human cells with genetic surgery in an editorial published in *Science*.[1] He recognized the formidable scientific challenges involved and the likely extended timelines required. Now, over 50 years later many of these challenges have been surmounted and the rapidly evolving gene therapy field is undergoing a major paradigm shift away from the predominance of gene addition approaches to the exploration of an increasing diversity of more elegant and precise genome editing approaches. This shift has been rapidly accelerated by the ingenious adaptation of a key component of the bacterial immune system, known as clustered regularly interspaced short palindromic repeats (CRISPR)-associated Cas9 (CRISPR/Cas9) nuclease. This in turn has led to dramatic improvements in the efficiency and accessibility of genome editing technology and the resultant initiation of early human clinical trials, most commonly focused on hematopoietic cell targets.[2]

2. The liver: A high value therapeutic target

The liver performs a host of complex functions.[3] These include (i) metabolism of fats, proteins and carbohydrates, (ii) storage of glycogen, vitamins and minerals, (iii) blood detoxification and purification, (iv) synthesis of plasma proteins, such as albumin and clotting factors, (v) bile production and excretion, and (vi) excretion of bilirubin, cholesterol, hormones and drugs. The predominant cell type in the liver is the hepatocyte, a cuboidal epithelial cell that performs most of the liver's functions and is affected in many inherited genetic/metabolic diseases. For many of these conditions contemporary therapies are inadequate and for a significant proportion liver transplantation is the best available therapeutic option. However, there is a lack of available donors and potential for irreversible damage before a transplant becomes available. Even when transplantation is possible, it brings associated risks of rejection and life-long immune suppression. Hepatocytes, therefore, are high value targets for liver-directed gene addition and genome editing strategies for the treatment of monogenetic disorders.

Gene therapy lies at the forefront of modern medicine and, with more than 30 years of incremental development, the field has reached an inflexion point where many difficult or impossible to treat diseases are increasingly falling within therapeutic reach.[4] In the liver, pioneering trials in adults with hemophilia B, and more recently hemophilia A, have demonstrated sustained clinical benefit following systemic delivery of adeno-associated virus (AAV) vectors encoding human factor IX or VIII, respectively.[5–7] On the back of these clinical success, AAV vectors have become the vector of choice for delivery of therapeutic transgenes to hepatocytes (see Table 1 for noteworthy clinical trials targeting the liver). Currently, there are in excess of 250 approved AAV-based trials,[12] using predominantly gene addition strategies, with at approximately 15% of these trials targeting the liver.[13–15]

Table 1 Selected clinical trials targeting the liver.

Delivery system	Gene addition or gene editing	Disorder	Clinical phase	Study start date	Trial sponsor	Recruitment status	Identifier/reference
Adenovirus	Addition	Ornithine transcarbamylase deficiency (OTC)	Phase I	October 1995	Children's National Research Institute	Terminated	NCT00004386
Adenovirus	Addition	Ornithine transcarbamylase deficiency (OTC)	Phase I	July 1998	University of Pennsylvania	Terminated	NCT00004498
AAV (scAAV2/8-LP1-hFIXco)	Addition	Hemophilia B (FIX deficiency)	Phase I	February 2010	St. Jude Children's Research Hospital	Active, not recruiting; clinical improvement noted	NCT00979238/ Ref. 5
AAV (rAAV2/5-PBGD)	Addition	Acute intermittent porphyria (AIP)	Phase I	November 2012	Digna Biotech S.L.	Completed; metabolic correction not achieved	NCT02082860/ Ref. 8
AAV (AAV5-hFVIII-SQ; BMN 270)	Addition	Hemophilia A (FVIII deficiency)	Phase I/II	August 2015	BioMarin Pharmaceutical	Active, not recruiting; clinical improvement noted	NCT02576795/ Ref. 6

Vector	Type	Disease	Phase	Date	Sponsor	Status	Identifier/Refs.
AAV (novel capsid; SPK-9001)	Addition	Hemophilia B (FIX deficiency)	Phase II	November 2015	Pfizer	Completed; clinical improvement noted	NCT02484092/Refs. 7,9
AAV (AAV8-hLDLR)	Addition	Homozygous familial hypercholesterolemia	Phase I/II	March 2016	Regenxbio Inc.	Active, not recruiting	NCT02651675
AAV (AAV2/6-ZFN[a]; SB-FIX)	Editing	Hemophilia B (FIX deficiency)	Phase I	November 2016	Sangamo Therapeutics	Active, not recruiting	NCT02695160
AAV (scAAV8OTC; DTX301)	Addition	Ornithine transcarbamylase deficiency (OTC)	Phase I	January 2017 (follow up August 2018, by invitation)	Ultragenyx Pharmaceutical Inc.	Recruiting; clinical improvement noted	NCT02991144 NCT03636438
AAV (AAV2/6-ZFN; SB-318)	Editing	Mucopolysaccharidosis I (MPSI)	Phase I/II	May 2017	Sangamo Therapeutics	Active, not recruiting	NCT02702115
AAV (AAV2/6-ZFN; SB-913)	Editing	Mucopolysaccharidosis II (MPSII)	Phase I/II	May 2017	Sangamo Therapeutics	Active, not recruiting	NCT03041324
AAV (AAV2/6-hF8; SB-525)	Addition	Hemophilia A (FVIII deficiency)	Phase III	June 2017	Pfizer	Active, not recruiting; clinical improvement noted	NCT03061201/ Ref. 10

Continued

Table 1 Selected clinical trials targeting the liver.—cont'd

Delivery system	Gene addition or gene editing	Disorder	Clinical phase	Study start date	Trial sponsor	Recruitment status	Identifier/reference
AAV (AAV5-hFIXco-Padua; AMT-061)	Addition	Hemophilia B (FIX deficiency)	Phase III	June 2018	UniQure Biopharma B.V.	Active, on hold; clinical improvement noted	NCT03569891/ Ref. 11
Lipid nanoparticle (CRISPR/Cas9)	Editing	Hereditary transthyretin amyloidosis	Phase I	October 2020	Intellia Therapeutics	Recruiting	NCT04601051

[a]ZFN, zinc-finger nuclease targeting the human albumin locus.

3. Liver biology and implications for genome editing

The most conceptually and clinically appealing gene therapy approach for monogenic disease is to perform precise molecular repair of the mutant locus in a sufficient number of somatic cells to confer therapeutic benefit. This has advantages over gene addition strategies that most commonly deliver a functional cDNA copy of the mutant gene under the transcriptional control of a heterologous promoter.[16] Here, the corrected cells retain the mutant locus, and the introduced transgene expression cassette remains either episomal or is inserted elsewhere into the host cell genome, in a semi-random manner, depending on the method of gene delivery used.[17,18] This limits application of gene addition approaches to recessive disease phenotypes, does not recapitulate physiological gene expression, and is at risk of loss of therapeutic efficacy over time as a consequence of either downregulation of heterologous promoter activity or the loss of vector genomes. Gene addition strategies employing vectors capable of integration also carry the added risk of insertional mutagenesis.[19,20] In contrast, gene repair offers the prospect of correcting both recessive and dominant disease phenotypes, retaining physiological control of gene expression, bypassing constraints imposed by the packaging capacity of the delivery system, and minimizing the risk of mutagenesis.

3.1 Metabolic zonation, blood flow and fenestration

The liver receives blood supply from two sources, systemic arterial blood from the hepatic artery and venous blood *via* the portal vein (Fig. 1A). Hepatic artery blood flow accounts for 25% of the cardiac output, despite the liver representing approximately 3% of total body mass.[3] The second source, portal venous blood, delivers a smaller volume of nutrient-rich blood from the internal organs, including the gut, pancreas and spleen.[3] Blood from both sources drains to the liver sinusoids that are lined by fenestrated endothelial cells with pores of between 100 and 200 nm depending on the species.[21] The high liver blood flow and small size of recombinant AAV vectors (20–25 nm) allows movement through the liver sinusoids and direct access to hepatocytes[3,22,23] (Fig. 1B).

Differential gene expression across the hepatic lobule results in zonation of metabolic functions (Fig. 1C). As a consequence, hepatocytes located in different hepatic zones have different capabilities and perform distinct and often opposing metabolic functions.[24,25] Because of this high degree of

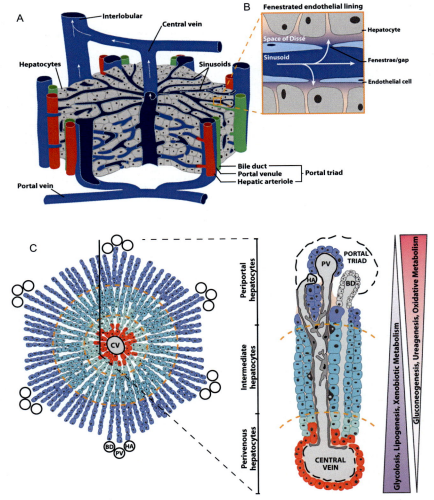

Fig. 1 Hepatic lobule structure and function. (A) View of the hepatic lobule showing vasculature and blood flow (white arrows) from the portal triad toward the central vein. (B) Magnified view of the sinusoid endothelia lining showing fenestration between cells, allowing for diffusion of large particles (∼0.5 μm) into the Space of Dissé surrounding hepatocytes. (C) Cross sectional view of hepatic lobule (left) and acinus (right) detailing zonation in hepatocytes relating to zonally graded metabolic function. BD, bile duct; CV, central vein; HA, hepatic arteriole; PV, portal venule.

specialization, the liver can efficiently fulfill a multitude of functions simultaneously. Patterns of zonation are generally dynamic and respond to metabolic demands. A notable example of metabolic zonation in the liver is ammonia detoxification. To ensure removal of this highly neurotoxic metabolite from the blood, ureagenesis (low affinity, high capacity) occurs in the periportal hepatocytes that express enzymes of the urea cycle most highly. This occurs in combination with scavenging by glutamine synthetase (high affinity, low capacity) in a single cell layer of the perivenous hepatocytes.[26,27] This zonated gene expression pattern is an important consideration, for both gene addition and genome editing technologies, when delivery to specific hepatocytes, determined by the disease target, may be required. For example, in both mice and dogs, recombinant AAV vectors pseudoserotyped with the AAV8 capsid demonstrate higher levels of transduction in hepatocytes located near the central veins. This, however, is not the case for non-human primates or mice following delivery of vectors serotyped with the AAV2 capsid.[28,29] These different transduction patterns have the potential to influence therapeutic efficacy when target genes show zonated expression or are integral to a zonated metabolic pathway (Fig. 1C). Beyond urea cycle disorders, an interesting contemporary example of therapeutic editing relevance is the albumin gene which is more highly expressed in the periportal zones of the hepatic lobule and is being targeted in gene-ride applications.[30] More efficient periportal targeting would therefore be expected to deliver higher therapeutic efficacy, however, such an effect has yet to be clinically demonstrated and would be target disease dependent. Never-the-less spatial patterns of target gene expression should clearly be considered and are becoming increasingly well documented for the liver transcriptome.[31]

3.2 Liver growth and implications for the pediatric population

The unique regenerative capacity of the liver was first captured in Greek mythology in the story of Prometheus.[32] Regeneration can occur by either proliferation of hepatocytes or differentiation of liver stem cells (LSCs) and oval cells,[33,34] and has the potential to impact therapeutic outcomes. Notably, to date, all AAV-based trials targeting the liver have been in adult patients. This largely reflects the biology of AAV vectors, which persist in the majority of target cells as extra chromosomal DNA known as episomes, with only a small fraction undergoing genomic integration.[35,36] These episomes are stable in the mostly quiescent adult liver. However, they are rapidly lost from the growing liver in concert with hepatocellular replication.[36] This may represent a major barrier to durable therapeutic effects in the pediatric liver following classical AAV-mediated gene addition, but there is currently no human clinical data addressing this point.

One key advantage of genome editing over gene addition is maintenance of durable therapeutic effects in replicating cells when non-integrating vector systems are used to achieve efficient delivery of the editing reagents and templates for repair. This is an important consideration for treating the pediatric population where the liver is estimated to undergo doublings by approximately 6 months, 2 years and 9 years of age.[37] Therefore, human hepatocytes in their native hepatic context will have lower patient age-dependent rates of replication which are likely to influence gene therapy outcomes. For example, low rates of cellular turnover in the adult liver facilitates maintenance of AAV episomes. In contrast, genome editing strategies that use classical homology-directed repair (HDR) require cell division for successful correction[38] and, therefore, would theoretically be more efficient in infants.

4. Genome editing technologies

In the early 1980s Capecchi and colleagues pioneered the use of homologous recombination (HR) to precisely modify complex mammalian genomes in cultured cells.[39–41] Building on these foundations Russell and Hirata were the first to exploit the greater gene transfer efficiency of recombinant AAV vectors, over earlier used non-viral approaches, to deliver DNA templates for HR.[42,43] In this seminal report, the level of homologous recombination achieved was unprecedented for the time and opened the door for studies exploring the therapeutic use of HR *in vivo* (reviewed in Alexander and Russell[43]). The knowledge that HR can be dramatically enhanced by the presence of double-strand DNA breaks subsequently lead to the combined use of AAV-mediated gene targeting technology and user-designed nucleases to introduce cleavage events at precisely defined loci.[44] This technological convergence has lifted genome editing efficiencies to contemporary levels, placing many disease targets in the liver and other organs within near term therapeutic reach.

During the early development stages of gene editing, zinc-finger nucleases (ZFNs) were the best tool available to perform genetic manipulation.[45,46] The ZFN architecture links a chain of DNA-binding transcription factors, zinc finger proteins (ZFPs), with the *FokI* non-specific nuclease. The ZFP region contains a tandem array of Cyc2-His2 fingers that each recognize ~3 bp of DNA, the combination of which can be engineered to target specific regions of DNA. To facilitate DNA cleavage, two ZFNs must be provided with ZFPs targeting top and bottom DNA strands on opposite sides of the desired cleavage site, allowing the two *FokI* nucleases to dimerize[47] (Fig. 2A). Although the small size of ZFNs makes them suitable for delivery in a variety of viral vectors, the complexity of predicting protein DNA interactions, as compared to Watson-Crick base pairing, has limited uptake of this technology.[48] Transcription activator-like effector nucleases (TALENs), the next developed gene editing tool, are also dependent on sequence-specific protein/DNA interactions and provide a system that functions similarly to ZFNs. They also confer higher DNA binding specificity, have fewer off-target effects, and more straightforward design of DNA-binding domains.[49,50] Compared to ZFNs, the DNA-binding system of TALENs target a longer sequence which is comprised of an array of TAL effector proteins that are each specific to a certain nucleotide (Fig. 2B). TALENs are also easier to engineer due to the lack of interfering effects between TAL effector proteins, however, synthesis is relatively expensive.[48]

Today, the CRIPSR/Cas system[51–53] is most commonly used to facilitate genome editing. This system represents a quantum advance because targeting is based on Watson-Crick base pairing between a readily designed 20-bp nucleotide single guide RNA (sgRNA) and the target DNA sequence. The sgRNA binds the target DNA sequence immediately upstream of a requisite protospacer adjacent motif (PAM) directing the Cas9 endonuclease to produce a double-stranded break (DSB) usually 3-bp upstream of the PAM (Fig. 2C). The CRISPR/Cas9 system from *Streptococcus pyogenes* has been widely used in viral gene delivery systems, with some Cas9 orthologs offering smaller coding sequences if packaging capacity is limiting.[54] Alternative Cas nucleases with unique features can also function in CRISPR editing systems, such as Cas12a which creates overhangs upon DNA cleavage with higher reported specificity.[55]

A concern with any system that edits the genome is the potential for off-target events, occurring at a non-desired location. The probability of such events is reduced by the careful selection of guide sequences that do not share homology with other locations in the genome. For CRISPR/Cas systems, the careful selection of optimal sgRNA sequences with the help of *in silico* prediction tools is essential for achieving high specificity, efficiency, and low off-target potential.[56] Repair of cleavage sites can also introduce errors at the site of cleavage, typically small base pair insertions and deletions known as indels that may be exploited for gene knock outs. However, repair can introduce unwanted changes into the genome in the form of either small indels or larger deletions and rearrangements that may be difficult to detect.[57]

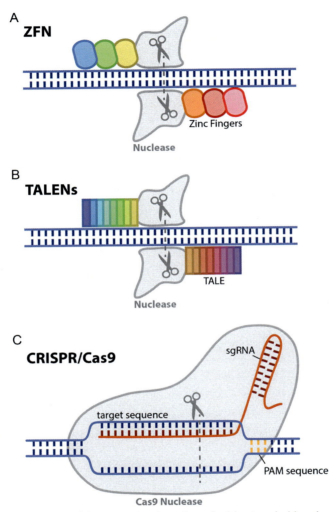

Fig. 2 Gene editing machinery used to introduce double-stranded breaks at targeted genomic locations. (A) Zinc-finger nuclease (ZFN) systems comprise of DNA-binding proteins (Zinc fingers) specific to the targeted DNA sequence, each binding ~3 base pairs, coupled with a non-specific nuclease that facilitates cleavage when two ZFNs bind complementary strands and align at the cleavage site. (B) Transcription activator-like effector nuclease (TALEN) systems comprise of a sequence of DNA-binding transcription activator-like effector (TALE) proteins targeted to a specific DNA sequence, each binding 1 base pair, coupled with a non-specific nuclease that facilitates cleavage when two TALENs bind complementary strands and align at the cleavage site. (C) CRISPR/Cas9 systems comprise of a guiding RNA (sgRNA) molecule which is reverse-complementary to a targeted sequence of DNA. The sgRNA directs the Cas9 nuclease to the target site where it recognizes a 2–6 base pair protospacer adjacent motif (PAM) and facilitates double-stranded cleavage.

When a DSB occurs, different DNA repair pathways can be activated depending on the cell cycle stage of the cell. The two main repair mechanisms used by mammalian cells to repair DSBs are non-homologous end joining (NHEJ) and homologous recombination (HR) also known as homology-directed repair (HDR)[58] which dictates the type of genome editing event observed and the population of cells that can be successfully edited (Fig. 3).

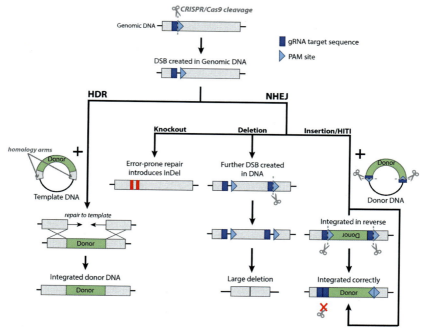

Fig. 3 Genome editing strategies mediated by double-stranded breaks (DSBs). Homology-directed repair (HDR) strategies rely on the homologous repair DNA repair pathway active in dividing cells for insertion of donor DNA sequences. A circular DNA template is provided containing the donor DNA with flanking regions of homology (homology arms) to the sequence surrounding the gnomic cut site. This facilitates recombination around the homologous regions and drives repair to the template DNA and integration of the donor segment. Methods utilizing the non-homologous end joining (NHEJ) DNA repair pathway allow for gene knockout, deletion, and insertion events. Repair of the DSB with the NHEJ pathway is error prone and introduces insertions/deletions (InDel) which can cause gene knockout. A further DSB can be introduced upstream or downstream of the primary DSB to facilitate a large deletion. An insertion facilitated by NHEJ, or homology-independent targeted integration (HITI), requires provision of donor DNA, where the donor segment is excised by two flanking CRISPR/Cas9 cleavage sites. The donor segment can then be integrated in reverse or correct directionality. Due to the alignment of guiding RNA (sgRNA) and protospacer adjacent motif (PAM) sites required for Cas9 cleavage, donor sequences integrated in reverse can be re-excised, while correctly integrated sequences cannot be cut further by Cas9.

Homology-directed repair occurs during the S/G2 phases of the cell cycle and at low efficiency in primary cells.[59] Genome editing approaches that exploit this pathway, therefore, can only be successfully applied when the target cell is dividing. In contrast, NHEJ is active throughout the cell cycle in both proliferating and post-mitotic cells.[60] Notably, the activity of NHEJ far exceeds the activity of HDR,[60] unless these pathways are deliberately altered to increase the likelihood of achieving HDR-mediated repair events.[61] New methods that effectively utilize the NHEJ pathway for accurate gene editing have been recently developed and include the use of donor templates. Among them homology-independent targeted integration (HITI) exploiting the NHEJ repair mechanism by creating CRISPR/Cas9-mediated DSBs in both the genome and on either side of the donor template using the same target sequences[62] and intercellular linearized Single homology Arm donor mediated intron-Targeting Integration (SATI) that combine both HDR and NHEJ.[63] Alternatively, Precise Integration into Target Chromosome (PITCh) can be used to perform targeted insertions in non-dividing cells.[64] This involves microhomologies at the end of the donor to direct it to specific target sites which are used by the microhomology-mediated end-joining (MMEJ) repair pathway, active during G1 and early S phase.[65] Finally, base editors are able to introduce precise point mutations into the genome, again without the requirement for HDR, and in the absence of DSBs or without the need for donor templates for repair.[66]

5. Disease specific challenges of genome editing in the liver

Genome editing will likely surpass gene therapy strategies based on gene addition in the future, however, challenges related to achieving therapeutic levels of editing, particularly *in vivo*, remain to be addressed. This is especially true for target organs such as the liver that are not inherently suited to *ex vivo* gene therapy approaches. Genome editing in the liver has been directed to a wide variety of therapeutic targets employing many different approaches including, but not limited to, gene disruption, targeted insertion, precise gene repair, and base editing.[62,66,67] However, HDR-mediated genome editing efficiency in adult animal livers is very low[68–70] unless undergoing rapid expansion.[71] Thus, given this limitation of HDR-based editing approaches and the current reach of gene delivery technology, to date the majority of studies investigating the utility of genome editing technologies to target primary hepatocytes *in vivo* have focused on either (i) locus-specific disruption, exploiting the more active non-homologous end joining (NHEJ) pathway,[44,72–75] (ii) targeting diseases where a selective growth advantage is conferred on the corrected cells[69,76,77] or, (iii) where the correction of only a small percentage of cells can confer therapeutic benefit.[78]

5.1 Cell autonomous and non-cell autonomous liver diseases

An important factor in determining the level of editing required for therapeutic success is whether the target disease is cell autonomous or non-cell autonomous. For cell autonomous diseases correction of individual cells does not cross-complement the defect in other cells. Conversely, for non-autonomous diseases correction of individual cells can cross-compliment the defect in other cells.

For non-cell autonomous diseases, commonly involving secreted proteins, gene transfer to only a modest number of hepatocytes has proven sufficient to confer the clinical benefit (Fig. 4). In contrast, cell autonomous diseases typically represent a greater therapeutic challenge. This higher threshold for therapeutic benefit varies in a disease-dependent manner and is a function of both the number of target cells that must be gene modified and the level of transgene expression required in individual cells. Liver-targeted *in vivo* genome editing represents an even greater therapeutic challenge and, although in its infancy, human trials have progressed to the clinic (Table 1). Notably, trials using ZFNs to target the endogenous and highly expressed albumin locus to treat hemophilia B and mucopolysaccharidosis (MPSI and MPSII) are open and recruiting patients.[79–81]

5.2 Hemophilia as an exemplar of a non-cell autonomous disease target

Hemophilia is a paradigm for non-cell autonomous disease targets, with gene addition strategies already showing success in the clinic (Table 1). This disease, however, has a low therapeutic threshold because an increase in plasma clotting factors of as little as 1% of wild-type levels can confer

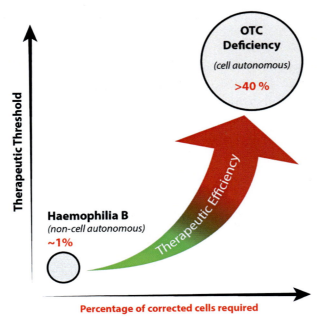

Fig. 4 Relationship between exemplar cell non-autonomous (Hemophilia B) and cell autonomous (OTC deficiency) diseases and their therapeutic reach. The percentage of corrected cells required for therapeutic benefit (red) is directly correlated with therapeutic threshold. Greater therapeutic efficiency is needed to reach diseases with a high therapeutic threshold.

significant phenotypic improvement (Fig. 4) by converting severe- to moderate-disease and for levels above 5%, a mild phenotype.[82] Notably, the transgene products for hemophilia treatment are secreted proteins and are usually expressed from a strong liver-specific promoter with resultant supraphysiological levels of expression. The highly active albumin promoter is a notable example and is being used in genome editing trials targeting hemophilia B to drive production of high levels of human factor IX.[79] Another option in some diseases contexts is the use of hyperactive factors. For example, a number of recent trials for hemophilia B have incorporated a naturally occurring single nucleotide variant of factor IX (R338L, FIX-Padua) with increased specific activity.[9,83–85]

5.3 Ornithine transcarbamylase deficiency as an exemplar of a cell autonomous disease target

Cell autonomous diseases, which can also be metabolically zonated, represent a greater therapeutic challenge. A classic example of a cell autonomous disease with a high therapeutic threshold is ornithine transcarbamylase (OTC) deficiency, the most common urea cycle disorder in humans with an estimated frequency of between 1:50,000 and 1:80,000. Urea cycle disorders are clear candidates for gene therapy because of the monogenic nature of the disorders, sub-optimal treatment options and severity of disease, particularly in neonates. Clinical insights, supported by mathematical modeling,[86] suggest that for OTC deficiency therapeutic efficacy of gene therapy interventions will correlate most strongly with the number of hepatocytes undergoing repair rather than the level of OTC activity achieved in individual cells (Fig. 4). This renders OTC deficiency a relatively more challenging therapeutic target, and accordingly, success in the treatment of this condition would have positive implications for a much broader set of liver diseases.

Proof-of-principal for precise gene repair has already been explored for OTC deficiency. Using a CRISPR/Cas9 system and template for repair, HDR has been used to permanently correct the *OTC* gene in mice,[87] however, the repair rates achieved would be unlikely to confer robust therapeutic benefit in humans. In a follow-on study from the same group, a similar HDR-based approach was used to introduce a functional *OTC* cDNA minigene, driven by the liver-specific human thyroxin binding globulin (TBG) promoter, into exon 4 of the murine *Otc* gene.[88] This is a more universal approach to restore OTC expression and reduce the need for personalized editing reagents for individual mutations. A more clinically relevant study, using human-specific editing reagents and chimeric human-mouse livers[89] populated with OTC-deficient patient-derived hepatocytes has also been described.[71] Again, using dual AAV-based vector delivery this time packaged into the highly human tropic NP59 capsid,[90] successful single nucleotide editing was observed in up to 29% of *OTC* alleles at clinically relevant vector doses (Fig. 5). This underscores the need for highly efficient gene delivery systems, particularly when co-transduction of multiple AAV vectors is required to deliver both the editing machinery and donor template for repair. This study performed a comprehensive analysis of events occurring at the target locus following Cas9-mediated cleavage and classical HDR-mediated template repair. In addition to precise *in vivo* gene repair of the patient mutation, unprecedented rates of cleavage at the *OTC* locus, identified by the presence of indels, single nucleotide substitutions and large insertions were also observed. Interestingly, the vast majority of insertions were of vector origin with a dominant contribution from the AAV inverted terminal repeat (ITR) sequences, an observation also reported by others.[91] Irrespective of the nature of these events, the key consequence is the likely

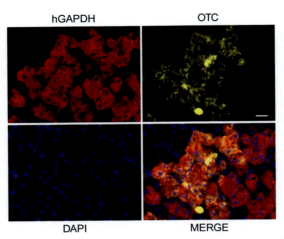

Fig. 5 Homology-directed repair restores OTC expression in patient-derived primary human hepatocytes *in vivo*. Representative immunohistochemistry images of chimeric human-mouse livers engrafted with patient-derived human hepatocytes 5-weeks post-treatment. Sections were labeled with antibodies against the human glyceraldehyde 3-phosphate dehydrogenase (hGAPDH, red) and OTC (yellow) proteins in liver sections. Nuclei were stained with DAPI. See Ginn et al.[71] for methodological details. Scale bar = 20 μm.

disruption of the reading frame and resultant loss of any residual OTC function resulting from the disease-causing mutated alleles. This consequence needs to be considered when targeting hypomorphic alleles that may confer clinical benefit to the patient.

These studies have all explored HDR-based genome editing approaches to correct the OTC deficiency. Given that classical HDR occurs only in cycling cells,[38] a major obstacle when targeting the human liver will be the rate of cellular replication. Notably, the adult liver is largely quiescent, with higher rates of mitotic activity during infancy and childhood when the liver is growing.[37] OTC deficiency is the most severe in its neonatal form, but also exists in late onset forms.[92] Gene therapies relying on HDR methods in the liver significantly decrease both the population of eligible patients based on age, and the population of hepatocytes that can be corrected at the time of treatment. Editing methods that are independent of the cell cycle and based on NHEJ for repair, such as HITI (Figs. 3 and 6), will be likely refinements to the technology, however, it remains plausible that strategies involving deliberate stimulation of the cell cycle might also prove useful.

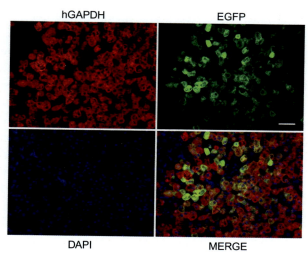

Fig. 6 Efficient integration and expression of a promoter-less EGFP-containing donor template into the *OTC* locus of primary human hepatocytes *in vivo* using homology-independent targeted integration (HITI). Representative immunohistochemistry images of chimeric human-mouse livers engrafted with wild-type human hepatocytes 4-weeks post-treatment. Sections were labeled with antibodies against the human glyceraldehyde 3-phosphate dehydrogenase (hGAPDH, red) and GFP (green) proteins in liver sections. Nuclei were stained with DAPI. Scale bar = 50 μm.

6. Strategies to achieve genome editing outcomes in the liver

6.1 Locus-specific gene disruption

The most conceptually simple application of genome editing technology to ameliorate a disease phenotype is to use locus-specific gene disruption to abolish an unwanted function, such as a mutant dominant allele. This can be achieved by disrupting the protein reading frame with the expected small insertions and deletions (indels) introduced into the genome when double-stranded breaks are repaired using the NHEJ pathway. The most notable example of this approach targets the therapeutically relevant gene *proprotein convertase subtilisin/kexin 9* (*PCSK9*) involved in cholesterol homeostasis.[93] This gene encodes a secreted serine protease that binds the low-density lipoprotein (LDL) receptor and targets it for lysosomal destruction. Thus, hyperactive gain-of-function mutations in *PCSK9* cause autosomal dominant hypercholesterolemia. In contrast, loss-of-function *PCSK9* mutants cause life-long low LDL-cholesterol levels and protection against cardiovascular disease. Using CRISPR-Cas9 editing technology, sgRNAs targeting the murine *Pcsk9* gene have been used to disrupt the protein coding sequence, successfully reduce serum Pcsk9 and total cholesterol levels.[44,72] Similar results targeting *Pcsk9* have been observed using both lipid nanoparticle delivery of the editing reagents to the murine liver[76] or a recombinant AAV vector expressing an engineered meganuclease to non-human primates.[94] These studies have established the therapeutic potential of targeting the *PCSK9* gene to prevent cardiovascular disease. However, the clinical translation of this approach requires validation of human-specific editing reagents unless a common sgRNA and PAM sequence exists between species. Using a humanized mouse model with chimeric mouse-human livers, human *PCSK9* has been successfully targeted with up 47% disruption observed at the locus. This resulted in a significant reduction of post-treatment levels of human PCSK9 protein secreted by the human hepatocytes in the serum.[74]

6.2 Selective expansion of genome engineered cells

Hereditary tyrosinemia type I (HT1) is a rare and severe autosomal recessive liver disease, with an incidence of approximately 1:100,000 worldwide, that is particularly suited to gene repair-based editing strategies. This is because the repaired hepatocytes have a selective replication advantage over non-repaired cells during liver growth and basal or disease-induced hepatocellular turnover. This replication advantage exists in only a limited number of diseases, making HT1 the most widely described disease model to use selective expansion to increase the percentage of corrected cells in the liver to therapeutic levels. In patients with HT1, mutation of fumarylacetoacetate hydrolase (FAH), the terminal enzyme of the tyrosine catabolic pathway, leads to accumulation of toxic metabolites and severe liver damage.[95] Interestingly, however, the liver of some patients shows a mosaic pattern of FAH activity due to the presence of clusters of healthy hepatocytes. These clusters originate from hepatocytes in which spontaneous reversion events have occurred, thereby conferring a selective growth advantage.[96,97] This strong *in vivo* selection is recapitulated in murine models of FAH deficiency[89,98] which has paved the way for investigating the utility of genome editing strategies for the treatment of HT1.[68,69,76,99–106]

The power of *in vivo* selection has been used successfully to correct a murine model of FAH deficiency using AAV-mediated gene repair *via* homologous recombination. In both neonatal and adult mice, treated with a vector comprising homology arms of *Fah* genomic sequence, FAH positive hepatocytes were identified and the underlying HT1 phenotype corrected despite a frequency of repair of approximately 1:1000 cells.[100] More recently, repair rates have been increased significantly using targeted nucleases, which although not strictly necessary in this model, will be likely required in the context of treating humans. This model has been exploited to investigate the utility of multiple genome editing strategies including the use of novel Cas9 proteins,[103] HDR,[69,99,106] base editors,[101,104] non-viral gene delivery[68,76] and *ex vivo* approaches.[102]

6.3 Targeted insertion of therapeutic transgenes into the liver

The liver exports up to one-third of the proteins it produces and plays a critical role in supporting homeostasis. The major protein synthesized by the liver is human albumin which is responsible for the transport of compounds including hormones and fatty acids and is important in maintaining oncotic pressure.[3] Clotting factors are other notable proteins produced in the liver that circulate in the blood stream.[107] Therefore, the unique potential of the liver to produce proteins that can be secreted by hepatocytes has been exploited in gene therapy strategies targeting diseases including hemophilia, lysosomal storage diseases and alpha-1-antitrypsin deficiency.[30,108–110] More recently, genome editing approaches using both ZFNs and the CRISPR/Cas9 system have been used to promote HDR in engineered murine models of hemophilia B.[78,111–113]

Genome editing strategies based on targeted insertion have also been explored to direct the event to a precise genomic location, usually the native, mutant locus or a characterized locus that represents a "safe harbor" with resulting high-level gene expression.[30,108–110,114] A seminal example of the later is the targeted insertion of exons 2–8 of the human coagulation factor IX (*F9*) into a F9 minigene placed into the murine ROSA26 locus.[114] In this humanized murine model of hemophilia B, ZFNs were designed to induce a site-specific double-strand break into the first intron of the human minigene to stimulate gene targeting. The ZFNs were co-delivered with a recombinant AAV vector containing a wild-type human *F9* donor template flanked by homology arms for repair.[114] Recombinant AAV gene delivery has also been used to treat murine models of hemophilia B by HDR.[78,111,113] However, targeting the albumin locus, which is highly expressed in the liver, as a novel method for *in vivo* protein replacement has been most widely described. Using AAV-mediated gene delivery, a promoter-less factor IX cDNA downstream of a 2A peptide coding sequence, both flanked by homology arms, was integrated immediately upstream of the stop codon of albumin in the absence of targeted nucleases.[30] Although only a low level of targeting at the albumin locus was achieved, reported at approximately 0.5%, between 7% and 20% of normal FIX expression was observed because the albumin locus is highly transcriptionally active. In addition, any potential off-target effects of nuclease delivery are avoided using this approach. This strategy, termed GeneRide, has also been used to correct murine models of Crigler-Najjar syndrome and methylmalonic acidemia.[115,116] In the latter, corrected cells had a selective growth advantage which mitigated the low efficiency of nuclease-free editing.[116] Using this nuclease-free approach, LogicBio Therapeutics has recently announced their SUNRISE trial, a multicenter, open label Phase I/II clinical trial that plans to assess the safety and tolerability of a single intravenous infusion of an AAV vector (LB-001) in pediatric patients with methylmalonic acidemia (MMA).[117]

Insertion rates can be increased with the use of targeted nucleases and this has been successful in treating murine models of hemophilia A with ectopic expression of factor VIII promoted by Cas9-mediated cleavage at the albumin locus.[118,119] To establish a general platform for liver-directed protein replacement therapies, expression of multiple therapeutic genes has been directed to the same location within the albumin locus.[120] Using ZFN-mediated site-specific integration, delivered *in vivo* by recombinant AAV vectors, long-term expression of human factors VIII and IX have been achieved in mouse models of hemophilia A and B at therapeutic levels. Furthermore, by using the same targeting reagents in wild-type mice, lysosomal enzymes were expressed that are deficient in Fabry and Gaucher diseases and in Hurler and Hunter syndromes[120] and has also been used to treat the neurological and systemic complications in mice with MPS I and MPS II deficiencies (Hurler and Hunter syndromes, respectively).[108,109] This strategy has now been adopted by Sangamo Therapeutics for the correction of indications including Hemophilia B (FIX TENDZ Study), Hurler syndrome (EMPOWERS Study, MPS I deficiency) and Hunter syndrome (CHAMPIONS Study, MPS II deficiency) in phase I/II clinical trials.[79–81] For the FIX TENDZ Study, clinical benefit has yet to be reported[121] and for the trials targeting MPS I and MPS II, although no drug-related adverse events have been observed, the levels of transgene expression were low.[122,123] This highlights the inherent challenges of translating novel gene therapies, including those that have successfully employed genome editing in animal models, to the clinical setting.

7. Translational considerations

Gene therapy clinical trials have a unique set of challenges that must be addressed to ensure the safety of human subjects. These include but are not limited to: ethical considerations and regulatory oversight, potential genotoxicity, unwanted immune responses to the gene delivery system or transgene,[124] the availability of animal models that faithfully recapitulate human disease for preclinical testing, and the ability to replicate these successes in large animals and human subjects (reviewed in Ginn and Alexander[125]). The translation of genome editing technologies have additional challenges that include pre-existing immunity against bacterial endonucleases and the unwanted introduction of mutations at both on-target and off-target sites which need to be carefully identified in model systems that will faithfully predict outcomes in humans.

7.1 Cellular responses to Cas proteins

Prior exposure to bacterial proteins such as Cas9 has the potential to influence the outcomes of therapeutic editing in humans. Recent reports have demonstrated a high seroprevalence of pre-existing immunity to Cas9 proteins from both *Streptococcus pyogenes* (SpCas9) and *Staphylococcus aureus* (SaCas9).[126–128] These represent, to date, the most commonly used Cas9 orthologues. In addition, questions have been raised about the efficacy and safety of CRISPR-Cas9-based *in vivo* editing in the liver. In a mouse model of liver-directed gene delivery, a cytotoxic T cell response was observed in animals with pre-existing SaCas9 immunity that resulted in apoptosis of hepatocytes and loss of both recombinant AAV genomes and edited cells.[129] It should be noted that these animals were deliberately vaccinated with SaCas9 one week prior to AAV-SaCas9 exposure. It will, therefore, be important to assess any impact of prior *S. aureus* exposure in non-human primates and patients. Toxicity associated with vector administration is also a consideration, particularly when high doses are used.[130] While this is not unique to gene therapy trials using genome editing, activation of p53-mediated DNA damage responses has been reported.[131,132] These have the potential to cause cell cycle arrest and apoptosis but can be mitigated by using highly specific sgRNA sequences with no detectable off-target cleavage sites[133] and transient delivery of the Cas9 editing machinery.[68,75,76]

7.2 Introducing unwanted mutations into the genome

One of the major disadvantages of a targeted correction strategy employing targeted nucleases, in contrast to locus-specific gene disruption, is the introduction of small insertions and deletions when double-stranded breaks are repaired by the cell using mechanisms such as NHEJ. These on-target mutations, in addition to the desired targeted repair, when present in exons or intronic sequences involved in splicing, are of particular concern due to the risk of converting hypomorphic alleles into null alleles.[134] Accordingly, inadvertent conversion of such alleles to a null state has the potential for therapeutic harm if there is not simultaneous conversion of a sufficient number of target alleles to a fully functional state. This critical balance will vary among disease phenotypes and be particularly important in cell autonomous diseases. Using urea cycle defects as an example, it is predicted that sub-physiological levels of enzymatic activity in a large number of hepatocytes will confer greater therapeutic benefit than physiological or supraphysiological activity in a smaller number of hepatocytes.[86] The use of nucleases can also introduce off-target mutations into the genome when the editing reagents cleave at non-specific sites with the potential to alter gene function, disrupt tumor suppressor genes or activate proto-oncogenes. Editing reagents, therefore, require careful selection and evaluation in clinically relevant models before moving to the clinic.[71] This concern has also led to the investigation of strategies such as GeneRide to avoid the use of nucleases altogether.[30] This relatively inefficient approach, however, will most likely be only applicable for a small number of diseases that have a low therapeutic threshold or where repair confers a positive selective advantage, such as hemophilia or tyrosinemia type I, respectively.

7.3 Logistic and commercial constraints

While genome editing for therapeutic purposes is without doubt a transformative paradigm, there are logistic and economic constraints to its application. The most conceptually appealing consequence of genome editing is the restoration of a mutant locus to its native state. In some disease contexts this might require the correction of a founder mutation carried by the majority of affected individuals, such that a single set of validated editing reagents would be broadly applicable. Conversely, a target disease with a high number of causative mutations in an effected population would require an extensive set of mutation-specific editing reagents to confer broad therapeutic benefit. This would increase development and manufacturing costs, add significant regulatory complexity, and reduce the incentive for commercialization, with the latter strongly influenced by target disease prevalence (market size). The consequence of these important translational considerations is likely to be a more widespread use of editing approaches where a small set of reagents have broad therapeutic utility irrespective of the nature and distribution of mutations in the target locus. Examples include the use of HITI or HDR to place a wild-type cDNA of the disease-causing gene under the transcriptional control of the native promoter for that gene[62,88] or, alternatively, under the control of a carefully selected promoter for another gene elsewhere in the genome,[30] depending on the desired pattern and level of gene expression.

8. Concluding remarks

With an increasing number of therapeutic successes being reported and investment in genome editing technology advancing rapidly, it is inevitable that "genetic surgery," as foretold by Nirenberg,[1] will revolutionize the gene therapy field and supplant conventional gene addition approaches. It is therefore essential that challenges related to patient safety and the requirement for efficient gene delivery are addressed as the field matures. This will be of particular importance when two-hit kinetics are required to achieve therapeutic endpoints. For example, systems utilizing lipid nanoparticle technology for transient delivery are being investigated to reduce potential off-target effects from prolonged nuclease expression in both preclinical models and in the clinic.[68,75,76,135–137] The near-term prospect of human clinical translation will also be critically dependent on the ongoing development and use of AAV capsids, with the capacity to target primary human hepatocytes *in vivo* with unprecedented efficiency.[90,138,139] This will ensure maximal editing rates are achieved using the lowest possible vector doses irrespective of the editing strategy that is ultimately used.

References

1. Nirenberg MW. Will society be prepared? *Science*. 1967;157(3789):633.
2. Frangoul H, Altshuler D, Cappellini MD, et al. CRISPR–Cas9 gene editing for sickle cell disease and beta-thalassemia. *N Engl J Med*. 2020;384(3):252–260.
3. Kuntz E, Kuntz H-D. *Hepatology: Textbook and Atlas 3rd Edition*. 3rd ed. Germany: Springer; 2009.
4. Ginn SL, Amaya AK, Alexander IE, Edelstein M, Abedi MR. Gene therapy clinical trials worldwide to 2017—an update. *J Gene Med*. 2018;20(5):e3015.
5. Nathwani AC, Reiss UM, Tuddenham EG, et al. Long-term safety and efficacy of factor IX gene therapy in hemophilia B. *N Engl J Med*. 2014;371(21):1994–2004.
6. Rangarajan S, Walsh L, Lester W, et al. AAV5-factor VIII gene transfer in severe hemophilia A. *N Engl J Med*. 2017;377(26):2519–2530.
7. Pasi KJ, Rangarajan S, Mitchell N, et al. Multiyear follow-up of AAV5-hFVIII-SQ gene therapy for hemophilia A. *N Engl J Med*. 2020;382(1):29–40.
8. D'Avola D, Lopez-Franco E, Sangro B, et al. Phase I open label liver-directed gene therapy clinical trial for acute intermittent porphyria. *J Hepatol*. 2016;65(4):776–783.
9. George LA, Sullivan SK, Giermasz A, et al. Hemophilia B gene therapy with a high-specific-activity factor IX variant. *N Engl J Med*. 2017;377(23):2215–2227.
10. Konkle BA, Stine K, Visweshwar N, et al. *Updated Follow-Up of the Alta Study, a Phase 1/2, Open Label, Adaptive, Dose Ranging Study to Assess the Safety and Tolerability of SB-525 Gene Therapy in Adult Patients With Severe Hemophilia A*. ISTH Academy (International Society on Thrombosis and Haemostasis), 2019.
11. Pipe SW, Recht M, Key NS, et al. LBA-6 first data from the phase 3 HOPE-B gene therapy trial: efficacy and safety of etranacogene dezaparvovec (AAV5-Padua hFIX variant; AMT-061) in adults with severe or moderate-severe Hemophilia B treated irrespective of pre-existing anti-capsid neutralizing antibodies. In: *Paper Presented at: 62nd ASH Annual Meeting and Exposition*; 2020.
12. *Gene Therapy Clinical Trials Worldwide Database*. John Wiley and Sons Ltd; 2015.
13. Baruteau J, Waddington SN, Alexander IE, Gissen P. Gene therapy for monogenic liver diseases: clinical successes, current challenges and future prospects. *J Inherit Metab Dis*. 2017;40(4):497–517.
14. Batty P, Lillicrap D. Advances and challenges for hemophilia gene therapy. *Hum Mol Genet*. 2019;28(R1):R95–R101.
15. Cozmescu AC, Counsell J, Gissen P. Gene therapies targeting the liver. *J Hepatol*. 2020;74(1):235–236.
16. Dunbar CE, High KA, Joung JK, Kohn DB, Ozawa K, Sadelain M. Gene therapy comes of age. *Science*. 2018;359(6372):eaan4672.
17. Miller DG, Trobridge GD, Petek LM, Jacobs MA, Kaul R, Russell DW. Large-scale analysis of adeno-associated virus vector integration sites in normal human cells. *J Virol*. 2005;79(17):11434–11442.
18. Gabriel R, Schmidt M, von Kalle C. Integration of retroviral vectors. *Curr Opin Immunol*. 2012;24(5):592–597.
19. Baum C, von Kalle C, Staal FJ, et al. Chance or necessity? Insertional mutagenesis in gene therapy and its consequences. *Mol Ther*. 2004;9(1):5–13.

20. Deyle DR, Russell DW. Adeno-associated virus vector integration. *Curr Opin Mol Ther.* 2009;11(4):442–447.
21. Wisse E, Jacobs F, Topal B, Frederik P, De Geest B. The size of endothelial fenestrae in human liver sinusoids: implications for hepatocyte-directed gene transfer. *Gene Ther.* 2008;15(17):1193–1199.
22. Jacobs F, Wisse E, De Geest B. The role of liver sinusoidal cells in hepatocyte-directed gene transfer. *Am J Pathol.* 2010;176(1):14–21.
23. Nguyen TH, Ferry N. Liver gene therapy: advances and hurdles. *Gene Ther.* 2004;11 (Suppl 1):S76–S84.
24. Jungermann K, Sasse D. Heterogeneity of liver parenchymal cells. *Trends Biochem Sci.* 1978;3(3):198–202.
25. Jungermann K, Katz N. Functional specialization of different hepatocyte populations. *Physiol Rev.* 1989;69(3):708–764.
26. Haussinger D, Sies H, Gerok W. Functional hepatocyte heterogeneity in ammonia metabolism. The intercellular glutamine cycle. *J Hepatol.* 1985;1(1):3–14.
27. Gebhardt R, Mecke D. Heterogeneous distribution of glutamine synthetase among rat liver parenchymal cells in situ and in primary culture. *EMBO J.* 1983;2(4):567–570.
28. Bell P, Wang L, Gao G, et al. Inverse zonation of hepatocyte transduction with AAV vectors between mice and non-human primates. *Mol Genet Metab.* 2011;104(3):395–403.
29. Dane AP, Wowro SJ, Cunningham SC, Alexander IE. Comparison of gene transfer to the murine liver following intraperitoneal and intraportal delivery of hepatotropic AAV pseudo-serotypes. *Gene Ther.* 2013;20(4):460–464.
30. Barzel A, Paulk NK, Shi Y, et al. Promoterless gene targeting without nucleases ameliorates haemophilia B in mice. *Nature.* 2015;517(7534):360–364.
31. Yu Y, Ping J, Chen H, et al. A comparative analysis of liver transcriptome suggests divergent liver function among human, mouse and rat. *Genomics.* 2010;96(5):281–289.
32. Chen TS, Chen PS. The myth of prometheus and the liver. *J R Soc Med.* 1994;87(12):754–755.
33. Fausto N, Campbell JS. The role of hepatocytes and oval cells in liver regeneration and repopulation. *Mech Dev.* 2003;120(1):117–130.
34. Duncan AW, Dorrell C, Grompe M. Stem cells and liver regeneration. *Gastroenterology.* 2009;137(2):466–481.
35. Schnepp BC, Jensen RL, Chen CL, Johnson PR, Clark KR. Characterization of adeno-associated virus genomes isolated from human tissues. *J Virol.* 2005;79(23):14793–14803.
36. Cunningham SC, Spinoulas A, Carpenter KH, Wilcken B, Kuchel PW, Alexander IE. AAV2/8-mediated correction of OTC deficiency is robust in adult but not neonatal Spf^{ash} mice. *Mol Ther.* 2009;17(8):1340–1346.
37. Coppoletta JM, Wolbach SB. Body length and organ weights of infants and children: a study of the body length and normal weights of the more important vital organs of the body between birth and twelve years of age. *Am J Pathol.* 1933;9(1):55–70.
38. Hustedt N, Durocher D. The control of DNA repair by the cell cycle. *Nat Cell Biol.* 2016;19(1):1–9.
39. Folger KR, Wong EA, Wahl G, Capecchi MR. Patterns of integration of DNA microinjected into cultured mammalian cells: evidence for homologous recombination between injected plasmid DNA molecules. *Mol Cell Biol.* 1982;2(11):1372–1387.
40. Smithies O, Gregg RG, Boggs SS, Koralewski MA, Kucherlapati RS. Insertion of DNA sequences into the human chromosomal beta-globin locus by homologous recombination. *Nature.* 1985;317(6034):230–234.
41. Capecchi MR. Altering the genome by homologous recombination. *Science.* 1989;244 (4910):1288–1292.

42. Russell DW, Hirata RK. Human gene targeting by viral vectors. *Nat Genet*. 1998;18(4): 325–330.
43. Alexander IE, Russell DW. The potential of AAV-mediated gene targeting for gene and cell therapy applications. *Curr Stem Cell Res*. 2015;1:16–22.
44. Ran FA, Cong L, Yan WX, et al. In vivo genome editing using Staphylococcus aureus Cas9. *Nature*. 2015;520(7546):186–191.
45. Beerli RR, Segal DJ, Dreier B, Barbas 3rd CF. Toward controlling gene expression at will: specific regulation of the erbB-2/HER-2 promoter by using polydactyl zinc finger proteins constructed from modular building blocks. *Proc Natl Acad Sci U S A*. 1998;95 (25):14628–14633.
46. Kim YG, Cha J, Chandrasegaran S. Hybrid restriction enzymes: zinc finger fusions to Fok I cleavage domain. *Proc Natl Acad Sci U S A*. 1996;93(3):1156–1160.
47. Urnov FD, Rebar EJ, Holmes MC, Zhang HS, Gregory PD. Genome editing with engineered zinc finger nucleases. *Nat Rev Genet*. 2010;11(9):636–646.
48. Li H, Yang Y, Hong W, Huang M, Wu M, Zhao X. Applications of genome editing technology in the targeted therapy of human diseases: mechanisms, advances and prospects. *Signal Transduct Target Ther*. 2020;5(1):1.
49. Moscou MJ, Bogdanove AJ. A simple cipher governs DNA recognition by TAL effectors. *Science*. 2009;326(5959):1501.
50. Boch J, Scholze H, Schornack S, et al. Breaking the code of DNA binding specificity of TAL-type III effectors. *Science*. 2009;326(5959):1509–1512.
51. Jinek M, Chylinski K, Fonfara I, Hauer M, Doudna JA, Charpentier E. A programmable dual-RNA-guided DNA endonuclease in adaptive bacterial immunity. *Science*. 2012;337(6096):816–821.
52. Cong L, Ran FA, Cox D, et al. Multiplex genome engineering using CRISPR/Cas systems. *Science*. 2013;339(6121):819–823.
53. Mali P, Yang L, Esvelt KM, et al. RNA-guided human genome engineering via Cas9. *Science*. 2013;339(6121):823–826.
54. Burstein D, Harrington LB, Strutt SC, et al. New CRISPR-Cas systems from uncultivated microbes. *Nature*. 2017;542(7640):237–241.
55. Chen JS, Ma E, Harrington LB, et al. CRISPR-Cas12a target binding unleashes indiscriminate single-stranded DNase activity. *Science*. 2018;360(6387):436–439.
56. Naeem M, Majeed S, Hoque MZ, Ahmad I. Latest developed strategies to minimize the off-target effects in CRISPR-Cas-mediated genome editing. *Cells*. 2020;9(7):1608.
57. Nelson CE, Wu Y, Gemberling MP, et al. Long-term evaluation of AAV-CRISPR genome editing for Duchenne muscular dystrophy. *Nat Med*. 2019;25(3):427–432.
58. Ceccaldi R, Rondinelli B, D'Andrea AD. Repair pathway choices and consequences at the double-strand break. *Trends Cell Biol*. 2016;26(1):52–64.
59. Iyama T, Wilson 3rd DM. DNA repair mechanisms in dividing and non-dividing cells. *DNA Repair*. 2013;12(8):620–636.
60. Lieber MR. The mechanism of double-strand DNA break repair by the non-homologous DNA end-joining pathway. *Annu Rev Biochem*. 2010;79:181–211.
61. Maruyama T, Dougan SK, Truttmann MC, Bilate AM, Ingram JR, Ploegh HL. Increasing the efficiency of precise genome editing with CRISPR-Cas9 by inhibition of nonhomologous end joining. *Nat Biotechnol*. 2015;33(5):538–542.
62. Suzuki K, Izpisua Belmonte JC. In vivo genome editing via the HITI method as a tool for gene therapy. *J Hum Genet*. 2018;63(2):157–164.
63. Suzuki K, Yamamoto M, Hernandez-Benitez R, et al. Precise in vivo genome editing via single homology arm donor mediated intron-targeting gene integration for genetic disease correction. *Cell Res*. 2019;29(10):804–819.
64. Nakade S, Tsubota T, Sakane Y, et al. Microhomology-mediated end-joining-dependent integration of donor DNA in cells and animals using TALENs and CRISPR/Cas9. *Nat Commun*. 2014;5:5560.

65. Taleei R, Nikjoo H. Biochemical DSB-repair model for mammalian cells in G1 and early S phases of the cell cycle. *Mutat Res.* 2013;756(1–2):206–212.
66. Anzalone AV, Koblan LW, Liu DR. Genome editing with CRISPR-Cas nucleases, base editors, transposases and prime editors. *Nat Biotechnol.* 2020;38(7):824–844.
67. Uddin F, Rudin CM, Sen T. CRISPR gene therapy: applications, limitations, and implications for the future. *Front Oncol.* 2020;10:1387.
68. Yin H, Song CQ, Dorkin JR, et al. Therapeutic genome editing by combined viral and non-viral delivery of CRISPR system components in vivo. *Nat Biotechnol.* 2016;34(3):328–333.
69. Shao Y, Wang L, Guo N, et al. Cas9-nickase-mediated genome editing corrects hereditary tyrosinemia in rats. *J Biol Chem.* 2018;293(18):6883–6892.
70. Zhang QS, Tiyaboonchai A, Nygaard S, et al. Induced liver regeneration enhances CRISPR/Cas9-mediated gene repair in tyrosinemia type 1. *Hum Gene Ther.* 2020. https://doi.org/10.1089/hum.2020.042.
71. Ginn SL, Amaya AK, Liao SHY, et al. Efficient in vivo editing of OTC-deficient patient-derived primary human hepatocytes. *JHEP Rep.* 2020;2(1):100065.
72. Ding Q, Strong A, Patel KM, et al. Permanent alteration of PCSK9 with in vivo CRISPR-Cas9 genome editing. *Circ Res.* 2014;115(5):488–492.
73. Jarrett KE, Lee CM, Yeh YH, et al. Somatic genome editing with CRISPR/Cas9 generates and corrects a metabolic disease. *Sci Rep.* 2017;7:44624.
74. Wang X, Raghavan A, Chen T, et al. CRISPR-Cas9 targeting of PCSK9 in human hepatocytes in vivo-brief report. *Arterioscler Thromb Vasc Biol.* 2016;36(5):783–786.
75. Finn JD, Smith AR, Patel MC, et al. A single administration of CRISPR/Cas9 lipid nanoparticles achieves robust and persistent in vivo genome editing. *Cell Rep.* 2018;22(9):2227–2235.
76. Yin H, Xue W, Chen S, et al. Genome editing with Cas9 in adult mice corrects a disease mutation and phenotype. *Nat Biotechnol.* 2014;32(6):551–553.
77. Borel F, Tang Q, Gernoux G, et al. Survival advantage of both human hepatocyte xenografts and genome-edited hepatocytes for treatment of alpha-1 antitrypsin deficiency. *Mol Ther.* 2017;25(11):2477–2489.
78. Ohmori T, Nagao Y, Mizukami H, et al. CRISPR/Cas9-mediated genome editing via postnatal administration of AAV vector cures haemophilia B mice. *Sci Rep.* 2017;7(1):4159.
79. *Ascending Dose Study of Genome Editing by Zinc Finger Nuclease Therapeutic SB-FIX in Subjects With Severe Hemophilia B*; 2019. https://clinicaltrials.gov/ct2/show/NCT02695160.
80. *Ascending Dose Study of Genome Editing by the Zinc Finger Nuclease (ZFN) Therapeutic SB-318 in Subjects With MPS I*; 2020. https://clinicaltrials.gov/ct2/show/NCT02702115.
81. *Ascending Dose Study of Genome Editing by the Zinc Finger Nuclease (ZFN) Therapeutic SB-913 in Subjects With MPS II*; 2019. https://clinicaltrials.gov/ct2/show/NCT03041324.
82. Mannucci PM, Tuddenham EG. The hemophilias—from royal genes to gene therapy. *N Engl J Med.* 2001;344(23):1773–1779.
83. Simioni P, Tormene D, Tognin G, et al. X-linked thrombophilia with a mutant factor IX (factor IX Padua). *N Engl J Med.* 2009;361(17):1671–1675.
84. Miesbach W, Meijer K, Coppens M, et al. Gene therapy with adeno-associated virus vector 5-human factor IX in adults with hemophilia B. *Blood.* 2018;131(9):1022–1031.
85. Monahan PE, Sun J, Gui T, et al. Employing a gain-of-function factor IX variant R338L to advance the efficacy and safety of hemophilia B human gene therapy: preclinical evaluation supporting an ongoing adeno-associated virus clinical trial. *Hum Gene Ther.* 2015;26(2):69–81.
86. Kok CY, Cunningham SC, Kuchel PW, Alexander IE. Insights into gene therapy for urea cycle defects by mathematical modeling. *Hum Gene Ther.* 2019;30(11):1385–1394.

87. Yang Y, Wang L, Bell P, et al. A dual AAV system enables the Cas9-mediated correction of a metabolic liver disease in newborn mice. *Nat Biotechnol*. 2016;34(3):334–338.
88. Wang L, Yang Y, Breton C, et al. A mutation-independent CRISPR-Cas9–mediated gene targeting approach to treat a murine model of ornithine transcarbamylase deficiency. *Sci Adv*. 2020;6(6):eaax5701.
89. Azuma H, Paulk N, Ranade A, et al. Robust expansion of human hepatocytes in Fah-/-/Rag2-/-/Il2rg-/- mice. *Nat Biotechnol*. 2007;25(8):903–910.
90. Paulk NK, Pekrun K, Zhu E, et al. Bioengineered AAV capsids with combined high human liver transduction in vivo and unique humoral seroreactivity. *Mol Ther*. 2018;26 (1):289–303.
91. Hanlon KS, Kleinstiver BP, Garcia SP, et al. High levels of AAV vector integration into CRISPR-induced DNA breaks. *Nat Commun*. 2019;10(1):4439.
92. Ah Mew N, Simpson KL, Gropman AL, et al. Urea cycle disorders overview. In: Adam MP, Ardinger HH, Pagon RA, et al., eds. *GeneReviews*® *[Internet]*. Seattle, WA: University of Washington; 2003:1993–2021. [Updated 2017 Jun 22].
93. Abifadel M, Varret M, Rabes JP, et al. Mutations in PCSK9 cause autosomal dominant hypercholesterolemia. *Nat Genet*. 2003;34(2):154–156.
94. Wang L, Smith J, Breton C, et al. Meganuclease targeting of PCSK9 in macaque liver leads to stable reduction in serum cholesterol. *Nat Biotechnol*. 2018;36(8):717–725.
95. Lindblad B, Lindstedt S, Steen G. On the enzymic defects in hereditary tyrosinemia. *Proc Natl Acad Sci U S A*. 1977;74(10):4641–4645.
96. Kvittingen EA, Rootwelt H, Brandtzaeg P, Bergan A, Berger R. Hereditary tyrosinemia type I. Self-induced correction of the fumarylacetoacetase defect. *J Clin Invest*. 1993;91(4):1816–1821.
97. Kvittingen EA, Rootwelt H, Berger R, Brandtzaeg P. Self-induced correction of the genetic defect in tyrosinemia type I. *J Clin Invest*. 1994;94(4):1657–1661.
98. Overturf K, Al-Dhalimy M, Tanguay R, et al. Hepatocytes corrected by gene therapy are selected in vivo in a murine model of hereditary tyrosinaemia type I. *Nat Genet*. 1996;12(3):266–273.
99. Krooss SA, Dai Z, Schmidt F, et al. Ex vivo/in vivo gene editing in hepatocytes using "all-in-one" CRISPR-adeno-associated virus vectors with a self-linearizing repair template. *iScience*. 2020;23(1), 100764.
100. Paulk NK, Wursthorn K, Wang Z, Finegold MJ, Kay MA, Grompe M. Adeno-associated virus gene repair corrects a mouse model of hereditary tyrosinemia in vivo. *Hepatology*. 2010;51(4):1200–1208.
101. Song CQ, Jiang T, Richter M, et al. Adenine base editing in an adult mouse model of tyrosinaemia. *Nat Biomed Eng*. 2020;4(1):125–130.
102. VanLith C, Guthman R, Nicolas CT, et al. Curative ex vivo hepatocyte-directed gene editing in a mouse model of hereditary tyrosinemia type 1. *Hum Gene Ther*. 2018;29 (11):1315–1326.
103. Agudelo D, Carter S, Velimirovic M, et al. Versatile and robust genome editing with Streptococcus thermophilus CRISPR1-Cas9. *Genome Res*. 2020;30(1):107–117.
104. Yang L, Wang L, Huo Y, et al. Amelioration of an inherited metabolic liver disease through creation of a de novo start codon by cytidine base editing. *Mol Ther*. 2020;28(7):1673–1683.
105. Junge N, Yuan Q, Vu TH, et al. Homologous recombination mediates stable fah gene integration and phenotypic correction in tyrosinaemia mouse-model. *World J Hepatol*. 2018;10(2):277–286.
106. Li N, Gou S, Wang J, et al. CRISPR/Cas9-mediated gene correction in newborn rabbits with hereditary tyrosinemia type I. *Mol Ther*. 2020. https://doi.org/10.1016/j.ymthe.2020.11.023.

107. Biggs R, Douglas AS, Macfarlane RG, Dacie JV, Pitney WR, Merskey. Christmas disease: a condition previously mistaken for haemophilia. *Br Med J.* 1952;2(4799):1378–1382.
108. Laoharawee K, DeKelver RC, Podetz-Pedersen KM, et al. Dose-dependent prevention of metabolic and neurologic disease in murine MPS II by ZFN-mediated in vivo genome editing. *Mol Ther.* 2018;26(4):1127–1136.
109. Ou L, DeKelver RC, Rohde M, et al. ZFN-mediated in vivo genome editing corrects murine hurler syndrome. *Mol Ther.* 2019;27(1):178–187.
110. Stephens CJ, Kashentseva E, Everett W, Kaliberova L, Curiel DT. Targeted in vivo knock-in of human alpha-1-antitrypsin cDNA using adenoviral delivery of CRISPR/Cas9. *Gene Ther.* 2018;25(2):139–156.
111. Anguela XM, High KA. Entering the modern era of gene therapy. *Annu Rev Med.* 2019;70:273–288.
112. Guan Y, Ma Y, Li Q, et al. CRISPR/Cas9-mediated somatic correction of a novel coagulator factor IX gene mutation ameliorates hemophilia in mouse. *EMBO Mol Med.* 2016;8(5):477–488.
113. Huai C, Jia C, Sun R, et al. CRISPR/Cas9-mediated somatic and germline gene correction to restore hemostasis in hemophilia B mice. *Hum Genet.* 2017;136(7):875–883.
114. Li H, Haurigot V, Doyon Y, et al. In vivo genome editing restores haemostasis in a mouse model of haemophilia. *Nature.* 2011;475(7355):217–221.
115. Porro F, Bortolussi G, Barzel A, et al. Promoterless gene targeting without nucleases rescues lethality of a Crigler-Najjar syndrome mouse model. *EMBO Mol Med.* 2017;9(10):1346–1355.
116. Chandler RJ, Venturoni LE, Liao J, et al. Promoterless, nuclease-free genome editing confers a growth advantage for corrected hepatocytes in mice with methylmalonic acidemia. *Hepatology.* 2020. https://doi.org/10.1002/hep.31570.
117. *LogicBio Therapeutics Announces SUNRISE Phase 1/2 Clinical Design for LB-001 for the Treatment of Methylmalonic Acidemia in Pediatric Patients [Press Release]*. LogicBio Website: LogicBio; 2020. August 10, 2020 at 6:35 AM EDT.
118. Chen H, Shi M, Gilam A, et al. Hemophilia A ameliorated in mice by CRISPR-based in vivo genome editing of human factor VIII. *Sci Rep.* 2019;9(1):16838.
119. Zhang JP, Cheng XX, Zhao M, et al. Curing hemophilia A by NHEJ-mediated ectopic F8 insertion in the mouse. *Genome Biol.* 2019;20(1):276.
120. Sharma R, Anguela XM, Doyon Y, et al. In vivo genome editing of the albumin locus as a platform for protein replacement therapy. *Blood.* 2015;126(15):1777–1784.
121. https://www.biospace.com/article/releases/sangamo-provides-clinical-development-update-including-early-phase-1-2-beta-thalassemia-gene-edited-cell-therapy-data/ (Published: Apr 02, 2019).
122. Muenzer J, Prada CE, Burton B, et al. CHAMPIONS: a phase 1/2 clinical trial with dose escalation of SB-913 ZFN-mediated in vivo human genome editing for treatment of MPS II (Hunter syndrome). In: *Paper Presented at: Molecular Genetics and Metabolism*; 2019.
123. Harmatz P, Lau HA, Heldermon C, et al. EMPOWERS: a phase 1/2 clinical trial of SB-318 ZFN-mediated in vivo human genome editing for treatment of MPS I (Hurler syndrome). In: *Paper Presented at: Molecular Genetics and Metabolism*; 2019.
124. Mingozzi F, High KA. Immune responses to AAV vectors: overcoming barriers to successful gene therapy. *Blood.* 2013;122(1):23–36.
125. Ginn SL, Alexander IE. Gene therapy. In: Balakrishnan N, Colton T, Everitt B, Piegorsch W, Ruggeri F, Teugels JL, eds. *Wiley StatsRef: Statistics Reference Online*; 2016. https://doi.org/10.1002/9781118445112.stat06926.pub2.
126. Charlesworth CT, Deshpande PS, Dever DP, et al. Identification of preexisting adaptive immunity to Cas9 proteins in humans. *Nat Med.* 2019;25(2):249–254.

127. Wagner DL, Amini L, Wendering DJ, et al. High prevalence of Streptococcus pyogenes Cas9-reactive T cells within the adult human population. *Nat Med.* 2019;25(2):242–248.
128. Simhadri VL, McGill J, McMahon S, Wang J, Jiang H, Sauna ZE. Prevalence of pre-existing antibodies to CRISPR-associated nuclease Cas9 in the USA population. *Mol Ther Methods Clin Dev.* 2018;10:105–112.
129. Li A, Tanner MR, Lee CM, et al. AAV-CRISPR gene editing is negated by pre-existing immunity to Cas9. *Mol Ther.* 2020;28(6):1432–1441.
130. Hinderer C, Katz N, Buza EL, et al. Severe toxicity in nonhuman primates and piglets following highdose intravenous administration of an AAV vector expressing human SMN. *Hum Gene Ther.* 2018;29(3):285–298.
131. Ihry RJ, Worringer KA, Salick MR, et al. p53 inhibits CRISPR–Cas9 engineering in human pluripotent stem cells. *Nat Med.* 2018;24(7):939–946.
132. Haapaniemi E, Botla S, Persson J, Schmierer B, Taipale J. CRISPR–Cas9 genome editing induces a p53-mediated DNA damage response. *Nat Med.* 2018;24(7):927–930.
133. Schiroli G, Conti A, Ferrari S, et al. Precise gene editing preserves hematopoietic stem cell function following transient p53-mediated DNA damage response. *Cell Stem Cell.* 2019;24(4):551–565 e558.
134. Maule G, Casini A, Montagna C, et al. Allele specific repair of splicing mutations in cystic fibrosis through AsCas12a genome editing. *Nat Commun.* 2019;10(1):3556.
135. Schuh RS, Poletto E, Pasqualim G, et al. In vivo genome editing of mucopolysaccharidosis I mice using the CRISPR/Cas9 system. *J Control Release.* 2018;288:23–33.
136. Intellia Therapeutics. *Intellia Therapeutics. Doses First Patient in Landmark CRISPR/Cas9 Clinical Trial of NTLA-2001 for the Treatment of Transthyretin Amyloidosis.* [press release]. 9th November, GlobeNewsWire, Intellia Therapeutics, Inc.; 2020.
137. Intellia Therapeutics. *Study to Evaluate Safety, Tolerability, Pharmacokinetics, and Pharmacodynamics of NTLA-2001 in Patients With Hereditary Transthyretin Amyloidosis With Polyneuropathy (ATTRv-PN).* U.S. National Library of Medicine; 2020. NCT04601051 ClinicalTrials.gov. Accessed 11 November 2020.
138. Lisowski L, Dane AP, Chu K, et al. Selection and evaluation of clinically relevant AAV variants in a xenograft liver model. *Nature.* 2014;506(7488):382–386.
139. Cabanes-Creus M, Hallwirth CV, Westhaus A, et al. Restoring the natural tropism of AAV2 vectors for human liver. *Sci Transl Med.* 2020;12(560):eaba3312.

CHAPTER NINE

Genome editing in lysosomal disorders

Luisa Natalia Pimentel-Vera[a,b], Edina Poletto[a,b], Esteban Alberto Gonzalez[a,b], Fabiano de Oliveira Poswar[c], Roberto Giugliani[a,b,c,d,e,f], and Guilherme Baldo[a,b,e,g],*

[a]Postgraduate Program in Genetics and Molecular Biology, UFRGS, Porto Alegre, Brazil
[b]Gene Therapy Center, HCPA, Porto Alegre, Brazil
[c]Medical Genetics Service, HCPA, Porto Alegre, Brazil
[d]INAGEMP, Porto Alegre, Brazil
[e]Biodiscovery Research Group, Experimental Research Center, HCPA, Porto Alegre, Brazil
[f]Department of Genetics, UFRGS, Porto Alegre, Brazil
[g]Department of Physiology, UFRGS, Porto Alegre, Brazil
*Corresponding author: e-mail address: gbaldo@hcpa.edu.br

Contents

1. Introduction: Lysosomal disorders	291
2. Physiopathology, clinical manifestations and natural history	292
3. Diagnosis and molecular genetics	293
4. Current treatments: Impact and limitations	297
4.1 Hematopoietic stem cell transplantation	297
4.2 Enzyme replacement therapy	297
4.3 Small molecules	300
5. Genome editing tools	301
6. Genome-edited cell models	303
7. Genome-edited animal models	306
8. Therapies based on genome editing	307
8.1 In vitro mutation-specific genome editing	307
8.2 Ex vivo genome editing targeting safe harbors loci	311
8.3 In vivo genome editing	314
9. Clinical trials for lysosomal disorders	317
10. Conclusions	319
References	319

Abstract

Lysosomal disorders are a group of heterogenous diseases caused by mutations in genes that encode for lysosomal proteins. With exception of some cases, these disorders still lack both knowledge of disease pathogenesis and specific therapies. In this sense, genome editing arises as a technique that allows both the creation of specific cell lines, animal models and gene therapy protocols for these disorders. Here we explain the main applications of genome editing for lysosomal diseases, with examples based on the literature. The ability to rewrite the genome will be of extreme importance to study and potentially treat these rare disorders.

Abbreviations

AAV	adeno associated virus
AAVS1	AAV integration site
ABE	adenine base editor
ALB	albumin
BBB	blood-brain barrier
CBE	cystine base editors
CCR5	chemokine (C-C motif) receptor
CHO	Chinese hamster ovary
CMT	Charcot-Marie-tooth disease
CNS	central nervous system
CRISPR	clustered regularly interspaced short palindromic repeats
DNA	deoxyribonucleic acid
DSB	double-strand break
EMA	European Medicines Agency
eNOS	endothelial nitric oxide synthase
ERT	enzyme replacement therapy
FDA	American Food and Drug Administration
FTDALS	frontotemporal dementia and/or amyotrophic lateral sclerosis
GAA	acid α-glucosidase
GALC	galactosylceramidase
GBA	glucocerebrosidase
GLA	α-galactosidase
gRNA	guide RNA
HDR	homology directed repair
HEK	human embryonic kidney
hESC	human embryonic stem cells
HEX	β-hexosaminidases
HGSNAT	heparan-alpha-glucosaminide *N*-acetyltransferase
HSC	hematopoietic stem cell
HSCT	hematopoietic stem cell transplantation
IDS	iduronate-2-sulfatase
IDUA	alpha-L-iduronidase
IL2RG	interleukin-2 receptor gamma chain
iPSCs	induced pluripotent stem cells
LAL	lysosomal acid lipase
LDs	lysosomal disorders
ML	mucolipidosis
MPS	mucopolysaccharidosis
NBIA	neurodegeneration with brain iron accumulation
NCL	neuronal ceroid lipofuscinosis
NHEJ	non-homologous end joining
Nick	single-strand break

NSCs	neural stem cells
PAM	protospacer adjacent motif
PC	pharmacological chaperone
PEG	polyethylene glycol
PME	progressive myoclonic epilepsy
RNA	ribonucleic acid
RNP	ribonucleoprotein
SNV	single nucleotide variant
SRT	substrate reduction therapy
TALENs	transcription activator-like effector nucleases
UCB	umbilical cord blood
ZFNs	zinc-finger nucleases

1. Introduction: Lysosomal disorders

The lysosomal disorders (LDs) comprise a group of over 70 monogenic disorders involving genes encoding for acid hydrolases, membrane proteins, transporters, enzyme modifiers and activators, or other proteins affecting the function of the lysosome. Many LDs result from an impairment in the catabolism of complex molecules, which as result tend to store in cells and tissues. For instance, a defect in any of the 11 enzymes related to the degradation of glycosaminoglycans (formerly denominated mucopolysaccharides) implicates in mucopolysaccharidoses I-IX (MPS I-IX). Likewise, an impairment in the catabolism of sphingolipids due to either the deficiency of lysosomal enzymes or their activators is the cause of sphingolipidoses, such as Fabry, Gaucher, Niemann Pick types A/B, Tay-Sachs and Sandhoff diseases, and a defective degradation of oligosaccharide chains of glycoproteins causes oligosaccharidoses (also known as glycoproteinoses), including sialidosis, fucosidosis, α-mannosidosis, β-mannosidosis, and aspartylglycosaminuria.[1] Defects in other lysosomal components may also result in lysosomal dysfunction with or without storage.

Almost all LDs have an autosomal recessive inheritance, except for Danon disease, mucopolysaccharidosis II and Fabry disease, which are X-linked traits. LDs are rare, with some conditions, like MPS IX and β-mannosidosis, having less than 100 cases described. Collectively, the global birth prevalence of the LDs is estimated to be around 1:7000.[2] However, this number is likely to be underestimated, considering the results of newborn screening studies that suggest the existence of a large number of undiagnosed patients with later-onset phenotypes, especially for Fabry disease, presently considered the commonest LD.[3,4] Furthermore, a much higher prevalence of certain LDs is observed in specific ethnicities, most likely as a result of bottleneck and founder effects.

In this chapter, we aim to discuss the clinical and molecular aspects of the LDs and how the use of genome editing tools may aid the development of better treatments, by overcoming the limitations of current therapeutical approaches.

2. Physiopathology, clinical manifestations and natural history

Lysosomes are involved in the degradation of several macromolecules through the action of more than 50 hydrolases. Those enzymes require the acidic environment of the lysosomes for their proper function and may be much less active in the cytosol. The final products (e.g., amino acids and monosaccharides) are released from the lysosomes and made available for other metabolic pathways outside the organelle. When a lysosomal enzyme is deficient, intra-lysosomal accumulation of undegraded or partially degraded substrates occurs, resulting in enlargement of lysosomes and promoting a cascade of secondary effects. Besides the storage of the substrate of the deficient enzyme, secondary storage may also occur in LDs, due to inhibition of other lysosomal degradation pathways,[5] as is the case of the accumulation of gangliosides in the mucopolysaccharidoses. Lysosomes are known to be involved in several cellular functions including autophagy,[6] cholesterol homeostasis[7] and cell death,[8] processes that may be either primarily or secondarily compromised in LDs.

Although lysosomes are ubiquitously present in the cells, the consequences of a deficiency of a lysosomal enzyme may result in different impacts in distinct cell types, according to the cell function and the tissue where a particular substrate is present. For instance, phagocytes play a major role in the engulfment and digestion of extracellular components and some lysosomal diseases may be accompanied by histopathological and functional changes in the cells of the mononuclear phagocyte system, particularly Gaucher and Niemann Pick diseases.[9] This results in a range of overlapping clinical manifestations including hepatosplenomegaly, osteolysis, interstitial lung disease and, in severe cases, neurological involvement.

For some LDs, clinical manifestations are nearly restricted to a few cells and tissues where the substrate is present, as is the case of Krabbe disease and metachromatic leukodystrophy. Both diseases result from impairments in the degradation of two major components of the myelin lipids, galactocerebroside and its sulfated derivative, sulfatide, resulting mainly in demyelinating disease of the central nervous system.[10]

Without treatment, most LDs will have a progressive course, with worsening of symptoms, disability and early death. There is a wide range of presentation for each of the LDs, which may span from prenatal presentations such as hydrops fetalis (as in some patients with Niemann Pick type C and Gaucher Disease type 2, and a large proportion of patients with MPS VII) to a much more attenuated phenotype presenting in adulthood and without the full range of manifestations (as in some patients with the cardiac variant of Fabry disease).

The rate of the disease progression is, in general, faster in patients who have more severe, infantile-onset presentations, while patients with the adult-onset disease may have a much slower progression. These characteristics of LDs may complicate the assessment of treatment efficacy in clinical trials if the intervention is not expected to result in improvement but only in stabilization or slower progression of the symptoms.[11] Table 1 summarizes the main lysosomal storage disorders, their clinical manifestations and the available treatments.

3. Diagnosis and molecular genetics

Considering the rarity of the LDs, their absence in most public newborn screening programs, and the fact that many patients do not have a family history of the disease (as expected for the autosomal recessive inheritance), diagnostic delays are common, with many patients being diagnosed several months or even decades after the onset of the symptoms.[13] LDs manifestations may be misattributed to more common conditions. For example, patients with Fabry disease and white matter involvement were misdiagnosed as multiple sclerosis[14] and patients with late-onset GM2 gangliosidosis and psychiatric manifestations were considered to suffer from schizophrenia.[15]

When a LD is suspected, the traditional approach involves assessing enzyme activity and/or the levels of informative biomarkers, which are typically derived from the undegraded substrates (e.g., the glycosphingolipids psychosine, glucosylsphingosine and globotriaosylsphingosine for Krabbe, Gaucher and Fabry diseases, respectively). The biochemical diagnosis is then confirmed through a molecular test, including DNA sequencing and deletion/duplication analysis. Enzyme activity assessments may be less reliable in

Table 1 Lysosomal diseases groups, manifestations and treatment.

Group of LDs	Diseases	Clinical manifestations	Treatment
Sphingolipidoses	Fabry, Gaucher, Krabbe and Farber diseases; GM1 and GM2 gangliosidoses; multiple sulfatases deficiency, saposin deficiencies	Hepatosplenomegaly, chronic kidney disease, anemia, thrombocytopenia, cognitive impairment, neurological regression, cherry-red spots, white matter disease, ataxia	ERT, SRT, HSCT, PC
Mucopolysaccharidoses	MPS I, II, IIIA, IIIB, IIIC, IIID, IVA, IVB, VI, VII, IX, MPS Plus	Coarse facial features, corneal clouding, joint stiffness, dysostosis multiplex, hepatomegaly, valve disease	ERT, HSCT
Oligossacharidoses	Sialidosis, fucosidosis, α-mannosidosis, β-mannosidosis, galactosialidosis and aspartylglycosaminuria	Coarse facial features, ataxia, myopathy, myoclonus, hepatosplenomegaly	ERT
Mucolipidoses	ML-II, ML-III, ML-IV	Coarse facial features, dysostosis multiplex, hepatomegaly, hyperparathyroidism, valve disease	None available
Neuronal ceroid lipofuscinoses (Batten diseases)	NCL1-14	Epilepsy, brain atrophy, ataxia, optic nerve atrophy	ERT
Disorders of lysosomal transport	Cystinosis, sialic acid storage disease, EPM4	Chronic kidney disease, Fanconi syndrome, coarse facial features, cardiomegaly, dysostosis multiplex, hypopigmented skin, myoclonus	SRT

Disorders of lysosomal cholesterol metabolism	Niemann Pick type C, lysosomal acid lipase deficiency	Hepatomegaly, splenomegaly, ataxia, neurodegeneration, steatosis, hypercholesterolemia	ERT, SRT
Disorders of autophagy	Autosomal recessive spastic paraplegia 11, 15, 48 and 49, Vici syndrome, NBIA 5, FTDALS 4, CMT type 2B	Spastic paraplegia, thin corpus callosum, ocular albinism, cardiomyopathy, extrapyramidal signs, fasciculations, peripheral neuropathy	None available
Disorders of lysosomal protein degradation	Pycnodysostosis, Papillon-Lefevre syndrome	Bone dysplasia, periodontitis, hyperkeratosis of palms and soles	None available

Classification as proposed by Ferreira.[12] The treatment column denotes if there is an available treatment for any of the conditions cited in each group. Disease abbreviations: CMT, Charcot-Marie-tooth disease; EPM, epilepsy progressive myoclonic; FTDALS, frontotemporal dementia and/or amyotrophic lateral sclerosis; ML, mucolipidosis; MPS, mucopolysaccharidosis; NBIA, neurodegeneration with brain iron accumulation; NCL, neuronal ceroid lipofuscinosis. Treatment abbreviations: ERT, enzyme replacement therapy; HSCT, hematopoietic stem cell transplantation; PC, pharmacological chaperone; SRT, substrate reduction therapy.

some situations, especially when investigating females for Fabry disease and when pseudo-deficient alleles are present. Furthermore, for conditions caused by the deficiency of activator proteins and other nonenzymatic proteins, molecular tests are the only available method for establishing a definitive diagnosis, although clinical and biochemical biomarkers may play a role in raising the level of suspicion.

Targeted mutation analysis is a feasible approach in many situations for selected LDs, considering the existence of frequent variants. For instance, the *HEXA* gene variants c.1421+1G>C, c.1274_1277dupTATC, and c.805G>A (p.G269S) are found in 98% of the Ashkenazi Jews with Tay-Sachs disease and 35% of non-Ashkenazi patients. On the other hand, in ~80% of Japanese patients with Tay-Sachs disease the *HEXA* gene variant c.571–1G>T is found.[16] For Gaucher disease, the variant c.1226A>G (N370S) is known to be highly represented in the Ashkenazi Jewish population and is associated with a later-onset phenotype.[17]

With the incorporation of massive parallel sequencing to the clinical practice, a simultaneous molecular testing approach for more than one condition is been advocated in some cases, especially considering the existence of significant clinical overlap among some LDs.[18] Targeted next-generation sequencing pipelines, however, may face the challenges of overcoming technical difficulties imposed by the presence of high GC content and pseudogenes and for some LDs, which may also need to be addressed when designing genome-editing tools.[19]

Pseudogenes have been described in different types of lysosomal diseases. The gene mutated in Gaucher disease, *GBA*, has a highly homologous pseudogene *GBAP1*, and nonreciprocal homologous recombination is a common mechanism of mutation for this disease. The two most common disease-causing recombinant variants are RecNcil and Recdelta55, which result in the incorporation of the sequence of *GBAP1* to the *GBA* gene resulting in the introduction of missense variants or the deletion of 55 nucleotides, respectively.[20] Similarly, in MPS type II, a pseudogene located 20 kb telomeric to the *IDS* gene is responsible for a range of gene rearrangements and deletions.[21]

With the increased availability of next-generation sequencing, unexpected phenotypes are being more and more described, as a clinical picture dominated by retinitis pigmentosa in some patients with mutations in *HGSNAT*, which is usually related to Mucopolysaccharidosis IIIC[22]. It is expected that with the growing use of genomic-based diagnosis, findings like this will become more common.

Besides the diagnosis, identification of causal variants may be valuable for establishing the prognosis and the most appropriate therapies, as further discussed in the next section. As LDs are caused by a loss of function mechanism, null variants are predicted to be associated with more severe phenotypes or even fetal death in some cases. However, genotype-phenotype correlations are not perfect, and patients with the same genotype were described as having variations in disease severity. Enzyme activity may also not have a clear correlation with the phenotype. Genetic, epigenetic and environmental factors are expected to modulate the phenotype, but the magnitude of the effects of those factors is not always defined.[5]

4. Current treatments: Impact and limitations

Despite significant advance in the comprehension of the pathophysiology of LDs, most conditions do not have an approved therapy, although important progress has been observed in the last two decades. Currently available treatments for lysosomal disorders include hematopoietic stem cell transplantation, enzyme replacement therapy and small molecules (Fig. 1).

4.1 Hematopoietic stem cell transplantation

The first established treatment modality for the LDs was the hematopoietic stem cell transplantation.[23] The cells, transplanted from a healthy donor will produce resident cells of hematopoietic origin, including glial cells and macrophages, which may act as permanent sources of the deficient enzyme in the central nervous system and other parts of the body.[5] HSCT is currently the first treatment choice for infants with severe MPS I (Hurler syndrome). This treatment can halt the neurodegeneration and attenuate the overall phenotype, especially when performed early in the disease course. HSCT is also recommended for some LD patients with metachromatic leukodystrophy and Krabbe disease.[24] However, morbidity and mortality associated with the immunosuppression and graft vs host disease preclude its more generalized use in attenuated presentations of LDs. Furthermore, HSCT has not been proven to be successful in some of the LDs with CNS involvement, including MPS III and Niemann Pick disease type C.[24]

4.2 Enzyme replacement therapy

Another treatment modality for LDs is enzyme replacement therapy (ERT), which consists of the periodical administration of a therapeutic enzyme. The

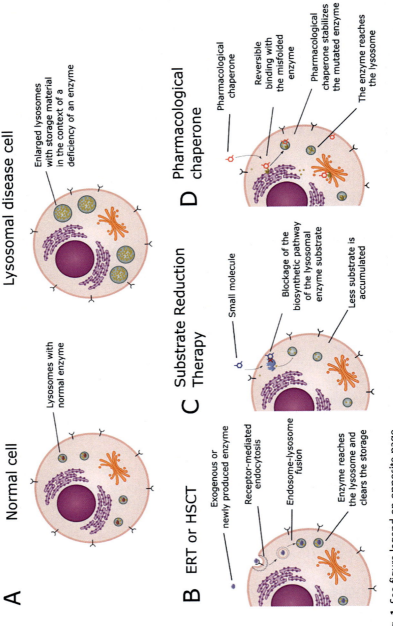

Fig. 1 See figure legend on opposite page.

first ERT for a LD, alglucerase, was approved in 1991 for Gaucher disease.[25] The product was extracted and purified from the human placenta, a method that was eventually replaced by the more scalable recombinant DNA technology. Enzyme replacement therapy is currently available for Gaucher, Fabry and Pompe diseases, as well as for MPS types I, II, IVA, VI and VII, lysosomal acid lipase deficiency, α-mannosidosis and CLN type 2. These biopharmaceuticals are produced in different cell or animal lineages, including Chinese hamster ovary (CHO) cells, human fibroblasts, carrot root cells and transgenic hens. In all cases, the enzymes are administered intravenously, except for CLN2, in which the approved product has an intracerebroventricular route of administration.

ERT was shown to promote a marked improvement in signs and symptoms of Gaucher disease type I, including anemia, thrombocytopenia, splenomegaly and bone pain, and this success was a key driver to develop new products for other LDs. Nevertheless, ERT does not seem to respond as well in other lysosomal diseases, where macrophages are not the target cells. For instance, worsening of cardiac and renal parameters has been observed for patients with Fabry disease on ERT, when treatment is started at older ages.[26] Likewise, some patients with late-onset Pompe disease on ERT have experienced worsening of strength and respiratory function after a period of improvement and stabilization of the disease.[27]

Intravenous ERT is, in general, of limited efficacy in addressing the CNS manifestations of LDs, since currently approved therapies are unable to cross the blood-brain barrier (BBB), a limitation that may be especially relevant for patients who have both somatic and neurological manifestations, such as those with Gaucher disease type III and severe MPS II. However, the demonstration of the efficacy of intracerebroventricular ERT for CLN type 2 indicates that intrathecal/intracerebroventricular ERT, either alone or in

Fig. 1 Mechanisms of action of the currently approved therapies for lysosomal diseases (LDs). (A) In the absence of a lysosomal enzyme, substrate accumulates in the lumen of the lysosomes with consequent lysosomal dysfunction. (B) Hematopoietic stem cell transplantation (HSCT) and enzyme replacement therapy (ERT) are aimed to provide the missing enzyme either through the production by donor cells (HSCT) or by regular infusion of recombinant enzymes (ERT). The newly produced or recombinant enzyme is captured by the LD cell through receptor-mediated endocytosis and directed to the lysosome. (C) Substrate reduction therapy (SRT) may be employed to inhibit the biosynthesis of the lysosomal substrate not degraded in LD cells. (D) Pharmacological chaperones (PCs) may bind transiently to the active site of the mutated enzyme, facilitating the enzyme targeting to the lysosome.

combination with intravenous ERT, may also be a promising approach for other LDs.[28]

ERT usually elicits an immune response with the production of immunoglobulins, especially in patients with null mutations, considering the lack of negative selection to the enzyme antigens. Although the presence of high titers of anti-drug antibodies may not always be directly associated with loss of efficacy or harm to the patient, adverse events may occur, including moderate to severe infusion-related reactions and worsening of clinical parameters and biomarkers.[29] ERT also requires the patient to commit to a strict schedule of periodical, and frequently hospital-based infusions that may be difficult to adhere to.

4.3 Small molecules

Besides cell transplantation and biotherapeutic proteins, small molecules are also an option for the treatment of some LDs, like Fabry and Gaucher diseases. For other LDs, such as cystinosis and Niemann Pick disease type C, small molecules are, currently, the only available treatment. The mechanisms of action of the small molecules used for LDs include pharmacological chaperone therapy (e.g., migalastat for Fabry disease) and substrate reduction therapy (e.g., miglustat and eliglustat for Gaucher disease).

Substrate reduction therapy (SRT) aims to halt the biosynthesis of the accumulated substrate and has also the potential to treat more than one LD, as the first step of a common pathway of synthesis may be inhibited. For instance, N-butyl-deoxynojirimycin (miglustat), which inhibits the synthesis of glucosylceramide, is approved for the treatment of Gaucher and Niemann Pick type C diseases.[30,31] As glucosylceramide is also a precursor of gangliosides in the biosynthetic pathway, miglustat has also been administrated to patients with GM1 and GM2 gangliosidoses in clinical trials or as an off-label treatment, although the evidence is not conclusive for its efficacy for those conditions.[32,33] Eliglustat, another SRT, has also demonstrated efficacy for Gaucher disease type 1, for which it was approved as a first-line treatment.[34] However, being unable to cross the blood-brain barrier, eliglustat is not applicable for neurodegenerative LDs.

Migalastat is currently the only approved pharmacological chaperone for a LD. It binds and provides stabilization for amenable mutant forms (misfolded functional variants) of α-galactosidase A, the enzyme deficient in Fabry disease. The medication has been shown to stabilize renal function, reduce cardiac mass, and improve gastrointestinal symptoms in its phase 3 clinical trial.[35] Due to its recent approval and the limited number of eligible patients with amenable variants, there is less available data regarding the long-term effectiveness of the therapy, but results to date have confirmed its potential as an important therapeutic option.[36]

5. Genome editing tools

In the past few years, some gene-based therapies have been approved by the American Food and Drug Administration (FDA) and the European Medicines Agency (EMA) for the treatment of blindness, cancer, viral infections, blood, and neuromuscular disorders.[37] Nevertheless, gene therapy has still been considered as an experimental therapy for most monogenic disorders, including LDs. Among the many hurdles that this approach holds, the most significant ones are immune responses to some viral vectors, random integration of lentiviral vectors in non-desired regions, or dilution of episomal vectors, such as adenovirus and adeno-associated virus, leading to loss of the transgene expression.[38] To overcome these drawbacks, researchers have investigated the potential of genome editing technologies for developing new therapies.

Genome editing tools offer the possibility of editing a specific genome region with precision in any organism. Overall, these tools rely on programmable nucleases that work along with customized DNA-binding motifs or RNA molecules that serve as guides to target the region of interest. Once the specific site is recognized by the nucleases, they introduce a double-strand break (DSB) or a single-strand break (nick), depending on the nucleases type.[39] The editing process begins when endogenous DNA repair machinery is activated after the break. There are two distinct repair mechanisms: non-homologous end joining (NHEJ) and homology-directed repair (HDR). In the first, the DNA molecule may undergo the introduction of indels, frequently knocking-out the gene. On the other hand, HDR is less common and requires the presence of a homologous DNA template to repair the break in a precise manner, copying the sequence at the break site from the template. Moreover, the HDR pathway is active mostly in proliferating cells.[40]

Zinc-finger nucleases (ZFNs), transcription activator-like effector nucleases (TALENs) and clustered regularly interspaced short palindromic repeats (CRISPR) are the genome editing tools most commonly employed. ZFNs are composed of a *Fok*I restriction endonuclease cleavage domain and an array of zinc-finger proteins. Each array is composed of three or four DNA zinc-finger domains each recognizing a DNA triplet. Zinc-finger domains are in charge of recognizing different specific DNA sequences—18 base pairs each in distal parts of the target sequence. Then, *Fok*I cuts DNA within a five- to seven-bp between the two flanking zinc-finger targets as the FokI enzyme requires dimerization for DNA cleavage.[41] The TALENs also consists of a *Fok*I endonuclease, but this is paired with transcription activator-like effectors (TALE) proteins. As ZFNs, TALENs also work as modules and undergo dimerization. TALE-binding domain has a series of repeat domains that recognize only a single base. In total, these effectors recognize 12- to 20-bps of DNA each, and FokI cuts within a 12- to 19-bp between them.[41] Currently, the most used method for gene editing is, however, based on CRISPR-Cas, which is more simple and versatile. CRISPR machinery involves only two components: the Cas9 enzyme, responsible for DNA cleavage, and a guide RNA (gRNA). The gRNA can be customized to virtually recognize any specific 20 bp sequence. Both components form a complex together and, after gRNA binds the target region by base-pairing, Cas9 makes the DSB. However, this cleavage is restricted to the presence of three NGG nucleotides (called PAM site) in the target sequence.[39]

More recently, different variations to CRISPR-Cas9 system have been proposed, mainly focusing on base editing. The first developed was the cytosine base editor (CBE). This system uses a cytidine deaminase enzyme linked to a Cas9 that underwent inactivation of one of its cleavage domains, becoming a nickase (Cas9n). This enzyme complex can catalyze the conversion of C-G base pairs to T-A in a target sequence, by converting cytidine to uridine which, in turn, is converted to thymidine by DNA repair mechanisms.[42] Another variation of the base editor is the adenine base editor (ABE), which can convert A-T base pairs to G-C by adenosine deaminase. The last CRISPR-derived tool, the prime editing system,[43] relies on the fusion of the Cas9n to an engineered viral reverse transcriptase enzyme to create a modified nuclease. The other component is a pegRNA, a longer gRNA that also serves as a template for the repair. After binding to the specific site, Cas9n makes a nick and the reverse transcriptase domain generates a complementary DNA sequence by copying the pegRNA, restoring a segment of the nicked strand. Thus, it mediates all 12 possible base-to-base conversions due to pegRNA carrying the desired sequence.[39]

Due to their potential, these tools have been used to develop several cell and animal models, aiming not only to contribute to understanding diseases' physiopathology but also to develop different strategies for gene therapy.

6. Genome-edited cell models

Patient-derived cells have been an important resource for understanding disease mechanisms and the development of new therapies. The majority of basic research or proof-of-concept studies for lysosomal disorders was done in fibroblasts, as they are easy to collect without the need for surgical intervention. With the development of genome editing technologies, cell models became more accessible as one can choose the most relevant cell type and induce the modification of interest, without the need for complex biopsies and patient availability. Currently, these models based on genome editing are mainly developed using induced pluripotent stem cells (iPSCs) or cell lines.

Innumerous cell models using iPSCs have been developed for lysosomal disorders, mainly for the ones with neurological involvement.[44] This is justifiable since little is known about the pathophysiological processes involved in neurodegeneration and access to relevant models is generally scarce. Differentiation in cardiomyocytes is also a big hit, as some lysosomal disorders have major cardiovascular pathology[45] that is not reverted with approved or even experimental therapies, like gene therapy.[46] Lastly, hematopoietic stem cells have been targeted for differentiation of iPSCs as well, as they can become a source of cells for autologous transplantation.[44]

For developing these iPSCs models, fibroblasts or blood cells are collected from patients, reprogrammed in vitro and, then, differentiated into the cell of interest. This way, the newly created model carries the patient's pathogenic variants and will present the disease phenotype. For controlled studies, however, an isogenic healthy control is required, as the cells behave differently depending on the genetic background. To address this, genome editing tools can be used to correct the disease-causing mutation in the iPSCs population, before the differentiation protocol.[47] In the absence of a patient's cells, healthy iPSCs can be edited to present various genotypes and to mimic the disease of interest.[48,49] A complete review about using iPSCs for LD studies was published by Borger and colleagues.[44]

Although the possibilities are virtually endless, the use of iPSCs is still dispendious, limiting its broad use. Cell lines, on the other hand, are much more affordable—they are easy to culture and have high reproducibility of results. They are generally easy to transfect, making the editing process easier, and many different cell types are available—including endothelial, myeloid, lymphoid and neuron-like cells, for example, in case specific morphologic characteristics are required. Here we reference some cell models for LD, created using genome editing tools, which became important resources for basic research and therapy development.

Fabry disease is caused by a deficiency of α-galactosidase A (GLA). The most common manifestations include renal and cardiovascular disease. Several cell lines for modeling Fabry disease have been developed using CRISPR-mediated genome editing. One of the first studies knocked-out *GLA* in the human embryonic kidney (HEK) 293 cells, to create a simple model for drug screening purposes.[50] In the same year, a more relevant cell type to study the disease was developed—immortalized podocytes were edited with CRISPR-Cas9 for microarray studies, identifying MAPK, VEGF and TGF-beta pathways as enriched in Fabry cells,[51] all of which are also involved in other glomerular diseases.

Fabry patients have increased vasculopathy incidence than healthy individuals due to the accumulation of substrate in endothelial cells. Thus, the endothelial cell line EA.hy926 was used to help understand some of the disease's mechanisms. GLA-deficient endothelial cells have lower endothelial nitric oxide synthase (eNOS) and treatment with substrate reduction therapy did not affect in this, suggesting that the absence of the enzyme rather than the accumulation of the substrate that affects eNOS activity.[52] Further investigation with the same model showed that decreased eNOS leads to increased secretion of von Willebrand factor and this is not altered neither by substrate reduction therapy nor administration of recombinant GLA.[53] Therefore, this aspect of the disease should be addressed separately. In a different approach, human embryonic stem cells (hESC) knocked-out for *GLA* were differentiated in cardiomyocytes, as cardiomyopathy is an important manifestation of Fabry disease. Proteome analysis of the new model showed downregulation of exocytotic vesicle release proteins, causing impaired autophagic flux and protein turnover and ultimately leading to increased apoptosis.[54] Altogether, these findings provide new insights about the disease pathophysiology and can help identify new targets for vasculopathy and cardiomyopathy in Fabry disease.

Macrophages are key cells in diseases of lipid metabolism, like atherosclerosis, acid lipase deficiency, Niemann-Pick and Gaucher diseases, in which they become foamy cells after an accumulation of lipids in their lysosomes. To explore the loss-of-function phenotype and accumulation of lipids in the macrophages, Zhang and colleagues[49] generated iPSCs knockout for lipase A and induced the differentiation of these cells to macrophages, observing the macrophage-specific response to lipase A deficiency. Alternatively, macrophage-like cells can be obtained using the hematopoietic cell line THP-1 and submitting them to a differentiation protocol, as developed to model Gaucher disease.[55] An advantage of using this last instead of other approaches is the possibility of high-throughput studies, mainly for drug testing and screening, as cell lines are better suited for large scale production.

Besides the macrophage model aforementioned, a microglia-like cell model for Gaucher disease using the human U87 cell line was also developed by Pavan and colleagues,[55] in an attempt to generate a more relevant model to study the neuronal component of the disease. U87 *GBA1* mutant cells showed retention of misfolded GBA in the endoplasmic reticulum, increased production of interleukin-1-beta, α-synuclein accumulation and increased cell death compared to non-edited cells. Accordingly, α-synuclein accumulation was also observed in GBA-knockout HEK293[56] and heterozygote mutant for Cathepsin D gene (CTSD), responsible for a lysosomal hydrolase that causes neuronal ceroid lipofuscinosis when absent.[57] Even though these models were created with different cell types and targeting different genes, they all conversely contributed to understanding part of lysosomal function. These are not useful only for lysosomal disorders; GBA-knockout HEK293 and A549 cells were used to study endocytic processes in viral infections. It was found that endocytic trafficking of viruses and cellular cargos are impaired in GBA-knockout cells.[58]

There are few Niemann-Pick type C (NPC) cell models generated with CRISPR-Cas9. The simplest is a near-haploid human cell line (HAP1) edited to carry new or private mutations in the *NPC1* gene that have been found in patients and lacked characterization.[59] Because it is a near-haploid line, editing is simplified, making this model useful mainly to resolve clinical interpretation of new variants. The classic HeLa and the CHO-Ldl-D cell lines were also knocked-out to generate different NPC models: the first was created to be used in drug screening studies, as HeLa cells are easy to culture.[60] The second was used to study glycosylation inhibition and its effect on cholesterol accumulation, in an attempt to better understand the biology of lysosomes.[61]

Using genome editing to create LDs cell models is a useful tool for basic research, initial screenings and proof-of-concept studies. For example, one can observe the effect of different pathogenic variants all at once in the same cell type; or the effect of a single variant in many cell types (Fig. 2). Lysosome biology can be explored and, this way, pathogenic mechanisms of lysosomal diseases can be elucidated.

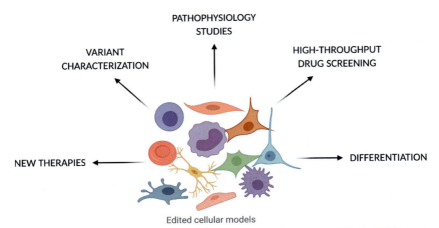

Fig. 2 Applications of genetically edited cell lines in disease modeling. Cell lines are easy to obtain and maintain in culture, thus being excellent models for initial pathophysiology studies, development of new therapies and high-throughput drug screening. They can also be used to characterize new variants observed in patients and, in cases where the line can be differentiated into other cell types, the process can be observed in the disease context.

7. Genome-edited animal models

As mentioned previously, cellular models play a role in several aspects, including proof-of-concept studies, for emergent therapeutic strategies. However, the use of multicellular transgenic model organisms is mandatory to study disease pathogenesis and treatment options, because they can more closely recapitulate the complex metabolic and cellular pathways occurring in patients. As a standard, the murine model has been the most used in this field; however, some of these do not manifest the clinical aspects of the LDs; as seen in the Gaucher, Fabry, Tay-Sachs, and cystinosis mouse models, compromising the understanding of experimental therapies.[62] Despite several mammalian species, including cats and dogs, naturally develop some of the LDs seen in humans, researchers have taken advantage of genome editing tools to develop models according to their research needs.

Zebrafish has become an interesting non-mammalian model, and its use in LDs studies has raised notably through the years. The reason is due to high comparable anatomy to humans, where the major systems and organs are present, and also its genetics—82% of all human disease-associated genes have orthologues in this model. Lately, a considerable number of stable mutants fish strains have been generated with these approaches (TALENs: 6, CRISPR-Cas9: 10; combined TALENs/CRISPR-Cas9: 1).[63] Mutants generated for Gaucher and Sandhoff diseases displayed sphingolipid accumulation, microglial alterations, neuron loss, apoptosis, and impaired locomotion. For mucolipidosis IV, pathological signs in the muscle and eye were observed in the models. The Niemann-Pick type C1 models displayed hepatic disease features as lipid storage, vacuolated hepatocytes and also Purkinje cell defects. Finally, the MPS II model showed most of the clinical presentations seen in patients, as impaired development, abnormal heart morphogenesis, and bone alterations, like craniofacial defects, scoliosis, and kyphosis. An extended review of Zebrafish for LD studies was recently published by Zhang and Peterson.[63]

Lastly, an ovine model for infantile neuronal ceroid lipofuscinosis (NCL1) was developed aiming to better translate the human condition. The sheep model presented biochemical, morphological, and neurological alterations including loss of vision and reduced lifespan.[64] With this, the use of genome editing tools to develop more accurate animal models that resemble the human disease condition holds promising potential.

8. Therapies based on genome editing

Moving to therapeutic approaches, unlike the classical gene therapies, genome editing came up with the possibility of a precise and one-time long-term treatment. That also includes a significant decrease in off-target activities and immunogenic risks related to the former ones.

Some LDs are better candidates for genome editing based therapies than others. As indicated previously, the majority of them rely on the lack of production of a secreted protein. Editing tools can be employed to develop a unified platform that allows the production of any of these proteins since only one gene is the target for each condition. From in vitro to in vivo studies, researchers have been working on the choice of the proper features for the desired approach, and the majority of efforts have brought positive and promising results. Herein below are presented the leading studies that comprise proof of concepts in a cell model, development of ex vivo stem cell therapy platforms, and in vivo preclinical studies (Table 2).

8.1 In vitro mutation-specific genome editing

Most LDs have a vast number of genetic variants reported, resulting in a broad range of phenotypes. Depending on the complexity of the patient

Table 2 Genome editing studies for lysosomal disorders.

Preclinical studies in cell models

Disease	Affected gene	Targeted gene/locus	Platform	Cell type	Delivery method	Genetic modification	References
MPS I	*IDUA*	*IDUA*	CRISPR–Cas9	Patient fibroblasts	Plasmid-Liposome complex	SNV correction	65,66
MPS I	*IDUA*	*IDUA*	CRISPR–Cas9	Mouse iPSCs	Plasmids	Precise deletion	67
Tay-Sach	*HEXA*	*HEXA*	Prime editing	HEK293T cell	Plasmids	SNV correction	43
Fabry	*GLA*	*GLA*	CRISPR–Cas9	Patient fibroblast	NS	SNV correction	68
Fabry	*GLA*	*GLA*	CRISPR–Cas9	Patient iPSCs	RNP	Deletion correction	69

Preclinical studies in murine models

Disease	Affected gene	Targeted gene/locus	Platform	In vivo/ex vivo	Delivery method	Genetic modification	References
Gaucher	*GBA*	CCR5	CRISPR–Cas9	Ex vivo	RNP/AAV6	Knock-in	70
Pompe	*GAA*	AAVS1	CRISPR–Cas9	Ex vivo	Plasmids	Knock-in	71
Krabbe	*GALC*	IL2RG	CRISPR–Cas9	Ex vivo	RNP/AAV6	Knock-in	72
Wolman disease	*LAL*	α-globin	CRISPR–Cas9	Ex vivo	RNP/AAV6	Knock-in	73
Fabry	*GLA*	α-globin	CRISPR–Cas9	Ex vivo	RNP/AAV6	Knock-in	73
MPS I	*IDUA*	α-globin	CRISPR–Cas9	Ex vivo	RNP/AAV6	Knock-in	73

Disease	Gene	Target	Editing tool	In vivo/Ex vivo	Delivery	Strategy	Ref
MPS I	IDUA	CCR5	CRISPR–Cas9	Ex vivo	RNP/AAV6	Knock-in	74
MPS I	IDUA	ROSA26	CRISPR–Cas9	In vivo	Liposome and plasmid vectors	Knock-in	75
MPS I	IDUA	ALB	ZFNs	In vivo	3 AAV2/8 vectors	Knock-in	76
MPS I	IDUA	ALB	CRISPR–Cas9	In vivo	2AAV 8 vectors	Knock-in	77
MPS I	IDUA	IDUA	CRISPR–cas9	In vivo	2AAV9 vectors	Allelic Exchange/SNV correction	78
MPS II	IDS	ALB	ZFNs	In vivo	3 AAV2/8 vectors	Knock-in	79
Gaucher	GBA	ALB	ZFNs	In vivo	2 AAV8 vectors	Knock-in	80
Fabry	GLA	ALB	ZFNs	In vivo	2 AAV8 vectors	Knock-in	80
Sandhoff/Tay-Sach	HEX	ALB	CRISPR–Cas9	In vivo	2 AAV8 vectors	Knock-in	81
Niemman Pick C	NPC1	NPC1	CBE	In vivo	AAV9 vector	SNV correction	82

Summary of preclinical and clinical studies using genome editing for LDs. AAVS1, AAV integration site; ALB, albumin; CBE, cystine base editors; CCR5, chemokine (C—C motif) receptor 5; GAA, acid α-glucosidase; GALC, galactosidase; GBA, glucocerebrosidase; GLA, α-galactosidase; HEX, β-hexosaminidases; IDS, iduronate 2 sulfatase; IDUA, iduronidase; IL2RG, interleukin-2 receptor gamma chain; LAL, lysosomal acid lipase; NS, not specified; RNP, ribonucleoprotein; SNV, single nucleotide variant.

variant, it is possible to "change" a single mismatched nucleotide or to correct a small indel with precision using genome editing tools. These edits have been assessed for different LDs using the CRISPR-Cas9 system in cellular models or even in patient's cells.

MPS I is one of the common LDs worldwide, with a prevalence of 1/100.000 births.[83] As well as other well-known MPS, it presents a broad spectrum of phenotypes that can be frequently associated with the genotype. The severe one (MPS I-H), is the most common among western patients and is the one with progressive neurologic impairment. This phenotype has been related to several pathogenic variants, but the most frequent is p. Trp402Ter.[83] Researchers evaluated the capability of the CRISPR-Cas9 system and a single-strand donor oligonucleotide complexed in commercial liposome to correct this variant in patient fibroblasts.[65] A similar approach was carried out later, but this time using cationic nanoemulsions as a non-viral delivery vector.[66] In both, IDUA activity was significantly increased (2–6% of normal) and the lysosomal mass was decreased in treated fibroblasts. The presence of correctly edited cells was confirmed by next-generation sequencing[65] and nanoemulsion/DNA complexes were well tolerated by cells (viability of about 80%),[66] demonstrating the potential of this approach for further studies. Nanomaterials such as nanoemulsions, liposomes, and nanoparticles have been studied in the past years as an alternative drug/gene delivery system for the treatment of many diseases and conditions.[84] The oil-water phase nanoemulsions have shown a notable cellular targeting ability due to their 1–100 nanometers (nm) droplet size, and their biocompatible and non-toxic materials that are commonly safe (GRAS) grade excipients approved by FDA.[85] Their nature allows them to protect the cargo (molecule to carry) and to bind and fuse with cell membranes since many of their components are lipids. Also, they have an increased half-life in circulation due to hydrophilic polymers such as PEG (polyethylene glycol) that avoid their interaction with plasma macrophages reducing their clearance.[86] Some nanoemulsion dosage forms, mostly carrying small molecules or active biopharmaceuticals, are already in the pharmaceutical market and several others are under development.[85] As well as we presented earlier for lysosomal storage diseases, nanoemulsions have been studied as nanocarriers to help the delivery of nucleic acids targeting brain tumors[86,87] and neuroinflammatory conditions,[88] with promising results, demonstrating reasonable transfection efficiency and lower cytotoxicity carrying RNA.

In the cardiac phenotype of Fabry disease, the vasculopathy is caused by the accumulation of Gb3 which, in turn, is caused by different pathogenic variants in *GLA*. Using CRISPR-Cas9 ribonucleoprotein (RNP) and single-stranded oligodeoxynucleotide (ssODN), the correction of a genomic deletion—c.803_806del, causing a frameshift in exon 6 and amino acid position 268—was performed in the patient's fibroblast-derived iPSCs. After differentiation into vascular endothelial cells, a significant improvement in GLA activity and enhanced expression of the angiogenic factors was observed when compared to non-differentiated iPSCs.[69] In a different study, a double sgRNA was employed to target the deletion IVS4 + 919 G > A in the *GLA* gene of FD patient's fibroblasts—a common variant in the cardiac type. This strategy succeeded in restoring the aberrant splicing, causing an increase in the enzyme activity of GLA and a decrease in the Gb3 accumulation.[68]

The CRISPR-Cas9 system was also studied to mediate gene insertion in cell models. The approach could be useful when complex pathogenic variants are present. As an example, the complete restoration of the exon VI in the *IDUA* gene was performed in iPSCs derived from knockout mouse embryonic fibroblasts.[67] The strategy consisted of removing the whole neomycin resistance gene placed into the exon VI of *IDUA* with double guide RNAs, and several donor templates, to allow complete coverage of the exon VI. The correction was successful without any undesired indels, and enzyme function was restored to levels equivalent to wild-type iPSC-derived fibroblasts; plus, the pluripotency of iPSCs was maintained.[67]

A more recent genome editing platform, called prime editing, is a more precise tool that allows modifying genes without the necessity of DSB and a long DNA donor.[43] As proof-of-concept, it was used to correct the most common mutation that causes Tay-Sachs disease, a 4-bp insertion in *HEXA* (*HEXA$^{1278 + TATC}$*), in a HEK293T cell line model that was generated previously using the same approach. As a result, prime editing allowed the correction of the mutation with 33% efficiency and 0.32% indel formation, showing its potential to correct genetic variants associated with human diseases.[43]

8.2 Ex vivo genome editing targeting safe harbors loci

Besides these in vitro studies, where the goal was editing specific disease-causing mutations, other researchers have searched for different genomic regions as the target for editing. They use "safe harbors" loci to integrate the entire cDNA of the functional gene to overcome the limitations of correcting each gene mutation individually. It is unlikely feasible that a specific therapeutic strategy targeting a dedicated mutation can reach the clinic when patients are very rare and there are a large number of disease-causing mutations. Moreover, the presence of pseudogenes hinders the design of gRNAs and donor templates and increase the risk of undesired genome recombination. To address these issues, studies using safe harbors, tissue-specific promoters and ex vivo or in vivo platforms have been conducted.

As mentioned previously in the introduction, allogeneic HSCT is to date the gold standard for the treatment of a few LDs. Ex vivo gene therapy using genome editing strategies allows for autologous transplantation, where the patient's cells are collected, edited outside the body and transplanted back into the patient. This approach removes the need for matching donors and reduces complications related to the procedure, like graft-vs-host disease. Besides hematopoietic stem cells (HSC), mesenchymal stromal cells and neural stem cells have been studied for gene therapy, due to their ability to engraft and migrate to the brain and bone.[89]

Pavani and colleagues proposed the use of the α-globin locus as a promising "safe harbor" candidate to integrate and express therapeutic transgenes.[73] The α-globin genes are present in four copies per cell and are expressed at an extremely high level by erythroid lineages. Also, it has been reported that the loss of three of the α-globin alleles does not represent a negative impact on the organism.[73] Since erythroid cells are the most abundant hematopoietic progeny, they hold the potential to secrete relevant amounts of therapeutic proteins as required for LDs. Using Cas9-gRNA ribonucleoprotein (RNP) and AAV6 donor vector, researchers targeted the α-globin locus in CD34+ cells, purified from human umbilical cord blood (UCB), to integrate lysosomal acid lipase (LAL), α-L-iduronidase (IDUA), or α-galactosidase (GLA) transgenes. As a result, mRNA and enzymes were detected post-transduction and at increased levels after erythroid differentiation. The enzyme produced by the erythroblasts derived from LAL-HSPCs presented cross-correction activity when co-cultured with patients' cells. Additionally, edited cells maintained their repopulation and multilineage potential after transplantation in immunodeficient NOD SCID gamma (NSG) mice.[73]

Another potential candidate as a safe harbor locus for LDs gene addition therapies is the chemokine (C-C motif) receptor 5 (CCR5), which is used in prospective therapies for HIV infection. This gene encodes an HIV-1 receptor, and when mutated, minimal consequences such as increased vulnerability to other virus infections (influenza or West Nile virus) have been described besides conferring resistance to HIV-1 infection in individuals.[90,91] However, as a matter of future concerns, it has been shown the potential risk of inducing a null CCR5 phenotype in bone development and immune system regulation in mice.[91]

For LDs emergent therapies, Gomez-Ospina and colleagues reported an efficient ex vivo genome editing approach using CRISPR-Cas9 and AAV6 to targeted α-L-iduronidase into CCR5 locus in human CD34+ cells under the control of synthetic promoters such as spleen focus-forming virus (SFFV) and 3-phosphoglycerate kinase (PGK). The rationale of the promoter choice was to exploit their constitutiveness nature and guarantee the expression of the transgene in all hematopoietic lineages to facilitate cross-correction.[74] After transplantation into immunodeficient NSG-MPS I mice, the engineered cells secreted supra-endogenous enzyme levels, improving phenotypical and biochemical features of MPS I, including neurological impairment. Edited cells maintained the self-renewal capacity and multilineage differentiation potential, but the engraftment of the edited cells was less efficient compared to unedited cells, 3.9% (0.8–9.7%) and 30.4% (7.7–48.2%) respectively. This lower engraftment was shown in cells modified at other loci as well.[92–94] Besides effective, this approach also proved to be safe, since the off-target activity of the CCR5 sgRNA was undetectable with the use of high-fidelity Cas9, p53 pathways were not altered and there was no evident tumorigenicity.[74]

More recently, the same platform was used with slight modifications. Human CD34+ cells were edited with a glucocerebrosidase expression cassette driven by a monocyte/macrophage-specific promoter (CD68) into the CCR5 locus.[70] The rationale of this strategy is related to the hallmark manifestations in Gaucher disease, which are mainly caused due to infiltration and inflammation by macrophages. While edited HSC did not express the enzyme, differentiated monocytes/macrophages had increased glucocerebrosidase activity both in vitro and 16 weeks post-transplantation, as measured in sorted cells from bone marrow, spleen and lung. The use of a specific promoter restricted enzyme expression to affected cells only and prevented ectopic (and possibly toxic) expression in hematopoietic stem cells.[70]

Different stem cells, such as neural stem cells (NSCs), can be a promising platform for the therapy of some LDs with CNS involvement. NSCs cells have shown a high capability of neural multilineage differentiation, migration, and proliferation when they are transplanted into the brain or spinal cord of NSG mice.[95] NSCs are suitable for in vitro manipulation as they continue to express NSC markers after been isolated, propagated, and banked.[72] To date, only one proof-of-concept study has demonstrated CRISPR-Cas9 genome editing in NSCs, using the Krabbe disease as model.[72] In this study, the authors were able to edit human brain-derived NSCs to produce the enzyme galactocerebrosidase (GALC) under the regulation of the IL2RG promotor. This locus was described as a safe harbor in NSCs due to the lack of functional role in these cells. Edited cells cross-corrected fibroblasts from Krabbe patients. Moreover, after transplantation into the cerebellum of juvenile mice, cells engrafted and maintained biological properties.[72]

Finally, iPSCs are also suitable candidates for ex vivo therapy due to their pluripotency and self-renewal characteristics. It was demonstrated by Van de wal et al. that iPSCs from Pompe disease patients can be edited in vitro to introduce the cDNA for the enzyme acid α-glucosidase (GAA) into the AAVS1 locus, another human safe harbor. Myogenic progenitors were generated from edited iPSCs and their differentiation capability into myotubes was assessed. In these cells, the correction of the disease was evident since GAA activity was threefold higher compared to healthy controls. Also, the restoration of the enzyme activity led to the normalization of glycogen accumulation. To assess their therapeutical potential, edited myogenic progenitors were transplanted in the tibia muscle of pre-injured NSG mice, where they engrafted robustly and contributed to muscle regeneration at the injured-site.[71]

8.3 In vivo genome editing

In vivo systemic genome editing strategies for the treatment of LDs have also been studied. Some of these strategies managed to reach clinical trials, but several approaches using CRISPR-Cas9, zinc-finger nucleases, and other newly emerging systems have only been used in preclinical studies aiming for reliable and efficient therapies.

For in vivo genome editing approaches, liposomes and viral vectors are the most promising strategies. Liposomes are highly utilized for delivering different cargoes, from small therapeutic molecules to gene therapy products such as plasmids. This non-viral delivery was studied aiming for the correction of Hurler syndrome in the MPS I mouse model.[96] As a proof of concept, cationic liposomes carrying CRISPR-Cas9 and cDNA donor plasmids targeting the ROSA26 locus were administrated in neonatal MPS I mice by a single hydrodynamic injection. At 6 months, treated mice presented an increase in enzyme activity in every visceral tissue, with the highest levels in the heart and lungs (~5% of normal activity). Indeed, the biodistribution of the complex showed a high affinity of the cationic liposome for these tissues.[96] The enzyme produced by edited cells also led to a significant reduction in the substrate (GAGs) accumulation and secretion in serum, urine and visceral organs, though still not to normalized levels. The effects of gene editing were also evidenced in the improvement of cardiovascular, respiratory and bone pathology.[75] Nevertheless, this approach failed to reach the CNS.

The albumin locus has also been shown to be a safe harbor in vivo. Researchers engineered a flexible platform where this locus is used for transgene expression without interrupting hepatic albumin production, exploiting the strong promoter activity.[97] Using AAV8 vectors (with a high tropism for hepatocytes) zinc-finger nuclease (ZFN), and cDNA donor templates were delivered systemically to target overexpression of α-galactosidase A, acid β-glucosidase, iduronate-2 sulfatase, or α-L-iduronidase in wild-type mice liver.[80] Donor DNA was designed to allow, after appropriate recombination, directed-splicing between intron 1 of the albumin locus and the transgene cDNA, to produce a functional hybrid therapeutic transcript. After treatment, the therapeutic enzymes were found in liver extracts and plasma, along with albumin protein.[80] To evaluate the clinical potential of the platform, more studies were held in MPS I and II mouse models, using AAV2/8 as delivery vectors.[76,79] In treated mice, supraphysiological levels of the enzyme were observed in serum, while GAGs content in urine was found significantly decreased in both disease models. The therapeutic enzymes were uptaken from the blood to other tissues with high enzyme activity levels detected in the spleen, heart, lung and muscle. As a remarkable finding, along with the correction of the metabolic disease, the strategy managed to prevent neurologic impairment in young MPS I and II mice.[76,79] These results suggested that supraphysiological levels of the enzyme produced by the liver, which is commonly achieved using this type of vector in animal models, allowed a fraction of IDUA/IDS to penetrate the BBB. Thus, even a small percentage of normal enzyme levels in the brain seems sufficient to prevent neurologic deficit.[79] This was different from what was observed previously from Schuh and colleagues,[96] where the levels of the enzyme compared to normal conditions were about 10% in liver and heart tissue and 2–5%

in the other visceral tissues. Thus, there was no sign of the enzyme in the brain.

This last strategy resulted in two clinical trials, which will be covered in the next section of this chapter. However, preliminary results showed a low expression of targeted enzymes in treated individuals. Aiming for an optimized outcome, the platform was recently redesigned using the CRISPR-Cas9 system instead of ZNF-nucleases.[77] Compared to the previous strategy, which required the co-transduction of three different AAV8 vectors to deliver the whole system, the new approach only uses two. The strategy succeeded in correcting both the metabolic and the neurological features of MPS I mice. Furthermore, no tumorigenesis or off-target activity was detected.[77] Additionally, the platform was tested recently to deliver a modified human Hex µ subunit (HEXM) into the albumin locus.[81] Since HEXM forms a homodimer that degrades GM2 gangliosides, the strategy represents a promising unified therapy for both Tay-Sachs and Sandhoff diseases.[81] Importantly, along with the increase in enzyme activity in the brain, memory was improved in treated mice when compared to untreated animals. Despite these promising results, more studies are needed before this new platform can reach clinical trials since this approach used a high volume/pressure injection to deliver the vectors, which is hard to scale up to humans.

Finally, the correction of some specific point mutations was also investigated in vivo using genome editing tools. In one study, allelic exchange was induced in a compound heterozygous MPS I mice model using the CRISPR-Cas9 system.[78] Here, a unique gRNA targeting an intronic site was used to create double-stranded DNA breaks in both chromosomal homologs aiming at switching arms between chromosomes in a chromosomal translocation-like mechanism. Researchers were able to restore IDUA levels (~0.5% of the wild-type level) in the heart of young treated mice accompanied by a considerable GAGs reduction. The results indicated the possibility of inducing allelic exchange in post-mitotic tissues. Nevertheless, chromosomal translocation occurs in low frequency compared to any other strategy based on NHEJ or HDR.[78] In another study, a full-length cytosine base editor was modified and split in a dual-AAV packaging vector.[82] After delivery into the organism, the components undergo reconstitution by trans-splicing machinery, overcoming the main limitation of reduced packaging capacity that AAV vectors have.[82] This dual strategy was tested in $Npc1^{I1061T}$ (c.3182 T > C) homozygous mice, a model that reproduces the neurological pathology and also has reduced lifespan. Low and high doses of the dual vectors were delivered by retro-orbital injections. Mice who received the higher dose survived longer than untreated and low-dose mice. Genomic DNA from the brain of surviving animals revealed that a total of 94% of alleles were edited—C-to-T—without any undesirable indels in the region.

9. Clinical trials for lysosomal disorders

Gene editing technology for LDs has been successfully applied in a variety of preclinical models, both ex vivo and in vivo.[75,77,80,98] The promising results from these studies have provided the basis for the development of clinical trials.

To date, two new therapies using gene editing, named SB-318 and SB-913, are being tested as a single treatment therapy for MPS I and MPS II patients, respectively. Both were designed using ZFN system to insert a therapeutic transgene into the albumin locus of the patient's liver cells.[77] Each one of these therapies consists of the intravenous infusion of two ZFNs (left and right arms), targeting the albumin locus and the corrective donor gene (*IDUA* or *IDS*) packaged into AAV2/6 vectors. This one-time therapy has the potential to provide permanent lifetime production of the impaired enzyme, to improve the current clinical outcome of ERT or HSCT, and the quality of life for patients.[99]

EMPOWERS and CHAMPIONS (clinicaltrials.gov, NCT02702115 and NCT03041324) are the two Phase 1/2 clinical trials, for respectively SB-318 and SB-913, that are ongoing in the U.S. They are multicenter, open-label, dose-escalating studies. Subjects received gene editing and initiated under protocol-specified schedule with monitoring of safety, IDS or IDUA activity and urinary glycosaminoglycan (GAG) levels, and functional assessments.[100,101] At present, the data analysis for both studies is still in progress.

The CHAMPIONS study started in 2017, being the first-ever clinical trial to attempt in vivo genome editing in patients with LDs. The purpose of this study is to assess the safety, tolerability and preliminary efficacy (changes from baseline in plasma IDS activity and urine GAG levels) of ascending doses of SB-913 in subjects with attenuated MPS II. MPS II patients receive a single-dose of SB-913 with 3 years of follow-up. A three-dose cohort with two subjects each is expected, along with three additional individuals who will receive the high dose, in an expanded cohort.[101] Interim data from the first three cohorts showed that SB-913 was generally well tolerated and no serious adverse events related to the drug were reported. Some adverse events related to the study drug were mild or moderate and were eventually resolved. The incidence of adverse events was assessed by Common Terminology Criteria for Adverse Events (CTCAE) and includes pruritus, flushing, erythema, increased serum transaminases, headache and pyrexia. Preliminary analysis of liver tissue biopsy, using RT-qPCR, showed evidence of albumin-IDS mRNA transcript in both subjects at the mid-dose, suggesting that genome editing had occurred. A substantial increase in plasma IDS activity was observed only in one patient (mid-cohort). Nevertheless, it decreased after the development of mild transaminitis, a known risk of AAV-based therapies.[101,102]

EMPOWERS is a clinical trial started in mid-2018 and designed, firstly, to assess the safety and tolerability of ascending doses of SB-318 in adult subjects with the mild form of MPS I and, secondly, to evaluate the changes from baseline in IDUA activity and urine GAG levels. The clinical design is similar to the CHAMPIONS study. Here, two-dose cohorts with two subjects each are conducted, along with an expansion of five additional subjects who will receive the high dose (total of nine subjects).[100] Interim data of the first 3 subjects across 2 dosing cohorts (1 patient with a low dose and 2 patients with high dose), showed that the administration of SB-318 was generally well tolerated without adverse events related to SB-318 treatment. Increases in leukocyte IDUA activity were observed in all three subjects and suggest a dose-dependent increase. On the other hand, plasma IDUA activity was not significantly changed from pre-treatment values.[100] The results provided so far indicate that efficacy needs to be improved since the levels of transgene expression are low.[77] This low expression seems to be the major obstacle and has already been described in other clinical trials for gene therapy.[77] As previously reported in animal studies, the low transgene expression issue may be improved by increasing the dose or through repeating administration, for example, using lipid nanoparticles delivery. Increasing the dose, however, could lead to a higher risk of toxicity, challenging vector production, and increased manufacture cost.[77] Preclinical studies indicated that CRISPR technology has higher efficiency than ZFN system with lower doses of AAV vector, which minimizes the risk of toxicity as well reduced complexities and challenges regarding viral vector manufacturing.[77] Otherwise, the use of lipid nanoparticle delivery enables repeated administrations, resulting in efficient in vivo genome editing, being a useful alternative tool.[103]

10. Conclusions

There are a myriad of applications for genome editing in LDs, including the development of cell and animal models to study disease pathogenesis, as well as permanent therapeutic approaches. Genome editing allows recreating the same mutation found in patients, which can be essential to develop individualized therapies. Therefore, we believe that the number of studies on LDs using techniques to rewrite the genome will grow exponentially in the next years.

References

1. Johnson WG. Disorders of glycoprotein degradation: sialidosis, fucosidosis, α-mannosidosis, β-mannosidosis, and aspartylglycosaminuria. In: *Rosenberg's Molecular and Genetic Basis of Neurological and Psychiatric Disease*. 5th ed. Elsevier; 2014:369–383. https://doi.org/10.1016/B978-0-12-410529-4.00033-4.
2. Waters D, Adeloye D, Woolham D, Wastnedge E, Patel S, Rudan I. Global birth prevalence and mortality from inborn errors of metabolism: a systematic analysis of the evidence. *J Glob Health*. 2018;8(2). https://doi.org/10.7189/jogh.08.021102, 021102.
3. Hsu T-R, Niu D-M. Fabry disease: review and experience during newborn screening. *Trends Cardiovasc Med*. 2018;28(4):274–281. https://doi.org/10.1016/j.tcm.2017.10.001.
4. Mechtler TP, Stary S, Metz TF, et al. Neonatal screening for lysosomal storage disorders: feasibility and incidence from a nationwide study in Austria. *Lancet*. 2012;379 (9813):335–341. https://doi.org/10.1016/S0140-6736(11)61266-X.
5. Marques ARA, Saftig P. Lysosomal storage disorders—challenges, concepts and avenues for therapy: beyond rare diseases. *J Cell Sci*. 2019;132(2):221739. https://doi.org/10.1242/jcs.221739.
6. Darios F, Stevanin G. Impairment of lysosome function and autophagy in rare neurodegenerative diseases. *J Mol Biol*. 2020;432(8):2714–2734. https://doi.org/10.1016/j.jmb.2020.02.033.
7. Pfeffer SR. NPC intracellular cholesterol transporter 1 (NPC1)-mediated cholesterol export from lysosomes. *J Biol Chem*. 2019;294(5):1706–1709. https://doi.org/10.1074/jbc.TM118.004165.
8. Petersen NHT, Olsen OD, Groth-Pedersen L, et al. Transformation-associated changes in sphingolipid metabolism sensitize cells to lysosomal cell death induced by inhibitors of acid sphingomyelinase. *Cancer Cell*. 2013;24(3):379–393. https://doi.org/10.1016/j.ccr.2013.08.003.
9. Platt FM, D'Azzo A, Davidson BL, Neufeld EF, Tifft CJ. Lysosomal storage diseases. *Nat Rev Dis Primers*. 2018;4(1):27. https://doi.org/10.1038/s41572-018-0025-4.
10. Maegawa GHB. Lysosomal leukodystrophies lysosomal storage diseases associated with white matter abnormalities. *J Child Neurol*. 2019;34(6):339–358. https://doi.org/10.1177/0883073819828587.
11. Lachmann RH. Treating lysosomal storage disorders: what have we learnt? *J Inherit Metab Dis*. 2020;43(1):125–132. https://doi.org/10.1002/jimd.12131.
12. Ferreira CR, van Karnebeek CDM, Vockley J, Blau N. A proposed nosology of inborn errors of metabolism. *Genet Med*. 2019;21(1):102–106. https://doi.org/10.1038/s41436-018-0022-8.

13. Reisin R, Perrin A, García-Pavía P. Time delays in the diagnosis and treatment of Fabry disease. *Int J Clin Pract.* 2017;71(1), e12914. https://doi.org/10.1111/ijcp.12914.
14. Berger JR. Misdiagnosis of multiple sclerosis in a female heterozygote with Fabry's disease. *Mult Scler Relat Disord.* 2019;30:45–47. https://doi.org/10.1016/j.msard.2019.01.040.
15. Cox TM. Lysosomal diseases and neuropsychiatry: opportunities to rebalance the mind. *Front Mol Biosci.* 2020;7:177. https://doi.org/10.3389/fmolb.2020.00177.
16. Mistri M, Mehta S, Solanki D, et al. Identification of novel variants in a large cohort of children with Tay–Sachs disease: an initiative of a multicentric task force on lysosomal storage disorders by Government of India. *J Hum Genet.* 2019;64(10):985–994. https://doi.org/10.1038/s10038-019-0647-8.
17. Beutler E, Nguyen NJ, Henneberger MW, et al. Gaucher disease: gene frequencies in the Ashkenazi Jewish population. *Am J Hum Genet.* 1993;52(1):85–88.
18. Málaga DR, Brusius-Facchin AC, Siebert M, et al. Sensitivity, advantages, limitations, and clinical utility of targeted next-generation sequencing panels for the diagnosis of selected lysosomal storage disorders. *Genet Mol Biol.* 2019;42(1):197–206. https://doi.org/10.1590/1678-4685-gmb-2018-0092.
19. Hanss Z, Boussaad I, Jarazo J, Schwamborn JC, Krüger R. Quality control strategy for CRISPR-Cas9-based gene editing complicated by a pseudogene. *Front Genet.* 2020;10:1297. https://doi.org/10.3389/fgene.2019.01297.
20. Zampieri S, Cattarossi S, Bembi B, Dardis A. GBA analysis in next-generation era. *J Mol Diagn.* 2017;19(5):733–741. https://doi.org/10.1016/j.jmoldx.2017.05.005.
21. Froissart R, Da Silva IM, Maire I. Mucopolysaccharidosis type II: an update on mutation spectrum. *Acta Paediatr.* 2007;96(455):71–77. https://doi.org/10.1111/j.1651-2227.2007.00213.x.
22. Haer-Wigman L, Newman H, Leibu R, et al. Non-syndromic retinitis pigmentosa due to mutations in the mucopolysaccharidosis type IIIC gene, heparan-alpha-glucosaminide N-acetyltransferase (HGSNAT). *Hum Mol Genet.* 2015;24(13):3742–3751. https://doi.org/10.1093/hmg/ddv118.
23. Beck M. Treatment strategies for lysosomal storage disorders. *Dev Med Child Neurol.* 2018;60(1):13–18. https://doi.org/10.1111/dmcn.13600.
24. Tan EY, Boelens JJ, Jones SA, Wynn RF. Hematopoietic stem cell transplantation in inborn errors of metabolism. *Front Pediatr.* 2019;7:433. https://doi.org/10.3389/fped.2019.00433.
25. Barton NW, Brady RO, Dambrosia JM, et al. Replacement therapy for inherited enzyme deficiency—macrophage-targeted glucocerebrosidase for Gaucher's disease. *N Engl J Med.* 1991;324(21):1464–1470. https://doi.org/10.1056/NEJM199105233242104.
26. Germain DP, Charrow J, Desnick RJ, et al. Ten-year outcome of enzyme replacement therapy with agalsidase beta in patients with Fabry disease. *J Med Genet.* 2015;52(5):353–358. https://doi.org/10.1136/jmedgenet-2014-102797.
27. Schoser B, Stewart A, Kanters S, et al. Survival and long-term outcomes in late-onset Pompe disease following alglucosidase alfa treatment: a systematic review and meta-analysis. *J Neurol.* 2017;264(4):621–630. https://doi.org/10.1007/s00415-016-8219-8.
28. Edelmann MJ, Maegawa GHB. CNS-targeting therapies for lysosomal storage diseases: current advances and challenges. *Front Mol Biosci.* 2020;7:291. https://doi.org/10.3389/fmolb.2020.559804.
29. Poswar F, Vairo F, Burin M, et al. Lysosomal diseases: overview on current diagnosis and treatment. *Genet Mol Biol.* 2019;42(1):165–177. https://doi.org/10.1590/1678-4685-gmb-2018-0159.
30. Giraldo P, Andrade-Campos M, Alfonso P, et al. Twelve years of experience with miglustat in the treatment of type 1 Gaucher disease: the Spanish ZAGAL project. *Blood Cells Mol Dis.* 2018;68:173–179. https://doi.org/10.1016/j.bcmd.2016.10.017.

31. Patterson MC, Mengel E, Vanier MT, Moneuse P, Rosenberg D, Pineda M. Treatment outcomes following continuous miglustat therapy in patients with Niemann-Pick disease Type C: a final report of the NPC Registry. *Orphanet J Rare Dis.* 2020;15(1):104. https://doi.org/10.1186/s13023-020-01363-2.
32. Leal AF, Benincore-Flórez E, Solano-Galarza D, et al. GM2 gangliosidoses: clinical features, pathophysiological aspects, and current therapies. *Int J Mol Sci.* 2020;21(17):6213. https://doi.org/10.3390/ijms21176213.
33. Fischetto R, Palladino V, Mancardi MM, et al. Substrate reduction therapy with Miglustat in pediatric patients with GM1 type 2 gangliosidosis delays neurological involvement: a multicenter experience. *Mol Genet Genomic Med.* 2020;8(10):1371. https://doi.org/10.1002/mgg3.1371.
34. Peterschmitt MJ, Cox GF, Ibrahim J, et al. A pooled analysis of adverse events in 393 adults with Gaucher disease type 1 from four clinical trials of oral eliglustat: evaluation of frequency, timing, and duration. *Blood Cells Mol Dis.* 2018;68:185–191. https://doi.org/10.1016/j.bcmd.2017.01.006.
35. Germain DP, Hughes DA, Nicholls K, et al. Treatment of Fabry's disease with the pharmacologic chaperone migalastat. *N Engl J Med.* 2016;375(6):545–555. https://doi.org/10.1056/NEJMoa1510198.
36. Germain DP, Nicholls K, Giugliani R, et al. Efficacy of the pharmacologic chaperone migalastat in a subset of male patients with the classic phenotype of Fabry disease and migalastat-amenable variants: data from the phase 3 randomized, multicenter, double-blind clinical trial and extension study. *Genet Med.* 2019;21(9):1987–1997. https://doi.org/10.1038/s41436-019-0451-z.
37. Ma CC, Wang ZL, Xu T, He ZY, Wei YQ. The approved gene therapy drugs worldwide: from 1998 to 2019. *Biotechnol Adv.* 2020;40:107502. https://doi.org/10.1016/j.biotechadv.2019.107502.
38. Colella P, Ronzitti G, Mingozzi F. Emerging issues in AAV-mediated in vivo gene therapy. *Mol Ther Methods Clin Dev.* 2018;8:87–104. https://doi.org/10.1016/j.omtm.2017.11.007.
39. Broeders M, Herrero-Hernandez P, Ernst MPT, van der Ploeg AT, Pijnappel WWMP. Sharpening the molecular scissors: advances in gene-editing technology. *iScience.* 2020;23(1), 100789. https://doi.org/10.1016/j.isci.2019.100789.
40. Ho BX, Loh SJH, Chan WK, Soh BS. In vivo genome editing as a therapeutic approach. *Int J Mol Sci.* 2018;19(9):2721. https://doi.org/10.3390/ijms19092721.
41. Gaj T, Sirk SJ, Shui SL, Liu J. Genome-editing technologies: principles and applications. *Cold Spring Harb Perspect Biol.* 2016;8(12):a023754. https://doi.org/10.1101/cshperspect.a023754.
42. Porto EM, Komor AC, Slaymaker IM, Yeo GW. Base editing: advances and therapeutic opportunities. *Nat Rev Drug Discov.* 2020;19(12):839–859. https://doi.org/10.1038/s41573-020-0084-6.
43. Anzalone AV, Randolph PB, Davis JR, et al. Search-and-replace genome editing without double-strand breaks or donor DNA. *Nature.* 2019;576(7785):149–157. https://doi.org/10.1038/s41586-019-1711-4.
44. Borger DK, McMahon B, Lal TR, Serra-Vinardell J, Aflaki E, Sidransky E. Induced pluripotent stem cell models of lysosomal storage disorders. *Dis Model Mech.* 2017;10(6):691–704. https://doi.org/10.1242/dmm.029009.
45. Nair V, Belanger EC, Veinot JP. Lysosomal storage disorders affecting the heart: a review. *Cardiovasc Pathol.* 2019;39:12–24. https://doi.org/10.1016/j.carpath.2018.11.002.
46. Poletto E, Pasqualim G, Giugliani R, Matte U, Baldo G. Effects of gene therapy on cardiovascular symptoms of lysosomal storage diseases. *Genet Mol Biol.* 2019;42(1):261–285. https://doi.org/10.1590/1678-4685-gmb-2018-0100.

47. Allende ML, Cook EK, Larman BC, et al. Cerebral organoids derived from Sandhoff disease-induced pluripotent stem cells exhibit impaired neurodifferentiation. *J Lipid Res*. 2018;59(3):550–563. https://doi.org/10.1194/jlr.M081323.
48. Latour YL, Yoon R, Thomas SE, et al. Human GLB1 knockout cerebral organoids: a model system for testing AAV9-mediated GLB1 gene therapy for reducing GM1 ganglioside storage in GM1 gangliosidosis. *Mol Genet Metab Rep*. 2019;21:100513. https://doi.org/10.1016/j.ymgmr.2019.100513.
49. Zhang H, Shi J, Hachet MA, et al. CRISPR/Cas9-mediated gene editing in human iPSC-derived macrophage reveals lysosomal acid lipase function in human macrophages-brief report. *Arterioscler Thromb Vasc Biol*. 2017;37(11):2156–2160. https://doi.org/10.1161/ATVBAHA.117.310023.
50. Song HY, Chiang HC, Tseng WL, et al. Using CRISPR/Cas9-mediated GLA gene knockout as an in vitro drug screening model for Fabry disease. *Int J Mol Sci*. 2016;17(12):2089. https://doi.org/10.3390/ijms17122089.
51. Pereira EM, Labilloy A, Eshbach ML, et al. Characterization and phosphoproteomic analysis of a human immortalized podocyte model of Fabry disease generated using CRISPR/CAS9 technology. *Am J Physiol Renal Physiol*. 2016;311(5):F1015–F1024. https://doi.org/10.1152/ajprenal.00283.2016.
52. Kaissarian N, Kang J, Shu L, Ferraz MJ, Aerts JM, Shayman JA. Dissociation of globotriaosylceramide and impaired endothelial function in α-galactosidase-A deficient EA.hy926 cells. *Mol Genet Metab*. 2018;125(4):338–344. https://doi.org/10.1016/j.ymgme.2018.10.007.
53. Kang JJ, Kaissarian NM, Desch KC, et al. α-galactosidase A deficiency promotes von Willebrand factor secretion in models of Fabry disease. *Kidney Int*. 2019;95(1):149–159. https://doi.org/10.1016/j.kint.2018.08.033.
54. Song C, Yarmishyn, et al. Generation of GLA-knockout human embryonic stem cell lines to model autophagic dysfunction and exosome secretion in Fabry disease-associated hypertrophic cardiomyopathy. *Cells*. 2019;8(4):327. https://doi.org/10.3390/cells8040327.
55. Pavan E, Ormazabal M, Peruzzo P, Vaena E, Rozenfeld P, Dardis A. Crispr/cas9 editing for Gaucher disease modeling. *Int J Mol Sci*. 2020;21(9):3268. https://doi.org/10.3390/ijms21093268.
56. Kim MJ, Jeon S, Burbulla LF, Krainc D. Acid ceramidase inhibition ameliorates α-synuclein accumulation upon loss of GBA1 function. *Hum Mol Genet*. 2018;27(11):1972–1988. https://doi.org/10.1093/hmg/ddy105.
57. Bae EJ, Yang NY, Lee C, Kim S, Lee HJ, Lee SJ. Haploinsufficiency of cathepsin D leads to lysosomal dysfunction and promotes cell-to-cell transmission of α-synuclein aggregates. *Cell Death Dis*. 2015;6(10), e1901. https://doi.org/10.1038/cddis.2015.283.
58. Drews K, Calgi MP, Harrison WC, et al. Glucosylceramidase maintains influenza virus infection by regulating endocytosis. *J Virol*. 2019;93(12):e00017–e00019. https://doi.org/10.1128/jvi.00017-19.
59. Erwood S, Brewer RA, Bily TMI, et al. Modeling Niemann-Pick disease type C in a human haploid cell line allows for patient variant characterization and clinical interpretation. *Genome Res*. 2019;29(12):2010–2019. https://doi.org/10.1101/gr.250720.119.
60. Du X, Lukmantara I, Yang H. CRISPR/Cas9-mediated generation of Niemann–pick C1 knockout cell line. *Methods Mol Biol*. 2017;1583:73–83. https://doi.org/10.1007/978-1-4939-6875-6_7.
61. Li J, Deffieu MS, Lee PL, Saha P, Pfeffer SR. Glycosylation inhibition reduces cholesterol accumulation in NPC1 protein-deficient cells. *Proc Natl Acad Sci U S A*. 2015;112(48):14876–14881. https://doi.org/10.1073/pnas.1520490112.

62. Haskins M. Gene therapy for lysosomal storage diseases (LSDs) in large animal models. *ILAR J.* 2009;50(2):112–121. https://doi.org/10.1093/ilar.50.2.112.
63. Zhang T, Peterson RT. Modeling lysosomal storage diseases in the zebrafish. *Front Mol Biosci.* 2020;7:82. https://doi.org/10.3389/fmolb.2020.00082.
64. Eaton SL, Proudfoot C, Lillico SG, et al. CRISPR/Cas9 mediated generation of an ovine model for infantile neuronal ceroid lipofuscinosis (CLN1 disease). *Sci Rep.* 2019;9:9891. https://doi.org/10.1038/s41598-019-45859-9.
65. de Carvalho TG, Schuh R, Pasqualim G, et al. CRISPR-Cas9-mediated gene editing in human MPS I fibroblasts. *Gene.* 2018;678:33–37. https://doi.org/10.1016/j.gene.2018.08.004.
66. Schuh RS, de Carvalho TG, Giugliani R, Matte U, Baldo G, Teixeira HF. Gene editing of MPS I human fibroblasts by co-delivery of a CRISPR/Cas9 plasmid and a donor oligonucleotide using nanoemulsions as nonviral carriers. *Eur J Pharm Biopharm.* 2018;122:158–166. https://doi.org/10.1016/j.ejpb.2017.10.017.
67. Miki T, Vazquez L, Yanuaria L, et al. Induced pluripotent stem cell derivation and ex vivo gene correction using a mucopolysaccharidosis type 1 disease mouse model. *Stem Cells Int.* 2019;2019:6978303. https://doi.org/10.1155/2019/6978303.
68. Chang S-K, Lu Y-H, Chen Y-R, et al. Correction of the GLA IVS4 + 919 G > A mutation with CRISPR/Cas9 deletion strategy in fibroblasts of Fabry disease. *Ann Transl Med.* 2017;5(suppl. 2), AB043. https://doi.org/10.21037/atm.2017.s043.
69. Do HS, Park SW, Im I, et al. Enhanced thrombospondin-1 causes dysfunction of vascular endothelial cells derived from Fabry disease-induced pluripotent stem cells. *EBioMedicine.* 2020;52:102633. https://doi.org/10.1016/j.ebiom.2020.102633.
70. Scharenberg SG, Poletto E, Lucot KL, et al. Engineering monocyte/macrophage − specific glucocerebrosidase expression in human hematopoietic stem cells using genome editing. *Nat Commun.* 2020;11:3327. https://doi.org/10.1038/s41467-020-17148-x.
71. van der Wal E, Herrero-Hernandez P, Wan R, et al. Large-scale expansion of human iPSC-derived skeletal muscle cells for disease modeling and cell-based therapeutic strategies. *Stem Cell Rep.* 2018;10(6):1975–1990. https://doi.org/10.1016/j.stemcr.2018.04.002.
72. Dever DP, Scharenberg SG, Camarena J, et al. CRISPR/Cas9 genome engineering in engraftable human brain-derived neural stem cells. *iScience.* 2019;15:524–535. https://doi.org/10.1016/j.isci.2019.04.036.
73. Pavani G, Laurent M, Fabiano A, et al. Ex vivo editing of human hematopoietic stem cells for erythroid expression of therapeutic proteins. *Nat Commun.* 2020;11(1):4146. https://doi.org/10.1038/s41467-020-17552-3.
74. Gomez-Ospina N, Scharenberg SG, Mostrel N, et al. Human genome-edited hematopoietic stem cells phenotypically correct Mucopolysaccharidosis type I. *Nat Commun.* 2019;10(1):4045. https://doi.org/10.1038/s41467-019-11962-8.
75. Schuh RS, Gonzalez EA, Tavares AMV, et al. Neonatal nonviral gene editing with the CRISPR/Cas9 system improves some cardiovascular, respiratory, and bone disease features of the mucopolysaccharidosis I phenotype in mice. *Gene Ther.* 2020;27(1–2):74–84. https://doi.org/10.1038/s41434-019-0113-4.
76. Ou L, DeKelver RC, Rohde M, et al. ZFN-mediated in vivo genome editing corrects murine hurler syndrome. *Mol Ther.* 2019;27(1):178–187. https://doi.org/10.1016/j.ymthe.2018.10.018.
77. Ou L, Przybilla MJ, Ahlat O, et al. A highly efficacious PS gene editing system corrects metabolic and neurological complications of mucopolysaccharidosis type I. *Mol Ther.* 2020;28(6):1442–1454. https://doi.org/10.1016/j.ymthe.2020.03.018.

78. Wang D, Li J, Song CQ, et al. Cas9-mediated allelic exchange repairs compound heterozygous recessive mutations in mice. *Nat Biotechnol.* 2018;36(9):839–842. https://doi.org/10.1038/nbt.4219.
79. Laoharawee K, DeKelver RC, Podetz-Pedersen KM, et al. Dose-dependent prevention of metabolic and neurologic disease in murine MPS II by ZFN-mediated in vivo genome editing. *Mol Ther.* 2018;26(4):1127–1136. https://doi.org/10.1016/j.ymthe.2018.03.002.
80. Sharma R, Anguela XM, Doyon Y, et al. In vivo genome editing of the albumin locus as a platform for protein replacement therapy. *Blood.* 2015;126(15):1777–1784. https://doi.org/10.1182/blood-2014-12-615492.
81. Ou L, Przybilla MJ, Tăbăran AF, et al. A novel gene-editing system to treat both Tay–Sachs and Sandhoff diseases. *Gene Ther.* 2020;27(5):226–236. https://doi.org/10.1038/s41434-019-0120-5.
82. Levy JM, Yeh WH, Pendse N, et al. Cytosine and adenine base editing of the brain, liver, retina, heart and skeletal muscle of mice via adeno-associated viruses. *Nat Biomed Eng.* 2020;4:97–110. https://doi.org/10.1038/s41551-019-0501-5.
83. Poletto E, Pasqualim G, Giugliani R, Matte U, Baldo G. Worldwide distribution of common IDUA pathogenic variants. *Clin Genet.* 2018;94(1):95–102. https://doi.org/10.1111/cge.13224.
84. Ventola CL. Progress in nanomedicine: approved and investigational nanodrugs. *P T.* 2017;42(12):742–755.
85. Ganta S, Talekar M, Singh A, Coleman TP, Amiji MM. Nanoemulsions in translational research—opportunities and challenges in targeted cancer therapy. *AAPS PharmSciTech.* 2014;15(3):694–708. https://doi.org/10.1208/s12249-014-0088-9.
86. Pan G, Shawer M, Øie S, Lu DR. In vitro gene transfection in human glioma cells using a novel and less cytotoxic artificial lipoprotein delivery system. *Pharm Res.* 2003;20(5):738–744. https://doi.org/10.1023/A:1023477317668.
87. Azambuja JH, Schuh RS, Michels LR, et al. Nasal administration of cationic nanoemulsions as CD73-siRNA delivery system for glioblastoma treatment: a new therapeutical approach. *Mol Neurobiol.* 2020;57(2):635–649. https://doi.org/10.1007/s12035-019-01730-6.
88. Yadav S, Gandham SK, Panicucci R, Amiji MM. Intranasal brain delivery of cationic nanoemulsion-encapsulated TNFα siRNA in prevention of experimental neuroinflammation. *Nanomed Nanotechnol Biol Med.* 2016;12(4):987–1002. https://doi.org/10.1016/j.nano.2015.12.374.
89. Gowing G, Svendsen S, Svendsen CN. Ex vivo gene therapy for the treatment of neurological disorders. *Prog Brain Res.* 2017;230:99–132. https://doi.org/10.1016/bs.pbr.2016.11.003.
90. Samson M, Libert F, Doranz BJ, et al. Resistance to HIV-1 infection in caucasian individuals bearing mutant alleles of the CCR-5 chemokine receptor gene. *Nature.* 1996;382(6593):722–725. https://doi.org/10.1038/382722a0.
91. Xie Y, Zhan S, Ge W, Tang P. The potential risks of C-C chemokine receptor 5-edited babies in bone development. *Bone Res.* 2019;7:4. https://doi.org/10.1038/s41413-019-0044-0.
92. Dever DP, Bak RO, Reinisch A, et al. CRISPR/Cas9 β-globin gene targeting in human hematopoietic stem cells. *Nature.* 2016;539:384–389. https://doi.org/10.1038/nature20134.
93. Wang J, Exline CM, Declercq JJ, et al. Homology-driven genome editing in hematopoietic stem and progenitor cells using ZFN mRNA and AAV6 donors. *Nat Biotechnol.* 2015;33(12):1256–1263. https://doi.org/10.1038/nbt.3408.

94. Genovese P, Schiroli G, Escobar G, et al. Targeted genome editing in human repopulating hematopoietic stem cells. *Nature.* 2014;510(7504):235–240. https://doi.org/10.1038/nature13420.
95. Uchida N, Chen K, Dohse M, et al. Human neural stem cells induce functional myelination in mice with severe dysmyelination. *Sci Transl Med.* 2012;4(165), 165er7. https://doi.org/10.1126/scitranslmed.3005469.
96. Schuh RS, Poletto E, Pasqualim G, et al. In vivo genome editing of mucopolysaccharidosis I mice using the CRISPR/Cas9 system. *J Control Release.* 2018;288:23–33. https://doi.org/10.1016/j.jconrel.2018.08.031.
97. Wechsler T, DeKelver R, Rohde M, et al. ZFN-mediated in vivo genome editing results in supraphysiological levels of lysosomal enzymes deficient in hunter and hurler syndrome and Gaucher disease. *Mol Ther.* 2015;27(1):178–187. https://doi.org/10.1038/mt.2015.74.
98. Leal AF, Espejo-Mojica AJ, Sánchez OF, et al. Lysosomal storage diseases: current therapies and future alternatives. *J Mol Med.* 2020;98(7):931–946. https://doi.org/10.1007/s00109-020-01935-6.
99. Harmatz P, Muenzer J, Burton BK, et al. Update on phase 1/2 clinical trials for MPS I and MPS II using ZFN-mediated in vivo genome editing. *Mol Genet Metab.* 2018;123(2):S59–S60. https://doi.org/10.1016/j.ymgme.2017.12.143.
100. Harmatz P, Lau HA, Heldermon C, et al. EMPOWERS: a phase 1/2 clinical trial of SB-318 ZFN-mediated in vivo human genome editing for treatment of MPS I (Hurler syndrome). *Mol Genet Metab.* 2019;126(2):S68. https://doi.org/10.1016/j.ymgme.2018.12.163.
101. Muenzer J, Prada CE, Burton B, et al. CHAMPIONS: a phase 1/2 clinical trial with dose escalation of SB-913 ZFN-mediated in vivo human genome editing for treatment of MPS II (Hunter syndrome). *Mol Genet Metab.* 2019;126(2):S104. https://doi.org/10.1016/j.ymgme.2018.12.263.
102. Poletto E, Baldo G, Gomez-Ospina N. Genome editing for mucopolysaccharidoses. *Int J Mol Sci.* 2020;21(2):500. https://doi.org/10.3390/ijms21020500.
103. Conway A, Mendel M, Kim K, et al. Non-viral delivery of zinc finger nuclease mRNA enables highly efficient in vivo genome editing of multiple therapeutic gene targets. *Mol Ther.* 2019;27(4):866–877. https://doi.org/10.1016/j.ymthe.2019.03.003.

CHAPTER TEN

Genome editing in mucopolysaccharidoses and mucolipidoses

Hallana Souza Santos, Edina Poletto, Roselena Schuh, Ursula Matte, and Guilherme Baldo*

Laboratório Células, Tecidos e Genes do Hospital de Clínicas de Porto Alegre, Porto Alegre, RS, Brazil
*Corresponding author: e-mail address: gbaldo@hcpa.edu.br

Contents

1. Clinical and molecular characteristics of mucopolysaccharidoses and mucolipidoses	328
1.1 Mucopolysaccharidoses	328
1.2 Mucolipidoses	331
2. Genome editing: General concepts	332
3. Genome editing for MPS and ML: In vitro studies	338
4. In vivo genome editing	341
5. Clinical trials	344
6. Conclusions	347
References	347

Abstract

Mucopolysaccharidoses (MPS) and mucolipidoses (ML) are disorders that alter lysosome function. While MPS are caused by mutation in enzymes that degrade glycosaminoglycans, the ML are disorders characterized by reduced function in the phosphotransferase enzyme. Multiple clinical features are associated with these diseases and the exact mechanisms that could explain such different clinical manifestations in patients are still unknown. Furthermore, there are no curative treatment for any of MPS and ML conditions so far. Gene editing holds promise as a tool for the creation of cell and animal models to help explain disease pathogenesis, as well as a platform for gene therapy. In this chapter, we discuss the main studies involving genome editing for MPS and the prospect applications for ML.

Abbreviations

AAVS	adeno-associated viruses
cDNA	complementary DNA
CNS	central nervous system
CRISPR	clustered regularly interspaced short palindromic repeat

DSBs	double-strand breaks
GAGs	glycosaminoglycans
GBA	glucosylceramidase beta
GNPTAB	N-acetylglucosamine-1-phosphotransferase
gRNA	guide RNA
HDR	homology-directed repair
hIDS	human IDS
IDS	iduronate-2-sulfatase
IDUA	alpha-L-iduronidase
IPSCs	induced pluripotent stem cells
LSDs	lysosomal storage disorders
ML	mucolipidoses
MPS	mucopolysaccharidoses
NHEJ	error-prone non-homologous end joining
NSCs	neural stem cells
PAM	protospacer adjacent motif
pegRNA	prime editing guide RNA
RVDs	repeat-variable diresidues
TALENs	transcription activator-like effector nucleases
ZFNs	zinc finger nucleases

1. Clinical and molecular characteristics of mucopolysaccharidoses and mucolipidoses

1.1 Mucopolysaccharidoses

The mucopolysaccharidoses (MPS) are a group of 11 lysosomal storage disorders caused by a deficiency in one of the hydrolases that participate in the multistep degradation of glycosaminoglycans (GAGs).[1] They all have autosomal recessive inheritance, with exception of MPS II which is X linked. Table 1 summarizes the different types of MPS, their respective gene and enzyme deficiency and the partially metabolized GAG.[2]

GAGs are highly sulfated, complex, linear polysaccharides composed of repeated disaccharide units. They are constituents of the extracellular matrix where they play a key role in cell signaling and modulate several biochemical processes, including regulation of cell growth and proliferation, promotion of cell adhesion and wound repair.[3] Therefore, depending on which GAG degradation pathway is affected by each MPS type, different organs and tissues are more preeminently affected.[4]

MPS patients present a set of common features that include progressive multisystemic involvement, infiltrated faces, skeletal and joint problems.

Table 1 General features and key clinical features of mucopolysaccharidoses.

MPS type	OMIM	Eponym	Enzyme	Gene	Glycosaminoglycan	Brain	Bone/joint	Viscera (Liver and spleen)
MPS I	#607014 / #607016	Hurler/ Scheie	alpha-L-iduronidase	*IDUA*	HS, DS	+++/++	++	+++
MPS II	#309900	Hunter	Iduronate sulfatase	*IDS*	HS, DS	+/−	++	++
						+++/−	++	++/+++
MPS III A	#252900	Sanfilippo A	Heparan N-sulfatase (sulfamidase)	*SGSH*	HS	+++	+	+
MPS III B	#252920	Sanfilippo B	alpha-*N*-acetyl-glucosaminidase	*NAGLU*	HS	+++	+	+
MPS III C	#252930	Sanfilippo C	Acetyl-CoA: alpha-glucosaminide acetyltransferase	*HGSNAT*	HS	+++	+	+
MPS III D	#252940	Sanfilippo D	*N*-acetylglucosamine 6-sulfatase	*GNS*	HS	+++	+	+
MPS IV A	#253000	Morquio A	Galactose 6-sulfatase	*GALNS*	KS, CS	−	+++	++
MPS IV B	#253010	Morquio B	beta-Galactosidase	*GLB1*	KC	−	+++	++
MPS VI	#253200	Maroteaux-Lamy	*N*-acetyl-galactosamine 4-sulfatase	*ARSB*	DS, CS	−	+++	+++
MPS VII	#253220	Sly	beta-glucuronidase	*GUSB*	HS, DS, CS	++	++	+++
MPS IX	#601492	Natowicz	Hyaluronidase	*HYAL1*	Hyaluronan	−	++	−

CS, chondroitin sulfate; DS, dermatan sulfate; HS, heparan sulfate; KS, keratan sulfate; MPS, mucopolysaccharidosis.
−, no involvement; +, mild; ++, moderate; +++, severe.

They also may present heart and respiratory abnormalities, hepatosplenomegaly, and neurological impairment, although this last characteristic is highly variable between and within MPS types. Nevertheless, neurocognitive issues are restricted to those MPS in which heparan sulfate degradation is impaired, whereas skeletal and joint manifestations are present when dermatan or keratan sulfate degradation pathways are involved. It is important to notice, however, that the type of skeletal disease in MPS IV A/B is markedly different from that seen in other MPS that have impaired dermatan sulfate degradation. Table 1 presents a summary of key clinical findings in the different MPS.

Age at onset may be as early as prenatal (in MPS VII) to late childhood (in MPS IX) and varies greatly depending on where in the clinical spectrum the patient is situated. For historical reasons MPS I is clearly recognized as having severe and attenuated forms (Hurler and Scheie syndromes, respectively), but there is a continuum of clinical presentations within any given MPS that correlates with residual enzyme activity (Fig. 1). This variation can be mostly explained by the genotype of the patients. Usually, patients homozygous for nonsense mutations such as p.Trp402Ter and p.Gln70Ter in MPS I have the most severe form, due to the fact that no residual enzyme activity is produced.[5]

The incidence of MPS varies widely according to type and geographic region. Khan et al. (2017) based on epidemiological data, estimated frequencies between 7.85 (for MPS VI) to 0.01 (for MPS IV B) per 100,000 live births.[6] Borges et al. (2020), using the frequency of disease-causing variants in populational databases estimated incidences of 7.10 (for MPS I) to 0.05 (for MPS III D) per 100,000 live births.[7] Nevertheless, incidences can be

Fig. 1 Mucopolysaccharidoses symptoms and onset. Continuum of clinical presentation in the MPS correlates with residual enzyme activity and age at onset of symptoms.

higher in specific regions for particular MPS types, such as the cluster of MPS VI in Northeast of Brazil.[8] Allelic heterogeneity is common in all MPS, with a few predominant alleles, such as p.Trp402Ter in MPS I or p.Arg245His in MPS III A, that account for about 50% of the pathogenic alleles in different populations. However, these are rather the exception and many different pathogenic variants have been described for these diseases.

1.2 Mucolipidoses

Lysosomal hydrolases require the addition of mannose-6-phosphate residues for a correct trafficking into the nascent lysosomes. This post-translational modification is performed by N-acetylglucosamine-1-phosphotransferase (GlcNAc-phosphotransferase; EC 2.7.8.17), an enzyme complex formed by six subunits: $a_2/b_2/g_2$. Both a and b subunits are encoded by the *GNPTAB* gene, located in chromosome 12q23.2, with 21 exons and spans over 85 kb. The non-catalytic G subunit is encoded by the *GNPTG* gene, located in chromosome 16p13.3 which has 11 exons and spans 11.44 kb. GlcNAc-phosphotransferase deficiency impairs trafficking and redirects lysosomal hydrolases from lysosomes to the extracellular milieu. This feature is used for diagnostic purposes, as patients will present reduced activity of several hydrolases in fibroblasts (but not in leukocytes) with increased enzyme activity in plasma.[9]

Mucolipidosis I is a distinct disorder caused by alterations in neuraminidase, being called Sialidosis, and therefore will not be discussed in this chapter. Mucolipidosis II (ML II disease, inclusion cell disease or I-cell disease) and III (ML III, pseudo-Hurler polydystrophy) are autosomal recessive disorders caused by defects in the GlcNAc-1-phosphotransferase complex. Although having a common biochemical basis, clinically there are three recognized types of ML. The severe phenotype is ML II (OMIM #252500), whereas milder forms are ML III α/β (OMIM #252600) and ML III gamma (OMIM #252605).

ML II is caused by mutations in *GNPTAB*, usually nonsense or frameshift, that cause complete loss of function of GlcNAc-1-phosphotransferase activity. The incidence is approximately 1:123,500 live births in Portugal[10] and an unusually high prevalence is found in the province of Quebec, Canada with 1:6184 live births and an estimated carrier rate of 1:39, probably due to a founder effect.[11] Symptoms are present at birth or even before, and death occurs in the first decade of life. Clinical features include

Table 2 General features of mucolipidoses (ML).

ML type	OMIM	Protein	Gene	Phenotype
ML II	#252500	GlcNAc-1-phosphotransferase (α/β subunits)	*GNPTAB*	Severe
ML III α/β	#252600	GlcNAc-1-phosphotransferase (α/β subunits)	*GNPTAB*	Intermediate
ML III δ	#252605	GlcNAc-1-phosphotransferase (δ subunit)	*GNPTG*	Attenuated

ML type I is Sialidosis, with different characteristics.

psychomotor retardation, coarse dysmorphic facial features, growth retardation, and restricted joint movement. Severe skeletal abnormalities, cardiac and pulmonary complications are also present and usually are the cause of death.[12]

ML III α/β is also due to mutations in *GNPTAB* but in this case it is usually associated to missense or splice site variants that retain some residual enzyme activity. Incidence is estimated to be similar to that of ML II and disease progression is slower, with clinical onset at approximately age 3 years and death in early-to-middle adulthood. Clinical features include joint stiffness, left and/or right ventricular hypertrophy in older individuals and death in early adulthood is often from cardiopulmonary causes.[13] Neuromotor development and intellect are the most variable features in ML IIIα/β.

ML III gamma is caused by mutations in *GNPTG*. No precise disease incidence is available, and it is considered an ultra-rare disease. Clinical onset is in early childhood and disease progress is slow (Table 2). It affects mainly the skeletal, joint, and connective tissues. Mild cardiac involvement is present.[14]

2. Genome editing: General concepts

Genome editing is a broadly used DNA engineering approach for biological research taking advantage of a new set of revolutionary technologies. The strategy involves making precise changes, such as insertion, deletion, or substitutions at a specific DNA locus. This technology can be used to edit diverse cell types and organisms, and it has been widely applied in pathophysiology studies or in the development of new gene-based therapies. Genome editing-based therapies are particularly promising for patients affected with genetic diseases,[15] especially monogenic diseases, like MPS and ML.

Currently, there is a wide variety of genome editing platforms. Among them, the most pertinent to treat diseases are the transcription activator-like effector nucleases (TALENs), zinc finger nucleases (ZFNs), clustered regularly interspaced short palindromic repeat (CRISPR)–associated (Cas) systems and the more recently described CRISPR-Cas9-based editors and prime editing.[15] To date, genome editing was performed in MPS using ZFN and CRISPR–Cas9. Although there are no reports of the application of genome editing tools in ML, the potential of the technique in this group of diseases is undeniable and will be discussed in this chapter as well.

Genome editing technologies are based on the use of engineered nucleases to create double-strand breaks (DSBs) at specific DNA sequences or genomic locations, to stimulate endogenous cellular DNA repair mechanisms. The error-prone non-homologous end joining (NHEJ) process incorporates random non-complementary nucleotides or delete nucleotides (indels). Depending on the type and position, it can change the open reading frame and generate non-functional proteins and, thus, gene knockouts. The homology-directed repair (HDR) needs a DNA donor template sequence to correct or insert the transgene. This template must have nucleotide sequences flanking the gene to be inserted that are homologous to those upstream and downstream of the break site[16] (Fig. 2).

The first programmable genome editing tools were ZFNs and TALENs (Fig. 3), which were designed by the fusion of a restriction enzyme and an unrelated DNA recognition domain. ZFNs consist of a nuclease domain of 196 amino acids, derived from the restriction enzyme FokI, and a DNA-binding domain made by consecutive Cys2His2 zinc finger units, each composed by ~30 amino acids. In the ZFN each zinc finger protein recognizes 3 base pairs, and the total recognized DNA sequence is about 9–18 nucleotides in length. TALENs are engineered by fusing an array of DNA-binding domain derived from transcription activator-like effectors (TALEs) with the FokI nuclease domain. The DNA recognition of TALENs is conferred by the repeat-variable diresidues (RVDs), which include multiple 33–35 amino acids repeat domains, each recognizes a single base pair for a total target site length between 15 and 30 nucleotides. For both systems, the recognition of two proximal DNA sequences is required for Fok1 dimerization and subsequent cleavage of the targeted region. Because DNA-binding domains derived from zinc fingers and TALE proteins can be customized they can recognize virtually any genomic DNA sequence.[17]

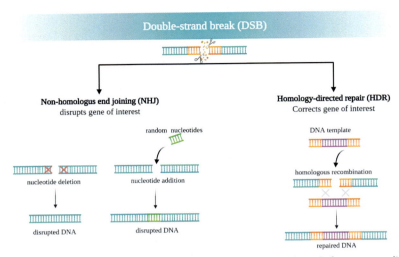

Fig. 2 Endogenous cellular DNA repair mechanisms exploited for gene editing. Following a DNA double-strand break (DSB), non-homologous end joining (NHEJ) joins two broken ends together, without using a homologous template for repair and leaving a disrupted DNA sequence with small insertion or deletions. This DNA repair pathway has been usually exploited to create gene knockouts. Homology-directed repair (HDR) is a template-dependent pathway for DNA DSB repair, which can be exploited for precise gene modification.

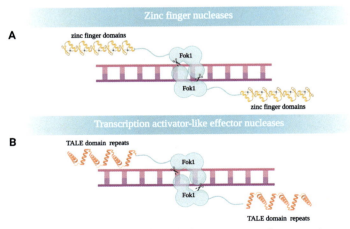

Fig. 3 Protein-based genome editing platforms. (A) Zinc finger nucleases (ZFNs) are formed by the DNA cleavage domain derived from the restriction enzyme *Fok*I and a DNA-binding domain zinc fingers. (B) Transcription activator-like effector nucleases (TALENs) are formed by the same FokI nuclease domain present in ZFNs while their DNA-binding domain is derived from transcription activator-like effectors (TALEs) proteins.

The most recent genome editing tools are based on the CRISPR-associated (Cas) system, which consists of an endonuclease (usually Cas9) and a guide-RNA. The Cas9 nuclease cleaves DNA guided by the guide-RNA (gRNA). This complex recognizes the target sequence (protospacer) through RNA and DNA base pairing, in the presence of short (3–7 bp) protospacer-adjacent motif (PAM) sequence bound directly by the Cas9 protein.[18] Targeting different sequences can be easily programmed by altering its gRNA, due to RNA-based recognition. New genome editing tools based on CRISPR have recently appeared. Base editing and Prime editors are CRISPR-mediated platforms that do not rely on DSBs. Base editors utilize a modified Cas9 such as inactive Cas9 or Cas9 Nickase, complexed with base-modifying enzymes (cytosine deaminase or adenosine deaminase) and a gRNA for converting one target base or base pair to another (for example, A-T to G-C or C-G to T-A). The prime editing utilizes a reverse transcriptase fused to a Cas9 Nickase and a prime editing guide RNA (pegRNA). The pegRNA serves as gRNA and as a template for reverse transcription[15,18] (Fig. 4).

Genome editing requires delivery of the components to target cells, and there are two approaches to this: in vivo or ex vivo. In vivo genome editing involves direct delivery in the body, either by intravenous delivery or in situ injection. On the other hand, ex vivo genome editing involves removal of the target cell population from the body, modifying it in vitro, and then transplanting it back into the original host.[19] These cells can be from the patient, from a suitable donor, or from a cell bank[15] (Fig. 5).

Regardless of the chosen approach (in vivo or ex vivo), it is necessary to have a vector that can access the nucleus of the cell, so that the editing occurs. Genome editing components cannot efficiently pass-through cell membranes, due to their molecular features. Therefore, the viral or non-viral delivery vectors are used to encapsulate and transport them. In some instances, a non-viral physical method is used. For ex vivo studies, the most common approaches use mechanical deformation or electroporation as a strategy to create transient holes in cell membranes, allowing nucleic acids and proteins to enter the cell. For in vivo administration of the gene editing components in animals, some common approaches include the application of high-volume hydrodynamic injection of nucleic acid into the tail vein or by direct injection into the embryo or zygote to create animal models (Fig. 6).

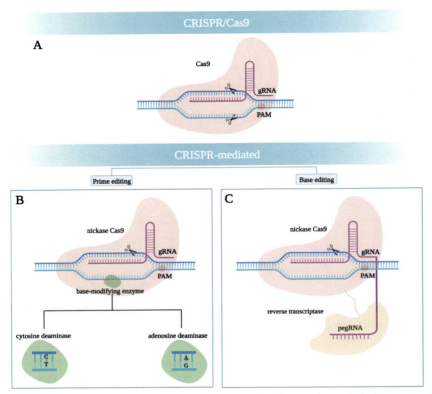

Fig. 4 CRISPR-based Genome Editing Platforms. (A) Clustered regularly interspaced short palindromic repeat (CRISPR)/Cas9 system: guide RNA (gRNA), Cas9 and protospacer adjacent motif (PAM). (B) Base editors: Cas9 Nickase complexed with base-modifying enzymes and a gRNA. (C) Prime editors: Cas9 Nickase attached to reverse transcriptase and complexed with prime editing guide RNA (pegRNA).

The main viral vectors used for genome editing are retroviruses, adenoviruses, and adeno-associated viruses (AAVs). Currently, AAVs have received special attention in research and clinical studies. Viral vectors are very efficient for gene delivery, but important immune responses have been reported in the history of viral gene therapy. On the one hand, non-viral vectors involve cationic materials to form nanoparticles that can be electrostatically complexed with DNA and other nucleic acids. The most successful classes of cationic materials used are naturally occurring and synthetic polymers and lipids. The vectors based on non-viral materials are biocompatible, less toxic, and immunogenic and are well-tolerated, but have reduced delivery efficiency.

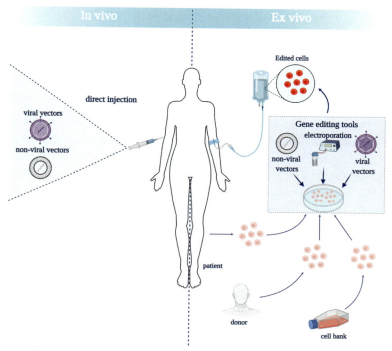

Fig. 5 Genome editing approaches: In vivo gene editing: direct injection of genome editing tools with viral or non-viral vectors. Ex vivo: removal of the target cell population from the body (patient, donor, or cell bank), in vitro modification (using non-viral, viral, or physical methods), and transplantation, usually back to the original host.

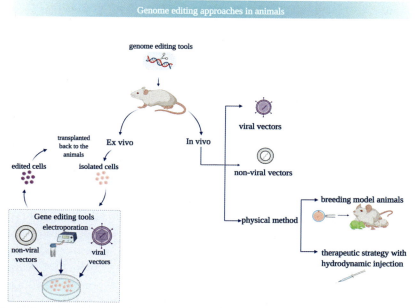

Fig. 6 Genome editing approaches in animal models. In vivo: direct injection of genome editing components with viral or non-viral vectors. Ex vivo: removal of the target cell population from the body, in vitro modification using non-viral, viral, or physical methods (hydrodynamic injection and direct injection) and injection of edited cells.

3. Genome editing for MPS and ML: In vitro studies

Preliminary studies involving pathophysiology or therapy development are traditionally done in patient-derived fibroblasts, as these cells present some disease features and are relatively easy to culture. When exposed to specific factors in the culture media, fibroblasts can even be differentiated into chondrocytes or osteoblast-like cells, becoming simple in vitro models of bone disease, for example. Stromal mesenchymal cells or hematopoietic stem cells also have the ability to differentiate into other cell types, as adipocytes or blood cells[20] (Fig. 7A). Lately, however, alternatives have been used to expand the possibilities of in vitro studies, mainly by means of using induced pluripotent stem cells (iPSCs), which can virtually be differentiated into many different cell types, as neural cells, cardiomyocytes or hematopoietic cells[21] (Fig. 7B).

In the last few years, several iPSCs models were created for Lysosomal storage disorders (LSDs) and MPS, including MPS I,[22] MPS II[23,24] (with a model from a heterozygote female patient with skewed X chromosome inactivation),[25] MPS IIIB,[26,27] MPS IVA[28] and MPS VII,[29] though none for ML. In 2011 the first iPSC MPS model reported was derived from keratinocytes and from mesenchymal stromal cells collected from two MPS I patients.[30] Once generated, alpha-L-iduronidase (IDUA)-deficient iPSCs were able to differentiate into hematopoietic and non-hematopoietic cells and the retroviral-mediated delivery of *IDUA* to correct the enzyme deficiency did not affect the differentiation pathway.

As the basis of MPS pathogenesis are still not completely understood in the central nervous system (CNS), patient-derived iPSCs are frequently differentiated into neural stem cells (NSCs). Indeed, despite MPS iPSCs maintain neural differentiation ability, MPS patient derived NSCs have a reduced self-renew capacity.[31] In MPS I NSCs, analysis of transcriptome revealed major alterations comparing to control cells, such as higher expression of autophagy genes and GAG biogenesis; in this study, cells derived from patients with the severe phenotypes present more pronounced alterations than cells from attenuated patients.[32] In the MPS II and MPS IIIB NSCs models, there are structural alterations in the Golgi complex and impaired autophagy.[32] When cultured in 3D organoids, MPS VII neural cells showed reduced neuronal activity and altered network connectivity.[29] Another interesting observation was done using MPS II cells, where recombinant enzyme administration did not normalize nor prevented GAG storage in

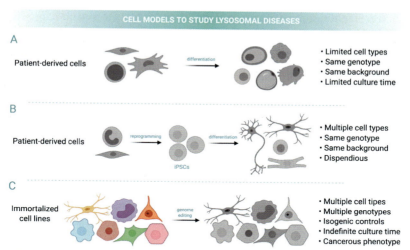

Fig. 7 Cell models to study lysosomal storage disorders (LSDs). (A) The obtention of patient-derived cells can be extremely invasive, thus not all cell types are available. Cells normally collected are fibroblasts, hematopoietic cells, and mesenchymal stromal cells. These cells can be differentiated into a small subset of cells, like osteoblasts, chondrocytes, adipocytes, and blood cells. (B) Alternatively, cells collected from patients can be reprogrammed into iPSCs cells, and afterwards be differentiated into virtually any cell type, like neurons or cardiomyocytes. In both patient-derived cells cases, all cells will harbor the patient genetic background and same disease-causing mutation and isogenic controls are only possible with genome editing. (C) Immortalized cell lines are derived from multiple individuals thus having different genetic marks. Multiple cell types can be chosen and edited to mimic the disease phenotype, with multiple genotypes. Gray cells: LSDs cells. Colored cells: wild-type cell lines.

neurons, astrocytes or oligodendrocytes, even when administered before GAG started to build up intracellularly[33]; these information from patient derived cells can give invaluable insights about therapeutic targets in the CNS.

An important aspect of using patient-derived iPSCs and healthy donor cells as controls is that these are not equivalent. One should be careful before drawing conclusions about what differences are due to the disease or simply due to distinct genetic background, epigenetic status or unmatched gender, age, and ethnicity. All these confounding factors can only be turned around by editing the cells—either by inducing pathogenic mutations in healthy cells or by correcting the disease-causing genotype in patient cells. In either way, both populations will harbor the same genome and epigenetic profile, except for the pathogenic variant in the specific gene of interest. Such approach was done using cells derived from MPS I mice, where isogenic iPSCs were generated from murine fibroblasts by CRISPR-mediated deletion of the exogenous sequence on exon 6, originally introduced to create the MPS I mouse model. Edited cells had a restored wild-type sequence and were efficiently differentiated into IDUA-secreting fibroblast-like cells.[34]

Despite the clear importance of creating iPSCs from LSDs patients, the process can be challenging and requires special strategies. In LSDs cells, the reprogramming process may be impaired by the very nature of the metabolic dysfunction. For example, the first model for MPS IIIB required co-culture of iPSCs with a feeder-layer of alpha-N-acetyl-glucosaminidase (NAGLU)-expressing cells during reprogramming as the lack of enzyme and/or accumulation of undegraded metabolites were impeding the procedure.[35] An additional barrier for iPSCs development of LSDs models is the cost: besides reagents, the reprogramming process is labor-intensive and requires experienced personnel.[21]

One way to overcome some limitations to iPSCs use is to generate cell models from immortalized cell lines (Fig. 7C). Although they might normally have cancerous features that can interfere in some studies, these are accessible cells, easy to culture and frequently easy to engineer.[36] There are currently multiple cell types available, from epithelial to neural cells—like the neuroblastoma line SH-SY5Y (ATCC CRL-2266). This cell line can be differentiated into neuron-like cells and be used as a model for neurodegenerative diseases. Such model was developed for MPS II, where SH-SY5Y cells were knocked-out in the iduronate-2-sulfatase (IDS) gene using specific gRNA targeting exon 3 and Cas9, transfected to the cells as a single plasmid vector.[37] The MPS II SH-SY5Y cells presented undetectable IDS activity and accumulation of GAGs, thus were used to validate observations made in the mouse model, which had overexpression of the protease cathepsin B in the brain. Increased protease expression, including cathepsin B, has been previously demonstrated in MPS I mouse models[38,39] and is thought to be part of the pathogenesis of lysosomal diseases. To confirm the data observed in the MPS II mice, cathepsin B was measured in knockout SH-SY5Y cells. The immunofluorescence data shows increased signal for cathepsin B and leakage to the cytoplasm, as this protease was not co-localized with lysosomal markers, suggesting alteration in the lysosomal membrane possibly caused by GAG storage or other unknown mechanism.[37]

There are no reports of genome editing or generation of cell models for ML, and for MPS, the field is still in its beginning. First reports of genome editing for MPS were in vitro studies using patient fibroblasts. In these proof-of-concept studies[40,41] cells were collected from a patient harboring the p.Trp402* mutation, the most common IDUA pathogenic variant.[5] Cationic nanoemulsions and liposomes were used to co-deliver a CRISPR-Cas9 and gRNA expression plasmid and a single-stranded oligonucleotide donor template containing a single nucleotide change (from tAg to tGg) to correct the p.Trp402Ter mutation. After 7 days of transfection, IDUA enzyme activity reached 5% of normal levels in the bulk population of edited cells; though seeming low, the achieved increased IDUA activity was enough to significantly reduce lysosome size in the whole cultured population, including non-edited cells, encouraging in vivo studies.[42,43]

Even though there are few studies using either iPSCs or genome editing for MPS, the applications of these two methodologies combined are huge. For example, it has been proposed the perspective of autologous neural stem cell transplantation, where iPSCs are edited, reprogrammed into NSCs and then transplanted back to the patient, in an attempt to serve as a permanent enzyme source in the brain.[44,45] While iPSCs can be efficiently edited,[46] human NSCs from biobanks can also be engineered to secrete lysosomal enzymes—as shown for Krabbe disease—and cross-correct enzyme-deficient cells.[47] A caveat of this approach, however, is the still remaining uncertainty regarding the safety of these cells once in vivo.

4. In vivo genome editing

Genome editing technologies have progressed to correct mutations that cause diseases through the use of nucleases such as ZFNs, TALENs, and more recently, CRISPR-Cas nucleases.[15] In vivo genome editing requires to tackle several critical aspects to achieve successful treatments. Indeed, the choice of gene delivery strategies[2,48] and biopharmaceutical aspects, such as route of administration, dose, therapeutic regimen, considering as well animal models or patient variables like age and other characteristics, lead to a very challenging path.[49–51]

Site-specific in vivo genome editing provides a promising approach for achieving long-term, stable therapeutic gene expression and has been successfully applied in a variety of preclinical models.[15] The first in vivo genome editing studies for MPS used ZFNs.[52] This first report describes a strategy for liver-directed protein replacement therapies using ZFN-mediated site-specific integration of either *IDUA* or *IDS* transgenes, delivered by an AAV vector, into the albumin locus. The study achieved long-term expression of lysosomal enzymes at therapeutic levels for both MPS I and MPS II.[52]

A similar in vivo genome editing approach was described for the murine model of MPS I syndrome, delivering ZFNs along with a corrective copy of the *IDUA* gene, which was inserted at the albumin locus in hepatocytes. In these studies they co-delivered mouse albumin specific ZFNs with a donor construct encoding a partial complementary DNA (cDNA) for either human *IDS* (mutated in MPS II), *IDUA* (MPS I), or *GBA* (glucosylceramidase beta, Gaucher disease) using AAV in wild type mice.[53] They demonstrated stable integration of the transgenes at the albumin locus, which resulted in liver-specific expression and secretion of the enzymes to plasma. For IDS and IDUA, this correlated well with increased enzymatic activity and protein expression in the liver, also detected in secondary tissues. *IDS*, *IDUA* or *GBA* expression remained stable in mice for 1–2 months, suggesting that this procedure is well-tolerated. Another approach from the same group led to sustained enzyme expression, secretion from the liver to circulation, and subsequent systemic uptake at levels sufficient for correction of metabolic disease and prevention of neurobehavioral deficits in MPS I mice.[54]

Laoharawee et al.[55] used a ZFN-targeting system to mediate insertion of the human *IDS* (hIDS) coding sequence into a "safe harbor" site, the intron 1 of the albumin locus in hepatocytes of an MPS II mouse model, the same locus used in the MPS I model. Three dose levels of recombinant AAV2/8 vectors encoding a pair of ZFNs and a *hIDS* cDNA donor were administered systemically. Suprahysiological vector dose-dependent levels of IDS enzyme were observed in the circulation and peripheral organs of treated mice, as well as reduction in GAG content. Surprisingly, they also demonstrated that ZFN-mediated genome editing prevented the development of neurocognitive deficit in young MPS II mice (6–9 weeks old) treated with a high vector dose. This ZFN-based platform for expression of therapeutic proteins from the albumin locus is a promising approach for treatment of MPS II and other lysosomal diseases.

The promising results observed in these first studies in animals[53–55] suggested that this approach could be effective in humans,[56] and the strategy is being tested in two clinical trials for MPS I (ClinicalTrials.gov, NCT02702115[57]) and MPS II (NCT03041324[58]). However, the initial results suggest that a very discreet increase in serum enzyme activity, possibly due to the low transgene expression level or an immune response to the vector. So, the researchers designed a proprietary system (PS) gene editing approach with CRISPR to insert a promoterless *IDUA* cDNA sequence in the albumin locus of hepatocytes. In this study, adeno-associated virus 8 (AAV8) vectors delivering the PS gene editing system were administered to newborn and adult MPS I mice.[59] As result, IDUA enzyme activity in the brain was significantly higher than control group, and neurobehavioral tests showed that treated mice had better memory and learning ability when compared to MPS I untreated mice. In addition, histological analysis showed efficacy and no vector-associated toxicity or increased tumorigenesis risk were observed. In summary, these results demonstrated the safety and efficacy of the PS in treating MPS I and paved the way for clinical studies. Additionally, as a therapeutic platform, the PS has the potential to treat other lysosomal diseases and new applications may rise from these studies.

Wang et al.[60] reported a genome-editing strategy to correct compound heterozygous mutations. The adeno-associated viral vector delivery of Cas9 gene and a guide RNA induced allelic exchange and rescued the disease phenotype in a mouse model of MPS I. This approach uses recombination of non-mutated genetic information present in compound heterozygous alleles into one functional allele without using donor DNA templates.[60] Neonatal MPS I mice were injected with a mixture of rAAV-SpCas9 and scAAV-sgRNA at 1:1 ratio in the facial vein. Six-week-old mice were injected with the same mixture of rAAV vectors in the tail vein. In the compound heterozygous MPS I mice, rAAV treatment at a young adult age restored IDUA activity to about 0.5% of the wild-type level, and substantially reduced GAG accumulation in the heart. Thus, Cas9-induced allelic exchange has the potential for gene correction in post-mitotic tissues. Several advantages distinguish this Cas9/sgRNA-mediated allelic exchange from other therapeutic in vivo genome-editing approaches such as Homologous Directed Repair, as no exogenous DNA repair template is needed. However, there are also important limitations to be considered, such as the low frequency of recombination events (lower than 1% of alleles) and the lack of information regarding safety of this approach, as no long-term study was published, and more than a single site in the genome needs to be cut, increasing the chance of off-target effects.

Another potential treatment approach is to engineer the patient's own hematopoietic system to express high levels of the deficient enzyme, thereby correcting the biochemical defect and halting disease progression. In this sense, Gomez-Ospina et al.[61] presented an efficient ex vivo genome editing approach using CRISPR/Cas9 that targeted alpha-L-iduronidase enzyme to the *CCR5* safe harbor locus in human CD34+ hematopoietic stem and progenitor cells. The modified cells secreted supraphysiological enzyme levels, maintained long-term repopulation and differentiation potential, and improved biochemical and phenotypic abnormalities in an immunocompromised mouse model of MPS I. These studies provide support for the development of genome edited CD34+ hematopoietic stem cells and may be applicable to other lysosomal storage disorders as well.

Recently, our group also tested gene editing using the CRISPR/Cas9 system in a MPS I mouse model.[42] Cationic liposomes were used to carry a CRISPR/Cas9 plasmid and a donor vector carrying the entire *IDUA* cDNA, aiming at the *ROSA26* locus. We compared our treated mice with animals treated with naked plasmids, and with untreated controls. A single hydrodynamic injection of the liposomal complex in newborn MPS I mice led to a significant increase in serum IDUA levels which were maintained for up to 6 months. The biodistribution of liposomal complexes after hydrodynamic injection was markedly detected in lungs and heart, corroborating the results of increased IDUA activity and decreased GAG storage especially in these tissues, while the group that received the naked plasmids presented increased IDUA activity especially in the liver. Furthermore, animals treated with the liposomal formulation presented improvement in cardiovascular parameters, one of the main causes of death observed in MPS I patients.[42] In another study, we also reported the effects of gene editing in cardiovascular, respiratory, bone, and neurologic functions in MPS I mice. Bone morphology showed partial improvement, while heart valves were still thickened, cardiac mass and aortic elastin breaks were reduced, with normalization of aortic diameter. Pulmonary resistance was normalized, suggesting improvement in respiratory function. In contrast, behavioral abnormalities and neuroinflammation still persisted, suggesting deterioration of the neurological functions.[43]

Genome editing is a potentially curative therapy for many genetic diseases, especially for MPS and ML. Ex vivo and in vivo genome editing platforms have been tested primarily on MPS, some approaches reaching clinical testing. The development of new genome editing platforms and the expansion to other diseases, may provide potential applications for both groups of disorders, which exceed the potential of current approaches.[15]

5. Clinical trials

Regarding MPS, the promising results in animal models suggested that gene editing could be effective in humans.[15] However, there are only two clinical trials involving gene editing for MPS to date, because clinical research in rare diseases has many challenges, such as carrying out trials in small populations, difficult identification of affected patients and the almost inevitable international scale of rare disease studies involving different regulatory agencies.[62] Furthermore, since gene editing is a relative new technique, several preclinical studies are still ongoing to gather enough data to translate to the clinic.

Despite these limitations, as briefly mentioned before, to date there are two clinical studies for MPS (clinicaltrials.gov). One for MPS I (NCT02702115) (Table 3)[57] and the other for MPS II (NCT03041324).[58] Both studies are mediated by ZFNs and delivered by AAV2/6 vectors. Moreover, it is important to highlight that enzymes produced by the liver are not expected to cross the blood-brain barrier. In view of, only patients with mild forms of the diseases that have little or no involvement of the CNS were recruited or studies.[15] Below we describe the main features and findings of the two clinical trials for MPS:

The study called "CHAMPIONS" is a clinical trial for MPS II (NCT03041324), the first trial to attempt to edit the genome in vivo in humans. A Phase 1/2 clinical trial evaluating the safety, tolerability, and effect on IDS enzyme activity of SB-913, designed to insert an IDS transgene into the hepatocyte albumin locus and, this way, to provide the long-term expression of IDS. Nine participants were distributed in four different cohorts with rising therapeutic doses of SB-913. The first results showed generally well-tolerated and no drug-related adverse events were reported at any dose with up to 10 months of exposure. Plasma IDS values no were detected and 1 subject in cohort 2 showed an increase.[15,58,63] Additional analysis of the trial data is ongoing.

The second trial is called "EMPOWERS": The clinical trial directed to individuals with MPS I (NCT02702115). A Phase 1/2 clinical trial evaluating the safety and tolerability of SB-318, designed to insert IDUA transgene into the hepatocyte albumin locus in doing so, provide the long-term expression of IDUA. Three participants were distributed in three different cohorts with increasing therapeutic doses of SB-318. The primary outcome presented was generally well-tolerated, even the highest dose of the vector.[57,59] No serious adverse events related to the study drug were reported with up to 3 months of exposure. However, the plasma IDUA values unchanged from pre-treatment levels. Two patients showed a decrease in the excretion of GAGs, while one patient remained with GAGS values above the normal range[64]. Additional analysis is ongoing.

There are no clinical trials based on genome editing for ML to date. The development of future therapies for ML, such as those based on genomic

Table 3 Clinical trials involving genome editing for Mucopolysaccharidoses.

NCT number	Title	Status	Conditions	Interventions	Phase	Sponsor/collaborators
NCT02702115	Ascending Dose Study of Genome Editing by the Zinc Finger Nuclease (ZFN) Therapeutic SB-318 in Subjects with MPS I	Active, not recruiting	MPS I	Biological: SB-318	I/II	Sangamo therapeutics
NCT03041324	Ascending Dose Study of Genome Editing by the Zinc Finger Nuclease (ZFN) Therapeutic SB-913 in Subjects with MPS II	Active, not recruiting	MPS II	Biological: SB-913	I/II	Sangamo therapeutics

MPS, mucopolysaccharidosis.
Adapted from clinicaltrials.gov.

editing, still needs to advance in experimental research and accumulate clinical experiences to learn about the natural history and pathophysiology of ML.

6. Conclusions

The mucopolysaccharidoses were one of the first group of diseases in which genome editing was employed. This is mainly due to the facts that (1) the mechanisms of disease are not completely understood and (2) patients with these disorders benefit from small increments in enzyme activity, making gene editing an alternative treatment approach. The escalation from animal models to clinical trials occurred fast in MPS disorders, but the initial results are still inconclusive regarding effectiveness of therapy, and better gene editing platforms are being tested. On the other hand, studies involving gene editing for mucolipidoses are still inexistent, also possibly due to the limitations of gene therapy for a disease which is not caused by a secreted hydrolase. However, the creation of cell lines for mucolipidoses can help elucidating mechanisms of disease and also allow drug screening and should be tested soon, especially considering the unmet medical needs for these diseases.

References

1. Giugliani R. Mucopolysacccharidoses: from understanding to treatment, a century of discoveries. *Genet Mol Biol.* 2012;4:924–931.
2. Baldo G, Giugliani R, Matte U. *Gene Delivery Strategies for the Treatment of Mucopolysaccharidoses.* vol 11. Taylor & Francis; 2014:449–459. https://doi.org/10.1517/17425247.2014.880689.
3. Sodhi H, Panitch A. Glycosaminoglycans in tissue engineering: a review. *Biomolecules.* 2021;11(1):1–22. https://doi.org/10.3390/biom11010029.
4. Giugliani R, Villarreal MLS, Araceli Arellano Valdez C, et al. Guidelines for diagnosis and treatment of Hunter syndrome for clinicians in Latin America. *Genet Mol Biol.* 2014;37(2):315–329. https://doi.org/10.1590/S1415-47572014000300003.
5. Poletto E, Pasqualim G, Giugliani R, Matte U, Baldo G. Worldwide distribution of common IDUA pathogenic variants. *Clin Genet.* 2018;94(1):95–102. https://doi.org/10.1111/cge.13224.
6. Khan SA, Peracha H, Ballhausen D, et al. Epidemiology of mucopolysaccharidoses. *Mol Genet Metab.* 2017;121(3):227–240. https://doi.org/10.1016/j.ymgme.2017.05.016.
7. Borges P, Pasqualim G, Giugliani R, Vairo F, Matte U. Estimated prevalence of mucopolysaccharidoses from population-based exomes and genomes. *Orphanet J Rare Dis.* 2020;15(1):324. https://doi.org/10.1186/s13023-020-01608-0.
8. Costa-Motta FM, Acosta AX, Abé-Sandes K, et al. Genetic studies in a cluster of mucopolysaccharidosis type VI patients in Northeast Brazil. *Mol Genet Metab.* 2011;104(4):603–607. https://doi.org/10.1016/j.ymgme.2011.09.017.

9. Velho RV, Harms FL, Danyukova T, et al. The lysosomal storage disorders mucolipidosis type II, type III alpha/beta, and type III gamma: update on GNPTAB and GNPTG mutations. *Hum Mutat.* 2019;40(7):842–864. https://doi.org/10.1002/humu.23748.
10. Pinto R, Caseiro C, Lemos M, et al. Prevalence of lysosomal storage diseases in Portugal. *Eur J Hum Genet.* 2004;12(2):87–92. https://doi.org/10.1038/sj.ejhg.5201044.
11. Plante M, Claveau S, Lepage P, et al. Mucolipidosis II: a single causal mutation in the N-acetylglucosamine-1-phosphotransferase gene (GNPTAB) in a French Canadian founder population. *Clin Genet.* 2008;73(3):236–244. https://doi.org/10.1111/j.1399-0004.2007.00954.x.
12. Khan SA, Tomatsu SC. Mucolipidoses overview: past, present, and future. *Int J Mol Sci.* 2020;21(18):1–20. https://doi.org/10.3390/ijms21186812.
13. Leroy JG, Cathey S, Friez MJ. In: Adam MP, Ardinger HH, Pagon RA, et al., eds. *Mucolipidosis II.* Seattle: University of Washington; 1993. Accessed November 13, 2020. https://www.ncbi.nlm.nih.gov/books/NBK1828/.
14. Raas-Rothschild A, Spiegel R. *Mucolipidosis III Gamma.* Seattle: University of Washington; 1993. Accessed November 13, 2020. http://www.ncbi.nlm.nih.gov/pubmed/20301784.
15. Poletto E, Baldo G, Gomez-Ospina N. Genome editing for mucopolysaccharidoses. *Int J Mol Sci.* 2020;21(2):1–20. https://doi.org/10.3390/ijms21020500.
16. Carroll D. Genome editing: past, present, and future. *Yale J Biol Med.* 2017;90(4):653–659. Accessed November 13, 2020. https://pubmed.ncbi.nlm.nih.gov/29259529/.
17. Zhang HX, Zhang Y, Yin H. Genome editing with mRNA encoding ZFN, TALEN, and Cas9. *Mol Ther.* 2019;27(4):735–746. https://doi.org/10.1016/j.ymthe.2019.01.014.
18. Zeballos CMA, Gaj T. Next-generation CRISPR technologies and their applications in gene and cell therapy. *Trends Biotechnol.* 2020, S0167-7799(20)30287-0. https://doi.org/10.1016/j.tibtech.2020.10.010.
19. Cox DBT, Platt RJ, Zhang F. Therapeutic genome editing: prospects and challenges. *Nat Med.* 2015;21:121–131. https://doi.org/10.1038/nm.3793.
20. Kingma SDK, Wagemans T, IJlst L, et al. Genistein increases glycosaminoglycan levels in mucopolysaccharidosis type I cell models. *J Inherit Metab Dis.* 2014;37(5):813–821. https://doi.org/10.1007/s10545-014-9703-x.
21. Luciani M, Gritti A, Meneghini V. Human iPSC-based models for the development of therapeutics targeting neurodegenerative lysosomal storage diseases. *Front Mol Biosci.* 2020;7:224. https://doi.org/10.3389/fmolb.2020.00224.
22. Suga M, Kondo T, Imamura K, et al. Generation of a human induced pluripotent stem cell line, BRCi001-A, derived from a patient with mucopolysaccharidosis type I. *Stem Cell Res.* 2019;36:101406. https://doi.org/10.1016/j.scr.2019.101406.
23. Varga E, Nemes C, Bock I, et al. Generation of mucopolysaccharidosis type II (MPS II) human induced pluripotent stem cell (iPSC) line from a 1-year-old male with pathogenic IDS mutation. *Stem Cell Res.* 2016;17(3):482–484. https://doi.org/10.1016/j.scr.2016.09.032.
24. Varga E, Nemes C, Kovacs E, et al. Generation of human induced pluripotent stem cell (iPSC) line from an unaffected female carrier of mucopolysaccharidosis type II (MPS II) disorder. *Stem Cell Res.* 2016;17(3):514–516. https://doi.org/10.1016/j.scr.2016.09.035.
25. Reboun M, Rybova J, Dobrovolny R, et al. X-chromosome inactivation analysis in different cell types and induced pluripotent stem cells elucidates the disease mechanism in a rare case of mucopolysaccharidosis type II in a female. *Folia Biol.* 2016;62(2):82–89. https://www.ncbi.nlm.nih.gov/pubmed/27187040.
26. Huang W, Xu M, Li R, et al. An induced pluripotent stem cell line (TRNDi006-A) from a MPS IIIB patient carrying homozygous mutation of p.Glu153Lys in the NAGLU gene. *Stem Cell Res.* 2019;37:101427. https://doi.org/10.1016/j.scr.2019.101427.

27. Vallejo-Diez S, Fleischer A, Martín-Fernández JM, et al. Generation of two induced pluripotent stem cells lines from a mucopolysaccharydosis IIIB (MPSIIIB) patient. *Stem Cell Res*. 2018;33:180–184. https://doi.org/10.1016/j.scr.2018.10.019.
28. Li R, Baskfield A, Beers J, et al. Generation of an induced pluripotent stem cell line (TRNDi005-A) from a mucopolysaccharidosis type IVA (MPS IVA) patient carrying compound heterozygous p.R61W and p.WT405del mutations in the GALNS gene. *Stem Cell Res*. 2019;36:101408. https://doi.org/10.1016/j.scr.2019.101408.
29. Bayo-Puxan N, Terrasso AP, Creyssels S, et al. Lysosomal and network alterations in human mucopolysaccharidosis type VII iPSC-derived neurons. *Sci Rep*. 2018;8(1):16644. https://doi.org/10.1038/s41598-018-34523-3.
30. Tolar J, Park IH, Xia L, et al. Hematopoietic differentiation of induced pluripotent stem cells from patients with mucopolysaccharidosis type I (Hurler syndrome). *Blood*. 2011;117(3):839–847. https://doi.org/10.1182/blood-2010-05-287607.
31. Kobolak J, Molnar K, Varga E, et al. Modelling the neuropathology of lysosomal storage disorders through disease-specific human induced pluripotent stem cells. *Exp Cell Res*. 2019;380(2):216–233. https://doi.org/10.1016/j.yexcr.2019.04.021.
32. Swaroop M, Brooks MJ, Gieser L, Swaroop A, Zheng W. Patient iPSC-derived neural stem cells exhibit phenotypes in concordance with the clinical severity of mucopolysaccharidosis I. *Hum Mol Genet*. 2018;27(20):3612–3626. https://doi.org/10.1093/hmg/ddy259.
33. Rybova J, Ledvinova J, Sikora J, Kuchar L, Dobrovolny R. Neural cells generated from human induced pluripotent stem cells as a model of CNS involvement in mucopolysaccharidosis type II. *J Inherit Metab Dis*. 2018;41(2):221–229. https://doi.org/10.1007/s10545-017-0108-5.
34. Miki T, Vazquez L, Yanuaria L, et al. Induced pluripotent stem cell derivation and ex vivo gene correction using a mucopolysaccharidosis type 1 disease mouse model. *Stem Cells Int*. 2019;2019:6978303. https://doi.org/10.1155/2019/6978303.
35. Lemonnier T, Blanchard S, Toli D, et al. Modeling neuronal defects associated with a lysosomal disorder using patient-derived induced pluripotent stem cells. *Hum Mol Genet*. 2011;20(18):3653–3666. https://doi.org/10.1093/hmg/ddr285.
36. Mirabelli P, Coppola L, Salvatore M. Cancer cell lines are useful model systems for medical research. *Cancers (Basel)*. 2019;11(8):1098. https://doi.org/10.3390/cancers11081098.
37. Azambuja AS, Pimentel-Vera LN, Gonzalez EA, et al. Evidence for inflammasome activation in the brain of mucopolysaccharidosis type II mice. *Metab Brain Dis*. 2020;35(7):1231–1236. https://doi.org/10.1007/s11011-020-00592-5.
38. Baldo G, Tavares AMV, Gonzalez E, et al. Progressive heart disease in mucopolysaccharidosis type I mice may be mediated by increased cathepsin B activity. *Cardiovasc Pathol*. 2017;27:45–50. https://doi.org/10.1016/j.carpath.2017.01.001.
39. Gonzalez EA, Martins GR, Tavares AMV, et al. Cathepsin B inhibition attenuates cardiovascular pathology in mucopolysaccharidosis I mice. *Life Sci*. 2018;196(January):102–109. https://doi.org/10.1016/j.lfs.2018.01.020.
40. de Carvalho TG, Schuh R, Pasqualim G, et al. CRISPR-Cas9-mediated gene editing in human MPS I fibroblasts. *Gene*. 2018;678:33–37. https://doi.org/10.1016/j.gene.2018.08.004.
41. Schuh RS, de Carvalho TG, Giugliani R, Matte U, Baldo G, Teixeira HF. Gene editing of MPS I human fibroblasts by co-delivery of a CRISPR/Cas9 plasmid and a donor oligonucleotide using nanoemulsions as nonviral carriers. *Eur J Pharm Biopharm*. 2018;122:158–166. https://doi.org/10.1016/j.ejpb.2017.10.017.
42. Schuh RS, Poletto É, Pasqualim G, et al. In vivo genome editing of mucopolysaccharidosis I mice using the CRISPR/Cas9 system. *J Control Release*. 2018;288(July):23–33. https://doi.org/10.1016/j.jconrel.2018.08.031.

43. Schuh RS, Gonzalez EA, Tavares AMV, et al. Neonatal nonviral gene editing with the CRISPR/Cas9 system improves some cardiovascular, respiratory, and bone disease features of the mucopolysaccharidosis I phenotype in mice. *Gene Ther.* 2020;27(1–2):74–84. https://doi.org/10.1038/s41434-019-0113-4.
44. Christensen CL, Choy FYM. A prospective treatment option for lysosomal storage diseases: CRISPR/Cas9 gene editing technology for mutation correction in induced pluripotent stem cells. *Diseases.* 2017;5(1):6. https://doi.org/10.3390/diseases5010006.
45. de Carvalho TG, Matte U, Giugliani R, Baldo G. Genome editing: potential treatment for lysosomal storage diseases. *Curr Stem Cell Rep.* 2015;1:9–15. (March). internal-pdf://228.60.152.87/Carvalho 2015—genome editing review.pdf.
46. Martin RM, Ikeda K, Cromer MK, et al. Highly efficient and marker-free genome editing of human pluripotent stem cells by CRISPR-Cas9 RNP and AAV6 donor-mediated homologous recombination. *Cell Stem Cell.* 2019;24(5):821–828 e5. https://doi.org/10.1016/j.stem.2019.04.001.
47. Dever DP, Scharenberg SG, Camarena J, et al. CRISPR/Cas9 genome engineering in engraftable human brain-derived neural stem cells. *iScience.* 2019;15:524–535. https://doi.org/10.1016/j.isci.2019.04.036.
48. Mashel TV, Tarakanchikova YV, Muslimov AR, et al. Overcoming the delivery problem for therapeutic genome editing: current status and perspective of non-viral methods. *Biomaterials.* 2020;258(July):120282. https://doi.org/10.1016/j.biomaterials.2020.120282.
49. Schuh R, Baldo G, Teixeira H. Nanotechnology applied to treatment of mucopolysaccharidoses. *Expert Opin Drug Deliv.* 2016;13(12):1709–1718. https://doi.org/10.1080/17425247.2016.1202235.
50. Nagree MS, Scalia S, McKillop WM, Medin JA. An update on gene therapy for lysosomal storage disorders. *Expert Opin Biol Ther.* 2019;19(7):655–670. https://doi.org/10.1080/14712598.2019.1607837.
51. Ho BX, Loh SJH, Chan WK, Soh BS. In vivo genome editing as a therapeutic approach. *Int J Mol Sci.* 2018;19(9):2721. https://doi.org/10.3390/ijms19092721.
52. Sharma R, Anguela XM, Doyon Y, et al. In vivo genome editing of the albumin locus as a platform for protein replacement therapy. *Blood.* 2015;126(15):1777–1784. https://doi.org/10.1182/blood-2014-12-615492.
53. DeKelver R, Rohde M, Tom S, et al. ZFN-mediated genome editing of albumin safe harbor in vivo results in supraphysiological levels of human IDS, IDUA and GBA in mice. *Mol Genet Metab.* 2016;114(2):S36. https://doi.org/10.1016/j.ymgme.2014.12.065.
54. Ou L, DeKelver RC, Rohde M, et al. ZFN-mediated in vivo genome editing corrects murine hurler syndrome. *Mol Ther.* 2019;27(1):178–187. https://doi.org/10.1016/j.ymthe.2018.10.018.
55. Laoharawee K, DeKelver RC, Podetz-Pedersen KM, et al. Dose-dependent prevention of metabolic and neurologic disease in murine MPS II by ZFN-mediated in vivo genome editing. *Mol Ther.* 2018;26(4):1127–1136. https://doi.org/10.1016/j.ymthe.2018.03.002.
56. Ernst MPT, Broeders M, Herrero-Hernandez P, Oussoren E, van der Ploeg AT, Pijnappel WWMP. Ready for repair? Gene editing enters the clinic for the treatment of human disease. *Mol Ther Methods Clin Dev.* 2020;18:532–557. https://doi.org/10.1016/j.omtm.2020.06.022.
57. Harmatz P, Lau HA, Heldermon C, et al. EMPOWERS: a phase 1/2 clinical trial of SB-318 ZFN-mediated in vivo human genome editing for treatment of MPS I (Hurler syndrome). *Mol Genet Metab.* 2019;126(2):S68. https://doi.org/10.1016/j.ymgme.2018.12.163.
58. Muenzer J, Prada CE, Burton B, et al. CHAMPIONS: a phase 1/2 clinical trial with dose escalation of SB-913 ZFN-mediated in vivo human genome editing for treatment of MPS II (Hunter syndrome). *Mol Genet Metab.* 2019;126(2):S104. https://doi.org/10.1016/j.ymgme.2018.12.263.

59. Ou L, Przybilla MJ, Ahlat O, et al. A highly efficacious PS gene editing system corrects metabolic and neurological complications of mucopolysaccharidosis type I. *Mol Ther.* 2020;28(6):1442–1454. https://doi.org/10.1016/j.ymthe.2020.03.018.
60. Wang D, Li J, Song C-Q, et al. Cas9-mediated allelic exchange repairs compound heterozygous recessive mutations in mice. *Nat Biotechnol.* 2018;36(9):839–842. https://doi.org/10.1038/nbt.4219.
61. Gomez-Ospina N, Scharenberg SG, Mostrel N, et al. Human genome-edited hematopoietic stem cells phenotypically correct mucopolysaccharidosis type I. *Nat Commun.* 2019;10(1):4045. https://doi.org/10.1038/s41467-019-11962-8.
62. Giugliani L, Vanzella C, Zambrano MB, et al. Clinical research challenges in rare genetic diseases in Brazil. *Genet Mol Biol.* 2019;42(1):305–311. https://doi.org/10.1590/1678-4685-gmb-2018-0174.
63. Sheridan C. Sangamo's landmark genome editing trial gets mixed reception. *Nat Biotechnol.* 2018;36(10):907–908. https://doi.org/10.1038/nbt1018-907.
64. Vera LNP, Baldo G. The potential of gene therapy for mucopolysaccharidosis type I. *Expert Opin Orphan Drugs.* 2020;8(1):33–41. https://doi.org/10.1080/21678707.2020.1715208.

CHAPTER ELEVEN

Gene and epigenetic editing in the treatment of primary ciliopathies

Elisa Molinari[a] and John A. Sayer[a,b,c,]*

[a]Translational and Clinical Research Institute, Faculty of Medical Sciences, Newcastle University, International Centre for Life, Central Parkway, Newcastle upon Tyne, United Kingdom
[b]Renal Services, The Newcastle Hospitals NHS Foundation Trust, Newcastle upon Tyne, United Kingdom
[c]NIHR Newcastle Biomedical Research Centre, Newcastle upon Tyne, United Kingdom
*Corresponding author: e-mail address: john.sayer@newcastle.ac.uk

Contents

1. Introduction — 355
2. Primary ciliopathies — 359
   2.1 Retinal ciliopathies — 364
   2.2 Renal ciliopathies — 364
3. CRISPR/Cas systems and their therapeutic applications — 365
   3.1 Challenges associated to the clinical implementation of CRISPR/Cas and related technological advances — 369
   3.2 *In vivo* delivery of CRISPR/Cas-based therapies — 371
4. CRISPR/Cas gene editing for the treatment of primary ciliopathies — 379
   4.1 Gene editing for the treatment of retinal ciliopathies — 379
   4.2 Gene editing for the treatment of renal ciliopathies — 382
5. CRISPR/Cas epigenetic editing for the treatment of primary ciliopathies — 383
   5.1 Epigenetic editing for the treatment of retinal ciliopathies: RP — 384
   5.2 Epigenetic editing for the treatment of renal ciliopathies: ADPKD — 386
6. Further considerations and concluding remarks — 387
Acknowledgments — 389
References — 389

Abstract

Primary ciliopathies are inherited human disorders that arise from mutations in ciliary genes. They represent a spectrum of severe, incurable phenotypes, differentially involving several organs, including the kidney and the eye. The development of gene-based therapies is opening up new avenues for the treatment of ciliopathies. Particularly attractive is the possibility of correcting *in situ* the causative genetic mutation, or pathological epigenetic changes, through the use of gene editing tools. Due to their versatility and efficacy, CRISPR/Cas-based systems represent the most promising gene editing toolkit for clinical applications. However, delivery and specificity issues have so far held back the translatability of CRISPR/Cas-based therapies into clinical practice, especially where systemic administration is required. The eye, with its characteristics of high accessibility and compartmentalization, represents an ideal target for *in situ* gene correction. Indeed, studies for the evaluation of a CRISPR/Cas-based therapy for *in vivo* gene correction to treat a retinal ciliopathy have reached the clinical stage. Further technological advances may be required for the development of *in vivo* CRISPR-based treatments for the kidney. We discuss here the possibilities and the challenges associated to the implementation of CRISPR/Cas-based therapies for the treatment of primary ciliopathies with renal and retinal phenotypes.

Abbreviations

5mC	5-methylcytosine
AAV	adeno-associated virus vectors
ADPKD	autosomal dominant polycystic kidney disease
AdV	adenovirus
ALMS	Alstrom syndrome
ARPKD	autosomal recessive polycystic kidney disease
ASO	antisense oligonucleotide
BBS	Bardet-Bield syndrome
Bes	base editors
Cas	CRISPR-associated protein
CRISPR	clustered regularly interspaced short palindromic repeats
crRNA	CRISPR RNA
DNMTs	DNA methyltransferases
DSB	double strand break
ESKD	end-stage kidney disease
HDACs	histone deacetylases
HDR	homology-directed repair
HKAc	histone lysine acetylation
HKme	histone lysine methylation
IFT	intraflagellar transport
iPSCs	induced pluripotent stem cells
JBTS	Joubert syndrome
LCA	leber congenital amaurosis
LV	lentivirus
MACS	magnetic-activated cell sorting
MKS	Meckel syndrome
NHEJ	non-homologous end joining
NPHP	nephronophthisis
OFD	orofacial-digital syndrome
PAM	protospacer adjacent motif
PERV	Porcine endogenous retrovirus
RP	retinitis pigmentosa
RPE	retinal pigment epithelium
sgRNA	single guide RNA
SLSN	Senior-Loken syndrome
SRTDs	short-rib thoracic dysplasias
TALENs	transcription activator-like effector nucleases
tracrRNA	trans-activating CRISPR RNA
ZFNs	zinc finger nucleases

1. Introduction

The primary cilium is a hair-like microtubule-based organelle that protrudes outside the cellular plasma membrane and can be found throughout the body, particularly on the apical surface of epithelial cells. The primary cilium consists of a microtubule scaffold, the axoneme, surrounded by the ciliary membrane that differs from the plasma membrane in protein and lipid composition. The structure of the ciliary axoneme is templated by the basal body, which is a modified mother centriole, connected to the daughter centriole by interconnecting fibers. Distal appendages, which are radial protrusions connecting the distal end of the basal body to the membrane, are also docking sites for the intraflagellar transport (IFT) machinery that mediates the bi-directional movement of ciliary cargoes along the axoneme, through the action of microtubule motors. Anterograde transport (IFT-B, toward the ciliary tip) is mediated by kinesin, while the retrograde transport (IFT-A) is mediated by dynein. The cilium is enriched in signaling molecules and no protein synthesis takes place within the cilium. Ciliary entry of proteins from the cytoplasm is controlled by the transition zone, a subcompartment of the cilium located at the proximal end of the axoneme and ultrastructurally characterized by Y-shaped densities that bridge the microtubule axoneme to the surrounding ciliary membrane.[1] The transition zone establishes a diffusion barrier between the cilium and the cytoplasm and is therefore particularly important for the regulation of ciliary molecular composition[2] (Fig. 1A). The tight regulation of its protein composition make the primary cilium a well-defined cellular sub-compartment, enriched in signaling molecules, such as components of the Hedgehog pathway and it is therefore viewed as the cell signaling hub.[3,4]

Specialized sensory functions characterize the outer segment of photoreceptor cells, a modified primary cilium which is responsible for light sensation. The outer segment is composed of hundreds of stacked membrane discs that accommodate the components of the phototransduction cascade, such as opsin and rhodopsin, which are synthesized within the photoreceptor inner segment. The inner segment is linked to the outer segment through the connecting cilium, which consists of an axoneme made of nine microtubule doublets nucleated at the apical surface of the inner segment.[5] The axoneme extends into the outer segment, converting to singlet microtubules toward its distal end. Proteins produced in the inner segment are actively transported through the connecting cilium by anterograde IFT-B,

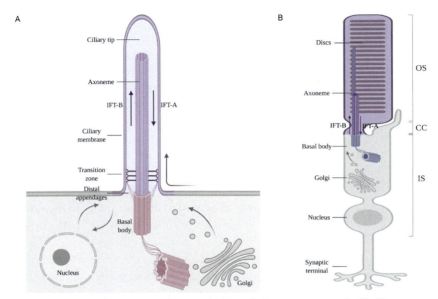

Fig. 1 Schematic of a primary cilium and of a rod photoreceptor cell. (A) Ciliary membrane proteins can be trafficked to the primary cilium by controlled lateral diffusion from the plasma membrane, by vesicular transport from the Golgi complex or by interaction with specific carrier proteins, in the case of peripheral membrane proteins. In all these cases, ciliary proteins need to surmount the transition zone, which acts as a sorting center at the base of the cilium. (B) In the IS, rhodopsin is trafficked in opsin-carrier vesicles that bud from the Golgi and reach the basal body, thanks to cytoplasmic dynein that moves along microtubules. Rhodopsin is then thought to be actively transported through the CC by anterograde IFT-B. The CC is the obligate passage for proteins moving between the IS and the OS and acts as a gatekeeper that controls the composition of the OS. In photoreceptor cells, distal to the IS, is the cell body that contains the nucleus and the synaptic terminal that extends into the outer plexiform layer of the retina. IFT, intraflagellar transport; OS, outer segment; CC, connecting cilium; IS, inner segment.

whereas movement of proteins from the outer segment back into the inner segment is mediated by retrograde IFT-A. The connecting cilium controls the transport of proteins to the outer segment and can be considered as a specialized transition zone[2,5] (Fig. 1B).

Gene mutations that disrupt primary cilium functions can result in a variety of disorders, collectively called primary ciliopathies.[6,7] They include a spectrum of overlapping and distinct clinical conditions involving different organs including the brain, eye, kidney and liver, as well as skeletal defects. Even primary ciliopathies with limited single organ phenotypes such as

Leber congenital amaurosis (LCA) or isolated nephronophthisis (NPHP) cause significant morbidity and a lifelong burden of disease. The use of innovative gene-based therapeutics strategies, such as antisense oligonucleotide (ASO) treatments, gene replacement and gene editing and reprogramming hold great potential for the treatment of these monogenic disorders. Genome editing to correct gene mutations and epigenetic editing to reverse pathological epigenetic modifications are particularly attractive as they act by directly correcting the disease-causing alterations at the DNA level, without introducing exogenous copies of the gene, whose spatiotemporal expression could be difficult to control. Moreover, the modifications introduced by gene editing are permanent, whereas the effects of ASOs, which act at the RNA level, are intrinsically temporary.

In recent years, various platforms for gene editing have been developed. They include zinc finger nucleases (ZFNs)[8] and transcription activator-like effector nucleases (TALENs).[9–11] ZFNs and TALENs systems are based on the activity of the *Fok*I nuclease domain, which is fused to the DNA-binding modules of zinc finger proteins in ZFNs and to transcription activator-like effector proteins (TALEs) in TALENs. Since these gene-editing tools function through DNA-protein interactions, they require laborious sequence-specific design and construction.

Clustered regularly interspaced short palindromic repeats (CRISPR) in combination with CRISPR-associated proteins (CRISPR/Cas) are natural adaptive immune systems of bacteria and archaea that provide sequence-specific resistance against bacteriophages and plasmids. The adaptation of this prokaryotic acquired immunity for gene editing has revolutionized modern biology.[12–14]

In contrast to ZFNs and TALENs, in the CRISPR/Cas system, the specificity of the genome recognition is determined by a short RNA sequence, which makes it a highly programmable tool.

CRISPR/Cas system is not just a fast and cheap gene editing tool, it is also extremely efficient and in the last years several alternative types of CRISPR/Cas systems have been discovered for genome editing.[15] Thanks to these properties, several CRISPR/Cas gene editing systems have been recently extensively exploited for the study of human diseases, including ciliopathies, where they have been deployed for the generation of animal models, for the study of the underlying molecular pathomechanisms and for the identification of novel therapeutic targets and causative genes.[16–20]

CRISPR/Cas *per se* merely introduce double strand breaks (DSBs) in the DNA in a site-specific manner, but the rational use and engineering of these systems has broadened their scope. Among the several possible applications of these systems, they can be used for the precise correction of disease-associated variants and for the modification of epigenetic marks, making CRISPR/Cas an extremely versatile tool for gene-based therapeutic purposes.

Here, we will discuss the current developments and the perspectives related to the employment of CRISPR/Cas-based systems in the treatment of human primary ciliopathies. In particular, we will focus on the possibility of using CRISPR/Cas to mediate genome editing and site-specific epigenetic reprogramming, to intervene on the retinal and renal degenerative manifestations of human primary ciliopathies. As opposed to skeletal and neurological defects, which are congenital and irreversible conditions, the degenerative nature of ciliopathy-associated retinal and renal phenotypes leaves a temporal window for therapeutic intervention, making it easier to envisage therapies that can halt or reverse these features.

We will also discuss the main challenges associated to the introduction of CRISPR/Cas-based therapies in the clinic. Some of the main limitations that are holding back the clinical application of CRISPR/Cas are the potentially difficult delivery of these relatively large molecules to the target tissue and safety issues associated to off-target effects and possible immunogenicity.

Indeed, so far, clinical studies on CRISPR/Cas-based therapies have mainly revolved around *ex vivo* gene editing and the only *in vivo* clinical trial to date involves the use of CRISPR/Cas to treat a retinal ciliopathy. This is likely due to the high accessibility, compartmentalization and immune specificities of the eye,[21] which overcome the main challenges and mitigate the risks associated to CRISPR/Cas applications *in vivo*.

A review of the current state of the research on the use of CRISPR/Cas for the treatment of retinal and renal ciliopathies showcases the relatively advanced stage of the implementation of CRISPR-based treatments for the eye, as opposed to a less accessible organ such as the kidney.

Hurdles in the delivery of CRISPR/Cas molecules to the kidney and the risks associated with systemic administration may argue in favor of *ex vivo* approaches in this case, such as the exploitation of CRISPR/Cas in xenogeneic kidney transplants, rather than *in situ* correction of the genetic alterations within the renal epithelium. Furthermore, the very efficient uptake of small nucleic acids by the kidney[22–24] may, where applicable, point toward the optimization of ASO- rather than CRISPR-based therapies for the treatment of renal ciliopathies.

Nevertheless, the technological advances that increase efficacy, safety and delivery of CRISPR/Cas systems are continuously being brought about and will likely lead to the expansion and the initiation of trials for the implementation of CRISPR-based therapies for retinal and renal ciliopathies, respectively.

2. Primary ciliopathies

Primary ciliopathies display extensive genetic and clinical heterogeneity. Phenotypical manifestations include neurodevelopmental defects such as cerebellar hypoplasia, skeletal defects such as abnormally short ribs and limbs and craniofacial defects, polydactyly, obesity, retinal degeneration and cystic and fibrocystic liver and kidney disease.[25] Some of these features can be found in isolation or, due to the widespread presence of the primary cilium in multiple tissues, they can combine to give rise to more complex syndromic phenotypes. These ciliopathy syndromes include Bardet-Bield syndrome (BBS, MIM # 209900) and Alstrom syndrome (ALMS, MIM # 203800), characterized by an obesity phenotype, the orofacial-digital syndrome (OFD, MIM # 311200) and short-rib thoracic dysplasias (SRTDs, MIM % 208,500), characterized by skeletal defects, Joubert syndrome (JBTS, MIM # 213300), which is a neurodevelopmental disorder, the severe, embryonic lethal Meckel syndrome (MKS, MIM # 249000), and Senior-Loken syndrome (SLSN, MIM # 266900), which features both renal and retinal involvement.

Linkage analysis and, more recently, next-generation sequencing have revealed the genetic complexity of primary ciliopathies identifying at least 150 different causal genes (Table 1).

The underlying mutated gene defect often explains only partially the resulting phenotype, which can vary depending on the gene, the mutation and possibly also on genetic modifiers.[40–43]

Some common ciliopathy causing mutations are human specific and so are the underpinning molecular mechanisms.[44,45] Therefore, the development of novel complex human models to investigate molecular pathomechanisms and novel therapies could be particularly useful. Advances have been made in recent years in the development of induced pluripotent stem cells (iPSCs)-derived retinal[46] and renal[47,48] organoids that recapitulate the complexity of these organs in a human setting and can be used to test novel therapeutics.

Table 1 Primary ciliopathies, associated genes and disease phenotypes.

Classification	Ciliopathy name	MIM	Gene(s)	Phenotype	References
Obesity	Bardet-Biedl syndrome (BBS)	# 209900	*CCDC28B, SDCCAG8, WDPCP, BBS5, LZTFL1, ARL6, BBS7, BBS12, PTHB1, TMEM67, C8orf37, IFT74, TRIM32, BBIP1, BBS1, BBS10, CEP290, TTC8, BBS4, BBS2, MKS1, MKKS, IFT27*	Usually AR. Cone-rod retinal dystrophy, obesity, postaxial polydactyly, cognitive impairment, hypogonadism and renal disease. Variable manifestation of hepatic fibrosis	26
	Alstrom syndrome (ALMS)	# 203800	*ALMS1*	AR. Cone-rod retinal dystrophy, sensorineural hearing loss and childhood obesity. Frequent involvement of dilated cardiomyopathy, renal, pulmonary, hepatic, and urologic dysfunctions	27
Skeletal	Short-rib thoracic dysplasias (SRTDs)	% 208500	*WDR35, IFT172, DYNC2LI1, TTC21B, IFT80, TCTEX1D2, EVC2, EVC, WDR19, INTU, NEK1, CEP120, WDR60, WDR34, DYNC2H1, IFT81, KIAA0586, IFT43, SRTD1, IFT140, IFT52, SRTD12*	AR. Constricted thoracic cage, short ribs, shortened tubular bones, and a "trident" appearance of the acetabular roof. Variable manifestation of polydactyly, cleft lip/palate and brain, eye, heart, kidneys, liver, pancreas, intestines, and genitalia anomalies. SRTDs include Ellis-van Creveld syndrome (EVC), asphyxiating thoracic dystrophy (ATD), short rib-polydactyly syndrome (SRPS), and Mainzer-Saldino syndrome (MZSDS)	28
	Orofacial-digital syndrome (OFD)	# 311200	*DDX59, IFT57, INTU, CPLANE1, TCTN3, C2CD3, KIAA0753, TMEM107, OFD1*	XL or AR. Facial anomalies, oral malformations (lobulated tongue and oral frenula) and digital abnormalities (polydactly). Variable brain and kidney involvement	29,30

Neurodevelopmental	Joubert syndrome (JBTS)	# 213300	INPP5E, TMEM216, AHI1, NPHP1, CEP290, TMEM67, RPGRIP1L, ARL13B, CC2D2A, OFD1, TTC21B, KIF7, TCTN1, TMEM237, TMEM138, CEP41, CPLANE1, TCTN3, ZNF423, TMEM231, CSPP1, PDE6D, TALPID3, TCTN2, CEP104, KIAA0556, B9D1, MKS1, TMEM107, ARMC9, CEP120, SUFU, PIBF1, B9D2, ARL3, FAM149B1	Mainly AR. Hypoplasia of the cerebellar vermis ("molar tooth sign") on brain MRI, ataxia, hypotonia, developmental delay, oculo-motor apraxia, dysregulation of breathing pattern. Variable manifestation of retinal dystrophy, renal disease, ocular coloboma, occipital encephalocele, hepatic fibrosis, polydactyly, oral hamartomas and endocrine abnormalities	31
	Meckel syndrome (MKS)	# 249000)	KIF14, NPHP3, CC2D2A, TMEM67, TMEM216, CEP290, TCTN2, RPGRIP1L, TMEM231, TMEM107, B9D1, MKS1, B9D2	AR. Cystic renal disease, posterior fossa abnormalities (occipital encephalocele) and hepatic ductal plate malformations. Often accompanied by polydactyly. Usually lethal in the prenatal or perinatal period	32

Continued

Table 1 Primary ciliopathies, associated genes and disease phenotypes.—cont'd

Classification	Ciliopathy name	MIM	Gene(s)	Phenotype	References
Retinal	Leber congenital amaurosis (LCA)	#204000	SPATA7, LCA5, RPGRIP1, CEP290, TULP1, CLUAP1	AR. Early-onset, severe retinal dystrophy. Poor visual function is accompanied by nystagmus, absent pupillary responses, photophobia, keratoconus and hyperopia	33
	Retinitis pigmentosa (RP)	#268000	RP1 RP2 RPGR PRPF31 PRPF8 TULP1 OFD1 FAM161A TOPORS USH2A PDE6B TTC8 PCARE ARL6 PRPF6 CLRN1 MAK C8orf37 IFT172 BBS2 IFT140 IFT43 ARL3 RP1L1 SPATA7	Variable inheritance. Progressive retinal degeneration, primarily affecting rod photoreceptors and retinal pigment epithelium and leading to night blindness, tunnel vision, and slowly progressive decreased central vision	34

Renal	Autosomal dominant polycystic kidney disease (ADPKD)	#173900	PKD1, PKD2, GANAB, DNAJB11	AD. Numerous, large, fluid filled renal cysts, with variable manifestation of liver cysts, and intracranial aneurysms. Typical age of onset is in middle life	35
	Autosomal recessive polycystic kidney disease (ARPKD)	#263200	PKHD1, DZIP1L	AR. Kidney cysts accompanied by biliary dysgenesis, hepatic fibrosis, portal hypertension and hypersplenism. Early *in utero* presentation: enlarged, echogenic kidneys, oligohydramnios and pulmonary hypoplasia. Can also manifest later in life with renal insufficiency accompanied by systemic and portal hypertension	36
	Nephronophthisis (NPHP)	#256100	NPHP1, INVS, NPHP3, NPHP4, IQCB1, CEP290, GLIS2, RPGRIP1L, NEK8, SDCCAG8, TMEM67, TTC21B, WDR19, ZNF423, CEP164, ANKS6, IFT172, CEP83, DCDC2, MAPKBP1, AHI1, CC2D2A, XPNPEP3, ATXN10, SLC41A1	AR. Renal tubular atrophy with tubular basement membrane disruption, progressive interstitial fibrosis, defects in urine concentration, loss of cortico-medullary differentiation and increased renal echogenicity at renal ultrasound. Early onset of ESKD (typically in the first two decades of life). Variable manifestation of hepatic fibrosis	37,38
Retinal and renal	Senior-Loken syndrome (SLSN)	#266900	NPHP4, SDCCAG8, NPHP1, TRAF3IP1, IQCB1, SLSN3, WDR19, CEP290	AR. NPHP + LCA	39

AR, autosomal recessive; XL, X-linked; AD, autosomal dominant; ESKD, end-stage kidney disease.

The accelerated discovery in the last two decades of many causal genes and the study of their function on appropriate models will be instrumental for the delivery of more efficacious treatments that act on the molecular basis of the disease, including targeted pharmacological treatments and gene-based therapies, such as gene replacement, editing and reprogramming and ASO-based interventions.

2.1 Retinal ciliopathies

Dysfunctions of the outer segment or the connecting cilium of photoreceptor cells of the retina can lead to dramatic consequences on these cells function and viability leading to retinal degeneration.

Retinal ciliopathies include LCA (MIM # 204000) and Retinitis pigmentosa (RP, MIM # 268000), which can present in isolation or as features of multi-systemic ciliopathies.

LCA is a severe retinal disease, with a typical onset in the first year of life. LCA is generally inherited as an autosomal recessive disease and its population frequency is around 1:50,000. Poor visual function is accompanied by nystagmus, absent pupillary responses and hyperopia. At least 20 different causative genes have been identified so far,[33] among which several ciliary genes (Table 1). Mutations in the *Centrosomal Protein* 290 (*CEP290*) are among the most common causes of LCA, accounting for 15%–20% of all known cases.[49,50]

RP is a condition that primarily affects rod photoreceptors and retinal pigment epithelium and becomes evident in the first or second decade of life. RP is the most common form of inherited retinal dystrophy, with a prevalence of 1:3500. RP manifests often with night blindness and peripheral vision loss as early symptoms and progresses to include central vision loss. Causative mutations have been identified in around 50 genes, several of which are genes that localize at the cilium and/or are important for ciliary functions (Table 1). The *RP GTPase Regulator* (*RPGR*) gene is the most common cause of X-linked RP.[34]

2.2 Renal ciliopathies

Primary cilia project from the apical surface of kidney epithelial cells into the lumen of the renal tubule, where they are proposed to mediate several fundamental processes including mechanosensation, chemosensation, differentiation, proliferation, cell cycle, and planar cell polarity. Ciliary dysfunctions can therefore affect renal tissue homeostasis in multiple ways, ultimately resulting in renal disease.[51]

Renal ciliopathies include autosomal dominant polycystic kidney disease (ADPKD, MIM # 173900), autosomal recessive polycystic kidney disease (ARPKD, MIM # 263200) and NPHP (MIM # 256100). They can be found in association with liver cysts and/or hepatic fibrosis. NPHP is also often part of more complex syndromic ciliopathies.

ADPKD is the most common genetic cause of chronic kidney disease and end-stage kidney disease (ESKD), with a prevalence of 1:2000. It is characterized by the formation of numerous, large, fluid filled renal cysts that can arise from all nephron segments but are predominately of collecting duct origin. Cysts slowly increase in number and expand in size, compressing the surrounding renal tissue and result in a decline in renal function that ultimately leads to ESKD, with a variable onset age (median age, 53 years). ADPKD is mainly caused by mutations in *Polycystic Kidney Disease 1* (*PKD1*) and *Polycystic Kidney Disease 2* (*PKD2*)[35] (Table 1).

ARPKD is rarer with a prevalence of 1:20,000 and it may present in neonates with massive kidney enlargement, intrauterine renal failure, oligohydramnios, and pulmonary hypoplasia or may present later in life with renal insufficiency accompanied by systemic and portal hypertension. ARPKD is primarily caused by mutations in *Polycystic Kidney and Hepatic Disease 1* (*PKHD1*)[36] (Table 1).

NPHP has an estimated incidence between 1:50,000 and 1:9,00,000 and represents one of the most frequent genetic causes of ESKD in children and young adults. NPHP is inherited as an autosomal recessive trait and up to 20 causative genes have been identified so far (Table 1). NPHP is typically characterized by normal or small-sized kidneys with loss of cortico-medullary differentiation and increased renal echogenicity at renal ultrasound. However, large polycystic kidneys may be observed in certain infantile NPHP forms.[37]

3. CRISPR/Cas systems and their therapeutic applications

CRISPRs are a family of repetitive DNA sequences within the bacterial or archaeal genome in which short sequences derived from invading foreign nucleic acids (spacers) can be incorporated and then transcribed as CRISPR RNA (crRNA) from the CRISPR locus. The crRNA can pair with the trans-activating CRISPR RNA (tracrRNA) to form a complex with the Cas proteins, which serve as RNA-guided nucleases in search for invading crRNA-complementary sequences (protospacers).[52,53]

CRISPR/Cas systems have been divided into 2 classes, 6 types and over 30 subtypes. Class 1 comprises multi-subunit Cas protein effectors and Class 2 features a single large effector protein.[54]

Due to their simplicity, Class 2 CRISPR/Cas systems have been largely repurposed as tools for gene editing. Among them Cas9 (type II) and Cas12a (type V) are suitable for DNA targeting, while Cas13 (type VI) for RNA targeting. The Cas9-based systems are so far the most commonly employed in genome editing; several Cas9 proteins from differing species have been identified and used for gene editing, including *Streptococcus pyogenes* Cas9 (SpCas9),[12–14] *Staphylococcus aureus* Cas9 (SaCas9),[55,56] *Neisseria meningitides* Cas9 (NmCas9),[57,58] *Francisella novicida* Cas9 (FnCas9),[59,60] *Streptococcus thermophilus* Cas9 (St1Cas9)[14,56] and *Streptococcus canis* Cas9 (ScCas9).[61,62]

For genome editing, tracrRNA:crRNA complexes can be engineered into a single guide RNA (sgRNA) for additional simplicity.[13] The Cas:sgRNA complex can be introduced in a cell, where it scans the genome, searching first for the appropriate protospacer adjacent motif (PAM), a short motif (1–4 nt) adjacent to the crRNA target sequence.[63] Different Cas orthologues possess different PAM specificities, allowing for a certain flexibility in the choice of the target sequence.[64,65] Additional flexibility is provided by the engineering of Cas9 variants with altered PAM specificities.[56,66–68] Upon recognition of the PAM sequence, the Cas:sgRNA complex unwinds the DNA in the PAM-proximal region (seed sequence).[13,69] If the sequence is complementary to the sgRNA target sequence, the Cas nuclease introduces a DSB.[13,70]

Generation of a DSB triggers host DNA repair pathways to re-ligate the cleaved DNA ends. These pathways include non-homologous end joining (NHEJ) and homology-directed repair (HDR)[71] (Fig. 2A).

NHEJ represents the primary DSB repair system in mammalian cells and it is active at any phase of the cell cycle. This repair process is error-prone and can introduce nucleotide insertions and/or deletions at the DSB site.[71]

HDR uses a homologous repair template to precisely repair the DSB. HDR activity is mainly restricted to the late S- or G2-phase, when a sister chromatid can serve as the repair template.[72] This repair mechanism can be leveraged to introduce the desired gene modifications by transfecting the cells with CRISPR/Cas together with a DNA repair template. HDR, although more precise, occurs at a low incidence rate compared to NHEJ. The low occurrence of HDR pathway and the fact that it is

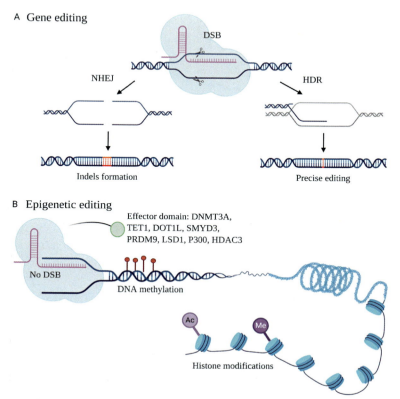

Fig. 2 Applications of CRISPR/Cas systems in gene and epigenetic editing. (A) Gene editing via DSB. The CRISPR/Cas system induces a DSB at the genomic locus specified by the sgRNA (in pink). The cell can repair the DSB through different mechanisms, including NHEJ or HDR. In NHEJ, random nucleotide insertions and deletions occur as the cell ligates the DSB, often resulting in gene knock out. In HDR, the DSB is repaired using a supplied template DNA (as plasmid or ssDNA) that presents homologous sequences to the target site. The sequence of the repair template is copied into the targeted site, resulting in precise gene correction. (B) Epigenetic editing. dCas is fused to the effector domain of an epigenetic modifier and recruited to a target location complementary to the sgRNA. This system does not produce a DSB, but allows the site-specific reprogramming of nearby sequences to control gene expression. Depending on the effector domain coupled to dCas (e.g., DNMT3A, TET1, DOT1L, SMYD3, PRDM9, LSD1, P300, HDAC3), the local epigenetic state, such as DNA methylation or histone post-translational modifications (methylation/ acetylation), is modified to revert pathogenic changes. sgRNA, single guide RNA; DSB, double strand break; NHEJ, non-homologous end joining; HDR, homology-directed repair; ssDNA, single-strand DNA; dCas, nuclease-dead Cas; Me, methylation; Ac, acetylation.

restricted to dividing cells pose a serious limit for the application of CRISPR/Cas-mediated gene editing for *in vivo* gene correction in the treatment of human disease.

The CRISPR/Cas toolbox is being continuously expanded and repurposed for other types of genetic interventions. Many of these new tools do not cut the DNA or introduce only a single strand break, which makes them safer for possible clinical applications.

Nuclease-dead Cas proteins (dCas, where both nuclease domains are inactivated *via* point mutations) can be exploited as RNA-guided DNA binding platforms, without eliciting DNA cleavage.[73] These systems can be used *per se* for transcriptional repression by sterically hindering the RNA polymerase activity on the target gene or on its promoter. dCas can also be fused to the catalytic domain of other proteins and exert their functions at the site specified by the sgRNA. For example, dCas9 has been fused to transcriptional effector domains to mediate transcriptional activation or inhibition.[73]

dCas9 has also been coupled to several epigenetic modifiers to mediate site-specific epigenetic editing (Fig. 2B). The dCas9 molecule has been fused to the catalytic domain of DNA methyltransferases (DNMTs)[74–76] and of the demethylase TET1,[74,77–79] allowing the modification of CpG islands around the dCas9 target site and influencing the expression of nearby genes. dCas9 has also been fused to histone-modifying enzymes catalytic cores, such as of the histone acetyltransferase P300[80] and the histone demethylase LSD1.[81] Other histone modification dCas9 tools include histone methyltransferases (DOT1L, SMYD3, and PRDM9),[82,83] and histone deacetylases (HDACs).[84]

Site-specific epigenetic editing can have important applications for the treatment of human disorders, such as the Fragile X syndrome, by reversing pathological epigenetic changes.[85]

Great clinical potential is also associated to catalytically impaired Cas nucleases fused to base-modification enzymes, to generate DNA base editors (BEs). The catalytically disabled nuclease does not generate a DSB at the target site, but modifies a nucleotide appropriately positioned relative to the PAM site within the target sequence and generates a nick in the opposite strand, to promote the repair of the non-edited strand.[86,87]

Base editing was shown to occur in post-mitotic cells that are resistant to DSB-stimulated HDR[88] and is therefore an extremely attractive alternative to HDR for therapeutic use in human.

Since the target nucleotide needs to be precisely positioned relative to the PAM site, it has been calculated that only ~26% of known pathogenic transitions can be targeted by SpCas9-based BEs and this warrants the development of BEs with additional PAM compatibilities.[89,90]

So far, BEs can mediate all four possible transition mutations (C to T, A to G, T to C, and G to A) and further work will be required to develop BEs to mediate transversions.[86,87]

Further current limitations for the clinical application of BEs are the fact that BEs can yield a low but detectable rate of indel formation and, as other CRISPR/Cas-mediated systems, it is associated to potential off-target effects.[86,87] Protein engineering efforts are being made to generate high-fidelity base editors and address these limitations.[91]

Recently, a new CRISPR-based gene editing strategy, called prime editing, has been developed with prospective applications in human disease therapy. A nicking Cas9 variant is fused to a reverse transcriptase for template-directed sequence alterations.[92] Further research will be required to reveal possibilities and limitations of this new tool.

3.1 Challenges associated to the clinical implementation of CRISPR/Cas and related technological advances

CRISPR/Cas systems hold great promise for the cure of several human diseases. The major challenges for the implementation of these tools in a clinical setting are efficacy, safety and delivery. All these aspects are strictly interdependent.

The irreversible nature of the modifications that can be introduced by most of CRISPR/Cas systems implies that the editing must occur at the target site with no off-target effects, to avoid unwanted mutagenesis. However, the relatively short length of the target sequence of the commonly used CRISPR/Cas9 system (~23 bp), coupled to the fact that these systems are relatively tolerant to single and even multiple mismatches, poses a serious issue for their specificity.[63,93,94]

Efficacy and specificity of CRISPR/Cas systems largely depend on target site selection and many computational tools are now available for rational design of the sgRNA.[95] In addition, Cas9 variants have been designed or evolved to increase specificity and reduce off-target effects.[89,96–102]

The chances of CRISPR/Cas off-target effects are also increased by its long-term expression in the target cells. Several approaches have therefore been developed to control the spatiotemporal activity of CRISPR/Cas systems, such as engineered switchable Cas nucleases[103] and self-limiting Cas circuits (SLiCES),[104] as well as the use of natural inhibitors for Cas9 and Cas12a.[105–108]

Another important aspect that influences the safety profile of these tools in a clinical setting is their interaction with the host immune system, especially considering that the species of origin of some CRISPR/Cas systems, such as *Staphylococcus aureus* and *Streptococcus pyogenes* are common human infectious agents.[109–112]

Some controversy regarding the safety of CRISPR/Cas systems also exist over recent reports that CRISPR/Cas gene editing may lead to a selective advantage of clones with oncogenic *p53* mutations, following induction of *p53*-dependent DNA damage response.[113–115]

Finally, for the use of CRISPR/Cas systems in the clinic, these tools need to be deliverable to the target cells. Target cell genome can be either engineered, or reprogrammed, outside the body (*ex vivo*); edited cells can bethen reinfused into the patient. Alternatively, target cell genome can be edited directly within the body (*in vivo*), through a local or systemic administration of CRISPR/Cas tools (Fig. 3).

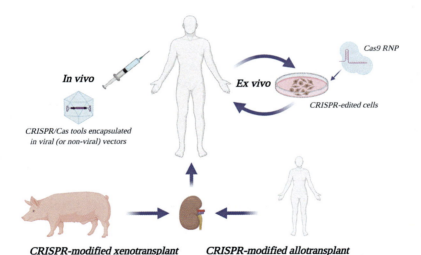

Fig. 3 Opportunities for *in vivo* and *ex vivo* CRISPR-based therapies in the treatment of human primary ciliopathies. For *in vivo* gene/epigenetic editing, CRISPR/Cas tools are delivered directly to the target tissue within the patient's body, using either viral or non-viral vectors, following systemic or local administration. In *ex vivo* editing, target cells or tissue are removed from the body, where they are modified using CRISPR/Cas tools in a controlled laboratory setting and then they are reintroduced into the patient's body. CRISPR-modified pigs for enhanced biosafety can also be used as organ donors in xenogeneic transplantations, to replace the malfunctioning kidney in renal ciliopathies. A similar approach could be envisaged to reduce the risk of rejection in allogenic organ transplantations. RNP, ribonucleoprotein.

While in *ex vivo* gene/epigenetic editing, CRISPR/Cas tools are delivered to the target cells in a controlled laboratory setting for enhanced efficacy and safety, *in situ* gene/epigenetic editing presents additional challenges associated with the *in vivo* delivery to the target tissue, especially when systemic administration is required. Indeed, the vast majority of ongoing clinical studies on CRISPR-based therapies focuses on *ex vivo* engineering of T cells for the treatment of HIV-1 infections or cancer and the only ongoing *in vivo* study involves the treatment of a retinal ciliopathy, following local administration to the eye[116] (Table 2).

Optimization of efficient and safe delivery methods is probably the main bottleneck in the development of *in vivo* CRISPR/Cas therapies.

3.2 *In vivo* delivery of CRISPR/Cas-based therapies

CRISPR-based editing systems can be delivered in the form of DNA with one or multiple plasmids encoding each component, as RNA (Cas-encoding mRNA and sgRNA) or as ribonucleoprotein complexes.

Whichever format is chosen, *in vivo* delivery of naked CRISPR/Cas system is problematic, due to potential interactions with serum nucleases and stimulation of the innate immune system, as well as to the large size and charge of these macromolecules, which normally prevent them from easily entering the target tissue from blood vessels.[117]

Hydrodynamic delivery is a physical delivery method that has been employed in preclinical models to deliver naked DNA plasmids and proteins *in vivo*. The hydrodynamic pressure is increased by pushing a large volume (8%–10% body weight) solution into the bloodstream.[118] This in turn increases the permeability of endothelial and parenchymal cells, allowing the entrance of gene editing tools. Delivery using this method is particularly efficient in the liver, kidney, lung, muscles and heart and it is particularly attractive as it avoids the deployment of delivery vectors.[119]

Indeed, using this method, it was possible to *in vivo* correct, albeit with very low efficiency, a mutation in the fumarylacetoacetate hydrolase (FAH) in mouse hepatocytes in a model of hereditary tyrosinemia.[120] However, it is hard to envisage application of hydrodynamic delivery of CRISPR systems in clinical practice, as it is highly traumatic for the tissues and can potentially lead to increased blood pressure, cardiac dysfunction and hepatomegaly.[118,121]

In the majority of cases is therefore necessary to use a vector to mediate the delivery of CRISPR/Cas to the target tissue. Vehicles for *in vivo* delivery of CRISPR-based systems can be divided into viral and non-viral.

Table 2 Interventional clinical studies on Leber congenital amaurosis 10 (LCA10).

NCT Number	Title	Acronym	Status	Interventions	Type of intervention	ROA	Phases	URL
NCT03872479	Single Ascending Dose Study in Participants With LCA10		Recruiting	Drug: AGN-151587	CRISPR/Cas	SRI	Phase 1|Phase 2	https://ClinicalTrials.gov/show/NCT03872479
NCT03913143	A Study to Evaluate Efficacy, Safety, Tolerability and Exposure After a Repeat-dose of Sepofarsen (QR-110) in LCA10	ILLUMINATE	Recruiting	Drug: sepofarsen|Other: Sham	ASO	IVT	Phase 2|Phase 3	https://ClinicalTrials.gov/show/NCT03913143
NCT03140969	Study to Evaluate QR-110 in Subjects With Leber's Congenital Amaurosis (LCA) Due to the c.2991+1655A>G Mutation (p.Cys998X) in the CEP290 Gene		Completed	Drug: QR-110	ASO	IVT	Phase 1|Phase 2	https://ClinicalTrials.gov/show/NCT03140969
NCT03913130	Extension Study to Study PQ-110-001	INSIGHT	Active, not recruiting	Drug: QR-110	ASO	IVT	Phase 1|Phase 2	https://ClinicalTrials.gov/show/NCT03913130

Data were downloaded from ClinicalTrials.gov on 09/10/2020. ROA, route of administration; SRI, subretinal injection; ASO, antisense oligonucleotide; IVT, intravitreal injection.

Viral platforms are probably the most efficient method for the introduction of exogenous genetic material into cells, but they also have several limitations, such as the high costs of their production at scale, the potential risk of insertional mutagenesis, immunogenicity and genotoxicity associated with viral integration.[122–125] Permanent transgene expression can also increase the risk of off-target effects. Indeed, historically the use of viral vectors for gene therapy has been occasionally associated to severe adverse events.[126–128]

The most widely applied vectors in gene-based therapies are Adeno-Associated Virus Vectors (AAVs). AAVs demonstrate high transduction efficiency and low immunogenicity and are characterized by a wide range of serotypes with broad tropism, allowing specific infection of different tissues.

The main limitation of AAVs is the low packaging capacity. This is particularly problematic when, together with the CRISPR/Cas duo, a repair template for precise gene correction following HDR needs to be delivered to the target cells. Moreover, dCas9 chimeras, which are particularly attractive for *in vivo* applications due to their potential safety, are large proteins and their encoding DNA represents a huge cargo for the small AAV vehicles. Several groups have shown *in vivo* gene editing by packaging the DNA coding for SpCas9 and sgRNA into two separate AAV particles.[129,130] Similarly, a split-intein Cas9 was developed, where two portions of the Cas9 protein were packaged into two separate vectors and joined when expressed in the cell.[131,132] However, both these dual vector strategies can add complexity to the system at the expense of editing efficiency.

A rationally designed truncated version of SpCas9 was also developed to overcome packaging limitations but displayed a loss of around 50% in activity when compared to wild-type Cas9.[133]

Alternatively, some SpCas9 (1368-1424 aa) orthologs are characterized by smaller size, such as SaCas9 (1053 aa), St1Cas9 (1122 aa), NmCas9 (1109 aa) and *Campylobacter jejuni* (CjCas9, 984 aa).[134] These smaller orthologs, however, tend to recognize longer PAM sequences, which limits the availability of target sequences.

Other nucleases belonging to the type V CRISPR/Cas systems (e.g., Cas12a) can display smaller size compared to SpCas9 and can represent a valid alternative to SpCas9 for AAV-mediated delivery. In addition, the type V CRISPR/Cas12a systems have a minimal sgRNA component (crRNA only with no tracrRNA)[135]; as a result, the sgRNA for Cas12a is about 40 nt long, which is less than half the size of a Cas9 sgRNA.

Alternative viral vectors such as Adenovirus (AdV) and Lentivirus (LV) vectors offer higher packaging capacity but also display higher immunogenicity compared to AAVs. In addition, LVs can integrate into host genomes, unless engineered to be produced as integrase deficient lentiviral vectors (Table 3).

Virus-like particles (VLPs) offer the opportunity to exploit the efficient viral transduction, without carrying any viral genomes. Indeed vesicles based on the fusogenic glycoprotein of the vesicular stomatitis virus (VEsiCas), were able to deliver RNPs and edit the genome of cardiomyocytes *in vivo*, when injected into the cardiac muscle of a mouse model.[139]

A very active area of research regards the development of new non-viral vector delivery systems. Non-viral delivery methods include lipid nanoparticles, gold nanoparticles, DNA "nanoclews" and cell-penetrating peptides. Non-viral delivery systems feature reduced immunogenicity, higher DNA packaging capacity and can be more easily produced at scale compared to viral vectors. The main limitation so far is the low delivery efficiency.[117]

3.2.1 Delivery of CRISPR/Cas to the retina

Currently, the most common delivery method used in gene therapy for eye disease is based on AAVs.

Type 2 AAV (AAV2) has been the first vector used in gene therapy and it is the most widely used serotype in ophthalmology.[140–142]

In terms of administration route, systemic administration is not recommended for retinal delivery, due to the presence of the blood-retinal barrier. However, the high accessibility of the eye makes local delivery a feasible approach. Local delivery also allows specific transfection of the target tissue, preventing dangerous off-target effects, as opposed to systemic delivery.

The two most common approaches of retinal delivery are intravitreal and subretinal injections. Although intravitreal injection is the least invasive procedure, viral vectors including AAV2 are unable to infect photoreceptors following this route.[143,144] Therefore, despite being more invasive, subretinal injection is commonly chosen for gene replacement and correction therapies aimed to treat photoreceptors. Development of new serotypes with improved photoreceptor transfection efficiency may be necessary to allow for less invasive administration routes.[145] It should be noted, however, that, due to the permanent nature of gene editing, deployment of non-invasive administration routes is less paramount than in those cases when repeated administration cycles are likely to be required.

Table 3 Comparison of the relevant features of adenoviral, lentiviral and adeno-associated viral vectors for clinical applications.

	AdV	LV	AAV
Genome	dsDNA	ssRNA	ssDNA/dsDNA
Host genome integration	No, 100% episomal	Yes, but integrase-deficient versions available	>90% episomal (recombinant AAV)
Packaging capacity	Up to 34 kb	8 kb	<5 kb
Tropism	Broad	Broad	Broad
Infects both dividing and non-dividing cells	Yes	Yes	Yes
Pathogenic	Yes	Yes	No
Immune response in target cells	High	Low	Very Low
Summary: main pros and cons for clinical applications	High packaging capacity, but can trigger severe inflammation and immune response	Higher packaging capacity than AAVs but features insertional mutagenesis potential and higher immunogenicity compared to AAVs	Not pathogenic and characterized by very low immunogenicity but low packaging capacity
Clinical trials	NCT03313596 (Hepatocellular carcinoma)	NCT02906202 (Beta-thalassemia) NCT03207009 (Beta-thalassemia)	NCT03370913 (Hemophilia A) NCT03641703 (Hemophilia B) NCT03612869 (Mucopolysaccharidosis type IIIA) NCT00999609 (Inherited retinal dystrophy due to RPE65 mutations; Leber congenital amaurosis) NCT03116113

Continued

Table 3 Comparison of the relevant features of adenoviral, lentiviral and adeno-associated viral vectors for clinical applications.—cont'd

	AdV	LV	AAV
			(X-linked retinitis pigmentosa) NCT03496012 (Choroideremia) NCT03392974 (Hemophilia A) NCT03293524 (Leber hereditary optic neuropathy) NCT03153293 (Leber hereditary optic neuropathy) NCT03569891 (Hemophilia B)
References	136	137	138

List of advanced interventional clinical studies (Phase II/III or Phase III) employing viral vectors for gene therapy as reported by ClinicalTrials.gov on 09/10/2020. For an exhaustive review of the therapeutic applications of each type of viral vector, you can refer to the reported references.

Promoter selection is also important to ensure specific expression of CRISPR/Cas tools in the target cell types. To drive the expression of CRISPR system components into photoreceptor cells but not in any other retinal cells, the human rhodopsin kinase (RK) promoter is commonly employed. RK is selectively active in rods and cones, but not in other retinal neurons or in the retinal pigment epithelium (RPE), moreover, the relatively small size make this promoter particularly attractive for AAV-mediated delivery.[146,147]

3.2.2 Delivery of CRISPR/Cas to the kidney

Renal anatomy and physiology make delivery of genetic material to this organ particularly difficult, especially through systemic administration.[148] Especially challenging is the targeting of tubular cells. The stringent molecular mass cut-off of the kidney glomerulus prevents particles above 50 kDa in mass to enter the kidney tubule, posing a limit even for the entrance of nano-scale non-viral vectors like lipid nanoparticles. The size of viral vectors is in the range of megadaltons and is therefore way above the glomerular cut-off size.[149]

Kidney uptake of small nucleic acids is extremely efficient,[22–24] but considering the average nucleotide size of 325 Da, only naked oligonucleotides <150 nt can reach tubule epithelia *via* systemic administration.

Hence, systemic administration of gene-based therapies, with the exception of small ASOs, is not a feasible approach to treat tubular disorders such as renal ciliopathies. Local delivery of CRISPR/Cas tools may therefore be required but, due to the low accessibility of the kidney, it implies more invasive practices compared to gene therapies for eye diseases.

Alternative administration routes of viral vectors, including retrograde injections into the ureter,[150] or direct injections into the renal interstitium,[151,152] were shown to mediate efficient transfection into kidney tubule cells in rodents. Transduction efficacy was compared for several AAV serotypes upon retrograde ureteral injections.[149] AAV8 was shown to be the most efficient, but also it mediated high off-target effects to the liver. AAV2, on the contrary, was associated to lower off-target activity to the liver, at the expense of on-target efficiency. Larger AdVs and LVs were also shown to transduce kidney cells upon local delivery with limited off-target tissue transduction.[149] It should be noted, however, that renal manifestations can be accompanied by liver involvement in some primary ciliopathies, including BBS, JBTS, ADPKD, ARPKD and NPHP (Table 1). In these cases, a partial leakage of AAV vectors from the kidney to the liver could be even seen as beneficial.

To partially overcome off-target delivery issues, CRISPR expression can be restricted through the use of kidney-specific promoters (e.g., the promoter of *Cadherin 16*, *CDH16*)[153,154] or nephron segment-specific promoters.[154]

Low tropism for kidney cells of currently available AAV serotypes may require higher injection doses, which can be associated to cytotoxicity. Efforts to engineer new AAV serotypes to increase targeted delivery to the kidney are therefore warranted.[150,155]

3.2.3 In vivo *delivery of ASOs: A better alternative for the kidney?*

ASOs as single stranded chemically modified oligonucleotides can be employed to treat human disease by several different mechanisms.[156] Chemical modifications (e.g., sugar and phosphodiester linkage modifications) allow ASOs to resist nuclease degradation, avoid or reduce immune responses and improve endocytotic cellular uptake.[157]

Naked ASOs can be delivered to cells in the central nervous system and to the eye through local administration.[158] Indeed, the development of sepofarsen (QR-110),[159] an ASO-based therapeutic administered *via* intravitreal injection to treat LCA10, has recently reached Phase II/III clinical trial (ILLUMINATE study, NCT03913143, Table 2).

Moreover, where endothelial fenestrations are large enough, the small size of ASOs allows passage through the vascular barrier and delivery through systemic administration to the target tissue. Upon systemic administration, ASOs accumulate primarily in the liver and kidney.[160] Oligonucleotides with a phosphorothioate backbone (the most widely used modification in medicine) were shown to be preferentially taken up in the kidney by proximal tubule cells upon intravenous injection in rats,[22] but were also shown to abundantly distribute into collecting duct-derived cysts in an ADPKD mouse model, following subcutaneous administration.[161]

Considering the superior ability to reach the kidney, even following systemic administration, ASOs represent therefore an excellent alternative to CRISPR/Cas systems for the *in vivo* gene-based treatment of renal ciliopathies.

Indeed, it was shown that the ASO-induced skipping of mutated exon 41 (carrying a nonsense mutation p.G1890*) from *CEP290* transcript restored near full-length protein expression and ameliorated ciliary phenotypes in urine-derived renal epithelial cells of a *CEP290* JBTS patient with cystic kidney disease. A similar approach also rescued ciliary and kidney phenotypes *in vivo* in a genetrap *Cep290* JBTS mouse model, following systemic administration of the ASO.[162]

Systemic administration of ASO to inhibit microRNA-17, which is a driver of cystogenesis, was also shown to be effective in ameliorating kidney phenotype in mouse models of ADPKD.[161]

Despite these encouraging results it should be noted that liver and kidney are preferential but not exclusive sites of ASOs biodistribution, which can also distribute to the adrenal gland, spleen, bone marrow and lymph nodes upon systemic administration,[160,161] where they could elicit off-target effects.

ASOs can be used in a wide range of different therapeutic strategies and therefore can be employed in the treatment of monogenic disorders with different pathomechanisms.[156] However, not all variants could be targetable by ASO therapeutics and the range of variants that can be tackled by gene editing remains wider at least in principle. For example, ASOs can be employed as splice modulators to correct the effects of splicing variants[159] but also to tackle dominant negative/gain of function variants, where the mutant allele leads to a toxic function, by mediating mRNA degradation or steric blocking of translation.[163] Gene upregulation through ASO treatment could also be used to tackle loss of function variants either by steric inhibition of 5′ UTR regulatory elements[164,165] or by mediating the skipping of truncating mutations from transcripts to elude nonsense-mediated decay.[166] Moreover, gene upregulation can be attained using ASOs to modulate non-productive alternative splicing in dominant haploinsufficiency diseases.[167] On the other hand, loss of function missense variants in a recessive context may not be generally suitable targets for an ASO-based therapeutic strategy, with the notable exception of the use of the ASO nusinersen in the treatment of spinal muscular atrophy.[168]

Another drawback of ASO therapeutics is that the likely non-permanent effects may require multiple and possibly life-long administrations.

Ten in-body ASO-based therapies have been so far approved by FDA,[158] while only one in-body CRISPR-based therapeutic is currently being tested in a Phase I/II clinical trial[116] (Table 2). While most likely this difference reflects the relatively recent history of CRISPR/Cas gene editing, in the next few years it will be clearer if the large size of CRISPR/Cas systems represents a critical liability for clinical in-body applications.

4. CRISPR/Cas gene editing for the treatment of primary ciliopathies

4.1 Gene editing for the treatment of retinal ciliopathies

The eye possesses peculiar features that make this organ an ideal target for gene-based therapies such as gene replacement and correction.

The eye is highly compartmentalized and accessible, it is small, allowing local delivery of vectors and genetic material concentrated in small volumes. It is also considered an immune-privileged site, in part due to the retina-blood barrier, which modulates the trafficking of immune cells to this organ, therefore reducing the risk of adverse immune response.[169]

Retinal degeneration in retinal ciliopathies is often severe and early onset. Gene correction to rescue photoreceptor function and viability in the early stages of the disease represents a promising therapeutic strategy.

Three-dimensional retinal organoids generated from patient-derived iPSCs can be an excellent pre-clinical *ex vivo* platform where to test the efficacy of gene correction therapies, allowing for a certain level of functional readout of the rescue in a human, patient-specific context.[170] A few proof-of-concept studies have already shown the efficacy of gene correction to rescue the functional defects observed in RP patient-specific retinal organoids.

Deng et al. generated retinal organoids from patients with *RPGR*-associated RP.[170] *RPGR* patient-derived retinal organoids displayed a decrease in the viability and morphology of photoreceptors compared to controls. Opsin mislocalization was also observed and the photoreceptor connecting cilium appeared shorter in RP organoids. At the gene expression level, patient organoids showed reduced retinal-specific gene expression and an upregulation of genes linked to necrosis and to G1/S and G2/M phase transitions. HDR-mediated gene correction of iPSCs prior to differentiation, using CRISPR/Cas9 tools and a DNA template, was able to partially rescue these defects. The rescue of the structural ciliary defects was accompanied by a partial restoration of the gene expression pattern. Moreover, gene correction was shown to mediate a functional rescue at the electrophysiological level,[170] providing a preliminary but powerful proof of concept for the efficacy of gene editing therapies in the treatment of RP.

Similar encouraging results were obtained for a different form of retinal ciliopathy. Retinal organoids were generated from patients with mutations in the *pre-mRNA processing factor 31* (*PRPF31*), which causes an autosomal dominant form of RP. CRISPR/Cas9-mediated gene correction restored well-aligned ciliary microtubules in corrected photoreceptors, as opposed to the aberrant ciliary morphologies observed in the uncorrected cells, and rescued protein expression.[171]

Although encouraging, these studies heavily rely on the exploitation of the HDR pathway for gene correction in iPSCs. Given the low efficiency of HDR in non-dividing cells, this approach may not be suitable for direct translation to the clinic, to treat postmitotic photoreceptor cells *in vivo*.[172] Different approaches, such as the use of BEs, could be envisaged to achieve gene correction in this type of disorders.

Another gene-based therapeutic paradigm that does not rely on the HDR pathway was also proposed to treat RP. Since RP is associated to early rod death followed by secondary cone degeneration, the conversion of adult rods into cones could make cells resistant to mutations in rod-specific genes while preventing secondary cone loss. The transcription factor neural retina-specific leucine zipper protein (NRL) acts during retinal development as a molecular switch that determines photoreceptor identity.[173,174] Using *rd1* mice (a severe retinal degeneration model mutated in the ciliary RP gene *Phosphodiesterase 6b*, *Pde6b*), Montana et al. showed that germline knockout of *Nrl* suppresses the retinal degenerative phenotype and demonstrated that reprogrammed rods are indeed resistant to degeneration.[175]

In order to translate this approach into a potential therapeutic strategy, Yu et al. studied the effects of knocking out *Nrl*, using CRISPR/Cas9 gene editing tools, in another *Pde6b* mouse (*rd10*). Gene editing of photoreceptor cells was attained upon subretinal injection of dual AAV8 vectors. Cone function was preserved after *Nrl* ablation and rod cells, although dysfunctional or dysmorphic and not efficiently transdifferentiated into cones, were relatively well maintained for at least 3 months post-injection.[176] It should be noted that mutations in *NRL* have been associated to human retinopathies.[177,178] A possible explanation for this apparent discrepancy may stem from the different functions or relevance of *NRL* in developing and mature postmitotic rods.

Another approach for the gene-based treatment of retinal ciliopathies that does not rely on the HDR pathway consists in the use of the CRISPR/Cas system for the correction of pathological splicing alterations. A paradigmatic example of this is the case of *CEP290* splicing correction for the treatment of LCA10.

The protein encoded by the *CEP290* gene localizes to the photoreceptor connecting cilium. The function of CEP290 within the photoreceptor cell is linked to the regeneration of the outer segment and to phototransduction, through the regulation of protein trafficking.[179]

The most common mutation associated to LCA10 is the intronic point mutation c.2991+1655A>G (IVS26 mutation), which creates a cryptic splice site, resulting in the inclusion of an aberrant exon of 128 bp that contains a premature stop codon. The removal (or inversion) of the IVS26 mutation could result in normal splicing and restoration of functional *CEP290* expression and consequently halt retinal degeneration.

As a proof of concept for the feasibility of this gene editing therapeutic strategy, the CRISPR/Cas9 system under the RK promoter was used to delete the corresponding intronic fragment in *Cep290* in wild-type mice *in vivo*. Upon subretinal injection of a dual AAV5 vector system, the desired gene editing of the wild-type *Cep290* gene was obtained with an average efficiency of around 20% in the retina.[180] Given the absence of good animal models that mimic the cellular and molecular features of *CEP290*-LCA10,[44] it is hard to establish whether the reached editing efficiency is indeed sufficient to produce a functional rescue. However, current estimations suggest that only a 10% of functional foveal cone photoreceptors could be sufficient for near-normal visual acuity.[181]

Following this rationale, a candidate genome-editing therapeutic, EDIT-101, based on the CRISPR/SaCas9 system, was developed to induce the removal or inversion of the IVS26 mutation.[181] Kinetics and dose response studies of subretinal delivery of EDIT-101 were performed in the *CEP290* IVS26 knock-in mouse model and on non-human primates and was well tolerated in all animals. The expression of SaCas9 was restricted to photoreceptors only and on-target *Cep290* gene editing reached an efficiency of around 30%, thus higher than the anticipated minimum threshold of 10% for clinical effectiveness.[181]

These successful preclinical investigations led to the initiation of the Phase I/II BRILLIANCE study (NCT03872479), to evaluate the safety, tolerability and efficacy of subretinal injections of EDIT-101 in participants with LCA10 caused by the IVS26 mutation. Notably, BRILLIANCE is the first clinical study to deploy the CRISPR/Cas gene editing system directly in the human body.[116]

It is worth noting that the BRILLIANCE study proceeds in parallel with the Phase II/III ILLUMINATE study on the ASO sepofarsen. These two studies are aimed at treating the same condition (LCA10 secondary to the IVS26 mutation) and follow a similar rationale, i.e., promoting the skipping of the cryptic exon. The ASO sepofarsen, which acts at the RNA level, is administered *via* intravitreal injection and the CRISPR-based EDIT-101, which acts at the DNA level, is administered *via* subretinal injection (Table 2). The results of these trials will be extremely informative to compare efficacy and safety profiles of ASO- and CRISPR-based therapies for the treatment of retinal disorders.

4.2 Gene editing for the treatment of renal ciliopathies

Despite the fact that renal and retinal degeneration are in principle both amenable to therapeutic intervention, the implementation of CRISPR/Cas gene editing therapies for renal ciliopathies is still in its infancy. Indeed, CRISPR/Cas gene editing as a means to treat renal ciliopathies has so far been investigated only *in vitro*, mainly for disease modeling purposes.

A remarkable phenotypic rescue was observed in gene corrected kidney organoids derived from ARPKD patients iPSCs. ARPKD kidney organoids partially mimic the cystic phenotype of human patients and display drastic cystogenesis when challenged by the upregulation of intracellular cAMP. HDR-mediated correction using CRISPR/Cas9 of the *PKHD1* mutation in one allele was able to rescue this cystic phenotype.[182]

Similar successful rescue was observed in kidney organoids derived from a syndromic ciliopathy patient with RP, skeletal defects and ESKD caused by compound heterozygous mutations in *Intraflagellar Transport 140 (IFT140)*. HDR-mediated gene correction using CRISPR/Cas9 of one *IFT140* allele rescued the clubbed ciliary morphology observed in uncorrected kidney organoids. Moreover, magnetic-activated cell sorting (MACS)-sorted EPCAM cells from gene corrected kidney organoids exhibited increased capacity to form spheroids with polarized epithelium, indicating a rescue of the apicobasal polarity defects of uncorrected cells. Gene corrected spheroids also presented a higher incidence of ciliated cells.[183]

Although the rescued phenotypes observed in gene corrected kidney organoids represent very promising results, we are still a long way from the translation to the clinic of *in vivo* CRISPR/Cas-based therapies for gene correction in the kidney, mainly because of delivery and safety issues.

A potential alternative to *in vivo* gene editing could be the *ex vivo* correction of kidney organoids derived from renal ciliopathy patients to be subsequently transplanted into the patient. This would overcome both graft rejection issues and the chronic shortage of transplantable organs. However, current limitations associated to kidney organoids such as reduced size and vasculature, immaturity and lack/underrepresentation of certain cell types prevents their usage as a replacement of human kidney for transplantation.

An additional application of gene editing for the treatment of kidney disease can be envisaged in the field of xenotransplants. Xenotransplatation of porcine kidneys has been considered for decades a viable alternative to the transplantation of human kidneys, but immunocompatibility issues between pig and human tissues have limited this application. The CRISPR/Cas system has been used to knock out three porcine surface antigens that elicit hyperacute rejection response by the immune system.[184]

Another safety issue associated to pig xenotransplants is the presence, in the pig genome, of porcine endogenous retrovirus (PERV) elements that could be transmitted to humans following organ transplantation. CRISPR/Cas gene editing has been used to disrupt critical gene sequences essential for PERV activity at multiple sites in the pig genome and produce animals in which no traces of PERV could be detected.[185] Coupling this approach to improve biosecurity of porcine kidneys to genetic modifications to diminish immunoreactivity could indeed lead to an unprecedented acceleration in the clinical translatability of xenotransplant research.

5. CRISPR/Cas epigenetic editing for the treatment of primary ciliopathies

Epigenetic modifications include chemical modifications of DNA and histone proteins and are critical regulators of gene expression. DNA methylation is catalyzed by DNMTs that post-replicatively add a methyl group to the fifth carbon of a cytosine residue to form 5-methylcytosine (5mC).[186] Such DNA modification often occurs in cytosines of the CpG dinucleotide sequence that tend to cluster in CpG-rich regions called CpG islands.[187] DNA methylation of gene promoters is usually associated to transcriptional silencing,[188] whereas gene body DNA methylation has been associated to both transcriptional activation and repression, depending on the context.[189] There is a complex interplay between DNA methylation and histone modifications, with the modification state of DNA influencing histone modification state and *vice-versa*.[190] Histone post-translational modifications regulate chromatin structure and the recruitment of regulators and therefore modulate gene expression. Histone modifications include histone lysine acetylation (HKAc) and methylation (HKme).[191] HKAc is mediated by histone acetyl transferases and is generally associated to relaxed chromatin and active gene expression, whereas histone deacetylation is generally associated to closed chromatin and repressed gene expression. HDACs are responsible for histone deacetylation and are subdivided into three main classes (HDAC I, II, and III).[192] HKme is mediated by histone methyltransferases and can be an active or repressive mark depending on the lysine residue modified and the extent of methylation (mono-, di-, or tri-methylation). Histone demethylases are responsible for demethylation.[191]

Disturbances in these processes have been associated to the pathogenesis of a broad variety of diseases, including cancer, autoimmune and neurological disorders[193] and pharmacological interventions on these processes are largely explored as therapeutic strategies.[194] CRISPR/Cas-mediated epigenetic editing can be exploited as a powerful treatment option in this context, as it can reverse aberrant epigenetic modifications at specific sites. Several lines of evidence indicate an involvement of epigenetic mechanisms into the pathogenesis of retinal and renal ciliopathies and could in principle be leveraged as new therapeutic targets.

5.1 Epigenetic editing for the treatment of retinal ciliopathies: RP

Farinelli et al. proposed that transcriptional silencing *via* cytosine methylation could represent a new pathological mechanism in RP.[195] Indeed, they detected an increase in nuclear DNA methylation as assessed *in situ* by 5mC staining in a sub-population of outer nuclear layer photoreceptors in several RP models, among which the retinal ciliopathy *rd1* mouse. The nuclei positive for 5mC were also labeled by TUNEL assay, suggesting an association between DNA methylation and photoreceptor degeneration. Genome-wide assessment of the genomic localization of DNA methylation and motif analysis demonstrated a hypermethylation enrichment at the binding sites of several important retina-specific transcription factors, including Yin Yang 1 (YY1) and NRL, which occurred alongside the transcriptional repression of the corresponding genes.

Importantly, pharmacological inhibition of DNA methylation with decitabine resulted in a significant reduction in photoreceptor cell death in *rd1* organotypic retinal explants.[195] Decitabine activity is not specific and can lead to unwanted effects at off-target sites. On the other hand, CRISPR/Cas fused to demethylases can be directed to specific sites involved in RP pathogenesis and can be envisaged as a more specific epigenetic intervention in this context. Further research will be required to evaluate the efficacy of this approach in the treatment of RP.

Another possible epigenetic mechanism that has recently emerged as a contributor to retinal degeneration is HKme, especially in the form of trimethylation of lysine 27 of histone H3 (H3K27me3). Increased H3K27me3, which is a repressive chromatin mark and mediates epigenetic silencing, was observed in the retinas of *rd1* mice. Importantly, subretinal injection of an inhibitor of methylation substantially delayed the loss of rod photoreceptors and improved electroretinography (ERG) response in mouse retinas.[196]

HDACs were also shown to be implicated in the pathogenesis of retinal ciliopathies. Using the *rd1* mouse model, it was shown that HDAC increase in activity in the retina precedes causally photoreceptors death.[197] Indeed, a global histone hypoacetylation would explain the significant changes that were observed in the gene expression profile of *rd1* retinas. Experiments on *rd1* retinal explants showed that pharmacological inhibition of class I/II HDAC exerts a prosurvival effect on photoreceptors.[197] However, no specific gene was reported to be the target of HDACs I/II in retinal degeneration and HDAC inhibition may prevent retinal degeneration by inducing the expression of neuroprotective genes. Hence, the global nature of histone modifications in the degenerating retina implies that site-specific CRISPR-mediated HKAc and histone demethylation may not be suitable therapeutic approaches in this case. Moreover, it was shown that the overexpression of class I/II HDAC4 and HDAC5 N-terminus prolongs photoreceptor survival in *rd1* mice.[198,199] Although the prosurvival effect of HDAC I/II overexpression was reported to be mainly mediated by the deacetylation of cytoplasmic substrates rather than histones,[198] these results are in sharp contrast with the neuroprotective effect of HDAC I/II inhibition in *rd1* retinas and warrant further studies to validate HDAC inhibition as a therapeutic target in retinal ciliopathies.

5.2 Epigenetic editing for the treatment of renal ciliopathies: ADPKD

Epigenetic interventions are currently largely investigated as therapeutic strategies in the cancer field. ADPKD shares several features with cancer, like the clonal origin of cysts and the proliferative phenotype of cyst-lining epithelium. Possibly due to this similarity, epigenetic modifications have been investigated as possible pathological mechanisms and therapeutic targets in ADPKD too.

The *Mucin-Like Protocadherin* (*MUPCDH*) gene was found to be hypermethylated in its promoter region, in ADPKD renal tissue and cells, where it is also significantly downregulated.[200] *MUPCDH* was shown to exert an anti-proliferative effect in kidney cells and its silencing by methylation was proposed to be a contributor of the cystic phenotype in ADPKD. Indeed, a certain correlation was observed between the methylation state of *MUPCDH* and cyst growth. Although this association was not sufficiently strong to make the methylation state of *MUPCDH* a solid biomarker of disease progression, interesting possibilities exist for the investigation of its potential as a target for epigenetic therapies.[200]

Moreover, genome-wide analyses detected global disturbances in the methylation profile of ADPKD kidney tissue.[201,202] Woo et al. reported a global hypermethylation in cystic renal cortex tissue of ADPKD patients.[201] Interestingly, comparison with microarray-based expression analysis showed that silencing by hypermethylation in ADPKD kidneys mainly occurred in highly expressed genes in normal kidneys. In particular, the *PKD1* gene body region was hypermethylated and hypermethylation correlated with gene downregulation. Treatment with decitabine increased the transcriptional expression of *PKD1* in human immortalized ADPKD cyst-lining epithelial cells and significantly repressed cyst growth in 3D cultures of Madin-Darby Canine Kidney (MDCK) cells.[201]

These results seem to point toward targeted demethylation of *PKD1* gene body as a promising treatment for ADPKD patients. However, DNA methylation analysis of similar ADPKD samples by a different group revealed that the *PKD1* gene, although hypermethylated, was overexpressed, rather than downregulated, in ADPKD kidneys, and a global hypomethylation in ADPKD tissues was reported rather than hypermethylation.[202]

These discrepancies may be due to the presence of different cell types in the tissues analyzed, which could confound the results or to differences in the methylation profiling methods used. It is also possible that intra- and inter-variability in ADPKD samples could contribute as a confounder. Indeed, analysis of the methylome of single cyst walls from a single ADPKD patient showed significant amount of variability across different cysts.[203]

Further investigation is therefore required to clarify the role of DNA methylation in ADPKD and to assess its suitability as a therapeutic target.

HDAC inhibition was also shown to be a very promising therapeutic strategy in ADPKD.

A drug screening first identified HDACs inhibitors as suppressors of polycystic kidney disease phenotypes in zebrafish.[204] Since then, different classes of HDACs were shown to be involved in the pathogenesis of ADPKD and HDACs inhibition has proven to be efficacious in several disease models.[205–208] However, the exact role of HDACs in the pathogenesis of ADPKD is not clear and the phenotypical rescue obtained through their inhibition may be mainly mediated by their effects on non-histone substrates (e.g., p53 and RB), rather than on a direct effect on chromatin remodeling. Therefore, CRISPR/Cas-mediated epigenetic editing may not be a successful strategy in the treatment of this pathogenic pathway.

Further efforts to clarify the exact meaning of epigenetic modifications in health and disease will be likely required before CRISPR/Cas-mediated site-specific epigenetic editing could be envisaged as a feasible therapeutic strategy in primary ciliopathies.

6. Further considerations and concluding remarks

CRISPR/Cas systems hold great promise as powerful tools for the treatment of human ciliopathies.

If it is hard to envisage an immediate utility of CRISPR/Cas targeted epigenetic editing as a therapeutic strategy in primary ciliopathies, CRISPR/Cas-mediated *in vivo* gene correction will be evaluated in clinical studies in the near future for the treatment of retinal ciliopathies.

The broad translation of CRISPR/Cas technology into clinical practice strongly depends on further improvements in its efficacy and safety profile.

Along with these improvements, the success of CRISPR-based therapies in the treatment of human ciliopathies is also dictated by the availability of a genetic diagnosis for a large proportion of ciliopathy patients, as precise information on the genetic alteration is needed to direct gene-based therapies.

Genetic diagnostic rates of primary ciliopathies are reported to be between 44% and 62%, depending on the sequencing method.[209] An improvement in this diagnostic rate is therefore a prerequisite for the implementation of more effective and targeted therapies, such as gene/epigenetic editing.

The time when the diagnosis is made is also important. In this chapter, we focused on two common manifestations of primary ciliopathies, retinal and renal degeneration. Due to their degenerative nature, retinal and renal disease in ciliopathies offer a temporal window for therapeutic intervention. The efficacy of the intervention very much depends on being able to hit this window. This highlights a further challenge, represented by an early diagnosis, which should ideally be made even before some of the defining disease phenotypes become evident. Moreover, the length of the therapeutic window may vary among different ciliopathies and even among different individuals, due to the high clinical heterogeneity of this class of disorders. The lack of a clear genotype-phenotype correlation in ciliopathies further impairs our ability to predict the temporal extent of such window.

Studies on a mouse model of ciliopathy show that the window for successful gene-base therapies of retinal degeneration is limited, with the efficacy and duration of the therapeutic rescue dictated not just by the number of viable photoreceptors but also by their state of terminal differentiation at the time of treatment.[210]

Early interventions such as gene editing in the germline or in the human embryo would overcome issues related to short therapeutic windows and would in principle allow the treatment of congenital phenotypes as well as degenerative ones. However, serious ethical issues limit the applicability of such early interventions.[211,212]

One further challenge for the implementation of CRISPR-based therapies is represented by the common hurdles that characterize the development and commercialisation pathway of every personalized medicine product,[213] including issues related to the high costs of clinical studies coupled to small market size, which require advances in the regulatory standards.

The challenges related to the development of CRISPR-based treatments for primary ciliopathies here described are not trivial but are being tackled by active research in the field. We can reasonably predict that technological and regulatory advances will be able to overcome such challenges within a decade and finally provide ciliopathy patients with very much needed effective therapies for these rare severe human disorders.

Acknowledgments

E.M. and J.A.S. are supported by Kidney Research UK (RP_006_20180227) and Northern Counties Kidney Research. Figures were created with BioRender.com.

References

1. Malicki JJ, Johnson CA. The cilium: cellular antenna and central processing unit. *Trends Cell Biol*. 2017;27(2):126–140. https://doi.org/10.1016/j.tcb.2016.08.002.
2. Nachury MV, Mick DU. Establishing and regulating the composition of cilia for signal transduction. *Nat Rev Mol Cell Biol*. 2019;20(7):389–405. https://doi.org/10.1038/s41580-019-0116-4.
3. Singla V. The primary cilium as the cell's antenna: signaling at a sensory organelle. *Science*. 2006;313(5787):629–633. https://doi.org/10.1126/science.1124534.
4. Wheway G, Nazlamova L, Hancock JT. Signaling through the primary cilium. *Front Cell Dev Biol*. 2018;6:8. https://doi.org/10.3389/fcell.2018.00008.
5. Bachmann-Gagescu R, Neuhauss SC. The photoreceptor cilium and its diseases. *Curr Opin Genet Dev*. 2019;56:22–33. https://doi.org/10.1016/j.gde.2019.05.004.
6. Hildebrandt F, Otto E. Cilia and centrosomes: a unifying pathogenic concept for cystic kidney disease? *Nat Rev Genet*. 2005;6(12):928–940. https://doi.org/10.1038/nrg1727.
7. Badano JL, Mitsuma N, Beales PL, Katsanis N. The ciliopathies: an emerging class of human genetic disorders. *Annu Rev Genomics Hum Genet*. 2006;7:125–148. https://doi.org/10.1146/annurev.genom.7.080505.115610.
8. Kim YG, Cha J, Chandrasegaran S. Hybrid restriction enzymes: zinc finger fusions to Fok I cleavage domain. *Proc Natl Acad Sci U S A*. 1996;93(3):1156–1160. https://doi.org/10.1073/pnas.93.3.1156.
9. Christian M, Cermak T, Doyle EL, et al. Targeting DNA double-strand breaks with TAL effector nucleases. *Genetics*. 2010;186(2):757–761. https://doi.org/10.1534/genetics.110.120717.
10. Li T, Huang S, Jiang WZ, et al. TAL nucleases (TALNs): hybrid proteins composed of TAL effectors and FokI DNA-cleavage domain. *Nucleic Acids Res*. 2011;39(1):359–372. https://doi.org/10.1093/nar/gkq704.
11. Mahfouz MM, Li L, Shamimuzzaman M, Wibowo A, Fang X, Zhu J-K. De novo-engineered transcription activator-like effector (TALE) hybrid nuclease with novel DNA binding specificity creates double-strand breaks. *Proc Natl Acad Sci U S A*. 2011;108(6):2623–2628. https://doi.org/10.1073/pnas.1019533108.
12. Gasiunas G, Barrangou R, Horvath P, Siksnys V. Cas9-crRNA ribonucleoprotein complex mediates specific DNA cleavage for adaptive immunity in bacteria. *Proc Natl Acad Sci U S A*. 2012;109(39):E2579–E2586. https://doi.org/10.1073/pnas.1208507109.
13. Jinek M, Chylinski K, Fonfara I, Hauer M, Doudna JA, Charpentier E. A programmable dual-RNA-guided DNA endonuclease in adaptive bacterial immunity. *Science*. 2012;337(6096):816–821. https://doi.org/10.1126/science.1225829.
14. Cong L, Ran FA, Cox D, et al. Multiplex genome engineering using CRISPR/Cas systems. *Science*. 2013;339(6121):819–823. https://doi.org/10.1126/science.1231143.
15. Anzalone AV, Koblan LW, Liu DR. Genome editing with CRISPR–Cas nucleases, base editors, transposases and prime editors. *Nat Biotechnol*. 2020;38(7):824–844. https://doi.org/10.1038/s41587-020-0561-9.
16. Van De Weghe JC, Rusterholz TDS, Latour B, et al. Mutations in ARMC9, which encodes a basal body protein, cause Joubert syndrome in humans and ciliopathy phenotypes in Zebrafish. *Am J Hum Genet*. 2017;101(1):23–36. https://doi.org/10.1016/j.ajhg.2017.05.010.

17. Frikstad K-AM, Molinari E, Thoresen M, et al. A CEP104-CSPP1 complex is required for formation of primary cilia competent in hedgehog signaling. *Cell Rep*. 2019;28 (7):1907–1922.e6. https://doi.org/10.1016/j.celrep.2019.07.025.
18. Latour BL, Van De Weghe JC, Rusterholz TDS, et al. Dysfunction of the ciliary ARMC9/TOGARAM1 protein module causes Joubert syndrome. *J Clin Invest*. 2020. https://doi.org/10.1172/JCI131656. Published online May 26.
19. Morbidoni V, Agolini E, Slep KC, et al. Biallelic mutations in the TOGARAM1 gene cause a novel primary ciliopathy. *J Med Genet*. 2020. https://doi.org/10.1136/jmedgenet-2020-106833. Published online August 3. jmedgenet-2020-106833.
20. Breslow DK, Hoogendoorn S, Kopp AR, et al. A CRISPR-based screen for Hedgehog signaling provides insights into ciliary function and ciliopathies. *Nat Genet*. 2018;50 (3):460–471. https://doi.org/10.1038/s41588-018-0054-7.
21. Taylor AW. Ocular immune privilege. *Eye*. 2009;23(10):1885–1889. https://doi.org/10.1038/eye.2008.382.
22. Oberbauer R, Schreiner GF, Meyer TW. Renal uptake of an 18-mer phosphorothioate oligonucleotide. *Kidney Int*. 1995;48(4):1226–1232. https://doi.org/10.1038/ki.1995.406.
23. Zhao Q, Zhou R, Temsamani J, Zhang Z, Roskey A, Agrawal S. Cellular distribution of phosphorothioate oligonucleotide following intravenous administration in mice. *Antisense Nucleic Acid Drug Dev*. 1998;8(6):451–458. https://doi.org/10.1089/oli.1.1998.8.451.
24. Amantana A, Iversen P. Pharmacokinetics and biodistribution of phosphorodiamidate morpholino antisense oligomers. *Curr Opin Pharmacol*. 2005;5(5):550–555. https://doi.org/10.1016/j.coph.2005.07.001.
25. Braun DA, Hildebrandt F. Ciliopathies. *Cold Spring Harb Perspect Biol*. 2017;9(3): a028191. https://doi.org/10.1101/cshperspect.a028191.
26. Forsyth R, Gunay-Aygun M. Bardet-Biedl syndrome overview. In: Adam MP, Ardinger HH, Pagon RA, et al., eds. *GeneReviews*®. Seattle: University of Washington; 1993. Accessed September 10, 2020 http://www.ncbi.nlm.nih.gov/books/NBK1363/.
27. Paisey RB, Steeds R, Barrett T, Williams D, Geberhiwot T, Gunay-Aygun M. Alström Syndrome. In: Adam MP, Ardinger HH, Pagon RA, et al., eds. *GeneReviews*®. Seattle: University of Washington; 1993. Accessed September 10, 2020 http://www.ncbi.nlm.nih.gov/books/NBK1267/.
28. Schmidts M. Clinical genetics and pathobiology of ciliary chondrodysplasias. *J Pediatr Genet*. 2015;03(02):049–064. https://doi.org/10.3233/PGE-14089.
29. Toriello HV, Franco B, Bruel A-L, Thauvin-Robinet C. Oral-facial-digital syndrome type I. In: Adam MP, Ardinger HH, Pagon RA, et al., eds. *GeneReviews*®. Seattle: University of Washington; 1993. Accessed September 10, 2020 http://www.ncbi.nlm.nih.gov/books/NBK1188/.
30. Franco B, Thauvin-Robinet C. Update on oral-facial-digital syndromes (OFDS). *Cilia*. 2016;5(1). https://doi.org/10.1186/s13630-016-0034-4.
31. Parisi M, Glass I. Joubert syndrome. In: Adam MP, Ardinger HH, Pagon RA, et al., eds. *GeneReviews*®. Seattle: University of Washington; 1993. Accessed September 10, 2020 http://www.ncbi.nlm.nih.gov/books/NBK1325/.
32. Hartill V, Szymanska K, Sharif SM, Wheway G, Johnson CA. Meckel-Gruber syndrome: an update on diagnosis, clinical management, and research advances. *Front Pediatr*. 2017;5:244. https://doi.org/10.3389/fped.2017.00244.
33. Kumaran N, Pennesi ME, Yang P, et al. Leber congenital amaurosis/early-onset severe retinal dystrophy overview. In: Adam MP, Ardinger HH, Pagon RA, et al., eds. *GeneReviews*®. Seattle: University of Washington; 1993. Accessed September 7, 2020 http://www.ncbi.nlm.nih.gov/books/NBK531510/.

34. Fahim AT, Daiger SP, Weleber RG. Nonsyndromic retinitis pigmentosa overview. In: Adam MP, Ardinger HH, Pagon RA, et al., eds. *GeneReviews®*. Seattle: University of Washington; 1993. Accessed September 7, 2020 http://www.ncbi.nlm.nih.gov/books/NBK1417/.
35. Harris PC, Torres VE. Polycystic kidney disease, autosomal dominant. In: Pagon RA, Adam MP, Ardinger HH, et al., eds. *GeneReviews(®)*. Seattle: University of Washington; 1993. http://www.ncbi.nlm.nih.gov/books/NBK1246/.
36. Sweeney WE, Avner ED. Polycystic kidney disease, autosomal recessive. In: Adam MP, Ardinger HH, Pagon RA, et al., eds. *GeneReviews®*. Seattle: University of Washington; 1993. Accessed April 24, 2020 http://www.ncbi.nlm.nih.gov/books/NBK1326/.
37. Stokman M, Lilien M, Knoers N. Nephronophthisis. In: Adam MP, Ardinger HH, Pagon RA, et al., eds. *GeneReviews®*. Seattle: University of Washington; 1993. Accessed September 7, 2020 http://www.ncbi.nlm.nih.gov/books/NBK368475/.
38. Srivastava S, Sayer JA. Nephronophthisis. *J Pediatr Genet*. 2014;3(2):103–114. https://doi.org/10.3233/PGE-14086.
39. Ronquillo CC, Bernstein PS, Baehr W. Senior–Løken syndrome: a syndromic form of retinal dystrophy associated with nephronophthisis. *Vision Res*. 2012;75:88–97. https://doi.org/10.1016/j.visres.2012.07.003.
40. Bachmann-Gagescu R, Dempsey JC, Phelps IG, et al. Joubert syndrome: a model for untangling recessive disorders with extreme genetic heterogeneity. *J Med Genet*. 2015;52(8):514–522. https://doi.org/10.1136/jmedgenet-2015-103087.
41. Zaki MS, Sattar S, Massoudi RA, Gleeson JG. Co-occurrence of distinct ciliopathy diseases in single families suggests genetic modifiers. *Am J Med Genet A*. 2011;155 (12):3042–3049. https://doi.org/10.1002/ajmg.a.34173.
42. Ramsbottom S, Miles C, Sayer J. Murine Cep290 phenotypes are modified by genetic backgrounds and provide an impetus for investigating disease modifier alleles. *F1000Res*. 2015;4:590. https://doi.org/10.12688/f1000research.6959.1.
43. Ramsbottom SA, Thelwall PE, Wood KM, et al. Mouse genetics reveals Barttin as a genetic modifier of Joubert syndrome. *Proc Natl Acad Sci U S A*. 2020;117 (2):1113–1118. https://doi.org/10.1073/pnas.1912602117.
44. Garanto A, van Beersum SEC, Peters TA, Roepman R, Cremers FPM, Collin RWJ. Unexpected CEP290 mRNA splicing in a humanized knock-in mouse model for leber congenital amaurosis. Li T, ed *PLoS One*. 2013;8(11), e79369. https://doi.org/10.1371/journal.pone.0079369.
45. Chen HY, Welby E, Li T, Swaroop A. Retinal disease in ciliopathies: Recent advances with a focus on stem cell-based therapies. Gunay-Aygun M, ed *Transl Sci Rare Dis*. 2019;4(1–2):97–115. https://doi.org/10.3233/TRD-190038.
46. Zhong X, Gutierrez C, Xue T, et al. Generation of three-dimensional retinal tissue with functional photoreceptors from human iPSCs. *Nat Commun*. 2014;5(1). https://doi.org/10.1038/ncomms5047.
47. Takasato M, Er PX, Chiu HS, et al. Kidney organoids from human iPS cells contain multiple lineages and model human nephrogenesis. *Nature*. 2015;526(7574):564–568. https://doi.org/10.1038/nature15695.
48. Morizane R, Lam AQ, Freedman BS, Kishi S, Valerius MT, Bonventre JV. Nephron organoids derived from human pluripotent stem cells model kidney development and injury. *Nat Biotechnol*. 2015;33(11):1193–1200. https://doi.org/10.1038/nbt.3392.
49. den Hollander AI, Koenekoop RK, Yzer S, et al. Mutations in the CEP290 (NPHP6) gene are a frequent cause of leber congenital amaurosis. *Am J Hum Genet*. 2006;79 (3):556–561. https://doi.org/10.1086/507318.
50. Sheck L, WIL D, Moradi P, et al. Leber congenital amaurosis associated with mutations in CEP290, clinical phenotype, and natural history in preparation for trials of novel therapies. *Ophthalmology*. 2018;125(6):894–903. https://doi.org/10.1016/j.ophtha.2017.12.013.

51. Pazour GJ, Quarmby L, Smith AO, Desai PB, Schmidts M. Cilia in cystic kidney and other diseases. *Cell Signal.* 2020;69:109519. https://doi.org/10.1016/j.cellsig.2019.109519.
52. Brouns SJJ, Jore MM, Lundgren M, et al. Small CRISPR RNAs guide antiviral defense in prokaryotes. *Science.* 2008;321(5891):960–964. https://doi.org/10.1126/science.1159689.
53. Hille F, Richter H, Wong SP, Bratovič M, Ressel S, Charpentier E. The biology of CRISPR-Cas: backward and forward. *Cell.* 2018;172(6):1239–1259. https://doi.org/10.1016/j.cell.2017.11.032.
54. Makarova KS, Wolf YI, Koonin EV. Classification and nomenclature of CRISPR-Cas systems: where from here? *CRISPR J.* 2018;1(5):325–336. https://doi.org/10.1089/crispr.2018.0033.
55. Ran FA, Cong L, Yan WX, et al. In vivo genome editing using Staphylococcus aureus Cas9. *Nature.* 2015;520(7546):186–191. https://doi.org/10.1038/nature14299.
56. Kleinstiver BP, Prew MS, Tsai SQ, et al. Engineered CRISPR-Cas9 nucleases with altered PAM specificities. *Nature.* 2015;523(7561):481–485. https://doi.org/10.1038/nature14592.
57. Hou Z, Zhang Y, Propson NE, et al. Efficient genome engineering in human pluripotent stem cells using Cas9 from Neisseria meningitidis. *Proc Natl Acad Sci U S A.* 2013;110(39):15644–15649. https://doi.org/10.1073/pnas.1313587110.
58. Esvelt KM, Mali P, Braff JL, Moosburner M, Yaung SJ, Church GM. Orthogonal Cas9 proteins for RNA-guided gene regulation and editing. *Nat Methods.* 2013;10(11):1116–1121. https://doi.org/10.1038/nmeth.2681.
59. Hirano H, Gootenberg JS, Horii T, et al. Structure and engineering of Francisella novicida Cas9. *Cell.* 2016;164(5):950–961. https://doi.org/10.1016/j.cell.2016.01.039.
60. Acharya S, Mishra A, Paul D, et al. *Francisella novicida* Cas9 interrogates genomic DNA with very high specificity and can be used for mammalian genome editing. *Proc Natl Acad Sci U S A.* 2019;116(42):20959–20968. https://doi.org/10.1073/pnas.1818461116.
61. Chatterjee P, Jakimo N, Jacobson JM. Minimal PAM specificity of a highly similar SpCas9 ortholog. *Sci Adv.* 2018;4(10), eaau0766. https://doi.org/10.1126/sciadv.aau0766.
62. Chatterjee P, Jakimo N, Lee J, et al. An engineered ScCas9 with broad PAM range and high specificity and activity. *Nat Biotechnol.* 2020;38(10):1154–1158. https://doi.org/10.1038/s41587-020-0517-0.
63. Sternberg SH, Redding S, Jinek M, Greene EC, Doudna JA. DNA interrogation by the CRISPR RNA-guided endonuclease Cas9. *Nature.* 2014;507(7490):62–67. https://doi.org/10.1038/nature13011.
64. Chylinski K, Le Rhun A, Charpentier E. The tracrRNA and Cas9 families of type II CRISPR-Cas immunity systems. *RNA Biol.* 2013;10(5):726–737. https://doi.org/10.4161/rna.24321.
65. Gasiunas G, Young JK, Karvelis T, et al. A catalogue of biochemically diverse CRISPR-Cas9 orthologs. *Nat Commun.* 2020;11(1):5512. https://doi.org/10.1038/s41467-020-19344-1.
66. Anders C, Bargsten K, Jinek M. Structural plasticity of PAM recognition by engineered variants of the RNA-guided endonuclease Cas9. *Mol Cell.* 2016;61(6):895–902. https://doi.org/10.1016/j.molcel.2016.02.020.
67. Legut M, Daniloski Z, Xue X, et al. High-throughput screens of PAM-flexible Cas9 variants for gene knockout and transcriptional modulation. *Cell Rep.* 2020;30(9):2859–2868.e5. https://doi.org/10.1016/j.celrep.2020.02.010.
68. Walton RT, Christie KA, Whittaker MN, Kleinstiver BP. Unconstrained genome targeting with near-PAMless engineered CRISPR-Cas9 variants. *Science.* 2020;368(6488):290–296. https://doi.org/10.1126/science.aba8853.

69. Szczelkun MD, Tikhomirova MS, Sinkunas T, et al. Direct observation of R-loop formation by single RNA-guided Cas9 and cascade effector complexes. *Proc Natl Acad Sci U S A*. 2014;111(27):9798–9803. https://doi.org/10.1073/pnas.1402597111.
70. Anders C, Niewoehner O, Duerst A, Jinek M. Structural basis of PAM-dependent target DNA recognition by the Cas9 endonuclease. *Nature*. 2014;513(7519):569–573. https://doi.org/10.1038/nature13579.
71. Chang HHY, Pannunzio NR, Adachi N, Lieber MR. Non-homologous DNA end joining and alternative pathways to double-strand break repair. *Nat Rev Mol Cell Biol*. 2017;18(8):495–506. https://doi.org/10.1038/nrm.2017.48.
72. Lisby M, Rothstein R. Cell biology of mitotic recombination. *Cold Spring Harb Perspect Biol*. 2015;7(3):a016535. https://doi.org/10.1101/cshperspect.a016535.
73. Xu X, Qi LS. A CRISPR–dCas toolbox for genetic engineering and synthetic biology. *J Mol Biol*. 2019;431(1):34–47. https://doi.org/10.1016/j.jmb.2018.06.037.
74. Amabile A, Migliara A, Capasso P, et al. Inheritable silencing of endogenous genes by hit-and-run targeted epigenetic editing. *Cell*. 2016;167(1):219–232.e14. https://doi.org/10.1016/j.cell.2016.09.006.
75. Vojta A, Dobrinić P, Tadić V, et al. Repurposing the CRISPR-Cas9 system for targeted DNA methylation. *Nucleic Acids Res*. 2016;44(12):5615–5628. https://doi.org/10.1093/nar/gkw159.
76. McDonald JI, Celik H, Rois LE, et al. Reprogrammable CRISPR/Cas9-based system for inducing site-specific DNA methylation. *Biol Open*. 2016;5(6):866–874. https://doi.org/10.1242/bio.019067.
77. Choudhury SR, Cui Y, Lubecka K, Stefanska B, Irudayaraj J. CRISPR-dCas9 mediated TET1 targeting for selective DNA demethylation at *BRCA1* promoter. *Oncotarget*. 2016;7(29):46545–46556. https://doi.org/10.18632/oncotarget.10234.
78. Morita S, Noguchi H, Horii T, et al. Targeted DNA demethylation in vivo using dCas9–peptide repeat and scFv–TET1 catalytic domain fusions. *Nat Biotechnol*. 2016;34(10):1060–1065. https://doi.org/10.1038/nbt.3658.
79. Xu X, Tao Y, Gao X, et al. A CRISPR-based approach for targeted DNA demethylation. *Cell Discov*. 2016;2(1). https://doi.org/10.1038/celldisc.2016.9.
80. Hilton IB, D'Ippolito AM, Vockley CM, et al. Epigenome editing by a CRISPR-Cas9-based acetyltransferase activates genes from promoters and enhancers. *Nat Biotechnol*. 2015;33(5):510–517. https://doi.org/10.1038/nbt.3199.
81. Kearns NA, Pham H, Tabak B, et al. Functional annotation of native enhancers with a Cas9–histone demethylase fusion. *Nat Methods*. 2015;12(5):401–403. https://doi.org/10.1038/nmeth.3325.
82. Cano-Rodriguez D, Gjaltema RAF, Jilderda LJ, et al. Writing of H3K4Me3 overcomes epigenetic silencing in a sustained but context-dependent manner. *Nat Commun*. 2016;7(1). https://doi.org/10.1038/ncomms12284.
83. Kim J-M, Kim K, Schmidt T, et al. Cooperation between SMYD3 and PC4 drives a distinct transcriptional program in cancer cells. *Nucleic Acids Res*. 2015;43(18):8868–8883. https://doi.org/10.1093/nar/gkv874.
84. Kwon DY, Zhao Y-T, Lamonica JM, Zhou Z. Locus-specific histone deacetylation using a synthetic CRISPR-Cas9-based HDAC. *Nat Commun*. 2017;8(1). https://doi.org/10.1038/ncomms15315.
85. Liu XS, Wu H, Krzisch M, et al. Rescue of fragile X syndrome neurons by DNA methylation editing of the FMR1 gene. *Cell*. 2018;172(5):979–992.e6. https://doi.org/10.1016/j.cell.2018.01.012.
86. Komor AC, Kim YB, Packer MS, Zuris JA, Liu DR. Programmable editing of a target base in genomic DNA without double-stranded DNA cleavage. *Nature*. 2016;533(7603):420–424. https://doi.org/10.1038/nature17946.

87. Gaudelli NM, Komor AC, Rees HA, et al. Programmable base editing of A•T to G•C in genomic DNA without DNA cleavage. *Nature.* 2017;551(7681):464–471. https://doi.org/10.1038/nature24644.
88. Yeh W-H, Chiang H, Rees HA, Edge ASB, Liu DR. In vivo base editing of post-mitotic sensory cells. *Nat Commun.* 2018;9(1). https://doi.org/10.1038/s41467-018-04580-3.
89. Hu JH, Miller SM, Geurts MH, et al. Evolved Cas9 variants with broad PAM compatibility and high DNA specificity. *Nature.* 2018;556(7699):57–63. https://doi.org/10.1038/nature26155.
90. Rees HA, Liu DR. Base editing: precision chemistry on the genome and transcriptome of living cells. *Nat Rev Genet.* 2018;19(12):770–788. https://doi.org/10.1038/s41576-018-0059-1.
91. Rees HA, Komor AC, Yeh W-H, et al. Improving the DNA specificity and applicability of base editing through protein engineering and protein delivery. *Nat Commun.* 2017;8(1). https://doi.org/10.1038/ncomms15790.
92. Anzalone AV, Randolph PB, Davis JR, et al. Search-and-replace genome editing without double-strand breaks or donor DNA. *Nature.* 2019;576(7785):149–157. https://doi.org/10.1038/s41586-019-1711-4.
93. Cradick TJ, Fine EJ, Antico CJ, Bao G. CRISPR/Cas9 systems targeting β-globin and CCR5 genes have substantial off-target activity. *Nucleic Acids Res.* 2013;41(20):9584–9592. https://doi.org/10.1093/nar/gkt714.
94. Fu Y, Foden JA, Khayter C, et al. High-frequency off-target mutagenesis induced by CRISPR-Cas nucleases in human cells. *Nat Biotechnol.* 2013;31(9):822–826. https://doi.org/10.1038/nbt.2623.
95. Liu G, Zhang Y, Zhang T. Computational approaches for effective CRISPR guide RNA design and evaluation. *Comput Struct Biotechnol J.* 2020;18:35–44. https://doi.org/10.1016/j.csbj.2019.11.006.
96. Slaymaker IM, Gao L, Zetsche B, Scott DA, Yan WX, Zhang F. Rationally engineered Cas9 nucleases with improved specificity. *Science.* 2016;351(6268):84–88. https://doi.org/10.1126/science.aad5227.
97. Kleinstiver BP, Pattanayak V, Prew MS, et al. High-fidelity CRISPR–Cas9 nucleases with no detectable genome-wide off-target effects. *Nature.* 2016;529(7587):490–495. https://doi.org/10.1038/nature16526.
98. Chen JS, Dagdas YS, Kleinstiver BP, et al. Enhanced proofreading governs CRISPR–Cas9 targeting accuracy. *Nature.* 2017;550(7676):407–410. https://doi.org/10.1038/nature24268.
99. Casini A, Olivieri M, Petris G, et al. A highly specific SpCas9 variant is identified by in vivo screening in yeast. *Nat Biotechnol.* 2018;36(3):265–271. https://doi.org/10.1038/nbt.4066.
100. Lee JK, Jeong E, Lee J, et al. Directed evolution of CRISPR-Cas9 to increase its specificity. *Nat Commun.* 2018;9(1). https://doi.org/10.1038/s41467-018-05477-x.
101. Vakulskas CA, Dever DP, Rettig GR, et al. A high-fidelity Cas9 mutant delivered as a ribonucleoprotein complex enables efficient gene editing in human hematopoietic stem and progenitor cells. *Nat Med.* 2018;24(8):1216–1224. https://doi.org/10.1038/s41591-018-0137-0.
102. Schmid-Burgk JL, Gao L, Li D, et al. Highly parallel profiling of Cas9 variant specificity. *Mol Cell.* 2020;78(4):794–800.e8. https://doi.org/10.1016/j.molcel.2020.02.023.
103. Richter F, Fonfara I, Gelfert R, Nack J, Charpentier E, Möglich A. Switchable Cas9. *Curr Opin Biotechnol.* 2017;48:119–126. https://doi.org/10.1016/j.copbio.2017.03.025.

104. Petris G, Casini A, Montagna C, et al. Hit and go CAS9 delivered through a lentiviral based self-limiting circuit. *Nat Commun.* 2017;8(1). https://doi.org/10.1038/ncomms15334.
105. Pawluk A, Amrani N, Zhang Y, et al. Naturally occurring off-switches for CRISPR-Cas9. *Cell.* 2016;167(7):1829–1838.e9. https://doi.org/10.1016/j.cell.2016.11.017.
106. Watters KE, Fellmann C, Bai HB, Ren SM, Doudna JA. Systematic discovery of natural CRISPR-Cas12a inhibitors. *Science.* 2018;362(6411):236–239. https://doi.org/10.1126/science.aau5138.
107. Jiang F, Liu J-J, Osuna BA, et al. Temperature-responsive competitive inhibition of CRISPR-Cas9. *Mol Cell.* 2019;73(3):601–610.e5. https://doi.org/10.1016/j.molcel.2018.11.016.
108. Uribe RV, van der Helm E, Misiakou M-A, Lee S-W, Kol S, Sommer MOA. Discovery and characterization of Cas9 inhibitors disseminated across seven bacterial phyla. *Cell Host Microbe.* 2019;25(2):233–241.e5. https://doi.org/10.1016/j.chom.2019.01.003.
109. Chew WL, Tabebordbar M, Cheng JKW, et al. A multifunctional AAV–CRISPR–Cas9 and its host response. *Nat Methods.* 2016;13(10):868–874. https://doi.org/10.1038/nmeth.3993.
110. Kim S, Koo T, Jee H-G, et al. CRISPR RNAs trigger innate immune responses in human cells. *Genome Res.* 2018;28(3):367–373. https://doi.org/10.1101/gr.231936.117.
111. Wagner DL, Amini L, Wendering DJ, et al. High prevalence of Streptococcus pyogenes Cas9-reactive T cells within the adult human population. *Nat Med.* 2019;25(2):242–248. https://doi.org/10.1038/s41591-018-0204-6.
112. Charlesworth CT, Deshpande PS, Dever DP, et al. Identification of preexisting adaptive immunity to Cas9 proteins in humans. *Nat Med.* 2019;25(2):249–254. https://doi.org/10.1038/s41591-018-0326-x.
113. Haapaniemi E, Botla S, Persson J, Schmierer B, Taipale J. CRISPR–Cas9 genome editing induces a p53-mediated DNA damage response. *Nat Med.* 2018;24(7):927–930. https://doi.org/10.1038/s41591-018-0049-z.
114. Ihry RJ, Worringer KA, Salick MR, et al. p53 inhibits CRISPR–Cas9 engineering in human pluripotent stem cells. *Nat Med.* 2018;24(7):939–946. https://doi.org/10.1038/s41591-018-0050-6.
115. Enache OM, Rendo V, Abdusamad M, et al. Cas9 activates the p53 pathway and selects for p53-inactivating mutations. *Nat Genet.* 2020;52(7):662–668. https://doi.org/10.1038/s41588-020-0623-4.
116. Ledford H. CRISPR treatment inserted directly into the body for first time. *Nature.* 2020;579(7798):185. https://doi.org/10.1038/d41586-020-00655-8.
117. Aghamiri S, Talaei S, Ghavidel AA, et al. Nanoparticles-mediated CRISPR/Cas9 delivery: recent advances in cancer treatment. *J Drug Deliv Sci Technol.* 2020;56: 101533. https://doi.org/10.1016/j.jddst.2020.101533.
118. Bonamassa B, Hai L, Liu D. Hydrodynamic gene delivery and its applications in pharmaceutical research. *Pharm Res.* 2011;28(4):694–701. https://doi.org/10.1007/s11095-010-0338-9.
119. Lino CA, Harper JC, Carney JP, Timlin JA. Delivering CRISPR: a review of the challenges and approaches. *Drug Deliv.* 2018;25(1):1234–1257. https://doi.org/10.1080/10717544.2018.1474964.
120. Yin H, Xue W, Chen S, et al. Genome editing with Cas9 in adult mice corrects a disease mutation and phenotype. *Nat Biotechnol.* 2014;32(6):551–553. https://doi.org/10.1038/nbt.2884.
121. Suda T, Gao X, Stolz DB, Liu D. Structural impact of hydrodynamic injection on mouse liver. *Gene Ther.* 2007;14(2):129–137. https://doi.org/10.1038/sj.gt.3302865.

122. Bessis N, GarciaCozar FJ, Boissier M-C. Immune responses to gene therapy vectors: influence on vector function and effector mechanisms. *Gene Ther.* 2004;11(S1):S10–S17. https://doi.org/10.1038/sj.gt.3302364.
123. Baum C, Kustikova O, Modlich U, Li Z, Fehse B. Mutagenesis and oncogenesis by chromosomal insertion of gene transfer vectors. *Hum Gene Ther.* 2006;17(3):253–263. https://doi.org/10.1089/hum.2006.17.253.
124. Deyle DR, Russell DW. Adeno-associated virus vector integration. *Curr Opin Mol Ther.* 2009;11(4):442–447.
125. van der Loo JCM, Wright JF. Progress and challenges in viral vector manufacturing. *Hum Mol Genet.* 2016;25(R1):R42–R52. https://doi.org/10.1093/hmg/ddv451.
126. Raper SE, Chirmule N, Lee FS, et al. Fatal systemic inflammatory response syndrome in a ornithine transcarbamylase deficient patient following adenoviral gene transfer. *Mol Genet Metab.* 2003;80(1–2):148–158. https://doi.org/10.1016/j.ymgme.2003.08.016.
127. Hacein-Bey-Abina S, Garrigue A, Wang GP, et al. Insertional oncogenesis in 4 patients after retrovirus-mediated gene therapy of SCID-X1. *J Clin Invest.* 2008;118(9):3132–3142. https://doi.org/10.1172/JCI35700.
128. Braun CJ, Boztug K, Paruzynski A, et al. Gene therapy for Wiskott-Aldrich syndrome—long-term efficacy and genotoxicity. *Sci Transl Med.* 2014;6(227):227ra33. https://doi.org/10.1126/scitranslmed.3007280.
129. Swiech L, Heidenreich M, Banerjee A, et al. In vivo interrogation of gene function in the mammalian brain using CRISPR-Cas9. *Nat Biotechnol.* 2015;33(1):102–106. https://doi.org/10.1038/nbt.3055.
130. Hung SSC, Chrysostomou V, Li F, et al. AAV-mediated CRISPR/Cas gene editing of retinal cells in vivo. *Investig Opthalmol Vis Sci.* 2016;57(7):3470. https://doi.org/10.1167/iovs.16-19316.
131. Truong D-JJ, Kühner K, Kühn R, et al. Development of an intein-mediated split–Cas9 system for gene therapy. *Nucleic Acids Res.* 2015;43(13):6450–6458. https://doi.org/10.1093/nar/gkv601.
132. Fine EJ, Appleton CM, White DE, et al. Trans-spliced Cas9 allows cleavage of HBB and CCR5 genes in human cells using compact expression cassettes. *Sci Rep.* 2015;5(1). https://doi.org/10.1038/srep10777.
133. Nishimasu H, Ran FA, Hsu PD, et al. Crystal structure of Cas9 in complex with guide RNA and target DNA. *Cell.* 2014;156(5):935–949. https://doi.org/10.1016/j.cell.2014.02.001.
134. Kim E, Koo T, Park SW, et al. In vivo genome editing with a small Cas9 orthologue derived from Campylobacter jejuni. *Nat Commun.* 2017;8(1):14500. https://doi.org/10.1038/ncomms14500.
135. Zetsche B, Gootenberg JS, Abudayyeh OO, et al. Cpf1 is a single RNA-guided endonuclease of a class 2 CRISPR-Cas system. *Cell.* 2015;163(3):759–771. https://doi.org/10.1016/j.cell.2015.09.038.
136. Gao J, Mese K, Bunz O, Ehrhardt A. State-of-the-art human adenovirus vectorology for therapeutic approaches. *FEBS Lett.* 2019;593(24):3609–3622. https://doi.org/10.1002/1873-3468.13691.
137. Milone MC, O'Doherty U. Clinical use of lentiviral vectors. *Leukemia.* 2018;32(7):1529–1541. https://doi.org/10.1038/s41375-018-0106-0.
138. Wang D, Tai PWL, Gao G. Adeno-associated virus vector as a platform for gene therapy delivery. *Nat Rev Drug Discov.* 2019;18(5):358–378. https://doi.org/10.1038/s41573-019-0012-9.
139. Montagna C, Petris G, Casini A, et al. VSV-G-enveloped vesicles for traceless delivery of CRISPR-Cas9. *Mol Ther–Nucleic Acids.* 2018;12:453–462. https://doi.org/10.1016/j.omtn.2018.05.010.

140. Bainbridge JWB, Smith AJ, Barker SS, et al. Effect of gene therapy on visual function in Leber's congenital amaurosis. *N Engl J Med*. 2008;358(21):2231–2239. https://doi.org/10.1056/NEJMoa0802268.
141. Hauswirth WW, Aleman TS, Kaushal S, et al. Treatment of Leber congenital amaurosis due to *RPE65* mutations by ocular subretinal injection of adeno-associated virus gene vector: short-term results of a phase I trial. *Hum Gene Ther*. 2008;19(10):979–990. https://doi.org/10.1089/hum.2008.107.
142. Maguire AM, Simonelli F, Pierce EA, et al. Safety and efficacy of gene transfer for Leber's congenital amaurosis. *N Engl J Med*. 2008;358(21):2240–2248. https://doi.org/10.1056/NEJMoa0802315.
143. Dalkara D, Kolstad KD, Caporale N, et al. Inner limiting membrane barriers to AAV-mediated retinal transduction from the vitreous. *Mol Ther*. 2009;17(12):2096–2102. https://doi.org/10.1038/mt.2009.181.
144. Yin L, Greenberg K, Hunter JJ, et al. Intravitreal injection of AAV2 transduces macaque inner retina. *Investig Opthalmol Vis Sci*. 2011;52(5):2775. https://doi.org/10.1167/iovs.10-6250.
145. Katada Y, Kobayashi K, Tsubota K, Kurihara T. Evaluation of AAV-DJ vector for retinal gene therapy. *PeerJ*. 2019;7, e6317. https://doi.org/10.7717/peerj.6317.
146. Khani SC, Pawlyk BS, Bulgakov OV, et al. AAV-mediated expression targeting of rod and cone photoreceptors with a human rhodopsin kinase promoter. *Investig Opthalmol Vis Sci*. 2007;48(9):3954. https://doi.org/10.1167/iovs.07-0257.
147. Sun X, Pawlyk B, Xu X, et al. Gene therapy with a promoter targeting both rods and cones rescues retinal degeneration caused by AIPL1 mutations. *Gene Ther*. 2010;17(1):117–131. https://doi.org/10.1038/gt.2009.104.
148. Davis L, Park F. Gene therapy research for kidney diseases. *Physiol Genomics*. 2019;51(9):449–461. https://doi.org/10.1152/physiolgenomics.00052.2019.
149. Rubin JD, Nguyen TV, Allen KL, Ayasoufi K, Barry MA. Comparison of gene delivery to the kidney by adenovirus, adeno-associated virus, and lentiviral vectors after intravenous and direct kidney injections. *Hum Gene Ther*. 2019;30(12):1559–1571. https://doi.org/10.1089/hum.2019.127.
150. Chung DC, Fogelgren B, Park KM, et al. Adeno-associated virus-mediated gene transfer to renal tubule cells via a retrograde ureteral approach. *Nephron Extra*. 2011;1(1):217–223. https://doi.org/10.1159/000333071.
151. Ortiz PA, Hong NJ, Plato CF, Varela M, Garvin JL. An in vivo method for adenovirus-mediated transduction of thick ascending limbs. *Kidney Int*. 2003;63(3):1141–1149. https://doi.org/10.1046/j.1523-1755.2003.00827.x.
152. Ortiz PA, Hong NJ, Wang D, Garvin JL. Gene transfer of eNOS to the thick ascending limb of eNOS-KO mice restores the effects of L-arginine on NaCl absorption. *Hypertension*. 2003;42(4):674–679. https://doi.org/10.1161/01.HYP.0000085561.00001.81.
153. Thomson RB, Igarashi P, Biemesderfer D, et al. Isolation and cDNA cloning of Ksp-cadherin, a novel kidney-specific member of the cadherin multigene family. *J Biol Chem*. 1995;270(29):17594–17601. https://doi.org/10.1074/jbc.270.29.17594.
154. Asico LD, Cuevas S, Ma X, Jose PA, Armando I, Konkalmatt PR. Nephron segment-specific gene expression using AAV vectors. *Biochem Biophys Res Commun*. 2018;497(1):19–24. https://doi.org/10.1016/j.bbrc.2018.01.169.
155. Chen S. Gene delivery in renal tubular epithelial cells using recombinant adeno-associated viral vectors. *J Am Soc Nephrol*. 2003;14(4):947–958. https://doi.org/10.1097/01.ASN.0000057858.45649.F5.
156. Quemener AM, Bachelot L, Forestier A, Donnou-Fournet E, Gilot D, Galibert M. The powerful world of antisense oligonucleotides: From bench to bedside. *WIREs RNA*. 2020;11(5). https://doi.org/10.1002/wrna.1594.

157. Crooke ST, Witztum JL, Bennett CF, Baker BF. RNA-targeted therapeutics. *Cell Metab.* 2018;27(4):714–739. https://doi.org/10.1016/j.cmet.2018.03.004.
158. Roberts TC, Langer R, Wood MJA. Advances in oligonucleotide drug delivery. *Nat Rev Drug Discov.* 2020;19(10):673–694. https://doi.org/10.1038/s41573-020-0075-7.
159. Dulla K, Aguila M, Lane A, et al. Splice-modulating oligonucleotide QR-110 restores CEP290 mRNA and function in human c.2991+1655A>G LCA10 models. *Mol Ther–Nucleic Acids.* 2018;12:730–740. https://doi.org/10.1016/j.omtn.2018.07.010.
160. Juliano RL. The delivery of therapeutic oligonucleotides. *Nucleic Acids Res.* 2016;44(14):6518–6548. https://doi.org/10.1093/nar/gkw236.
161. Lee EC, Valencia T, Allerson C, et al. Discovery and preclinical evaluation of anti-miR-17 oligonucleotide RGLS4326 for the treatment of polycystic kidney disease. *Nat Commun.* 2019;10(1). https://doi.org/10.1038/s41467-019-11918-y.
162. Ramsbottom SA, Molinari E, Srivastava S, et al. Targeted exon skipping of a CEP290 mutation rescues Joubert syndrome phenotypes in vitro and in a murine model. *Proc Natl Acad Sci U S A.* 2018;115(49):12489–12494. https://doi.org/10.1073/pnas.1809432115.
163. Aslesh T, Yokota T. Development of antisense oligonucleotide gapmers for the treatment of Huntington's disease. In: Yokota T, Maruyama R, eds. *Gapmers.* Springer US; 2020:57–67. Methods in Molecular Biology; vol 2176. https://doi.org/10.1007/978-1-0716-0771-8_4.
164. Liang X-H, Sun H, Shen W, et al. Antisense oligonucleotides targeting translation inhibitory elements in 5' UTRs can selectively increase protein levels. *Nucleic Acids Res.* 2017;45(16):9528–9546. https://doi.org/10.1093/nar/gkx632.
165. Sasaki S, Sun R, Bui H-H, Crosby JR, Monia BP, Guo S. Steric Inhibition of 5' UTR regulatory elements results in upregulation of human CFTR. *Mol Ther.* 2019;27(10):1749–1757. https://doi.org/10.1016/j.ymthe.2019.06.016.
166. Aartsma-Rus A, Ferlini A, Goemans N, et al. Translational and regulatory challenges for exon skipping therapies. *Hum Gene Ther.* 2014;25(10):885–892. https://doi.org/10.1089/hum.2014.086.
167. Lim KH, Han Z, Jeon HY, et al. Antisense oligonucleotide modulation of non-productive alternative splicing upregulates gene expression. *Nat Commun.* 2020;11(1):3501. https://doi.org/10.1038/s41467-020-17093-9.
168. Hua Y, Vickers TA, Baker BF, Bennett CF, Krainer AR. Enhancement of SMN2 exon 7 inclusion by antisense oligonucleotides targeting the exon. Misteli T, ed *PLoS Biol.* 2007;5(4):e73. https://doi.org/10.1371/journal.pbio.0050073.
169. DiCarlo JE, Mahajan VB, Tsang SH. Gene therapy and genome surgery in the retina. *J Clin Invest.* 2018;128(6):2177–2188. https://doi.org/10.1172/JCI120429.
170. Deng W-L, Gao M-L, Lei X-L, et al. Gene correction reverses ciliopathy and photoreceptor loss in iPSC-derived retinal organoids from retinitis pigmentosa patients. *Stem Cell Rep.* 2018;10(4):1267–1281. https://doi.org/10.1016/j.stemcr.2018.02.003.
171. Buskin A, Zhu L, Chichagova V, et al. Disrupted alternative splicing for genes implicated in splicing and ciliogenesis causes PRPF31 retinitis pigmentosa. *Nat Commun.* 2018;9(1). https://doi.org/10.1038/s41467-018-06448-y.
172. Bakondi B. *In vivo* versus *ex vivo* CRISPR therapies for retinal dystrophy. *Expert Rev Ophthalmol.* 2016;11(6):397–400. https://doi.org/10.1080/17469899.2016.1251316.
173. Mears AJ, Kondo M, Swain PK, et al. Nrl is required for rod photoreceptor development. *Nat Genet.* 2001;29(4):447–452. https://doi.org/10.1038/ng774.
174. Daniele LL, Lillo C, Lyubarsky AL, et al. Cone-like morphological, molecular, and electrophysiological features of the photoreceptors of the *Nrl* knockout mouse. *Investig Opthalmol Vis Sci.* 2005;46(6):2156. https://doi.org/10.1167/iovs.04-1427.
175. Montana CL, Kolesnikov AV, Shen SQ, Myers CA, Kefalov VJ, Corbo JC. Reprogramming of adult rod photoreceptors prevents retinal degeneration. *Proc Natl Acad Sci U S A.* 2013;110(5):1732–1737. https://doi.org/10.1073/pnas.1214387110.

176. Yu W, Mookherjee S, Chaitankar V, et al. Nrl knockdown by AAV-delivered CRISPR/Cas9 prevents retinal degeneration in mice. *Nat Commun.* 2017;8 (1). https://doi.org/10.1038/ncomms14716.
177. Bessant DAR, Payne AM, Mitton KP, et al. A mutation in NRL is associated with autosomal dominant retinitis pigmentosa. *Nat Genet.* 1999;21(4):355–356. https://doi.org/10.1038/7678.
178. Nishiguchi KM, Friedman JS, Sandberg MA, Swaroop A, Berson EL, Dryja TP. Recessive NRL mutations in patients with clumped pigmentary retinal degeneration and relative preservation of blue cone function. *Proc Natl Acad Sci U S A.* 2004;101 (51):17819–17824. https://doi.org/10.1073/pnas.0408183101.
179. Anand M, Khanna H. Ciliary transition zone (TZ) proteins RPGR and CEP290: role in photoreceptor cilia and degenerative diseases. *Expert Opin Ther Targets.* 2012;16 (6):541–551. https://doi.org/10.1517/14728222.2012.680956.
180. Ruan G-X, Barry E, Yu D, Lukason M, Cheng SH, Scaria A. CRISPR/Cas9-mediated genome editing as a therapeutic approach for leber congenital amaurosis 10. *Mol Ther.* 2017;25(2):331–341. https://doi.org/10.1016/j.ymthe.2016.12.006.
181. Maeder ML, Stefanidakis M, Wilson CJ, et al. Development of a gene-editing approach to restore vision loss in Leber congenital amaurosis type 10. *Nat Med.* 2019;25 (2):229–233. https://doi.org/10.1038/s41591-018-0327-9.
182. Low JH, Li P, Chew EGY, et al. Generation of human PSC-derived kidney organoids with patterned nephron segments and a de novo vascular network. *Cell Stem Cell.* 2019;25(3):373–387.e9. https://doi.org/10.1016/j.stem.2019.06.009.
183. Forbes TA, Howden SE, Lawlor K, et al. Patient-iPSC-derived kidney organoids show functional validation of a ciliopathic renal phenotype and reveal underlying pathogenetic mechanisms. *Am J Hum Genet.* 2018;102(5):816–831. https://doi.org/10.1016/j.ajhg.2018.03.014.
184. Estrada JL, Martens G, Li P, et al. Evaluation of human and non-human primate antibody binding to pig cells lacking GGTA1/CMAH/β4GalNT2 genes. *Xenotransplantation.* 2015;22(3):194–202. https://doi.org/10.1111/xen.12161.
185. Niu D, Wei H-J, Lin L, et al. Inactivation of porcine endogenous retrovirus in pigs using CRISPR-Cas9. *Science.* 2017;357(6357):1303–1307. https://doi.org/10.1126/science.aan4187.
186. Okano M, Xie S, Li E. Cloning and characterization of a family of novel mammalian DNA (cytosine-5) methyltransferases. *Nat Genet.* 1998;19(3):219–220. https://doi.org/10.1038/890.
187. Illingworth RS, Bird AP. CpG islands—'A rough guide.'. *FEBS Lett.* 2009;583 (11):1713–1720. https://doi.org/10.1016/j.febslet.2009.04.012.
188. Bird AP, Wolffe AP. Methylation-induced repression—belts, braces, and chromatin. *Cell.* 1999;99(5):451–454. https://doi.org/10.1016/S0092-8674(00)81532-9.
189. Aran D, Toperoff G, Rosenberg M, Hellman A. Replication timing-related and gene body-specific methylation of active human genes. *Hum Mol Genet.* 2011;20 (4):670–680. https://doi.org/10.1093/hmg/ddq513.
190. Rose NR, Klose RJ. Understanding the relationship between DNA methylation and histone lysine methylation. *Biochim Biophys Acta Gene Regul Mech.* 2014;1839 (12):1362–1372. https://doi.org/10.1016/j.bbagrm.2014.02.007.
191. Kouzarides T. Chromatin modifications and their function. *Cell.* 2007;128 (4):693–705. https://doi.org/10.1016/j.cell.2007.02.005.
192. Haberland M, Montgomery RL, Olson EN. The many roles of histone deacetylases in development and physiology: implications for disease and therapy. *Nat Rev Genet.* 2009;10(1):32–42. https://doi.org/10.1038/nrg2485.
193. Moosavi A, Motevalizadeh AA. Role of epigenetics in biology and human diseases. *Iran Biomed J.* 2016;(5):246–258. https://doi.org/10.22045/ibj.2016.01.

194. Ganesan A, Arimondo PB, Rots MG, Jeronimo C, Berdasco M. The timeline of epigenetic drug discovery: from reality to dreams. *Clin Epigenetics*. 2019;11(1). https://doi.org/10.1186/s13148-019-0776-0.
195. Farinelli P, Perera A, Arango-Gonzalez B, et al. DNA methylation and differential gene regulation in photoreceptor cell death. *Cell Death Dis*. 2014;5(12):e1558. https://doi.org/10.1038/cddis.2014.512.
196. Zheng S, Xiao L, Liu Y, et al. DZNep inhibits H3K27me3 deposition and delays retinal degeneration in the rd1 mice. *Cell Death Dis*. 2018;9(3). https://doi.org/10.1038/s41419-018-0349-8.
197. Sancho-Pelluz J, Alavi MV, Sahaboglu A, et al. Excessive HDAC activation is critical for neurodegeneration in the rd1 mouse. *Cell Death Dis*. 2010;1(2):e24. https://doi.org/10.1038/cddis.2010.4.
198. Chen B, Cepko CL. HDAC4 regulates neuronal survival in normal and diseased retinas. *Science*. 2009;323(5911):256–259. https://doi.org/10.1126/science.1166226.
199. Guo X, Wang S-B, Xu H, et al. A short N-terminal domain of HDAC4 preserves photoreceptors and restores visual function in retinitis pigmentosa. *Nat Commun*. 2015;6(1). https://doi.org/10.1038/ncomms9005.
200. Woo YM, Shin Y, Hwang J-A, et al. Epigenetic silencing of the MUPCDH gene as a possible prognostic biomarker for cyst growth in ADPKD. *Sci Rep*. 2015;5(1). https://doi.org/10.1038/srep15238.
201. Woo YM, Bae J-B, Oh Y-H, et al. Genome-wide methylation profiling of ADPKD identified epigenetically regulated genes associated with renal cyst development. *Hum Genet*. 2014;133(3):281–297. https://doi.org/10.1007/s00439-013-1378-0.
202. Bowden SA, Rodger EJ, Bates M, Chatterjee A, Eccles MR, Stayner C. Genome-scale single nucleotide resolution analysis of DNA methylation in human autosomal dominant polycystic kidney disease. *Am J Nephrol*. 2018;48(6):415–424. https://doi.org/10.1159/000494739.
203. Bowden SA, Stockwell PA, Rodger EJ, et al. Extensive inter-cyst DNA methylation variation in autosomal dominant polycystic kidney disease revealed by genome scale sequencing. *Front Genet*. 2020;11. https://doi.org/10.3389/fgene.2020.00348.
204. Cao Y, Semanchik N, Lee SH, et al. Chemical modifier screen identifies HDAC inhibitors as suppressors of PKD models. *Proc Natl Acad Sci U S A*. 2009;106(51):21819–21824. https://doi.org/10.1073/pnas.0911987106.
205. Xia S, Li X, Johnson T, Seidel C, Wallace DP, Li R. Polycystin-dependent fluid flow sensing targets histone deacetylase 5 to prevent the development of renal cysts. *Development*. 2010;137(7):1075–1084. https://doi.org/10.1242/dev.049437.
206. Fan LX, Li X, Magenheimer B, Calvet JP, Li X. Inhibition of histone deacetylases targets the transcription regulator Id2 to attenuate cystic epithelial cell proliferation. *Kidney Int*. 2012;81(1):76–85. https://doi.org/10.1038/ki.2011.296.
207. Zhou X, Fan LX, Sweeney WE, Denu JM, Avner ED, Li X. Sirtuin 1 inhibition delays cyst formation in autosomal-dominant polycystic kidney disease. *J Clin Invest*. 2013;123(7):3084–3098. https://doi.org/10.1172/JCI64401.
208. Cebotaru L, Liu Q, Yanda MK, et al. Inhibition of histone deacetylase 6 activity reduces cyst growth in polycystic kidney disease. *Kidney Int*. 2016;90(1):90–99. https://doi.org/10.1016/j.kint.2016.01.026.
209. Wheway G, Genomics England Research Consortium, Mitchison HM. Opportunities and challenges for molecular understanding of ciliopathies—the 100,000 genomes project. *Front Genet*. 2019;10. https://doi.org/10.3389/fgene.2019.00127.
210. Datta P, Ruffcorn A, Seo S. Limited time window for retinal gene therapy in a preclinical model of ciliopathy. *Hum Mol Genet*. 2020;29(14):2337–2352. https://doi.org/10.1093/hmg/ddaa124.

211. Lanphier E, Urnov F, Haecker SE, Werner M, Smolenski J. Don't edit the human germ line. *Nature*. 2015;519(7544):410–411. https://doi.org/10.1038/519410a.
212. Doudna JA. The promise and challenge of therapeutic genome editing. *Nature*. 2020;578(7794):229–236. https://doi.org/10.1038/s41586-020-1978-5.
213. Knowles L, Luth W, Bubela T. Paving the road to personalized medicine: recommendations on regulatory, intellectual property and reimbursement challenges. *J Law Biosci*. 2017;4(3):453–506. https://doi.org/10.1093/jlb/lsx030.

CHAPTER TWELVE

Genome editing in stem cells for genetic neurodisorders

Claudia Dell' Amico[a,†], Alice Tata[c,†], Enrica Pellegrino[a,b], Marco Onorati[a,*], and Luciano Conti[c,*]

[a]Unit of Cell and Developmental Biology, Department of Biology, University of Pisa, Pisa, Italy
[b]Host-Pathogen Interactions in Tuberculosis Laboratory, The Francis Crick Institute, London, United Kingdom
[c]Department of Cellular, Computational and Integrative Biology—CIBIO, University of Trento, Trento, Italy
*Corresponding authors: e-mail address: marco.onorati@unipi.it; luciano.conti@unitn.it

Contents

1. Introduction: The basics of genome editing technologies — 405
   1.1 Programmable nucleases: ZFNs, TALENs, and CRISPR/Cas9 technology — 406
2. Pluripotent stem cells and neural stem cells: An overview — 408
   2.1 Modeling human neurodevelopment with pluripotent stem cells — 409
   2.2 *In vitro* neural stem cell systems — 412
   2.3 New frontiers in disease modeling and cell replacement — 414
3. Genome editing for neurodevelopmental disorders — 416
   3.1 Exploring neurodevelopmental programs with PSCs and NSCs — 416
   3.2 *In vitro* modeling of neurodevelopmental disorders — 418
4. Genome editing of neurodegenerative disorders — 423
   4.1 Dissecting neurodegenerative processes with human PSCs and NSCs — 423
   4.2 *In vitro* modeling of neurodegenerative diseases — 424
5. Future directions and concluding remarks — 429
Acknowledgments — 430
References — 431

Abstract

The recent advent of genome editing techniques and their rapid improvement paved the way in establishing innovative human neurological disease models and in developing new therapeutic opportunities. Human pluripotent (both induced or naive) stem cells and neural stem cells represent versatile tools to be applied to multiple research needs and, together with genomic snip and fix tools, have recently made possible the creation of unique platforms to directly investigate several human neural affections. In this chapter, we will discuss genome engineering tools, and their recent improvements, applied to the stem cell field, focusing on how these two technologies may be pivotal

[†] Co-first authors.

instruments to deeply unravel molecular mechanisms underlying development and function, as well as disorders, of the human brain. We will review how these frontier technologies may be exploited to investigate or treat severe neurodevelopmental disorders, such as microcephaly, autism spectrum disorder, schizophrenia, as well as neurodegenerative conditions, including Parkinson's disease, Huntington's disease, Alzheimer's disease, and spinal muscular atrophy.

Abbreviations

AD	Alzheimer's disease
aGSK-3β	active glycogen synthase kinase-3β
APP	amyloid precursor protein
ASD	autism spectrum disorder
ASO	antisense oligonucleotide
Aβ	amyloid beta
Cas	CRISPR-associated
Chr	chromosome
CNS	central nervous system
CRISPR	clustered regularly interspaced short palindromic repeats
crRNA	CRISPR-RNA
DA	dopaminergic
DSB	double-stranded breaks
ESCs	embryonic stem cells
fAD	familial Alzheimer's disease
FDA	Food and Drug Administration
GABA	gamma-aminobutyric acid
GWAS	genome wide-association studies
HD	Huntington's disease
HDR	homology-directed repair
hiPSCs	human induced pluripotent stem cells
HTT	Huntingtin
IGF1	insulin-like growth factor 1
InDels	insertions/deletions
IPC	intermediate progenitor cell
iPSCs	induced pluripotent stem cells
iSVZ	inner subventricular zone
MCPH	microcephaly primary hereditary
mDA	midbrain dopaminergic
MN	motor neuron
NCX	neocortex
NES cells	neuroepithelial stem cells
NHEJ	non-homologous end-joining
NSCs	neural stem cells
oRGCs	outer radial glia cells
oSVZ	outer subventricular zone
pcSN	*Substantia Nigra Pars Compacta.*
PD	Parkinson's disease
PSCs	pluripotent stem cells

PSEN	presenilin
RGCs	radial glia cells
RGENs	RNA-guided engineered nucleases
RNAi	RNA interference
sAD	sporadic Alzheimer's disease
scRNA-SEQ	single-cell RNA-sequencing
SCZ	schizophrenia
SMA	spinal muscular atrophy
SMN	survival motor neuron
TALENs	transcription activator like effector nucleases
TFs	transcription factors
tracrRNA	trans-activating crispr RNA
vRGCs	ventricular radial glia cells
VZ	ventricular zone
ZFNs	zinc-finger nucleases

1. Introduction: The basics of genome editing technologies

Genome engineering encompasses the broad palette of recombination technologies, to which programmable nucleases are adding a new dimension. Classical genome engineering involves the introduction of large modifications, both through homologous or random recombination, for gene targeting or transgenesis. In contrast, the term "genome editing" emerged to specify the use of programmable nucleases, which have been almost exclusively linked to the introduction of small changes in the genome, *i.e.*, "edits."

Programmable nucleases, including meganucleases, zinc-finger nucleases (ZFNs), transcription activator like effector nucleases (TALENs) and RNA-guided engineered nucleases (RGENs), generate double-stranded breaks (DSB) in specific DNA sequences. Taking advantage of natural cellular pathways of DNA repair, programmable nucleases are largely used for the introduction of targeted sequence changes.

Initially demonstrated with meganucleases,[1] the introduction of a DSB at a genomic site can be repaired *via* a variety of cellular DNA repair mechanisms,[2] broadly subdivided into non-homologous end-joining (NHEJ) and homology-directed repair (HDR).

The most frequent outcome of DSB repair is the generation of small insertions/deletions (InDels) through the error prone NHEJ pathway. While this process is not effective at perfectly repairing the targeted or cut DNA, it offers a highly efficient system to disrupt genes. This method is referred to as "non-homologous" since the broken ends are directly ligated without the need for a homologous template, and instead exploits short homologous DNA sequences termed microhomologies to guide the repairing process after end resection.

On the other hand, in the presence of an exogenous DNA repair template, containing sequences homologous to the region close to the DSB, it is possible to introduce precise genetic changes by making use of other, although less efficient, HDR pathways.

The HDR mechanisms, in contrast to the NHEJ pathway, can only occur during the late S and G_2 phases[3] when sister chromatids are more easily available as a DNA template, thus allowing for a more faithful duplication of the genome by providing critical support for DNA replication and telomere maintenance.

1.1 Programmable nucleases: ZFNs, TALENs, and CRISPR/Cas9 technology

1.1.1 Zinc finger nucleases (ZFNs)

The first targetable nucleases are represented by ZFNs. ZFNs were created in 1996[4] by fusion of zinc-finger DNA-binding modules derived from natural transcription factors (TFs) and a nuclease domain of the Type IIS restriction enzyme, *FokI*. Each zinc-finger domain can recognize a trinucleotide DNA sequence and multiple fingers combined together can be used to bind longer DNA sequences, providing desired on-target specificity.[5,6] Because the nuclease domain must dimerize to cut DNA, the system uses two ZFNs engineered to recognize different closely spaced nucleotide sequences within the target site. Simultaneous recognition and binding of both ZFNs on the opposite DNA strands[7] allow for the activation of *FokI* cleavage domain and the creation of a DSB. Nonetheless, the major drawback of this system is that zinc-finger motifs aligned in an array influence the specificity of neighboring zinc-fingers, making the design and selection of modified zinc-finger arrays quite time-consuming and costly. Also, it can result in more erratic specificity of the final arrangement.

1.1.2 Transcription activator-like effector nucleases (TALENs)

The second site-specific nucleases are the TALENs. Similar to ZFNs, TALENs employ DNA-binding modules from bacterial TALEs linked to the same *FokI* cleavage domain.[8] In nature, these TALEs, encoded by *Xanthomonas* spp. proteobacteria, are injected into host plant cells *via* a type III secretion system and bind to genomic DNA to alter transcription in the host cells, thereby facilitating pathogenic bacterial colonization.[9] In the case of TALENs, a single TALE motif composed of 33–35 conserved amino acid repeated motifs recognizes only one nucleotide, while an array of TALEs can associate with a longer sequence (typically up to 18 bp, the maximum length for ZFN recognition). This does not affect the binding specificity of neighboring TALEs, making the engineering of TALENs definitively much easier than ZFNs.

1.1.3 Clustered regularly interspaced short palindromic repeats (CRISPR)/Cas9

In the last decade, the discovery of the CRISPR/CRISPR-associated (Cas) microbial adaptive immune system has completely revolutionized the genome engineering field, emerging as an easy and effective alternative to induce targeted genetic modifications.[10]

CRISPR-associated enzymes, such as Cas9 and Cas12a (previously termed Cpf1[11]) are RGENs that can precisely target any region of the genome *via* the complementarity of the RNA-guide to the DNA sequence of interest.

More than 20 years of research have given life to the most powerful genome editing technique and its use has been reported in many organisms—including yeast, nematode, fruit-fly, zebrafish, mouse, monkey and human.[12] To date, CRISPR/Cas9 represents the most accurate, efficient, and cost-effective genome-editing tool. The system is comprised of two major elements[10]:

- Single-guide RNA, *i.e.*, an artificial molecule combined by the linkage between the CRISPR-RNA (crRNA) and the trans-activating crispr RNA (tracrRNA) at the 5′ end and 3′ end, respectively. In nature, crRNAs are molecules that are 42 nt in length and consist of a guide sequence of 20-bases complementary to the target DNA. The tracrRNA is a 75 nt length molecule that anneals with crRNA forming a dual crRNA:tracrRNA complex. Moreover, the tracrRNA enables activity of the crRNA favoring crRNA maturation, stem-loop formation and binding to the Cas9 protein.
- Cas9 nuclease that binds to the crRNA:tracrRNA duplex or engineered single-guide RNA and that acts by cleaving the DNA. Cas9 has two nuclease domains: one HNH-like and one RuvC-like, each of which cleaves one strand of the target. Cas9 cleaves DNA at the site complementary to the 20 bp guide sequences of crRNA, which is three nucleotides upstream of the protospacer adjacent (PAM) sequence of the

target DNA. PAM is a short sequence in the target DNA that lies adjacent to the 3′ end of the 20 nt guide RNA sequence and it is essential in target recognition by CRISPR/Cas9. For *S. pyogenes* Cas9, the most used type II CRISPR/Cas9 system, the PAM is 5′-NGG-3′, where N stands for any nucleotides.

2. Pluripotent stem cells and neural stem cells: An overview

Since the pioneering Evans and Martin's breakthrough discoveries in 1981,[13,14] several scientific discoveries have fueled our knowledge about pluripotent stem cells (PSCs), helping us to harness their incredible potential for research and therapeutic applications.[15] These progressive advancements have opened up the possibility to effectively derive embryonic stem cells (ESCs) and create undifferentiated cell lines from the inner cell mass of several mammalian blastocysts, including humans.[13,16,17] The essential attribute of ESCs is their ability to grow indefinitely while maintaining a pluripotent status, *i.e.*, the remarkable potential to differentiate into all cell types of the post-implantation stage embryo and adult body. Under defined *in vitro* conditions, ESCs are capable of *ad infinitum* self-renewing. In the absence of growth factors and/or in presence of morphogens, ESCs can differentiate to form trophoblasts and derivatives of all three embryonic germ layers, *i.e.*, endoderm, mesoderm and ectoderm.[18] These unique features have opened up the exploitation of ESCs as a powerful tool in therapeutic application, too.

However, the difficulty to generate patient- or disease-specific ESCs and ethical and technical controversies stemming from the collection and derivation from supernumerary human blastocysts hinder their applications in the research and medical field. These considerations have propelled the search for alternative paths to generate pluripotent cells starting from easily available sources. Indications that pluripotency is a status that can be reacquired/induced in somatic cells came from the revolutionizing results on animal cloning attained by Sir John Gurdon in 1962.[19] Indeed, he generated cloned tadpoles by nuclear transfer of modified eggs in *Xenopus*, thus providing the first example of cellular reprogramming.[19] This experiment demonstrated that after transplantation of the somatic nucleus into an enucleated egg, the mature nucleus is able to reactivate early embryonic molecular programs and to establish all types of cells, proving that all the necessary genetic information is present in the nuclei of somatic cells and that the egg cytoplasm contains the necessary factors to bring about such reprogramming. This groundbreaking discovery unveiled the prospect to challenge the dogma that the acquisition of cell fate could occur only unidirectionally, *i.e.*, from an immature or pluripotent to a mature or differentiated state, as suggested by the epigenetic landscape formulation, with a ball rolling down from the top of Waddington's mountain to the bottom of a valley.[20]

Thanks to Gurdon's contribution and pursuing the idea that somatic cells can be experimentally reversed to a pluripotent status, in 2006 Shinya Yamanaka generated the first induced pluripotent stem cells (iPSCs). He showed that reprogramming of somatic cells into a pluripotent ESC-like status is achievable by means of the delivery of a combination of transcription factors.[21,22] iPSCs are very close to ESCs for morphology, proliferation, gene expression, and exhibit functional pluripotency, including the capability to generate adult chimeras competent for germline transmission.[23–25] For these reasons, iPSCs are currently recognized as a major breakthrough in stem cell research.

2.1 Modeling human neurodevelopment with pluripotent stem cells

Human PSCs, which embrace both ESCs and iPSCs, represent an extraordinary *ex vivo* source with the potential to mirror the cellular complexity existing during Central Nervous System (CNS) development. The uniqueness of the human brain is reflected in its complex, yet highly organized, cellular circuitries and in the cognitive and behavioral repertoire that defines us as human. The human brain is the product of an evolutionary and developmental history that resulted in its progressive enlargement and specialization.

Human CNS develops through a dynamic and protracted process in which myriad cell types are generated by neural stem cells (NSCs) and assembled into an intricate synaptic circuitry. Deviations from the regular path of development can lead to an assortment of pathologies, including neurological and psychiatric disorders that affect some of the most distinctly human aspects, *i.e.*, cognition and behavior.[26–29] During early CNS development, the neural tube is made of a pseudostratified layer of neuroepithelial cells lining the central cavity.[28,30–32] These cells constitute the ventricular zone (VZ) of the neural tube and are the founders of all neural progenies (*i.e.*, neurons and glial cells) of the adult CNS (Fig. 1). Later on, neuroepithelial cells convert into a distinct class of NSCs known as radial glia cells

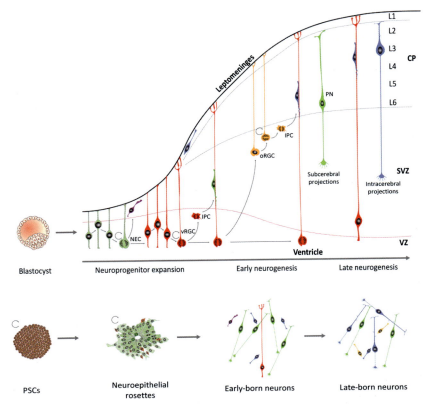

Fig. 1 Schematic illustration of neocortical development and comparison with *in vitro* recapitulation from pluripotent stem cells (PSCs). (*Top*) Neuroepithelial cells (NECs) undergo symmetric cell division to expand the initial pool and later transition into ventricular radial glia cells (vRGCs). vRGCs begin asymmetric cell division to generate another vRGC and a nascent projection neuron (PN). Neurons then migrates radially from the ventricular zone (VZ) along the RGC basal processes into the cortical plate (CP). Early-born projection neurons settle into the deep layers (Layers 5 and 6), while late-born neurons in upper layers. Additionally, some populations of RGC daughter cells convert themselves into intermediate progenitor cells (IPCs) or outer radial glial cells (oRGCs) in the subventricular zone (SVZ). (*Bottom*) PSCs can recapitulate the main events and timing of cortical development *in vitro*. Modified from Baggiani M, Dell'Anno MT, Pistello M, Conti L, Onorati M. Human neural stem cell systems to explore pathogen-related neurodevelopmental and neurodegenerative disorders. Cell 2020;9 (8):1–31. doi:10.3390/cells9081893.

(RGCs) which reside in the VZ and in the inner and outer parts of the subventricular zone (iSVZ and oSVZ, respectively) (Fig. 1). These cell populations serve as progenitors to generate neurons and macroglia (*i.e.*, astrocytes and oligodendrocytes) and to provide scaffold for migrating nascent neurons.[28,30,31] RGCs divide, but differently from neuroepithelial cells, the divisions of RGCs are mostly asymmetric, giving rise to either a daughter RGC, an intermediate progenitor cell (IPC, also known as a transit amplifying cell), or a nascent neuron that subsequently migrates out of the VZ or the SVZ to its final location near the pial surface. In the human neocortex, neuroepithelial cells give rise first to ventricular RGCs (vRGCs), which later transit into outer RGCs (oRGCs).[33–35] During early corticogenesis, at the beginning of the second trimester, the vRGCs and IPCs give rise to neurons present in the deep layers. On the other hand, oRGCs give rise to later born IPCs, and differentiate predominantly into upper layer neurons.

Since human PSCs hold the potential to generate any kind of committed or mature cell, they have been exploited as a unique platform to investigate neural development and its deviations (Fig. 1).

In this paragraph, we will briefly describe how, taking advantage of developmental biology knowledge, different worldwide laboratories have set diverse and specific protocols that recapitulate in a dish the main hallmarks of *in vivo* CNS development. A plethora of different protocols for converting PSC into mature neuronal populations has been proposed in the last decade since stem cell-based neuroscience has become a broad and rapid-expanding field. Molecular cues and timing are fundamental components of the differentiation process and these issues become particularly true when referring to neuronal differentiation. In fact, thanks to well defined protocols, it is now possible to derive different defined neuronal subtypes with region-specific identity from PSCs.[36] Generally, these protocols consist of an initial neural induction phase which brings cells from a pluripotent state to a neural lineage. In fact, the addition of supplements like N2 and heparin and small molecules that inhibit BMP and Lefty/Activin/TGFβ pathways (*e.g.*, dual SMAD inhibition protocol[37]) lead to neuroectodermal cell emergence. By adding other neural supplements and neurotrophins (*e.g.*, BDNF, NGF, NT3/4, *etc.*) neuroectodermal cells can be then terminally differentiated into mature neurons within 40–80 days.[37] Different neuronal fates can be generated through molecular patterning, thanks to the action of morphogens.

For example, to generate GABAergic medium-spiny neurons, a population that originates in the lateral ganglionic eminence (LGE)[38] and firstly affected in Huntington's disease, a recent protocol[39] proposes to firstly apply dual SMAD inhibition and then an intermediate step that produces LGE precursors thanks to exposure to Sonic Hedgehog (SHH) and DKK-1. At the end, a minimal supportive medium added with neurotrophins, such as BDNF, sustains the terminal differentiation of mature striatal neurons within 80 days. Optimizations of the protocol also include ascorbic acid (that promotes differentiation and neuronal network formation), Activin A (enhancing lateral-forebrain identity[40]) and other supplements. Similar protocols that include the administration of the different morphogens can give rise to neurons with diverse regional specifications, like cortical, midbrain, hindbrain or spinal cord. Beyond these positional identities, it is possible to tightly adjust neuronal fate obtaining specific sub-populations. For example, in order to generate midbrain dopaminergic (mDA) neurons, selectively degenerating in Parkinson's disease, floor plate progenitors (FOXA2-positive) are required. This trait can be reached by exposing the cultures to SHH and FGF8 (necessary for midbrain/hindbrain border definition) during neural induction phase. Maturation of mDA neurons progresses with neurotrophins in 50 days.[41]

Differently, for spinal cord neuron derivation,[42] after neural induction, progressive specification of caudal fates is reached by adding FGFs, retinoic acid, TGF-β and WNTs. Thanks to these molecules, motor neuron precursors with a distinctly spinal identity are generated and then terminally differentiated in 45/50 days.

Noteworthy, many of the CNS pathologies often have a neurodevelopmental component which means that the clinical outcome represents the result of defects occurring at neural progenitor level. In this case, NSCs represent the population to interrogate *in vivo* and *in vitro* in order to unravel disease-relevant altered mechanisms.

2.2 *In vitro* neural stem cell systems

NSCs represent a unique multipotent stem cell entity that can be established *in vitro*. Since Reynolds and Weiss (1992)[43] made the landmark discovery that NSCs could be maintained in culture *via* propagation as free-floating aggregates (named neurospheres), several works have progressively improved the systems for NSC long-term and homogeneous culture. In 2005, it has been shown that NSCs can be expanded as monolayer cultures, named NS cells, with full preservation of their neurogenic potential.[44] NS cells were derived from fetal and adult mouse CNS, but also from neural-committed mouse ESCs and iPSCs.[44–49] Additionally, NS cells were also derived from post-mortem human fetal tissue.[50,51] Although NS cells showed features of neurogenic RGCs, they emerged to be restricted to the generation of GABAergic neurons, independently of the different sources they were derived from.[47–49] On the other hand, embryonic neuroepithelial (NES) cells possess a great self-renewing potential and wide multilineage differentiation potential. Thanks to these unique properties, neuroepithelial cells represent an ideal candidate for *in vitro* studies related to NSC biology, neuronal and glial differentiation, and various neurodevelopmental diseases.[32]

Recently, the derivation and characterization of neocortical (NCX) NES cells has been described.[52] NCX-NES cell lines were derived from primary neuroepithelium of human post-mortem specimens ranging from 5 to 8 post-conceptional weeks (pcw). After derivation, NCX-NES cells form neural rosettes, reminiscent of the radial arrangement in the native neural tube. NES cells exhibit stem/progenitor cell characteristics as they express the neuroepithelial marker SOX1 and the pan-neural stem cell markers Nestin, SOX2, and Vimentin. NCX-NES cells could be expanded *in vitro* for more than 1 year and around 38 passages with no evidence of chromosomal instability.[52] NES cells retain regional identity after long-term expansion, as demonstrated by the expression of key transcription factors (*i.e.*, FOXG1 and OTX2) demarcating proliferative zones of the early human forebrain. NES cells show great neurogenic potentials, giving rise to mature neurons with extended complex neurites. Furthermore, they generate GFAP-positive astrocytes, thus demonstrating their multipotentiality. Single-cell RNA-sequencing (scRNA-seq) on expanded NCX-NES cells and cells from donor-matched brains demonstrated that the majority of cells from the donor-matched brain tissue samples express canonical marker genes of neuroepithelial cells and RGCs of the dorsal forebrain.[52] Remarkably, NCX-NES cells exhibit a close transcriptional signature of early NSCs as their donor-matched genetically identical NCX cells. Together, these data establish NES cell lines as a consistent model of early human brain development. A similar NES cell population derived from developing human spinal cord has been described, able to maintain regional identity and neuronal commitment of the caudal CNS.[53] Of note, spinal-cord NES cells were successfully tested in cell grafting approaches after spinal cord injury in mice.[53]

Human PSCs represent an extraordinary *ex vivo* source of neural progenitors that can be captured and maintained *ad infinitum* in a self-renewing state. The development of neural induction protocols has provided the possibility to generate *in vitro*-derived expandable NSC systems as a platform for studying basic human neurodevelopment, disease mechanisms, and potential therapeutics.[54] Seminal studies have identified rosette-type NSCs that resemble neural tube-stage progenitors, capable of responding to patterning instructions, but not long-term expandable.[55] On the other hand, a long-term population of NES cells (named lt-hESNSCs or lt-NES) was described by Koch and colleagues.[56] However, *in vitro* culture conditions bias lt-NES regional identity from rostral (first five passages) to more caudal midbrain-hindbrain identity (later passages).[56] In this direction, Li et al.[57] described a small molecule-based neural induction method to derive primitive neural progenitors from human ESCs. However, also in these conditions, the NSC population was biased toward a midbrain/hindbrain neural fate.[57] A more recent elegant work improved the isolation and propagation paradigm of neuroepithelial and RG-like cells,[58] but the true identity and physiological relevance of human PSC-derived NSCs are open to interrogations because cells could acquire transcriptional and epigenetic programs that diverge from the cell state *in vivo*.

2.3 New frontiers in disease modeling and cell replacement

As described in the previous paragraphs, human iPSCs (hiPSCs) and their neuroderivatives—NSCs and mature neurons—can recapitulate in a dish the main hallmarks of human neurodevelopment. For this reason, they offer a unique insight on human CNS development otherwise scarcely accessible for ethical and practical reasons. Moreover, the use of patient-derived cells allows to develop coherent models of a specific human pathology. This becomes particularly useful when a disease affects neural progenitor features poorly recapitulated by common animal models (*i.e.*, rodents). In fact, lissencephalic (smooth brain) rodent species lack a prominent population of oRGCs, responsible for cortical expansion and gyrification. Moreover, in the occurrence of polygenic disorders with great variability and complex genetic background, the use of patient-derived cells allows to study the disease in the same genetic context of the affected subject.

Conventional 2D-cell cultures sometimes cannot precisely replicate human pathophysiology because of the absence of complex interactions among different cell types—as occurs *in vivo*. In fact, the lack of 3D environmental cues limits analyses related to the structural development of neural tissues as well as neurodegeneration, and crucial questions cannot be properly addressed by using 2D cell culture systems.[59]

To bridge this gap, 3D human cell culture models, including spheroids and organoids, have been developed to closely replicate some aspects of the brain environment, like neuronal and glial cell interactions, and incorporating the effects of blood stream.[60] Despite their experimental relevance, also these 3D cultures present some limitations like shaping, lack of nutrient and gaseous exchange that confine organoid development to prenatal brain equivalent. Moreover, restricted uniformity and reproducibility—due to inadequate engineering of the cellular microenvironment and the extra-cellular matrix—make organoids difficult for screenings or high-throughput testing. Notwithstanding, several approaches have been proposed to overcome organoid current limitations in the attempt to reproduce *in vitro* more realistic scenarios and enable their application in drug testing and clinical trials and for personalized medicine applications.[61]

Another field of application concerns cell-based therapies for brain diseases, which find their roots back in the 70' when, for the first time, it was demonstrated that neurons implanted into the adult brain could reverse behavioral deficits of Parkinson's disease. This pivotal experiment - carried out in rats with lesions of nigrostriatal dopaminergic system - paved the way for a later series of experiments that led to the first clinical trial with neural transplantation in 1987. Later on, in the 90s, a similar approach was applied to Huntington's disease patients with the implant of fetal striatal grafts.[62]

The principle of cell-replacement relies on the replacement of damaged/dead cells—as a result of a trauma or disease—with healthy ones. Nowadays, the perspective to address severe neurological disorders is gaining increasing interest thanks to the advent of iPSC technology which has provided a huge step forward, since they represent a source of defined neural populations for transplantation and open to personalized therapeutic approaches.

Parkinson's disease represents a paradigmatic candidate for cell replacement because the loss of dopaminergic neurons—which is responsible for the symptomatic motor deficit of this pathology—is ascribed to a small nucleus, the *substantia nigra pars compacta*, unlike other neurodegenerative diseases that affect larger areas and/or multiple neuronal types. Moreover, dopaminergic neurons of the nigrostriatal pathway specifically innervate the dorsal striatum, making the connections reestablishment more feasible. For instance, Parmar and Takahashi groups in 2014[63] successfully transplanted hiPSC-derived dopaminergic neural precursors into the putamen of Parkinson's rat models. Grafted cells successfully integrated into the rodent's putamen and differentiated in a pure population of dopaminergic neurons, improving animals behavior and highlighting the broad potential of cell replacement in Parkinson's disease treatment.

Despite safe autologous transplant of hiPSC-derivatives has been tested in Parkinson's disease patients, careful evaluation of the strategy needs to be carried out in the direction of personalized medicine.

3. Genome editing for neurodevelopmental disorders
3.1 Exploring neurodevelopmental programs with PSCs and NSCs

Human brain development begins during the first weeks of pregnancy and lasts after the twentieth year of age.[64] Neurodevelopmental events that occur after birth concern mainly synaptic pruning and myelination in the cerebral cortex.[65] In this section, we will describe how deviations from the normal program can originate severe disorders. We will analyze some paradigmatic examples on how genome editing, and stem cells, can assist in building models useful for understanding of human brain development and related disorders.

During human embryonic development, the first evidence of neural tissue shows up around the third pcw when neuro-inductive processes begin. This phenomenon, borne by dorsal ectoderm, represents the steppingstone for CNS genesis.[66] The occurrence of neural induction leads to the generation of the so-called neural plate which—folding over itself—will give rise to the neural tube and, lastly, to the CNS.[67] Neural tube is constituted by neuroepithelial cells that in early neurodevelopmental stages proliferate in order to expand their pool and differentiate into other populations of NSCs, more restricted and specialized. These progenitors will further expand their pool and undergo terminal differentiation into early- and late-born neurons.[34] Human neurodevelopment departs in many aspects from the ones of the usual animal models. Hence, devising human specific cell-based systems represent a critical point.

Human NSCs are the most used cell system for neurodevelopmental studies because they are self-renewing and recapitulate the features of earliest neural progenitors. Indeed, when cultured, they spontaneously arrange in polarized rosette-like structures, resembling the neural tube organization during embryonic neurodevelopment. Furthermore, by providing the right conditions, they are able to mature into neurons and glial cells.

Hence, focusing on the different types of NSCs and neural progenitors that populate the brain during neurogenesis, researchers can dissect the developmental and molecular mechanisms that guide progenitors through their differentiative program[20] and provide a broad view on neurogenetic events.[68]

Moreover, hiPSC-based systems offer pathological-relevant patient-specific models that open a distinctive experimental window to inspect neurodevelopmental diseases (Fig. 2), a subject that will be reviewed in the following paragraph.

Genome editing tools for CNS disease modeling 417

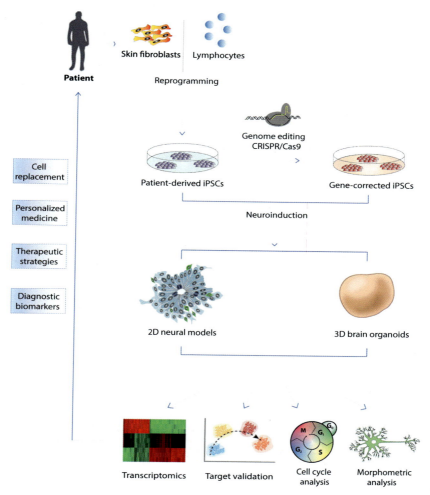

Fig. 2 Possible applications of genome editing on patient-derived hiPSCs to investigate and treat neurodevelopmental/neurodegenerative diseases. hiPSCs can be reprogrammed from patient's somatic cells. Representative cell types are shown. Once established, patient-derived hiPSCs can be genetically corrected (*e.g.*, with CRISPR/Cas9 technology) to set isogenic hiPSC lines as gold-standard controls. To model neurodevelopmental and neurodegenerative conditions, these lines can be differentiated into a neural fate both in 2D or 3D systems (*e.g.*, neural stem cells or brain organoids). The wide spectrum of analyses that can be carried out on these neural cultures opens the road to developmental studies, mechanisms of the disease, discovery of diagnostic biomarkers, cell therapy and, more broadly, to personalized medicine.

3.2 In vitro modeling of neurodevelopmental disorders

As mentioned before, iPSCs can be reprogrammed from somatic patient's cells (*e.g.*, fibroblasts or blood cells) and re-differentiated into NSCs and mature neurons. These neuro-derivatives possess the same genotype of the patient carrying all the mutations that led to the patient's pathology.

This offers the unique possibility to model *in vitro* patient's affections. Thus, through patient's somatic cell reprogramming, it is possible to generate pluripotent cell cultures genetically matching to patients. Then, iPSCs can be directed toward neural fate to follow neurogenetic processes and to study their defects; these purposes can be pursued through 2D-adherent cultures or 3D brain organoids.[69]

In this section, we will analyze three pathologies with different etiology and different outcomes, all sharing neurodevelopmental roots. We will discuss some evidence coming from the application of human ESC/iPSCs and NSC-based models, important to understand the molecular and cellular mechanisms of these diseases.

3.2.1 Microcephaly

Microcephaly is a rare (ranging from 1/2.000.000 to 1/10.000; an increased incidence in consanguineous population has been observed[70]) neurodevelopmental disorder characterized by the reduction of the occipito-frontal circumference of 2–3 standard deviations below the mean for sex, age, and ethnicity.[71] The different causes of this pathology can be broadly distinguished into genetic and environmental roots. Infections by neurotropic agents transmitted from the mother to the fetus lead to CNS development alterations that, in many cases, result in microcephaly.[32,72] On the other hand, genetic causes associated with microcephaly are often linked to autosomal recessive mutations (MCPH: Microcephaly Primary Hereditary) that affect genes encoding for centrosomal/mitotic spindle proteins. To date, 25 genetic loci related to MCPH have been identified; the most common are MCPH5 (*ASPM*; 68.6%), MCPH2 (*WDR62*; 14.1%) and MCPH1 (*Microcephalin*; 8%).[73]

Human PSC-based models are instrumental to study MCPH. Both 2D and 3D cell systems involve PSC commitment toward neural fate passing through NSC populations until neocortical mature neurons. This allows to analyze the impact of the mutations at diverse differentiative stages of neuronal development.

One of the first pioneering studies, involving cerebral organoids made from microcephalic patient-derived hiPSCs (carrying *CDK5RAP2* mutation), identified in premature NSC differentiation—at the expense of NSC proper proliferation—the cause leading to patient's microcephaly.[69] This work paved the way for MCPH investigation through hiPSC-brain organoids.

In this scenario, genome editing techniques are pivotal to deeply investigate gene contribution to neurodevelopmental disorder etiology (Fig. 2). In fact, the synergy of hiPSCs and gene editing allows investigators to introduce known pathologic mutations in healthy hiPSCs—mimicking a disrupted phenotype—to investigate neurodevelopment affections.[74] A recent work based on ablated *WDR62* hiPSC-derived brain organoids showed a reduction of organoid size due to neural progenitor disruption. Moreover, authors observed delayed primary cilium disassembly and deferred cell cycle progression leading to decreased proliferation and premature differentiation of NSCs,[75] thus linking the microcephalic phenotype to perturbation of NSC biology.

Another remarkable paradigmatic study in this field comes from Tungadi and colleagues. By using CRISPR-based *ASPM* knockout, they demonstrated that ASPM alone does not affect spindle morphology or mitotic progression because of its redundancy with another MCPH-associated protein (CDK5RAP2). In fact, only when also the pericentriolar material protein CDK5RAP2 is depleted in *ASPM KO* cells, spindle poles become unfocused throughout prometaphase resulting in anaphase delayed onset, thus advocating a link between spindle pole disruption and microcephaly.[76]

Together, these examples demonstrate the role of NSC disruption in microcephalic phenotypes linked to different mutations. The application of patient derived-hiPSCs allows to recapitulate the altered neurogenetic events that occurred in that precise genetic context and *in vitro* gene correction importantly contribute to define the pathogenetic landscape.

Nowadays, a fundamental application of genome editing consists in gene correction to generate isogenic lines. This approach entails in reverting existing mutations in patient hiPSCs to produce cell lines that possess the same genetic heritage of patient cells except for the mutation, thus representing the gold standard for *in vitro* disease modeling (Fig. 2).

3.2.2 Autism spectrum disorder
Autism spectrum disorder (ASD) is a group of psychiatric conditions that occur in as much as 0.6% of the population (more commonly in males).[77]

ASD are characterized by language deficits, social communication difficulties and repetitive behaviors with an early onset in childhood. In particular, significant impairments in the motor, complex language, and memory occur. Concurrently, intact or superior performance in attention, sensory perception, simple language and visual–spatial domain has been observed.[78] This picture is the result of many years of controversy concerning the basis of the pathology. In fact, in the past, nearly every aspect of neurophysiological functioning was addressed to underlie the abnormal behavior typical of this syndrome.

While many aspects of ASD remain poorly understood, recent findings have pointed out the genetic and developmental origins of the condition[79]; for this reason, ASD is ascribed to neurodevelopmental disorders.[80]

Many rare mutations linked to ASD have been identified but they are extremely heterogeneous and often lead to ASD related to other inherited and non-inherited risk factors.[81] To date, about 80% of the cases have no clear etiology. In this scenario, employing patient-derived hiPSC-neuroderivatives is providential. For example, Mariani and colleagues produced iPSC-telencephalic organoids from 4 donor families with an ASD-affected child. Despite the heterogeneity in genotypes, authors found perturbations in coherent programs of gene expression and common features of neurodevelopmental alterations. In particular, cell-cycle timing, synaptic growth and GABA/glutamate neuronal differentiation imbalance has been observed.[81]

Another interesting study employed 2D culture of hiPSC-derived from a cohort of ASD patients who display macrencephaly (20–30% of idiopathic ASD[82]). Cells were induced toward neural fate and analyzed at defined differentiative stages revealing that ASD-neural progenitors exhibit higher proliferation rate compared to controls, supporting a cell cycle dysregulation hypothesis, proposed to explain macrocephaly. Moreover, an imbalance of excitatory/inhibitory neuronal ratio has been also found in 2D cultures where a significant reduction in the amount of spontaneous activity in ASD neurons has also been reported. Patient-derived neuronal cultures also allowed to investigate IGF1 involvement—a target proposed in the last years for ASD treatment—revealing a potential cellular mechanism for its therapeutic effect. For instance, IGF1 seems to ameliorate the pronounced deficiency in network connectivity displayed by ASD-neurons.[83] The latter example is emblematic in demonstrating the efficacy of 2D models to investigate drug response on a patient-specific model, paving the way for personalized medicine also in the neurodevelopmental disease context (Fig. 2).

Since the heterogeneity and complex contribution of genomic variants to the disease, genome editing technologies, applied to stem cell-based models, represent an instrumental tool to dissect the etiology.

Disruptive mutations in the *CHD8* gene represent an important genetic risk factor. Noteworthy, previous studies showed that this gene regulates multiple cell processes critical for neural functions, and its targets are enriched in ASD-associated genes. Wang and colleagues applied CRISPR/Cas9 technology to target *CHD8* ($CHD8^{+/-}$) in hiPSCs to mimic the loss-of-function status in humans. The authors performed transcriptomic analyses of neural progenitors and neurons derived from $CHD8^{+/-}$ hiPSCs, highlighting the disruption of several genes already associated with ASD and schizophrenia. Among them, many are implicated in brain size control, underlying the correlation between ASD and macrocephaly.[84]

3.2.3 Schizophrenia

Schizophrenia (SCZ) is a chronic, severe psychiatric disorder that affects the way a patient thinks, acts, expresses emotions, perceives reality, and relates to others. SCZ patients display both positive symptoms—like hallucinations, delusion and repetitive movements—and negative symptoms, including apathy, alogia and withdrawal from social situations and relationships.[85]

This pathology, which shows a complex and diverse genetic architecture,[86] has been pointed as a familial disorder with complex and non-Mendelian mode of transmission. Moreover, through genome wide-association studies (GWAS), 108 independent genomic loci associated with SCZ have been identified.[87]

Concerning SCZ etiology, even though multiple hypotheses have been proposed, many evidences point to neurodevelopmental involvement. For example, according to the "two hit model" proposed by Kashavan, anomalous development during two critical time periods (early brain development and adolescence) combines to produce the symptoms associated with SCZ. In particular, early developmental insults may lead to dysfunction of specific neural networks. On the other hand, during adolescence, loss of plasticity and excessive elimination of synapses may trigger the emergence of symptoms.[88] Considering neural network dysfunction, the great variability between symptoms and implicated genetic loci is reflected into a plethora of molecular and functional alterations.[89] Indeed, *post-mortem* studies and animal models indicate that the balance of excitatory and inhibitory activity of cortical circuits is altered.[88]

Additionally, several cytoarchitectural studies underline early abnormal laminar organization and orientation of neurons in affected subjects.[89]

Given this complex scenario, application of patient-derived hiPSCs to build 2D and 3D *in vitro* models becomes a promising approach to dissect several aspects of this multifaceted and changeable pathology. An iconic example of how hiPSCs—and their neuroderivatives—allow to deepen the knowledge about such complex disease is related to one of the first reports of a SCZ family.[90] Back in 1990, Clair and colleagues described a large Scottish pedigree with schizoaffective disorder owed to a balanced chr(1;11) translocation, with 16 of the 23 family members that displayed major mental affections.[91] Later on, Millar and colleagues found that the translocation led to disruption of two genes (*DISC1* and *DISC2*—disrupted in schizophrenia 1 and 2, respectively) thus supporting the hypothesis that this alteration could have been linked to psychiatric illness.[92]

In 2015, taking advantage of a *DISC1* model generated in hiPSCs through CRISPR/Cas9 technology, it has been found that disruption of *DISC1*, near the site of the translocation, results in decreased DISC1 protein levels leading to an increased canonical Wnt signaling in neural progenitor cells. Moreover, authors highlighted altered expression of neural fate markers such as FOXG1 and EOMES (also known as TBR2) that influenced the distribution of cell types generated during cortical development.[93] The same authors proposed a model based on isogenic unaffected and *DISC1* mutant hiPSCs (*DISC1 ex8 wt/μ*). The generated cerebral organoids were examined for morphology and gene expression. Results pointed out disorganized organoid structures and altered gene expression such as downregulation of *BRN2*—coding for a transcription factor that upregulates pro-neural genes already associated with SCZ diagnosis—*TBR2*, and *DCX* (implicated in IPC determination and neuronal migration, respectively). Moreover, decreased expression of deep-layer cortical markers and increased expression of upper-layer markers suggest an unbalance during neurogenesis.[74] This example is also emblematic in underlying the benefits of *in vitro* gene editing, useful to introduce known pathologic mutations in hiPSCs to mimic a phenotype for investigation of neurodevelopmental affections.[74]

Precise gene editing/correction *in vitro* allows to insert/revert existing mutations in hiPSCs to generate isogenic cell lines, thus allowing the dissection of the real contribution of the putative mutation, polymorphism or combination of multiple polymorphisms to the disease. This is an instrumental tool in particular for diseases with a complex genetic architecture, such as SCZ. For example, a thorough study published in 2014 reported the generation of iPSCs from four members of a family in which a frameshift mutation in *DISC1* co-segregated with major psychiatric disorders. In this circumstance, the authors produced different isogenic hiPSC lines through genome editing technique (TALEN nuclease). By correcting the 4-bp deletion—linked to *DISC* mutation—they generated isogenic control lines in order to address potential effects of family genetic background. Forebrain neurons derived from mutant iPSCs showed synaptic vesicle release and synaptic boutons density deficits while isogenic hiPSC-derived neurons displayed a rescued phenotype. Finally, authors highlighted a dysregulated expression of many genes related to synapses and psychiatric disorders into affected neurons.[94]

4. Genome editing of neurodegenerative disorders

4.1 Dissecting neurodegenerative processes with human PSCs and NSCs

The CNS is one of the most vulnerable elements of the body and has a limited capacity to regenerate itself from both acute injuries and neurodegenerative diseases. The latter occur as a result of processes that drive the progressive loss of CNS neuronal subtypes that could arise from sporadic or familial/genetic triggers. Current treatments are predominantly focused on alleviating symptoms, so there is a need for the development of novel therapeutic interventions.

Notably, compared with other organs, progress in the molecular basis of neurodegenerative disorders has evolved slowly. This is mainly due to the poor availability of patients' brain tissue samples that often is only evocative of the final phase of the disease. These limitations have impeded advancements in the knowledge of the mechanisms underlying neurodegenerative diseases onset and progression.[95] Although genetically engineered animal models have led to promising advances, these systems are still insufficient and do not faithfully mirror the disease mechanisms due to species differences and genetic backgrounds. Both ESCs and iPSCs are now widely used in studies for various applications in the neurodegenerative disease field, such as autologous cell therapy, monogenic and multigenic diseases modeling, and as substrates for drug and genetic screens.

4.2 In vitro modeling of neurodegenerative diseases

Over the past decade, several patient-specific hiPSCs have been employed to model neurodegenerative diseases. Patient-derived hiPSCs can be in fact differentiated *in vitro* into disease-relevant neuronal subtypes or corrected using gene targeting to repair disease-causing mutations. iPSC technology has opened to the production of large-scale CNS cells with specific genotypes, providing an unlimited amount of affected neuronal subtypes to the potential development of promising drugs that are most suitable for the patient both in terms of efficacy and toxicity profile. The additional prospective is to develop sources of autologous-repaired hiPSC-derived progeny for cell therapy. In this direction, Rubio and colleagues described a new platform where neurons can be generated *in vitro* and manipulated using CRISPR/Cas9 to inactivate specific genes associated with different neuropathologies in humans.[96] Interestingly, the advent of the CRISPR-Cas9 gene technology has improved the efficiency of genome editing and accelerated the generation of isogenic controls[97] opening to precise establishment of genotype-phenotype correlations.[98]

4.2.1 Huntington's disease

Huntington's disease (HD) is a progressive neurodegenerative genetic disease associated with a wide variety of motor impairment (choreic symptoms) and psychiatric manifestations caused by abnormal CAG trinucleotide expansion in the 5' of the huntingtin gene (*HTT*). HD is an autosomal dominant pathology with a combination of gain- and loss-of-function mechanisms.[99] It is a fatal disorder at present and the few available pharmacological therapies represent only palliative treatments. The life expectancy of HD patients is usually about 10–15 years after the onset of the symptoms. At tissue level, the main pathological hallmarks of HD are represented by the initial loss of PPP1R1B (also known as DARPP-32)-positive striatal neurons and the presence of intracellular aggregates of mutant HTT protein in neuronal cells whose role in the onset/progression of the disease is still not fully understood.[100]

HiPSC-based models from HD patients carrying a CAG repeat expansion were originally developed by Zhang and colleagues.[101] They reported that the HD hiPSC-derived neurons exhibit enhanced caspase activity upon growth factor deprivation compared to normal cells. Jeon and colleagues in two different works illustrated that though showing no evident HD phenotype, HD hiPSC-derived NSCs were able to sharply respond to proteasome inhibition and express several markers of HD pathology after long-term transplantation in the mouse brain.[102,103] Also, a comprehensive study by the HD Consortium reported hundreds of distinct phenotypic differences between HD and control hiPSC lines, showing clear signs of HD-related pathology in several lineages. Notably, the authors found that lines with the longer CAG repeat expansions had more severe pathological phenotypes and ultimately lead to enhanced neuronal death.[104] Noteworthy, An and colleagues in another study showed that several HD-phenotypes, including altered mitochondrial bioenergetics and susceptibility to cell death, accompanied with pathogenic HD-associated signaling pathways, could be reversed by genetic correction with a homologous recombination approach to reconstitute a normal repeat for the expanded CAG repeat trait.[105] These findings show that HD hiPSCs could be a well-characterized and unique resource to elucidate disease mechanisms and to test genomic correction approaches (Fig. 2).

At another level, RNA interference (RNAi) and antisense oligonucleotide (ASO) agents targeting *Htt* messenger RNA have shown efficacy in HD preclinical models[106,107] and a recent Phase 1/2a trial showed that multiple intrathecal ASO administrations reduced normal and mutant huntingtin (mHTT) protein levels in participants' cerebrospinal fluid.[108] While partial reduction (~45%) of normal HTT expression is tolerated in the non-human primate striatum,[109] the long-term reductions are unclear given its involvement in myriad biological functions.[110] In mice, for example, *Htt* knockout is embryonically lethal[111] and perinatal loss leads to motor dysfunction and other neuropathology.[112] Since reduction of normal HTT below 50% concomitant with mHTT poses an unknown risk, the development of an allele-selective, mHTT-lowering approach suitable for clinical use has been a long-sought goal. To this end, Zaitler and colleagues have recently reported[113] the exploitation of engineered zinc finger protein transcription factors (ZFP-TFs) to target the pathogenic CAG repeat and selectively lower mHTT as a therapeutic strategy. Using patient-derived fibroblasts and iPSC-neurons, they demonstrate that ZFP-TFs selectively repress >99% of HD-causing alleles over a wide dose range while preserving expression of >86% of normal alleles. Noteworthy, other CAG-containing genes are minimally affected. Using three HD mouse models, they also demonstrated improvements in a range of molecular, histopathological, electrophysiological and functional endpoints.

4.2.2 Spinal muscular atrophy

Spinal muscular atrophy (SMA) is an autosomal recessive genetic disease characterized by the loss of the survival motor neuron (SMN) protein, leading to progressive death of motor neurons (MNs) and causing weakness and wasting of muscles required for movements.[114] There are two *SMN* genes in humans, *SMN1* and *SMN2*. Although the genetic cause of SMA has been mapped to the *SMN1* gene, the mechanisms underlying selective MN degeneration in SMA remain poorly understood.[115]

To model SMA, Ebert and co-workers have reported the generation of iPSC lines from skin fibroblast samples from a child with Type 1 SMA. These were subsequently differentiated into MNs that, at later stages in culture, exhibited hallmarks of neuronal degeneration when compared to hiPSC-derived MNs from the child's unaffected mother.[116] Noteworthy, SMA-derived hiPSCs treated with valproate or tobramycin showed a significant increase in SMN levels compared with untreated hiPSCs indicating that SMA hiPSCs respond to drug treatment in a similar fashion to SMA fibroblasts. This result elevates hiPSCs as a platform useful to screen for drugs specifically active in preserving MN viability. Two years later, Chang et al. reported the generation of hiPSC lines from fibroblasts of five SMA patients and found that SMA hiPSCs exhibited a reduced capacity to produce MNs coupled to a delayed neurite outgrowth. Ectopic SMN expression in these hiPSC lines was able to restore normal MN differentiation and rescued the phenotype of aberrant neurite outgrowth.[117] Another study from Corti et al. focused on genetic correction for potential therapeutic approaches by generating hiPSCs from SMA patients (SMA hiPSCs) using non-viral, non-integrating episomal vectors and used a targeted gene correction approach based on single-stranded oligonucleotides to convert *SMN2* gene into an *SMN1*-like gene. Uncorrected SMA hiPSC-derived MNs manifested disease-specific features, including reduction in MN survival and size. These phenotypes were ameliorated in genetically corrected cells. The transplantation of corrected MNs derived from SMA hiPSCs into an SMA mouse model extended the life span of the animals and improved the disease phenotype, suggesting that genetically corrected hiPSC-derived MNs could represent a source for future therapeutic strategies.[118]

In the gene therapy direction, in 2019 the US Food and Drug Administration (FDA) has approved Zolgensma for the treatment of pediatric patients less than 2 years of age with SMA with biallelic mutations in *SMN1*. Zolgensma is a gene replacement therapy comprising an adeno-associated viral vector containing the human *SMN* gene under control of the chicken beta-actin promoter. This therapy addresses the genetic root of the disease by increasing functional SMN protein in MNs and preventing neuronal cell death, thus resulting in improved neuronal and muscular function as previously demonstrated in transgenic animal models.[119]

4.2.3 Parkinson's disease

Parkinson's disease (PD) is a long-term, progressive neurodegenerative disorder that mainly affects the motor system. Death of dopaminergic (DA) neurons in the *pars compacta* of the *Substantia nigra* (pcSN) results in the deficiency of dopamine delivery to the striatum and represents the major defect leading to the disease. Symptoms include postural instability, resting tremors, rigidity, and bradykinesia. Currently, dopamine agonists, anticholinergic agents and L-DOPA, a dopamine precursor, are available for the treatment. However, these pharmacological treatments have been shown to be scarce in the long-term managing of PD symptomatology. As such, alternative therapies are urgently needed.[120]

PD is mainly idiopathic, nonetheless some studies conducted on patients with family histories of PD have revealed the presence of causative mutations in the genes *LRRK2, PARK2, DJ-1, PINK1*, or *SNCA*.[121] Inclusions of Lewy bodies filled with mutant α-synuclein protein in the dying neurons have been observed in *postmortem* analysis of the brains of PD patients. To date, *SNCA* and *LRRK2* mutations were identified to be responsible for the autosomal dominant form of PD.[122] Initial PD hiPSCs generated by Soldner and colleagues in 2009 failed to reveal relevant phenotypes in differentiated DA neurons.[123] Later, Nguyen and colleagues generated hiPSC lines from a patient harboring the G2019S mutation in *LRRK2* gene and showed that DA neurons derived from these lines exhibit defects partially consistent with early PD phenotype, *i.e.*, enhanced susceptibility to oxidative stress.[124] Similarly, Sanchez-Danes and co-workers reported that hiPSC-derived DA neurons from a patient carrying the same *LRRK2* mutation manifest morphological alterations when compared to control lines, including reduced neurite complexity.[125] Mutations in either *PINK1* or *PARK2* have been described to cause recessive forms of inherited PD characterized by impaired mitophagy. In studies involving patient-derived hiPSCs with *PARK2* mutations, it was shown that DA neurons exhibited morphological and functional mitochondria abnormalities along with increased oxidative stress. Noteworthy, these defects were not present in the original patients' fibroblasts, neither in the hiPSCs.[126] Jiang and colleagues reported similar alterations in *PARK2* mutant hiPSC-derived DA neurons, including enhanced sensibility to oxidative stress and monoamine oxidases activity, increased spontaneous dopamine release, and decreased dopamine uptake. Notably, these phenotypes were reversed by expression of normal PARK2.

In the modeling of complex diseases like PD it is of primary importance to differentiate hiPSCs into cell subtypes mainly affected in the disease to better reveal phenotypes that could otherwise be silent. To this regard, Devine and colleagues showed that hiPSC lines carrying triplication of *SNCA* locus do express double amount of α-synuclein only when differentiated into DA neurons, while expression was not found in the original fibroblasts and hiPSCs.[127]

Gene editing approaches have been exploited by several studies for correcting PD associated mutations in hiPSCs. ZFN-mediated genome editing technology was used by researchers to correct the underlying point mutations (A53T) in *SNCA* PD hiPSCs. However, the phenotypes of mutant and corrected hiPSC-derived DA neurons in this work were not assessed.[128] A more recent study[129] reported the correction by ZFN-mediated homologous recombination in hiPSCs carrying *GBA1* mutation, showing that corrected hiPSC-derived DA neurons exhibit rescued disease-linked lysosomal and autophagic phenotypes and restoration of non-pathological levels of α-synuclein.

4.2.4 Alzheimer's disease

Alzheimer's disease (AD) is the most common form of age-related dementia, characterized by progressive cognitive disturbance and loss of memory.[130] It is distinguished by the presence of major hallmarks, including extracellular accumulation of amyloid beta (Aβ) plaques and intracellular aggregation of the microtubule associated protein, tau (also known as MAPT). AD exists in only 1–5% in its familial form (fAD), characterized by autosomal dominant inheritance of Presenilin (*PSEN*) 1 or 2 and Amyloid Precursor Protein (*APP*)[131,132] while the majority of AD cases to date are sporadic (sAD) and multifactorial with suspected role of epigenetics and viral infection involved in the course of progression. List of suspected genes include *MAPT, BACE1, BACE2, ADAM10, ADAM17*, and others.[133,134]

The establishment of fAD patient-derived hiPSCs carrying *PSEN1* (A246E) and *PSEN2* (N141I) mutations has been very important in the field to model the disease *in vitro*. fAD iPSC-derived neurons have been shown to have significantly increased Aβ secretion and to sharply respond to γ-secretase modulators and inhibitors.[135] Indeed, these neuronal cells showed AD-like biochemical hallmarks, including an increase in Aβ ratio and reduced capability to secrete Aβ when treated with β-secretase inhibitors, γ-secretase inhibitors, and a nonsteroidal anti-inflammatory drug. Additional hiPSC lines have been established by Israel and colleagues by reprogramming of primary fibroblasts from two fAD patients, both with duplication of the *APP* gene, two with sAD and two non-demented control individuals. Neurons from differentiated cultures were purified with fluorescence-activated cell sorting and characterized. Relative to controls, hiPSC-derived purified neurons from the two fAD patients and sAD patients exhibited significantly higher levels of the pathological marker amyloid-β,$^{1-40}$ phospho-tau(Thr 231) and active glycogen synthase kinase-3β (aGSK-3β). Interestingly, these alterations were alleviated by β-secretase, not γ-secretase, inhibitor treatment both in fAD and in sAD samples.[136]

CRISPR/Cas9 can be applied in the AD drug-discovery process (Fig. 2). For example, researchers are focusing on new ways to decrease β-amyloid production. CRISPR/Cas9 can be applied to knock-down the precursors of the γ-secretase to predict and anticipate the adverse effects that could be encountered with the use of this new drug, before going through the actual clinical trial.[137]

5. Future directions and concluding remarks

Human brain disorders can be poorly recapitulated in mice models; in fact, mice possess a smooth cerebral cortex that is 1000 times smaller than the gyrified human one. In particular, lissencephalic species show reduced proliferation of oSVZ-progenitor cells during human neurodevelopment, a population that has been found fundamental for cortical expansion and gyrification.[138] For these reasons, genome editing technique applied to patient-specific hiPSCs and their neuroderivatives has provided a novel and complementary approach to interrogate cellular and molecular phenotypes in 2D and 3D systems, with the potential to uncover developmental and pathological mechanisms and giving a deep insight into human neurodevelopment and neurodegenerative diseases (Fig. 2).

However, cellular models possess some limitations due to the lack of complex physiological conditions. Thus, investigators have attempted different approaches to address this issue. One of them is represented by hiPSC-neuron transplant in humanized mice, thereby integrating the human disease genetic background with *in vivo* physiology of an organism.[139]

Advances in the treatment of neurodegenerative diseases can be expected in the near future, especially for those affected by a known genetic component such as HD and SMA. To date, as previously mentioned, for these two diseases, drugs based on gene correction[119] or clinical trials are under development.[108] Of primary importance will be to further study the effective therapeutic window and the safety of the constructs. In parallel, important clinical trials and *in vivo* studies are aiming to better understand the possibility of using neuronal populations derived from PSCs for transplantation approaches. This will be of great importance especially for pathologies with more heterogeneous etiology such as PD and HD. In this direction, it is necessary to highlight a clinical trial carried out by the TRANSNEURO Consortium and still in progress, which uses human fetal ventral mesencephalic tissue as transplant material.[140] At a preclinical level, several authors have demonstrated how transplantation of stem cell-derived neuronal populations have positive effects on neurological, cognitive, and electrophysiological alterations in rodent models of HD such as the works of Reidling et al., Jeon et al.[102,141] Other studies involving hiPSC-derived cholinergic neuronal precursors transplanted in the hippocampus of transgenic AD mice showed how precursors successfully differentiated into mature cholinergic neurons and a reversal of spatial memory impairment.[142] One limitation is that hiPSCs and their early progeny have a significant risk of tumor formation because of their highly proliferative features. Recent studies demonstrated that adult cells could be reprogrammed into mature neurons without taking them back to their stem cell phenotypic state, thereby eliminating the risk of tumor formation. In a recent study, also Wernig et al. demonstrated that reprogramming fibroblasts into DA neurons improved the performance on behavioral tests in PD rat models.[143] To sum up, for degenerative diseases such as HD, SMA, AD and PD, the restoration of lost neural tissue could be a successful treatment option. This goal could be reached in combination with precise gene-editing tools. Risks are being minimized thanks to technological and procedural development and regeneration therapy in synergy with genetic tools may soon be translated into the clinical setting.

Acknowledgments

We apologize for not citing many original papers due to the space limit. The research in the authors' laboratories was supported by 2017 NARSAD Young Investigator Grant from the Brain & Behavior Research Foundation (#26565 to M.O.), "PRA—Progetti di Ricerca di Ateneo" from University of Pisa (Institutional Research Grants, Project no. PRA_2018_68 to M.O.), Ricerca Finalizzata from Ministero della Salute (GR-2018-12367290 to M.O.) and Fondazione Cariplo (Prot. 2019-3396 to L.C.).

References

1. Rouet P, Smih F, Jasin M. Introduction of double-strand breaks into the genome of mouse cells by expression of a rare-cutting endonuclease. *Mol Cell Biol.* 1994;14:8096–8106. Published online https://doi.org/10.1128/mcb.14.12.8096.
2. Carroll D. Genome engineering with targetable nucleases. *Annu Rev Biochem.* 2014;83:409–439. Published online https://doi.org/10.1146/annurev-biochem-060713-035418.
3. Mjelle R, Hegre SA, Aas PA, et al. Cell cycle regulation of human DNA repair and chromatin remodeling genes. *DNA Repair (Amst).* 2015;30:53–67. https://doi.org/10.1016/j.dnarep.2015.03.007.
4. Kim YG, Cha J, Chandrasegaran S. Hybrid restriction enzymes: zinc finger fusions to Fok I cleavage domain. *Proc Natl Acad Sci U S A.* 1996;93(3):1156–1160. https://doi.org/10.1073/pnas.93.3.1156.
5. Pavletich NP, Pabo CO. Zinc finger-DNA recognition: crystal structure of a Zif268-DNA complex at 2.1 A. *Science.* 1991;252(5007):809–817. https://doi.org/10.1126/science.2028256.
6. Smith J, Bibikova M, Whitby FG, Reddy AR, Chandrasegaran S, Carroll D. Requirements for double-strand cleavage by chimeric restriction enzymes with zinc finger DNA-recognition domains. *Nucleic Acids Res.* 2000;28(17):3361–3369. https://doi.org/10.1093/nar/28.17.3361.
7. Bitinaite J, Wah DA, Aggarwal AK, Schildkraut I. Foki dimerization is required for DNA cleavage. *Proc Natl Acad Sci U S A.* 1998;95(18):10570–10575. https://doi.org/10.1073/pnas.95.18.10570.
8. Joung JK, Sander JD. TALENs: a widely applicable technology for targeted genome editing. *Nat Rev Mol Cell Biol.* 2013;14(1):49–55. https://doi.org/10.1038/nrm3486.
9. Miller JC, Tan S, Qiao G, et al. A TALE nuclease architecture for efficient genome editing. *Nat Biotechnol.* 2011;29(2):143–148. https://doi.org/10.1038/nbt.1755.
10. Doudna JA, Charpentier E. Genome editing. The new frontier of genome engineering with CRISPR-Cas9. *Science.* 2014;346(6213). https://doi.org/10.1126/science.1258096, 1258096.
11. Zetsche B, Gootenberg JS, Abudayyeh OO, et al. Cpf1 is a single RNA-guided endonuclease of a class 2 CRISPR-Cas system. *Cell.* 2015;163(3):759–771. https://doi.org/10.1016/j.cell.2015.09.038.
12. Lander ES. The heroes of CRISPR. *Cell.* 2016;164(1):18–28. https://doi.org/10.1016/j.cell.2015.12.041.
13. Evans MJ, Kaufman MH. Establishment in culture of pluripotential cells from mouse embryos. *Nature.* 1981;292(5819):154–156. https://doi.org/10.1038/292154a0.
14. Martin GR. Isolation of a pluripotent cell line from early mouse embryos cultured in medium conditioned by teratocarcinoma stem cells. *Proc Natl Acad Sci U S A.* 1981;78(12 II):7634–7638. https://doi.org/10.1073/pnas.78.12.7634.
15. Avior Y, Sagi I, Benvenisty N. Pluripotent stem cells in disease modelling and drug discovery. *Nat Rev Mol Cell Biol.* 2016;17(3):170–182. https://doi.org/10.1038/nrm.2015.27.
16. Bradley A, Evans M, Kaufman MH, Robertson E. Formation of germ-line chimaeras from embryo-derived teratocarcinoma cell lines. *Nature.* 1984;309(5965):255–256. https://doi.org/10.1038/309255a0.
17. Capecchi MR. Gene targeting in mice: functional analysis of the mammalian genome for the twenty-first century. *Nat Rev Genet.* 2005;6(6):507–512. https://doi.org/10.1038/nrg1619.
18. Knoblich JA. Mechanisms of asymmetric stem cell division. *Cell.* 2008;132:P583–P597. Published online https://doi.org/10.1016/j.cell.2008.02.007.

19. Gurdon JB. Adult frogs derived from the nuclei of single somatic cells. *Dev Biol.* 1962;4(2):256–273. https://doi.org/10.1016/0012-1606(62)90043-X.
20. Waddington CH. *The Strategy of the Genes: A Discussion of Some Aspects of Theoretical Biology.* London: Routledge; 2014. https://doi.org/10.4324/9781315765471.
21. Okita K, Ichisaka T, Yamanaka S. Generation of germline-competent induced pluripotent stem cells. *Nature.* 2007;448(7151):313–317. https://doi.org/10.1038/nature05934.
22. Robertson E, Bradley A, Kuehn M, Evans M. Germ-Line transmission of genes introduced into cultured pluripotential cells by retroviral vector. *Nature.* 1986;323(6087):445–448. https://doi.org/10.1038/323445a0.
23. Skarnes WC, Rosen B, West AP, et al. A conditional knockout resource for the genome-wide study of mouse gene function. *Nature.* 2011;474(7351):337–344. https://doi.org/10.1038/nature10163.
24. Takahashi K, Tanabe K, Ohnuki M, et al. Induction of pluripotent stem cells from adult human fibroblasts by defined factors. *Cell.* 2007;131(5):861–872. https://doi.org/10.1016/j.cell.2007.11.019.
25. Takahashi K, Yamanaka S. Induction of pluripotent stem cells from mouse embryonic and adult fibroblast cultures by defined factors. *Cell.* 2006;126(4):663–676. https://doi.org/10.1016/j.cell.2006.07.024.
26. Diaz AL, Gleeson JG. The molecular and genetic mechanisms of neocortex development. *Clin Perinatol.* 2009;36(3):503–512. https://doi.org/10.1016/j.clp.2009.06.008.
27. Rakic P. Specification of cerebral cortical areas. *Science.* 1988;241(4862):170–176. https://doi.org/10.1126/science.3291116.
28. Silbereis JC, Pochareddy S, Zhu Y, Li M, Sestan N. The cellular and molecular landscapes of the developing human central nervous system. *Neuron.* 2016;89(2):248. https://doi.org/10.1016/j.neuron.2015.12.008.
29. Sousa AMM, Meyer KA, Santpere G, Gulden FO, Sestan N. Evolution of the human nervous system function, structure, and development. *Cell.* 2017;170(2):226–247. https://doi.org/10.1016/j.cell.2017.06.036.
30. Kriegstein A, Alvarez-Buylla A. The glial nature of embryonic and adult neural stem cells. *Annu Rev Neurosci.* 2009;32:149–184. https://doi.org/10.1146/annurev.neuro.051508.135600.
31. Breunig JJ, Haydar TF, Rakic P. Neural stem cells: historical perspective and future prospects. *Neuron.* 2011;70(4):614–625. https://doi.org/10.1016/j.neuron.2011.05.005.
32. Baggiani M, Dell'Anno MT, Pistello M, Conti L, Onorati M. Human neural stem cell systems to explore pathogen-related neurodevelopmental and neurodegenerative disorders. *Cell.* 2020;9(8):1–31. https://doi.org/10.3390/cells9081893.
33. Fietz SA, Kelava I, Vogt J, et al. OSVZ progenitors of human and ferret neocortex are epithelial-like and expand by integrin signaling. *Nat Neurosci.* 2010;13(6):690–699. https://doi.org/10.1038/nn.2553.
34. Hansen DV, Lui JH, Parker PRL, Kriegstein AR. Neurogenic radial glia in the outer subventricular zone of human neocortex. *Nature.* 2010;464(7288):554–561. https://doi.org/10.1038/nature08845.
35. Reillo I, Borrell V. Germinal zones in the developing cerebral cortex of ferret: ontogeny, cell cycle kinetics, and diversity of progenitors. *Cereb Cortex.* 2012;22(9):2039–2054. https://doi.org/10.1093/cercor/bhr284.
36. Corti S, Faravelli I, Cardano M, Conti L. Human pluripotent stem cells as tools for neurodegenerative and neurodevelopmental disease modeling and drug discovery. *Expert Opin Drug Discovery.* 2015;10(6):615–629. https://doi.org/10.1517/17460441.2015.1037737.
37. Chambers SM, Fasano CA, Papapetrou EP, Tomishima M, Sadelain M, Studer L. Highly efficient neural conversion of human ES and iPS cells by dual inhibition of SMAD signaling. *Nat Biotechnol.* 2009;27(3):275–280. https://doi.org/10.1038/nbt.1529.

38. Onorati M, Castiglioni V, Biasci D, et al. Molecular and functional definition of the developing human striatum. *Nat Neurosci*. 2014;17(12):1804–1815. https://doi.org/10.1038/nn.3860.
39. Delli Carri A, Onorati M, Lelos MJ, et al. Developmentally coordinated extrinsic signals drive human pluripotent stem cell differentiation toward authentic DARPP-32+ medium-sized spiny neurons. *Development*. 2013;140(2):301–312. https://doi.org/10.1242/dev.084608.
40. Arber C, Precious SV, Cambray S, et al. Activin a directs striatal projection neuron differentiation of human pluripotent stem cells. *Development*. 2015;142(7): 1375–1386. https://doi.org/10.1242/dev.117093.
41. Wang M, Ling KH, Tan JJ, Lu CB. Development and differentiation of midbrain dopaminergic neuron: from bench to bedside. *Cells*. 2020;9(6):1489. https://doi.org/10.3390/cells9061489.
42. Trawczynski M, Liu G, David BT, Fessler RG. Restoring motor neurons in spinal cord injury with induced pluripotent stem cells. *Front Cell Neurosci*. 2019;13:369. https://doi.org/10.3389/fncel.2019.00369.
43. Reynolds BA, Weiss S. Generation of neurons and astrocytes from isolated cells of the adult mammalian central nervous system. *Science*. 1992;255(5052):1707–1710. https://doi.org/10.1126/science.1553558.
44. Conti L, Pollard SM, Gorba T, et al. Niche-independent symmetrical self-renewal of a mammalian tissue stem cell. *PLoS Biol*. 2005;3(9):1594–1606. https://doi.org/10.1371/journal.pbio.0030283.
45. Pollard SM, Conti L, Sun Y, Goffredo D, Smith A. Adherent neural stem (NS) cells from fetal and adult forebrain. *Cereb Cortex*. 2006;16(SUPPL. 1):i112–i120. https://doi.org/10.1093/cercor/bhj167.
46. Goffredo D, Conti L, Di Febo F, et al. Setting the conditions for efficient, robust and reproducible generation of functionally active neurons from adult subventricular zone-derived neural stem cells. *Cell Death Differ*. 2008;15(12):1847–1856. https://doi.org/10.1038/cdd.2008.118.
47. Albieri I, Onorati M, Calabrese G, et al. A DNA transposon-based approach to functional screening in neural stem cells. *J Biotechnol*. 2010;150(1):11–21. https://doi.org/10.1016/j.jbiotec.2010.07.027.
48. Onorati M, Camnasio S, Binetti M, Jung CB, Moretti A, Cattaneo E. Neuropotent self-renewing neural stem (NS) cells derived from mouse induced pluripotent stem (iPS) cells. *Mol Cell Neurosci*. 2010;43(3):287–295. https://doi.org/10.1016/j.mcn.2009.12.002.
49. Onorati M, Binetti M, Conti L, et al. Preservation of positional identity in fetus-derived neural stem (NS) cells from different mouse central nervous system compartments. *Cell Mol Life Sci*. 2011;68(10):1769–1783. https://doi.org/10.1007/s00018-010-0548-7.
50. Sun Y, Pollard S, Conti L, et al. Long-term tripotent differentiation capacity of human neural stem (NS) cells in adherent culture. *Mol Cell Neurosci*. 2008;38(2):245–258. https://doi.org/10.1016/j.mcn.2008.02.014.
51. Hook L, Vives J, Fulton N, et al. Non-immortalized human neural stem (NS) cells as a scalable platform for cellular assays. *Neurochem Int*. 2011;59(3):432–444. https://doi.org/10.1016/j.neuint.2011.06.024.
52. Onorati M, Li Z, Liu F, et al. Zika virus disrupts phospho-TBK1 localization and mitosis in human neuroepithelial stem cells and radial glia. *Cell Rep*. 2016;16 (10):2576–2592. https://doi.org/10.1016/j.celrep.2016.08.038.
53. Dell'Anno MT, Wang X, Onorati M, et al. Human neuroepithelial stem cell regional specificity enables spinal cord repair through a relay circuit. *Nat Commun*. 2018;9 (1):3419. https://doi.org/10.1038/s41467-018-05844-8.

54. Conti L, Cattaneo E. Neural stem cell systems: physiological players or in vitro entities? *Nat Rev Neurosci.* 2010;11(3):176–187. https://doi.org/10.1038/nrn2761.
55. Elkabetz Y, Panagiotakos G, Al Shamy G, Socci ND, Tabar V, Studer L. Human ES cell-derived neural rosettes reveal a functionally distinct early neural stem cell stage. *Genes Dev.* 2008;22(9):1257. https://doi.org/10.1101/gad.1616208.1995.
56. Koch P, Opitz T, Steinbeck JA, Ladewig J, Brüstle O. A rosette-type, self-renewing human ES cell-derived neural stem cell with potential for in vitro instruction and synaptic integration. *Proc Natl Acad Sci U S A.* 2009;106(9):3225–3230. https://doi.org/10.1073/pnas.0808387106.
57. Li W, Sun W, Zhang Y, et al. Rapid induction and long-term self-renewal of primitive neural precursors from human embryonic stem cells by small molecule inhibitors. *Proc Natl Acad Sci U S A.* 2011;108(20):8299–8304. https://doi.org/10.1073/pnas.1014041108.
58. Edri R, Yaffe Y, Ziller MJ, et al. Analysing human neural stem cell ontogeny by consecutive isolation of Notch active neural progenitors. *Nat Commun.* 2015;6:6500. https://doi.org/10.1038/ncomms7500.
59. Yu F, Hunziker W, Choudhury D. Engineering microfluidic organoid-on-a-chip platforms. *Micromachines.* 2019;10(3):165. https://doi.org/10.3390/mi10030165.
60. Slanzi A, Iannoto G, Rossi B, Zenaro E, Constantin G. In vitro models of neurodegenerative diseases. *Front Cell Dev Biol.* 2020;8:328. https://doi.org/10.3389/fcell.2020.00328.
61. Pellegrini L, Bonfio C, Chadwick J, Begum F, Skehel M, Lancaster MA. Human CNS barrier-forming organoids with cerebrospinal fluid production. *Science.* 2020;369 (6500):eaaz5626. https://doi.org/10.1126/science.aaz5626.
62. Lindvall O, Björklund A. Cell replacement therapy: helping the brain to repair itself. *NeuroRx.* 2004;1(4):379–381. https://doi.org/10.1602/neurorx.1.4.379.
63. Doi D, Samata B, Katsukawa M, et al. Isolation of human induced pluripotent stem cell-derived dopaminergic progenitors by cell sorting for successful transplantation. *Stem Cell Rep.* 2014;2(3):337–350. https://doi.org/10.1016/j.stemcr.2014.01.013.
64. Smith JL, Schoenwolf GC. Neurulation: coming to closure. *Trends Neurosci.* 1997;20 (11):510–517. https://doi.org/10.1016/S0166-2236(97)01121-1.
65. Huttenlocher PR. Synaptic density in human frontal cortex—developmental changes and effects of aging. *Brain Res.* 1979;163(2):195–205. https://doi.org/10.1016/0006-8993(79)90349-4.
66. Hamburger V, Hamilton HL. A series of normal stages in the development of the chick embryo. *J Morphol.* 1951;88(1):49–92. https://doi.org/10.1002/jmor.1050880104.
67. Sanes DH, Reh TA, Harris WA, Landgraf M. *Development of the Nervous System.* Elsevier; 2019.
68. Rifes P, Isaksson M, Rathore GS, et al. Modeling neural tube development by differentiation of human embryonic stem cells in a microfluidic WNT gradient. *Nat Biotechnol.* 2020;38(11):1265–1273. https://doi.org/10.1038/s41587-020-0525-0.
69. Lancaster MA, Renner M, Martin CA, et al. Cerebral organoids model human brain development and microcephaly. *Nature.* 2013;501(7467):373–379. https://doi.org/10.1038/nature12517.
70. Woods CG, Bond J, Enard W. Autosomal recessive primary microcephaly (MCPH): a review of clinical, molecular, and evolutionary findings. *Am J Hum Genet.* 2005;76 (5):717–728. https://doi.org/10.1086/429930.
71. De Araújo JSS, Regis CT, Gomes RGS, et al. Microcephaly in north-east Brazil: a retrospective study on neonates born between 2012 and 2015. *Bull World Health Organ.* 2016;94(11):835–840. https://doi.org/10.2471/BLT.16.170639.
72. Neu N, Duchon J, Zachariah P. TORCH infections. *Clin Perinatol.* 2015;42(1):77–103. https://doi.org/10.1016/j.clp.2014.11.001.

73. Naveed M, Kazmi SK, Amin M, et al. Comprehensive review on the molecular genetics of autosomal recessive primary microcephaly (MCPH). *Genet Res (Camb)*. 2018;100:e7. Published online https://doi.org/10.1017/S0016672318000046.
74. Srikanth P, Lagomarsino VN, Muratore CR, et al. Shared effects of DISC1 disruption and elevated WNT signaling in human cerebral organoids. *Transl Psychiatry*. 2018;8(1):77. https://doi.org/10.1038/s41398-018-0122-x.
75. Zhang W, Yang SL, Yang M, et al. Modeling microcephaly with cerebral organoids reveals a WDR62–CEP170–KIF2A pathway promoting cilium disassembly in neural progenitors. *Nat Commun*. 2019;10(1):2612. https://doi.org/10.1038/s41467-019-10497-2.
76. Tungadi EA, Ito A, Kiyomitsu T, Goshima G. Human microcephaly ASPM protein is a spindle pole-focusing factor that functions redundantly with CDK5RAP2. *J Cell Sci*. 2017;130(21):3676–3684. https://doi.org/10.1242/jcs.203703.
77. Xiao Z, Qiu T, Ke X, et al. Autism spectrum disorder as early neurodevelopmental disorder: evidence from the brain imaging abnormalities in 2-3 years old toddlers. *J Autism Dev Disord*. 2014;44(7):1633–1640. https://doi.org/10.1007/s10803-014-2033-x.
78. Minshew NJ, Goldstein G, Siegel DJ. Neuropsychologic functioning in autism: profile of a complex information processing disorder. *J Int Neuropsychol Soc*. 1997;3(4):303–316. https://doi.org/10.1017/s1355617797003032.
79. Elsabbagh M, Divan G, Koh YJ, et al. Global prevalence of autism and other pervasive developmental disorders. *Autism Res*. 2012;5(3):160–179. https://doi.org/10.1002/aur.239.
80. Wang H. Modeling neurological diseases with human brain organoids. *Front Synaptic Neurosci*. 2018;10(June):1–14. https://doi.org/10.3389/fnsyn.2018.00015.
81. Mariani J, Coppola G, Zhang P, et al. FOXG1-dependent dysregulation of GABA/glutamate neuron differentiation in autism spectrum disorders. *Cell*. 2015;162(2):375–390. https://doi.org/10.1016/j.cell.2015.06.034.
82. Bodurtha J, Duis J. *Signs and Symptoms of Genetic Conditions—A Handbook*. vol. 03. Thieme Medical Publishing; 2015.
83. Marchetto MC, Belinson H, Tian Y, et al. Altered proliferation and networks in neural cells derived from idiopathic autistic individuals. *Mol Psychiatry*. 2017;22(6):820–835. https://doi.org/10.1038/mp.2016.95.
84. Wang P, Mokhtari R, Pedrosa E, et al. CRISPR/Cas9-mediated heterozygous knockout of the autism gene CHD8 and characterization of its transcriptional networks in cerebral organoids derived from iPS cells. *Mol Autism*. 2017;8(1):1–17. https://doi.org/10.1186/s13229-017-0124-1.
85. U.S. Department of Health and Human Services, National Institutes of Health NI of MH. *Schizophrenia*; 2020. Published https://www.nimh.nih.gov/health/topics/schizophrenia/index.shtml#part_152034.
86. Gao R, Penzes P. Common mechanisms of excitatory and inhibitory imbalance in schizophrenia and autism spectrum disorders. *Curr Mol Med*. 2015;15(2):146–167. https://doi.org/10.2174/1566524015666150303003028.
87. Ripke S, Neale BM, Corvin A, et al. Biological insights from 108 schizophrenia-associated genetic loci. *Nature*. 2014;511(7510):421–427. https://doi.org/10.1038/nature13595.
88. Fatemi SH, Folsom TD. The neurodevelopmental hypothesis of Schizophrenia, revisited. *Schizophr Bull*. 2009;35(3):528–548. https://doi.org/10.1093/schbul/sbn187.
89. Demjaha A, MacCabe JH, Murray RM. How genes and environmental factors determine the different neurodevelopmental trajectories of schizophrenia and bipolar disorder. *Schizophr Bull*. 2012;38(2):209–214. https://doi.org/10.1093/schbul/sbr100.
90. St Clair D, Blackwood D, Muir W, et al. Association within a family of a balanced autosomal translocation with major mental illness. *Lancet*. 1990;336(8706):13–16. https://doi.org/10.1016/0140-6736(90)91520-K.

91. Fletcher JM, Evans K, Baillie D, et al. Schizophrenia-associated chromosome 11q21 translocation: identification of flanking markers and development of chromosome 11q fragment hybrids as cloning and mapping resources. *Am J Hum Genet*. 1993;52(3):478–490.
92. Millar JK, Wilson-Annan JC, Anderson S, et al. Disruption of two novel genes by a translocation co-segregating with schizophrenia. *Hum Mol Genet*. 2000;9(9):1415–1423. https://doi.org/10.1093/hmg/9.9.1415.
93. Srikanth P, Han K, Callahan DG, et al. Genomic DISC1 disruption in hiPSCs alters Wnt signaling and neural cell fate. *Cell Rep*. 2015;12(9):1414–1429. https://doi.org/10.1016/j.celrep.2015.07.061.
94. Wen Z, Nguyen HN, Guo Z, et al. Synaptic dysregulation in a human iPS cell model of mental disorders. *Nature*. 2014;515(7527):414–418. https://doi.org/10.1038/nature13716.
95. Robinton DA, Daley GQ. The promise of induced pluripotent stem cells in research and therapy. *Nature*. 2012;481(7381):295–305. https://doi.org/10.1038/nature10761.
96. Rubio A, Luoni M, Giannelli SG, et al. Rapid and efficient CRISPR/Cas9 gene inactivation in human neurons during human pluripotent stem cell differentiation and direct reprogramming. *Sci Rep*. 2016;6(1). https://doi.org/10.1038/srep37540, 37540.
97. Skarnes WC, Pellegrino E, McDonough JA. Improving homology-directed repair efficiency in human stem cells. *Methods*. 2019;164–165:18–28. https://doi.org/10.1016/j.ymeth.2019.06.016.
98. Poon A, Zhang Y, Chandrasekaran A, et al. Modeling neurodegenerative diseases with patient-derived induced pluripotent cells: possibilities and challenges. *N Biotechnol*. 2017;39:190–198. https://doi.org/10.1016/j.nbt.2017.05.009.
99. Ha AD, Fung VSC. Huntington's disease. *Curr Opin Neurol*. 2012;25(4):491–498. https://doi.org/10.1097/WCO.0b013e3283550c97.
100. Saudou F, Humbert S. The biology of Huntingtin. *Neuron*. 2016;89(5):910–926. https://doi.org/10.1016/j.neuron.2016.02.003.
101. Zhang N, An MC, Montoro D, Ellerby LM. Characterization of human Huntington's disease cell model from induced pluripotent stem cells. *PLoS Curr*. 2010;2(OCT):1–11. https://doi.org/10.1371/currents.RRN1193.
102. Jeon I, Lee N, Li JY, et al. Neuronal properties, in vivo effects, and pathology of a Huntington's disease patient-derived induced pluripotent stem cells. *Stem Cells*. 2012;30(11):2602. https://doi.org/10.1002/stem.1245.
103. Jeon I, Choi C, Lee N, et al. In vivo roles of a patient-derived induced pluripotent stem cell line (HD72-iPSC) in the YAC128 model of huntington's disease. *Int J Stem Cells*. 2014;7(1):43–47. https://doi.org/10.15283/ijsc.2014.7.1.43.
104. Mattis VB, Svendsen SP, Ebert A, et al. Induced pluripotent stem cells from patients with huntington's disease show CAG repeat expansion associated phenotypes. *Cell Stem Cell*. 2012;11(2):264–278. https://doi.org/10.1016/j.stem.2012.04.027.
105. An MC, Zhang N, Scott G, et al. Genetic correction of huntington's disease phenotypes in induced pluripotent stem cells. *Cell Stem Cell*. 2012;11(2):253–263. https://doi.org/10.1016/j.stem.2012.04.026.
106. Kordasiewicz HB, Stanek LM, Wancewicz EV, et al. Sustained therapeutic reversal of Huntington's disease by transient repression of Huntingtin synthesis. *Neuron*. 2012;74(6):1031–1044. https://doi.org/10.1016/j.neuron.2012.05.009.
107. Stanek LM, Sardi SP, Mastis B, et al. Silencing mutant huntingtin by adeno-associated virus-mediated RNA interference ameliorates disease manifestations in the YAC128 mouse model of Huntington's Disease. *Hum Gene Ther*. 2014;25(5):461–474. https://doi.org/10.1089/hum.2013.200.
108. Rodrigues FB, Ferreira JJ, Wild EJ. Huntington's Disease Clinical Trials Corner: June 2019. *J Huntingtons Dis*. 2019;8(3):363–371. https://doi.org/10.3233/JHD-199003.

109. Grondin R, Kaytor MD, Ai Y, et al. Six-month partial suppression of Huntingtin is well tolerated in the adult rhesus striatum. *Brain*. 2012;135(4):1197–1209. https://doi.org/10.1093/brain/awr333.
110. Zuccato C, Valenza M, Cattaneo E. Molecular mechanisms and potential therapeutical targets in Huntington's disease. *Physiol Rev*. 2010;90(3):905–981. https://doi.org/10.1152/physrev.00041.2009.
111. Duyao MP, Auerbach AB, Ryan A, et al. Inactivation of the mouse huntington's disease gene homolog Hdh. *Science*. 1995;269(5222):407–410. https://doi.org/10.1126/science.7618107.
112. Dragatsis I, Levine MS, Zeitlin S. Inactivation of Hdh in the brain and testis results in progressive neurodegeneration and sterility in mice. *Nat Genet*. 2000;26(3):300–306. https://doi.org/10.1038/81593.
113. Zeitler B, Froelich S, Marlen K, et al. Allele-selective transcriptional repression of mutant HTT for the treatment of Huntington's disease. *Nat Med*. 2019;25(7):1131–1142. https://doi.org/10.1038/s41591-019-0478-3.
114. Hamilton G, Gillingwater TH. Spinal muscular atrophy: going beyond the motor neuron. *Trends Mol Med*. 2013;19(1):40–50. https://doi.org/10.1016/j.molmed.2012.11.002.
115. Edens BM, Ajroud-Driss S, Ma L, Ma YC. Molecular mechanisms and animal models of spinal muscular atrophy. *Biochim Biophys Acta Mol basis Dis*. 2015;1852(4):685–692. https://doi.org/10.1016/j.bbadis.2014.07.024.
116. Ebert AD, Yu J, Rose FF, et al. Induced pluripotent stem cells from a spinal muscular atrophy patient. *Nature*. 2009;457(7227):277–280. https://doi.org/10.1038/nature07677.
117. Chang T, Zheng W, Tsark W, et al. Brief report: phenotypic rescue of induced pluripotent stem cell-derived motoneurons of a spinal muscular atrophy patient. *Stem Cells*. 2011;29(12):2090–2093. https://doi.org/10.1002/stem.749.
118. Corti S, Nizzardo M, Simone C, et al. Genetic correction of human induced pluripotent stem cells from patients with spinal muscular atrophy. *Sci Transl Med*. 2012;4(165). https://doi.org/10.1126/scitranslmed.3004108, 165ra162.
119. Al-Zaidy SA, Mendell JR. From clinical trials to clinical practice: practical considerations for gene replacement therapy in SMA type 1. *Pediatr Neurol*. 2019;100:3–11. https://doi.org/10.1016/j.pediatrneurol.2019.06.007.
120. Zahoor I, Shafi A, Haq E. Pharmacological treatment of Parkinson's disease. In: *Parkinson's Disease: Pathogenesis and Clinical Aspects*. Codon Publications; 2018:129–144.
121. Schulte C, Gasser T. Genetic basis of Parkinson's disease: inheritance, penetrance, and expression. *Appl Clin Genet*. 2011;4:67–80. https://doi.org/10.2147/TACG.S11639.
122. Sundal C, Fujioka S, Uitti RJ, Wszolek ZK. Autosomal dominant Parkinson's disease. *Parkinsonism Relat Disord*. 2012;18(SUPPL. 1):S7–S10. https://doi.org/10.1016/s1353-8020(11)70005-0.
123. Soldner F, Hockemeyer D, Beard C, et al. Parkinson's disease patient-derived induced pluripotent stem cells free of viral reprogramming factors. *Cell*. 2009;137(7):1356. https://doi.org/10.1016/j.cell.2009.06.017.
124. Nguyen HN, Byers B, Cord B, et al. LRRK2 mutant iPSC-derived da neurons demonstrate increased susceptibility to oxidative stress. *Cell Stem Cell*. 2011;8(3):267–280. https://doi.org/10.1016/j.stem.2011.01.013.
125. Sánchez-Danés A, Richaud-Patin Y, Carballo-Carbajal I, et al. Disease-specific phenotypes in dopamine neurons from human iPS-based models of genetic and sporadic Parkinson's disease. *EMBO Mol Med*. 2012;4(5):380–395. https://doi.org/10.1002/emmm.201200215.
126. Imaizumi Y, Okada Y, Akamatsu W, et al. Mitochondrial dysfunction associated with increased oxidative stress and α-synuclein accumulation in PARK2 iPSC-derived

127. Devine MJ, Ryten M, Vodicka P, et al. Parkinson's disease induced pluripotent stem cells with triplication of the α-synuclein locus. *Nat Commun.* 2011;2(1):440. https://doi.org/10.1038/ncomms1453.
128. Soldner F, Laganière J, Cheng AW, et al. Generation of isogenic pluripotent stem cells differing exclusively at two early onset parkinson point mutations. *Cell.* 2011;146(2):318–331. https://doi.org/10.1016/j.cell.2011.06.019.
129. Schöndorf DC, Aureli M, McAllister FE, et al. IPSC-derived neurons from GBA1-associated Parkinson's disease patients show autophagic defects and impaired calcium homeostasis. *Nat Commun.* 2014;5:4028. https://doi.org/10.1038/ncomms5028.
130. Alzahimer's Association. 2010 Alzheimer's disease facts and figures. *Alzheimers Dement.* 2010;6(2):158–194. https://doi.org/10.1016/j.jalz.2010.01.009.
131. Goate A. Segregation of a missense mutation in the amyloid β-protein precursor gene with familial Alzheimer's disease. *J Alzheimers Dis.* 2006;9(SUPPL. 3):341–347. https://doi.org/10.3233/jad-2006-9s338.
132. Schellenberg GD, Payami H, Wijsman EM, et al. Chromosome 14 and late-onset familial Alzheimer disease (FAD). *Am J Hum Genet.* 1993;53(3):619–628.
133. Karch CM, Goate AM. Alzheimer's disease risk genes and mechanisms of disease pathogenesis. *Biol Psychiatry.* 2015;77(1):43–51. https://doi.org/10.1016/j.biopsych.2014.05.006.
134. Jiang T, Yu J-T, Tian Y, Tan L. Epidemiology and etiology of Alzheimer's disease: from genetic to non-genetic factors. *Curr Alzheimer Res.* 2013;10(8):852–867. https://doi.org/10.2174/15672050113109990155.
135. Yagi T, Ito D, Okada Y, et al. Modeling familial Alzheimer's disease with induced pluripotent stem cells. *Hum Mol Genet.* 2011;20(23):4530–4539. https://doi.org/10.1093/hmg/ddr394.
136. Israel MA, Yuan SH, Bardy C, et al. Probing sporadic and familial Alzheimer's disease using induced pluripotent stem cells. *Nature.* 2012;482(7384):216–220. https://doi.org/10.1038/nature10821.
137. Kolli N, Lu M, Maiti P, Rossignol J, Dunbar GL. Application of the gene editing tool, CRISPR-Cas9, for treating neurodegenerative diseases. *Neurochem Int.* 2018;112:187–196. https://doi.org/10.1016/j.neuint.2017.07.007.
138. Zaqout S, Morris-Rosendahl D, Kaindl AM. Autosomal recessive primary microcephaly (MCPH): an update. *Neuropediatrics.* 2017;48(3):135–142. https://doi.org/10.1055/s-0037-1601448.
139. Cohen MA, Wert KJ, Goldmann J, et al. Human neural crest cells contribute to coat pigmentation in interspecies chimeras after in utero injection into mouse embryos. *Proc Natl Acad Sci U S A.* 2016;113(6):1570–1575. https://doi.org/10.1073/pnas.1525518113.
140. Barker RA, Farrell K, Guzman NV, et al. Designing stem-cell-based dopamine cell replacement trials for Parkinson's disease. *Nat Med.* 2019;25(7):1045–1053. https://doi.org/10.1038/s41591-019-0507-2.
141. Reidling JC, Relaño-Ginés A, Holley SM, et al. Human neural stem cell transplantation rescues functional deficits in R6/2 and Q140 Huntington's disease mice. *Stem Cell Rep.* 2018;10(1):58–72. https://doi.org/10.1016/j.stemcr.2017.11.005.
142. Fujiwara N, Shimizu J, Takai K, et al. Restoration of spatial memory dysfunction of human APP transgenic mice by transplantation of neuronal precursors derived from human iPS cells. *Neurosci Lett.* 2013;557(PB):129–134. https://doi.org/10.1016/j.neulet.2013.10.043.
143. Wernig M, Zhao JP, Pruszak J, et al. Neurons derived from reprogrammed fibroblasts functionally integrate into the fetal brain and improve symptoms of rats with Parkinson's disease. *Proc Natl Acad Sci U S A.* 2008;105(15):5856–5861. https://doi.org/10.1073/pnas.0801677105.

CHAPTER THIRTEEN

Reprogramming translation for gene therapy

Chiara Ambrosini[*], Francesca Garilli, and Alessandro Quattrone

Laboratory of Translational Genomics, CIBIO—Department of Cellular, Computational and Integrative Biology, University of Trento, Trento, Italy
[*]Corresponding author e-mail address: chiara.ambrosini@unitn.it

Contents

1. Introduction	440
2. Translational regulation in eukaryotes	442
2.1 Protein synthesis in eukaryotes	442
2.2 Translational regulation	444
3. Kozak consensus sequence	445
3.1 The Kozak sequence influences translation	445
3.2 Kozak sequences and disease	450
3.3 Kozak as a genome editing target	453
4. uORFs	455
4.1 uORFs impact on translation	455
4.2 Diseases caused by uORF mutations	458
4.3 uORFs as genome editing targets	460
5. Internal ribosomal entry sites (IRESs)	463
5.1 IRES elements and cap-independent translation initiation	463
5.2 IRESs as therapeutic targets	465
6. Concluding remarks	466
Acknowledgments	466
References	467

Abstract

Translational control plays a fundamental role in the regulation of gene expression in eukaryotes. Modulating translational efficiency allows the cell to fine-tune the expression of genes, spatially control protein localization, and trigger fast responses to environmental stresses. Translational regulation involves mechanisms acting on multiple steps of the protein synthesis pathway: initiation, elongation, and termination. Many cis-acting elements present in the 5′ UTR of transcripts can influence translation at the initiation step. Among them, the Kozak sequence impacts translational efficiency by regulating the recognition of the start codon; upstream open reading frames (uORFs) are associated with inhibition of translation of the downstream protein; internal ribosomal entry sites (IRESs) can promote cap-independent translation. CRISPR-Cas technology is a revolutionary gene-editing tool that has also been applied to the regulation of gene expression. In this chapter, we focus on the genome editing approaches developed to modulate the translational efficiency with the aim to find novel therapeutic approaches, in particular acting on the *cis*-elements, that regulate the initiation of protein synthesis.

Abbreviations

ABE	adenine base editor
CBE	cytosine base editor
DSB	double-strand break
HDR	homology-directed repair
IC	initiation complex
IRESs	internal ribosomal entry sites
mORF	main ORF
NHEJ	non-homologous end-joining repair
PIC	preinitiation complex
TC	ternary complex
TIS	translation initiation sequence
uAUG	upstream AUG
uORFs	upstream open reading frames

1. Introduction

Protein synthesis is a crucial, energy-consuming process for a cell, and it must be tightly regulated to ensure cell homeostasis and survival. Understanding the complex processes that underlie translational regulation could provide new targets for genome editing approaches aimed at modifying the levels of gene expression with therapeutic purposes, by acting on post-transcriptional steps. The CRISPR-Cas system has revolutionized the field of genome editing in the last decade. This system works thanks to a single-guide RNA which base-pairs with a target sequence in the DNA and allows an endonuclease (Cas) to perform a double-strand break (DSB) on the locus of interest. The DSB activates the cellular DNA repair mechanisms, which more often will lead to non-homologous end-joining repair (NHEJ), causing gene disruption because of insertions or deletions of usually few random nucleotides. After the DSB, insertion of a precise edit in the DNA can also be achieved thanks to homology-directed repair (HDR), adding a donor DNA to the system.[1] The CRISPR-Cas gene-editing tool has been increasingly applied in the regulation of gene expression, mainly acting on transcriptional efficiency.[2,3] Nevertheless, several examples of applications of CRISPR-Cas technology on the translational layer of regulation have been reported and will be discussed in this chapter.

Eukaryotic translation is a multi-step process that consists of initiation, elongation, termination, and ribosome recycling. Each step is promoted by several *cis* and *trans* factors necessary for the process to take place and/or able to modulate its efficiency. Among them, translation initiation can be tuned by the presence of *cis*-acting elements in the 5′ UTR of the mRNA, such as (i) the Kozak sequence, which impacts translational efficiency by regulating the recognition of the start codon; (ii) upstream open reading frames (uORFs), associated with inhibition of translation of the downstream protein; (iii) internal ribosomal entry sites (IRESs), that promote cap-independent translation (Fig. 1). In this chapter, we describe the importance of these translation regulatory elements and their targeting with genome editing approaches for the development of innovative therapeutics for genetic diseases.

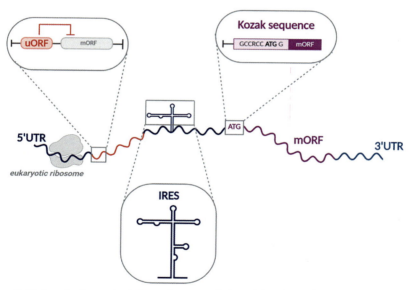

Fig. 1 Main *cis*-elements influencing translation initiation of eukaryotic mRNA. Upstream open reading frames (uORFs) are inhibitory open reading frames able to hamper translation of the downstream protein encoded by the main ORF (mORF); internal ribosome entry sites (IRESs) are complex RNA structures commonly found in a subset of cellular mRNAs, but originally discovered in viruses, that can promote cap-independent translation; the Kozak sequence is the most favorable nucleotide context around the starting codon of a protein.

2. Translational regulation in eukaryotes
2.1 Protein synthesis in eukaryotes

Eukaryotic translation can be described as a four-step process: initiation, elongation, termination, and ribosome recycling. Initiation is a rate-limiting step of translation that starts with the formation of a complex between the 40S small ribosomal subunit and the initiator Methionyl-tRNA$_i$ and ends with the assembly of the 80S ribosome, ready for the elongation phase.[4] The main points of translation initiation are depicted in Fig. 2. The first step is characterized by the formation of the 43S preinitiation complex (PIC): the 40S small ribosomal subunit binds to the ternary complex (TC), which is composed of the eukaryotic initiation factor eIF2 (a GTPase), GTP, and Met-tRNA$_i$, in a reaction favored by eIF1, 1a, 3 and 5.[5] The resulting 43S PIC subsequently loads onto an activated mRNA thanks to the eIF4F cap-binding complex. eIF4F is composed of eIF4G, a scaffold protein, which in turn binds to mRNA, eIF4E (a cap-binding protein), eIF4A (a DEAD-box RNA helicase), and poly-A binding protein (PABP). As a result, the mRNA is stabilized in a circularized closed-loop structure and the complex is ready for the next step. Most eukaryotic mRNAs are translated according to the scanning mechanism, in which the so formed PIC scans the 5′ UTR in an energy-consuming process until it finds the start codon, AUG. In particular, this process involves the ATP-dependent unwinding of secondary structures in the 5′ UTR hampering translation, which allows the movement of the 43S complex along the mRNA.[6] The RNA unwinding is operated by eIF4A (stimulated by eIF4B) and other helicases such as Dhx29 and Ddx3/Ded1.[5] To scan each base in the 5′ UTR, consecutive triplets enter the peptidyl (P) decoding site of the small ribosomal subunit searching for complementarity with the anticodon in the Met-tRNA$_i$. Although according to the "position rule"[7] the AUG closest to the 5′ end is generally used as the translation initiation codon, the ribosome can also skip it if the nucleotide context (Kozak sequence, see Section 3) is not optimal, in a process called leaky scanning.[8] When the suitable AUG codon is decoded, the scanning process ends and the 48S initiation complex (IC) is formed. Most initiation factors are released, including eIF1. This allows inorganic phosphate release, and consequently eIF2-GDP dissociation and PIC conversion into a closed conformation. Finally, eIF5B-GTP promotes the joining of the large 60S ribosomal subunit and the formation of the elongation-competent 80S IC. eIF5 is the factor considered to be

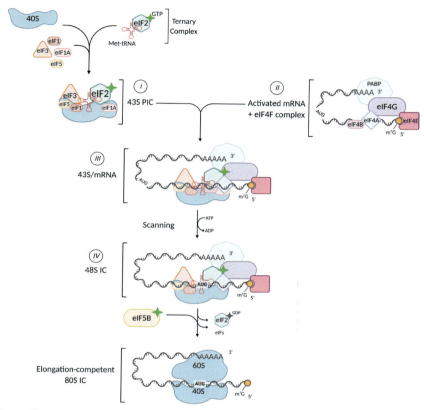

Fig. 2 The main steps of translation initiation in eukaryotes. The first step (I) is the formation of the 43S preinitiation complex (PIC), by the ternary complex (TC), composed of eIF2-GTP and Met-tRNA, and the 40S ribosomal subunit with eIF1, 1A, 3 and 5; (II) the mRNA is activated by the eIF4F complex, comprising eIF4A, B, E, G and polyA-binding protein (PABP), and reaches a circularized closed-loop structure; (III) the 43S PIC is loaded onto the mRNA and scans the 5' UTR until the recognition of the start codon AUG; (IV) when the AUG is decoded, most eIFs are released, 48S initiation complex (IC) is formed, and eIF5B-GTP favors the joining of the 60S ribosomal subunit to reach the 80S elongation-competent IC.[9,10]

responsible for the discrimination of the correct AUG by the scanning ribosome. It is indeed able to activate eIF2-driven ATP hydrolysis only if the scanning is paused for a sufficiently long time. This enables the machinery to distinguish a real initiator codon from other triplets (such as UUG).[9,10]

The second and central step of protein synthesis in eukaryotes is elongation, in which the encoded protein is progressively synthesized, from the first codon after the AUG until the 80S ribosome encounters the termination codon. Elongation can be divided into three main steps that lead to the inclusion of each subsequent amino acid in the nascent peptide: decoding, peptide bond formation, and mRNA-tRNA complex translocation.[11,12] The process starts with the Met-tRNA$_i$ bound to the peptidyl (P) site of the ribosome, the start codon AUG base-paired with its anticodon, and the second codon of the mRNA ORF in the aminoacyl (A) site of the ribosome. In the decoding phase, the correct aminoacyl-tRNA, charged with the amino acid corresponding to the codon in the A site of the ribosome, is selected. Each aminoacyl-tRNA is delivered to the A site in a ternary complex with the eukaryotic elongation factor eEF1A bound to GTP. The recognition of the codon present in the A site by base-pairing with the anticodon of the aminoacyl-tRNA causes GTP hydrolysis and release of eEF1A-GDP. The aminoacyl-tRNA is then accommodated into the A site of the ribosome.[13,14] The exchange factor eEF1B recycles eEF1A-GDP to eEF1A-GTP for the next round of elongation.

When a stop codon (UAG, UGA, or UAA) reaches the A site of the ribosome, termination takes place. Eukaryotic termination is carried out by two factors: the class I release factor, eRF1 catalyzes stop-codon recognition and hydrolysis of the ester bond of the peptidyl-tRNA; the class II factor, eRF3, is a GTPase that stimulates the release of the peptide from eRF1.[14,15]

The recycling is the final step of translation, in which the mRNA and the deacetylated tRNA are released, the ribosomal subunits are dissociated and the system is again ready for the subsequent round of initiation.[16]

2.2 Translational regulation

Contrary to the mechanisms of gene expression regulation acting at the level of transcription, translational control allows the cell to change more rapidly the levels of the expressed proteins. Consequently, modulating protein translation can be used to answer rapidly to potential environmental stress stimuli, or to control the distribution of the proteins spatially and temporally to rapidly re-shape the proteome[17,18]; moreover, translation regulation acquires a pivotal role in embryonic development, in which gene transcription is mostly inactive, and in other development and differentiation processes such as stem-cell proliferation.[19,20]

A large number of mechanisms have evolved to regulate translation.[21] We can generally distinguish between two types of controls, those acting on the global translation of the cell, and those acting at the level of specific mRNAs. Most general translation regulatory mechanisms concentrate on the step of translation initiation. The phosphorylation of several factors involved in this phase is a common way to quickly regulate the rate of the global process. One of the most well-studied examples, especially under stress conditions, is the phosphorylation of the α subunit of the heterotrimer eIF2, which causes an inhibition of its ability to bind to Met-tRNA$_i$ in the ternary complex, a crucial step for translation initiation.[4] An opposite example is the phosphorylation of eIF4E-binding proteins (4E-BPs, namely 4E-BP1, 4E-BP2, and 4E-BP3[22]) by the mTOR kinase, inhibiting 4E-BPs ability to bind to eIF4E. As a consequence, eIF4E binds to eIF4G and the m7G cap, collectively promoting initiation.[21,23,24] In the case of mechanisms acting at the level of specific mRNAs, translation regulation often relies on *cis*-elements and *trans*-acting factors that can enhance or hinder the recruitment of ribosomes on the mRNA. These factors are mostly present in both the 5′ and 3′ untranslated regions (UTRs) of the mRNA. The elements present in the 3′ UTR are usually *trans*-acting factors instrumental to the modulation of the transcript localization, degradation and translation through the binding of RNA-binding proteins (RBPs) or by ribonucleoprotein (RNP) particles containing micro-RNAs (miRNAs).[20,25] The *cis*-acting elements are present in the 5′ UTR of the transcripts and are, instead, mostly responsible for the processes of ribosomal scanning on the mRNA or the recognition of the AUG start codon by the translational machinery.[26] Here, we will focus on these latter elements, their impact on translation, and the opportunities offered by their regulation with genome editing tools to influence translation of specific genes.

3. Kozak consensus sequence
3.1 The Kozak sequence influences translation

In 1981 Marylin Kozak defined the Kozak consensus sequence, after decades of deep studies on eukaryotic mRNA, as the optimal translation initiation sequence (TIS).[27] The researcher investigated the pivotal role of the AUG codon and other key sequences located around it. The AUG sequence had been already recognized as the start site for protein biosynthesis, even though just a few AUG triplets in a single mRNA were really recognized as ribosome binding sites. Some years before, Kozak came up with the idea that ribosomes are able to bind the 5′ end of mRNA and then start to scan the entire sequence until the recognition of the first AUG codon.[7] In this model, the only variable that allows the ribosomes to recognize the starting codon is its position along the mRNA sequence. This process was defined by Kozak as the "scanning mechanism". To confirm this hypothesis, the researcher analyzed the sequences of more than 100 different eukaryotic messengers, finding out that the model was respected in 90% of the cases. On the other hand, in 10% of the analyzed sequences, the model did not reflect the real case and the translation started from a downstream AUG codon that was not the first in line. Furthermore, as the AUG flanking regions had been analyzed in a growing number of cases, it had become clear that the sequence located around the starting codon is not random. Kozak defined the region flanking the initiator codon as the 3 nucleotides before the AUG (positions −3, −2, −1), and a single nucleotide after the AUG (position +4), where the Adenine of the starting codon represents the position +1.[27] As a matter of fact, the researcher reported that the positions −3 and +4 are highly conserved. To further investigate the role of the nucleotides located in these crucial positions, *in vitro* ribosome binding studies were performed. Kozak observed that placing a purine in position −3 or +4 enhances the binding efficiency of ribosomes, in particular an Adenine (A) in −3 position and a Guanine (G) in +4 position. This conclusion was reinforced by the fact that those specific purines were never found in non-functional AUG sites in eukaryotic messengers.[27] Anyway, the distinction between functional and non-functional translation initiation sites was still not completely clarified, since there are some internal AUGs in the eukaryotic mRNAs that possess the favorable flanking sequence, but are not recognized by the ribosome to initiate translation. In addition, in 1980 Sherman showed that translation appears to start at the most proximal AUG codon and that in selected mRNAs in yeast, when the physiological initiation codon was disrupted by mutagenesis, the introduction of a new AUG triplet downstream is able to restore the translation, without any dependency on the presence of an optimal flanking region.[28]

Based on these previous results and the last studies, apparently in contrast with the "functional and non-functional" AUG starting sites theory, Kozak formulated a reviewed scanning mechanism: the first in line AUG codon is favored by its position, and in most of the cases it also presents purines in −3 and +4 hotspot positions, even though ribosomes are able to start the translation also at an internal AUG start codon if that site presents a favorable flanking region and the upstream AUG triplets show pyrimidine in position −3. The presence of additional AUG triplets can somehow inhibit translation, recruiting part of the 40S subunits to the non-functional starting codons. Additionally, the presence of multiple favored AUGs in a single mRNA can give rise to the translation of more than one protein per messenger.[27] For example, SV40 mRNA was reported to direct synthesis of either VP2 or VP3 viral protein by two different AUGs starting codons.[29,30] Another example, reported also by Kozak, is the case of herpes thymidine kinase protein, in which the synthesis of a second small protein is mediated by the same mRNA.[31]

With a big compilation of mRNAs of lower and higher organisms, Kozak took advantage of the increased number of variants and the stringent criteria of evaluation to confirm the observations postulated in 1981 and to find new variables and information about the crucial role of the AUG flanking region in eukaryotes. The distribution of positions from -1 to -6 has been described as non-random, as well as the presence of a purine in -3 position, more precisely, the predominance of a Cytosine in positions -1, -2, -4, and -5 has been reported for a large number of transcripts.[32] It has been clearly shown that the AUG initiation codons were characterized by the presence of a purine in position -3 (most often A), and in a few cases (3%) a pyrimidine was found in that position, as was already remarked in the previous survey. Interestingly, the flanking region of internal downstream AUGs, in most of the cases, did not show the same features. This can be easily explained by the presence of methionine in the coding sequence that is codified by the AUG codon as well.

In conclusion, the sequence CC(A,G)CC$\underline{AUG}$(G) (the initiation codon is underlined) emerges as a consensus sequence for eukaryotic initiation sites. Furthermore, Kozak confirmed that between two different AUG codons, located in an equally favorable flanking sequence, the first in line at the 5′ terminus is always the favored one, because 40S ribosomal subunits likely scan the 5′ end of the mRNA linearly. Instead, in the cases in which the first AUG codons found in the 5′ terminus do not show a favorable flanking region, mostly with a pyrimidine in position -3, ribosomes are able to recognize the initiator codon, but due to the non-appropriate sequence around the AUG triplet, the 40S subunit skips that initiator, scanning up to the next AUG codon with a more favorable consensus sequence. However, how this mechanism influences protein translation was still not clear.[33]

Kozak then investigated the conservation of the motif around AUG codons using a larger and more diversified database. The AUG flanking region was expanded considering the 12 nucleotides stretch that precedes the AUG initiator codon. It was found that the expanded consensus region in vertebrates is GCCGCC(A,G)CC$\underline{AUG}$G. The positions -3 and $+4$ remain crucial and are the most important positions in the sequence, defining the "strength" of the consensus region: pyrimidines in those positions determine a "weak" consensus sequence, while purines signal a favorable region and a "strong" AUG initiator codon. Indeed, this explanation was also useful in defining the role of upstream AUG codons with weak consensus sequences, which seem to have an inhibitory effect on translation.[34] Last but not least, a G repetition in positions $-3, -6, -9$ is quite noticeable in the vertebrate consensus sequence context, although its role in those positions was not completely clarified.[35]

Further studies on 340 different genes showed a non-uniform distribution of Guanine in the triplets around AUGs, whose usual position is the first one in the triplet.[36] Although it can be assumed that the primary sequence flanking AUG codons has an important function in ribosome recognition and binding, Kozak did not exclude the possibility that other sequences located around the AUG might have a role as well. Indeed, in 1986 it has been described by Kozak that sequences located downstream but adjacent to AUG initiators can assume a loop structure that facilitates ribosome binding.[37] The researcher also noticed that the space between the formation of the loop and AUG codons was relevant in modulating the translation initiation efficiency by slowing the 40S subunit scanning and stopping it in the correct position with respect to the functional AUG initiator. This hypothesis was confirmed in 1997 by Huang and colleagues, who reported the crucial contribution of 40S subunits slowing during translation. A longer pause occurs only when the subunit encounters a functional AUG triplet.[38]

The Kozak model is still valid nowadays. Nevertheless, in more recent years many different studies were performed on the Kozak sequence, its characteristics, and functions (Fig. 3). For example, in 2014 Noderer and colleagues introduced a genome-wide-study approach to study translation initiation efficiency of 65,536 different TIS sequences, selected by randomizing eight positions in the Kozak sequence.[39] The researchers combined fluorescence-activated cell sorting (FACS) with high-throughput DNA sequencing. In this study, a genetic reporter system was adapted for analyzing all the different TIS sequences of interest. In this reporter, the translation of green fluorescent protein (GFP) was under the control of the library of TIS sequences. Red fluorescent protein (RFP) was translated from the same transcript using an IRES and used for normalization. The entire library was cloned into lentiviral vectors and then the viruses containing the TIS variants were used to transduce the cells. After 3 days, FACS was used to isolate cell subpopulations measuring GFP/RFP levels within 20 different ranges. All the TIS sequences in each group were amplified and sequenced by the

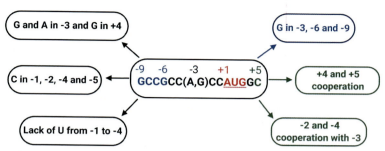

Fig. 3 The Kozak sequence. The central panel illustrates the favorable AUG flanking region for mRNA translation initiation defined by Kozak M. in vertebrates. The three peripheral panels in black summarize the characteristics of the conserved motif of the Kozak sequence found in lower eukaryotes. After the studies performed in higher eukaryotes and in vertebrates, Kozak M. confirmed the previous results and defined the final AUG flanking region. The panel in blue on the right shows the last characteristics found by Kozak M. The two panels in dark green represent, instead, the discoveries about Kozak sequences of the last decade.

Illumina next-generation platform. The researchers called this innovative system FACS-Seq approach. The importance of positions −3 and +4 was confirmed by this study. However, the analysis also found the −2, −4, and +5 positions to influence protein expression. Interestingly, Noderer and colleagues also reported significant cooperation between combinations of nucleotides present in hotspot positions: +4 and +5 positions have been reported to be strongly correlated. On the contrary, positions −2 and −4, previously reported as not so important in TIS recognition, were found to finely correlate with the −3 position[39] (Fig. 3).

Starting from Noderer's work, other libraries were designed by the researchers in order to study new Kozak sequence variants. Indeed, later a competitive-growth-assay-based library was developed in yeast. This technology used the measurement of the His3 proteins in yeasts transformed with a library of plasmids harboring the HIS3 reporter gene and random Kozak variants. The cells were grown in a selection histidine depleted media, the Kozak sequence variants were analyzed and the enrichment or depletion of each variant was correlated to the respective His3 expression over time. Compared to the limited number of cells that can be screened with FACS, this method achieved the analysis of half-million variants and new *cis*-regulatory motifs were identified as decisive in ribosome mRNA recognition and binding.[40] For example, G quadruplex motif was found to influence expression but other novel motifs are still not linked with specific mechanisms. These latter motifs could represent specific target sites for RNA-binding proteins.

All these studies corroborated the notion that the Kozak sequence is a predominantly conserved feature among lower and higher eukaryotes, it has a crucial role in the regulation of translation, and each nucleotide in AUG flanking region can contribute to the overall ribosomal recognition of the translation initiation sequence.

As a matter of fact, the interest in the primary role of the Kozak sequence in translation regulation is increasing as is demonstrated by the growing number of new studies published during the last 10 years.[39–42] From 2014 until the present day, many groups worldwide investigated the functional roles of the region flanking the TIS. The researchers applied different high-throughput technologies and computational approaches to study a constantly increasing number of Kozak and 5′ UTR sequences. An example is the development of a new 5′ UTR calculator in yeast where 2041 different 5′ UTR sequence variants with their respective protein abundances have been screened.[41] The crucial role of purine in the −3 position but also the combinatorial effect of the −2 and −3 positions have been confirmed. Moreover, the formation of mRNA secondary structures close to the AUG initiator codons was shown as highly important as well[41] (Fig. 3).

The increasingly recognized role and understanding of the Kozak sequence impact on translation have led to attempts to regulate it for the modulation of gene expression. In a recent work Blanco and colleagues aimed at improving the yield and product quality of bispecific antibodies (bsAbs) by tuning translational strength.[42] BsAbs are antibodies able to recognize two different epitopes on one or two different antigens. The authors screened a library of Kozak sequences created by randomizing 4 positions before and one after the AUG to identify variants able to enhance the expression of heavy chains (HCs) and light chains (LCs). They found that by regulating the expression of HCs and LCs, bispecific assembly increased by more than two-fold over control, accompanied by a higher product purity.[42]

3.2 Kozak sequences and disease

In the last two decades, it has been shown that the conservation of the consensus Kozak sequence is crucial to maintain the rate of translation initiation and how disruption of this motif can lead to impairment in ribosome mRNA recognition and binding.

For this reason, several diseases were found out to be correlated to mutations located in the Kozak sequence. In 2010 Xu and colleagues performed an NCBI database screening on SNPs located in Kozak sequences or directly in AUG codons, naming them kSNPs and sSNPs respectively.[43] The researchers listed all the reported human (k,s)SNPs, trying to provide an explanation of how diseases are associated with (k,s)SNP-containing genes via haplotype analysis. They highlighted some cases where (k,s)SNPs were located in the same gene linked to the disease-associated SNPs already reported to increase the susceptibility to the pathology. This correlation could be explained by the fact that translation modulation caused by SNPs located in the Kozak region can impact protein production and for this reason, they can influence disease risks. For example, a SNP in position −3 in the diazepam-binding inhibitor (*DBI*) gene has been reported to cause a decrease in protein production *in vitro*.[43] Indeed, *DBI* mutations have been already signaled as disease-associated SNPs in anxiety disorder and type II diabetes.[44,45] Other disorders have been found to be caused by polymorphisms that affect crucial Kozak sequence residues. Coronary artery disease (CAD) and its most acute manifestation, myocardial infarction (MI), are complex diseases highly diffused in the population. In both of these acute coronary syndromes (ACS), platelets play a crucial role in the development and progression of the disease. Multiple genes and a series of respective SNPs have been associated with these pathologies. The GPIb/IX/V complex is the major platelet surface receptor and GPIbα is the largest component of this complex. Thymidine to Cytosine single nucleotide substitution at position −5 from the start codon is one of the Kozak sequence polymorphisms that characterize *GP1BA* gene. It has been shown that a sample of individuals developing a stroke was enriched in the T/C genotype (32.2%) compared to the control. Thanks to flow cytometry, the researchers showed that the amount of GPIbα on platelets surface increases proportionally to the −5C allele presence in the cell population.[46] Thus, the presence of a higher number of the protein receptors on platelets surface could be caused by an increased translation of GPIbα due to the Kozak SNP.[47] This event could be crucial in the pathogenesis of ischemic stroke. GPIbα expression could make platelets more adhesive and more rapidly activated when stress occurs. In the end, the overactivation of platelets might cause thrombosis and vessel occlusion. This mechanism could explain the reason why patients that harbor this specific SNP, with ischemic stroke, are more prone to develop chronic platelet activation, usually associated with poststroke mortality.[48]

Another clear evidence of the strict correlation between Kozak modulation by SNPs is the polymorphism that affects the CD40 protein in Graves' disease (GD). This is an antibody-mediated autoimmune disease characterized by hyperthyroidism, diffuse goiter, and lymphocytic infiltration of the thyroid. CD40 is a surface-expressed molecule involved in B-cells regulation and activation and it is expressed in non-terminally differentiated B cells. Due to its pivotal role in immune response, it has been found that CD40 deregulation is involved in auto-immune pathologies. In the case of Graves' disease, a SNP located in position −1 in the Kozak sequence of the *CD40* codifying gene has been reported by the researchers as a risk factor for GD. Similar to the GPIbα association to the risk of stroke, *CD40* translation could increase because of a specific −1 SNP genotype, and overexpression of the protein in B-cell surface could influence the etiology of the disease, boosting the immune response.[49]

Concerning these three cases, it can be concluded that modifications in the Kozak sequence and 5' UTR region are capable of increasing translation of disease-associated genes. Individuals harboring Kozak SNPs are more prone to develop disease or to have a worse prognosis.

On the other hand, other cases where 5' UTR SNPs have been reported, show a decrease in translation due to a negative modulation of the Kozak sequence. In 2000, Usuki and Maruyama found that in the Ataxia with vitamin E deficiency (AVED) pathology, one of the two disease-associated mutations affected the Kozak sequence.[50] AVED is an autosomal recessive neurodegenerative disease caused by mutations in α-tocopherol transfer protein (α-TTP). This protein is a receptor able to bind Vitamin E with high affinity, and impairment in function or in production of this gene causes problems in binding and handling of Vitamin E in specific tissues.[50] The researchers showed that a C to T modification in position −1 of the Kozak region of the *α-TTP* gene leads to a strong reduction in protein production due to a non-favorable Kozak.[51]

Another example reported the year later regards a study on the *BRCA1* gene, already well known and characterized as the gene involved in most cases of breast and ovarian cancer.[52] Usually, hereditary mutations on this onco-suppressor gene involved in DNA replication and maintenance, cause inactivation of the BRCA1 codified protein. In less frequent cases of sporadic breast cancer, the mutations in the *BRCA1* gene are reported to be somatic, dysregulating protein function only in target tissues. In a screening performed on a group of individuals with sporadic *BRCA1* mutations, it has been found out a G to C transversion falling in position −3 of Kozak sequence of the onco-suppressor. The researchers investigated by *in vitro* and *in vivo* experiments the role of this mutation on translation efficiency and reported that the mutated form is less efficiently translated compared to the wild-type.[53]

In conclusion, the Kozak consensus sequence has a key role in the regulation of translation and just a single nucleotide modification in specific target genes could lead to disease associate risk.

3.3 Kozak as a genome editing target

The Kozak sequence can be seen as a target for genome editing approaches aimed at modulating translational efficiency (Fig. 4). The first and more straightforward application is to induce gene silencing of target genes by disrupting the Kozak sequence. Such an effort has been attempted in 2020 by two groups simultaneously.[54,55] Both groups used the technology of CRISPR-Cas base editors, a recently developed CRISPR-Cas tool that allows introducing specific point mutations in the desired site by delivering

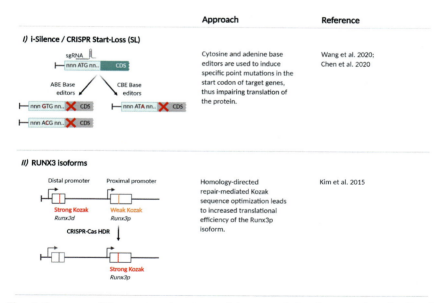

Fig. 4 Genome editing approaches to modulate translation by targeting the Kozak sequence. (*I*) Base editors have been used to insert point mutations in the Kozak sequence of target genes to impair protein translation; (*II*) CRISPR-Cas-mediated HDR has been exploited to modify the Kozak sequence of a target gene to understand the biological role of different isoforms.

a deaminase fused with a defective Cas (nCas, able to nick only one strand of the DNA) to the target locus.[56] The two originally developed types of base editors are cytosine base editors (CBE), which allow converting C-G base pair into T-A base pair,[57] and adenine base editors (ABE), which convert an A-T into a G-C base pair.[58] More recently developed base editors are able to perform transversion mutations[59] or to install concurrently both adenine and cytosine editing.[60,61] Wang and colleagues applied base editing to the Kozak sequence creating i-Silence, an ABE-mediated strategy to mutate the start codon of desired genes from ATG to GTG or ACG. The authors demonstrated the efficacy of this method by targeting four endogenous loci in HEK293T cells *(HDAC1, SEC61B, PIGH,* and *FTL)*, reaching an efficiency of substitution ranging from 60% to 80% for all the target genes, as assessed by deep sequencing. Moreover, they used i-Silence to target the programmed cell death protein 1 (PD-1) *in vivo*, injecting the mRNA of the base editor and the sgRNA in mouse embryos, and obtaining a PD-1 knockout mouse.[54] The major concern that can be pointed out about this approach is the base editors-induced off-targets. Despite the authors analyzed off-targets genome-wide, it is now known that it is important to also evaluate transcriptome-wide off-target effects induced by these newly developed tools. It has been shown that the base editors can cause off-target deaminase activity in a guide-RNA independent manner and this promiscuous reactivity can lead to undesired off-targets on the genome and the transcriptome of the cells.[62] This issue has been addressed in recent works creating new deaminase variants able to reduce the off-target effects by, for example, introducing mutations that decrease the enzyme ability to bind or edit RNA.[63–65] Besides the therapeutic potential, Wang and colleagues point out the applicability of i-Silence as a tool for investigating the functions of different isoforms of target genes by mutating the different start codons.[54,66]

In the second aforementioned work, Chen and colleagues propose CRISPR Start-Loss (CRISPR-SL), an analogous approach that takes advantage of both types of base editors (CBE and ABE). The authors tested the system on endogenous loci in HEK293T, as well as in rabbits, and they underline that cytosine base editors can expand the repertoire of genes targetable by this approach.[55]

In a work by Kim and colleagues, CRISPR-Cas-mediated HDR was used to demonstrate the importance of the Kozak sequence in the regulation of gene expression, in particular in fine-tuning the translation of different isoforms from a single gene (i.e. *RUNX3*).[67] The transcription factor

RUNX3 regulates the differentiation of T lymphocytes, specifically promoting the development of CD8+ cytotoxic T cells and CD4+ TH1 cells.[68] In these cells, *RUNX3* is expressed at high levels and translated as an isoform from a distal promoter (Runx3d). RUNX3 protein is instead undetectable in naive CD4+ T cells and TH2 cells, where a different isoform is derived by a proximal promoter (Runx3p). In naive CD4+ T cells and TH2 cells, the expression of the RUNX3 protein must be restricted to allow CD4+ T cells to differentiate in TH2.[69] The authors demonstrated that Runx3p is regulated by a sub-optimal Kozak sequence, which is, at least in part, responsible for the non-permissive expression of the protein in some subsets of T cells. By using CRISPR-Cas mediated HDR, they modified the Runx3p Kozak with a predicted optimal nucleotide context in a T cell line and observed a 3-fold increase in translational efficiency (i.e., protein to mRNA ratios). Even if their data demonstrate that additional layers of regulation influence the expression of this protein, such as feedback mechanisms at the level of the transcripts, this work proves that the Kozak sequence has a pivotal role also in the spatial and temporal differential gene expression.[67]

Collectively, these examples show a growing interest in the translation initiation site as a target for genome editing approaches, which could help to find therapeutic solutions but also to unravel previously unknown biological functions of target genes.

4. uORFs
4.1 uORFs impact on translation

Upstream open reading frames (uORFs) are short elements within the 5′ UTR of the mRNA able to inhibit translation from the main ORF (mORF).[70] More specifically, uORFs are defined as short potential reading frames with an upstream AUG (uAUG) in the 5′ UTR and an in-frame stop codon upstream or downstream of the starting codon of the mORF.[71] Among the *cis*-acting elements that can affect translation initiation, uORFs play a major role. It is in fact known from recent bioinformatic analyses that about 50% of human mRNAs contain at least one predicted functional uORF[72] and we know from ribosome footprinting that uORFs can be also translated.[73] Moreover, it has been demonstrated that uORFs are broadly conserved over evolution, indicating a crucial biological role.[74]

Several factors influence the ability of a uORF to repress translation from the downstream mORF:
- *The nucleotide context*: a strong Kozak sequence around the AUG of the uORF is more likely to be recognized and to trigger the translation of the uORF encoded peptide, negatively affecting the mORF[75];
- *The uORF position*: a long distance from the 5′ cap and a short intercistronic distance (i.e. the distance between the termination codon of the uORF and the AUG of the mORF) are positively correlated to the repressive activity[76–78];
- *The number of uORFs*: some transcripts bear multiple uORFs. This can lead to combinatorial effects that are difficult to predict on the mORF.[79] Interestingly, the mRNAs that present multiple uORFs in their 5′ UTR often belong to the classes of transcription factors, growth factors, and oncogenes, all transcripts that are scarcely translated under basal conditions.[80]

Usually, uORFs are able to downregulate translation of the mORF by 30%–80%.[81,82] Their inhibitory activity is exerted by two prevalent mechanisms: ribosome stalling and ribosome dissociation. In the first case, the ribosome gets stalled while translating the uORF-encoded peptide, either during the elongation or the termination phase. This effect is mainly driven by the nucleotide context encountered by the ribosome (such as the presence of rare codons), by RNA secondary structures, or by interaction with *trans*-acting factors.[83,84] If the ribosomal stalling takes place at the uORF termination, its stop codon could be recognized as a premature termination and trigger the activation of the nonsense-mediated mRNA decay (NMD) pathway.[85,86] Ribosome dissociation from the mRNA after the translation of uORFs, instead, is usually caused by peptides encoded by the uORFs, and is more likely to happen after long uORFs followed by short intercistronic distances.[87–89]

In particular cases, uORFs can also function as translational modulators, and, for example, promote the expression of certain mORFs from a subset of genes. This happens predominantly through the integrated stress response (ISR) pathway, which is activated in response to several different extrinsic stresses, such as amino acid starvation, hypoxia or UV radiation, as well as intrinsic ones, such as the accumulation of unfolded proteins in the endoplasmic reticulum (ER).[90,91] ISR central step is the phosphorylation of eIF2 on the serine 51 of its α subunit (eIF2α-P) by a family of four serine/threonine kinases, each activated by different stimuli. eIF2α phosphorylation hampers the activity of guanine nucleotide exchange factor eIF2B, resulting in decreased GDP to GTP exchange, reduced formation of the 43S preinitiation complex, and therefore in lowering of initiation of general translation. This global translational repression allows cells to save the energy spent in protein synthesis and to reprogram rapidly their gene expression in response to the stress. eIF2α phosphorylation allows the expression of specific subsets of proteins that can help to ameliorate the cellular stress or, if homeostasis is not restored and the stress stimulus is too strong or prolonged in time, will ultimately lead to apoptosis. Many of these factors are translated by transcripts containing one or more uORFs.[92] Among the genes preferentially expressed during ISR that help the recovery of the cell, there is the activating transcription factor 4 (*ATF4*), transcription factors CCAAT/enhancer-binding protein-α (*C/EBPα*) and -β (*C/EBPβ*)[90,93,94]; transcription factors such as CCAAT/enhancer-binding protein homologous protein *CHOP*, instead, induce the apoptosis in case of a stress stimulus too severe to overcome[95]; finally, growth arrest and damage-inducible protein 34 (*GADD34*) operates by negative feedback on ISR, leading to dephosphorylation of eIF2α.[96]

This modulation and the consequent relief of the uORF inhibitory activity on the mORF is possible thanks to two main mechanisms: ribosome leaky scanning and reinitiation.

Leaky scanning is a mechanism in which the ribosome bypasses the uORF and starts translation at the subsequent downstream AUG. The main factors that influence leaky scanning are the vicinity of the uORF to the 5′ end of the mRNA and the Kozak sequence of the uORF. The Kozak sequence seems to have a pivotal role also in ISR. It appears, in fact, that mRNAs that are preferentially translated in basal conditions present the uORF in a poor Kozak sequence and the mORF in a strong one, while genes repressed under stress conditions have a strong nucleotide context around the AUG of the uORF (uAUG).[71,97] An example of the influence of the nucleotide context can be found in *GADD34*. There are 2 uORFs in the 5′ UTR of *GADD34*, both in a poor Kozak context. In basal conditions, uORF1 is bypassed while uORF2 is translated. The peptide encoded by uORF2 contains a Pro-Pro-Gly sequence next to the stop codon of the uORF, which allows ribosome dissociation and leads to repression of the mORF. During stress conditions and in response to eIF2α-P, however, both uORFs are bypassed allowing the preferential expression of *GADD34*.[87,96]

Translation reinitiation takes place when the 40S ribosome subunit, after the translation of a uORF, remains associated with the mRNA and continues scanning until a new ribosome is ready to start translation at a downstream AUG. Reinitiation is more likely to happen after a very short uORF, if the available TC is sufficient, and if the intercistronic distance is adequately extended to allow a new competent ribosome to assemble.[70,98,99] A good example of how the mechanism of reinitiation can influence gene expression is provided by *ATF4*, which is highly conserved among vertebrates.[100] The 5′ UTR of *ATF4* contains 2 uORFs, both in a strong Kozak sequence, of which uORF1 is only 3 amino acid long, while uORF2 is 59 amino acid long and overlaps with the mORF. In basal conditions, the ribosome translates uORF1, the 60S subunit and the tRNA are recycled, while the 40S subunit remains attached to the mRNA. The available TC allows the ribosome to reconstitute and reinitiate translation at uORF2. The final result is translational repression of the mORF. In stress conditions, instead, the 40S subunit requires more time to resume translation due to the lower availability of TC, and this causes the bypass of uORF2 and the reinitiation of translation at the mORF.[101]

These examples underline the importance of uORFs, which, as the other main translational regulators, allow the cell to respond to a potential stress stimulus by reprogramming its gene expression quickly translating peptides from different ORFs of a specific subset of transcripts.

4.2 Diseases caused by uORF mutations

The impact of uORFs on translation regulation suggests that mutations altering existing uORFs could give rise to or predispose to human diseases. In the last years, indeed, the work of many laboratories has highlighted the existence of a large number of mutations able to create or eliminate uORFs, leading to neurologic, metabolic disorders, rare diseases, and even cancer susceptibility.[71,94,102] In 2009, Calvo and colleagues coined the term puORF (polymorphic uORF) to identify a uORF created or deleted by a SNP.[82] The authors, analyzing the millions of SNP present in the dbSNP database,[103] identified puORFs in 509 human genes.[82]

In 2015 Ye and colleagues analyzed the ClinVar, COSMIC, and TCGA databases, and identified 3751 mutations in the 5′ UTR that alter the number of uORFs.[77] Wethmar and colleagues used the dbSNP database to map all the human SNPs listed to the known uORF sequences and found more than 1300 SNPs affecting the start codon of uORFs and > 2000 affecting the uORF Kozak sequences in 2610 genes, leading to the consideration that up to 14.6% of human genes could be influenced in their translation by SNPs altering uORF-initiation.[104]

More recently, Whiffin and colleagues performed a genome-wide study using 15,708 whole-genome sequences from the Genome Aggregation Database (gnomAD).[105] The authors first identified a class of "high-impact uORF perturbing variants" in the genome (containing more than 145,000 SNPs), and then found that 2.2% of them are observed at least once in the gnomAD database.[106]

A list of the known examples of diseases caused by mutations affecting the uORFs can be found in reviews such as[107] and.[71] For example, a SNP or a point mutation can have the ability to create a new uORF. This is the case of the interferon regulatory factor 6 (IRF6), in which a -48 A to T mutation creates an overlapping uORF which decreases the translational efficiency of the mORF and causes the Van der Woude syndrome.[108] The inhibitory potential of this newly formed uORF has been detected and experimentally validated in the work of Calvo and colleagues, where the authors observed a significant reduction in expression (70–100%) when testing this patient mutation in a luciferase reporter construct.[82] An example of a mutation that disrupts an existing uORF can be found in the thrombopoietin gene (TPO).[107] There are seven uORFs in the 5′ UTR of this gene, among which the last overlapping one has been validated as the strongest inhibitor of expression from the mORF.[109,110] Pathologically increased levels of TPO are associated with hereditary thrombocythemia, characterized by high levels of platelets and risk of thrombosis.[111] Several different mutations linked to this disease disrupt the seventh uORF and up-regulate TPO translation.[112,113] For example, a -31 G to T mutation creates a novel stop codon downstream of the uORF AUG. This variation truncates 42 amino acids from the uORF which does not overlap with the mORF anymore, reducing its inhibitory effect.[110]

Finally, some mutations can indirectly affect uORF-mediated translational control. For example, an amino acid substitution in the uORF coding sequence can impair its ability to repress translation, if the uORF exerts its inhibitory effect through the peptide it encodes.[107] This is what happens for the dopamine D3 receptor gene[114] and for the WDR46 gene,[115] in which a single amino acid change increases the susceptibility to schizophrenia and aspirin-exacerbated respiratory disease, respectively. Another indirect impairment of uORF function is exemplified by the Shwachman-Bodian-Diamond syndrome gene (SBDS), mutated in the Shwachman-Diamond syndrome. The diseased state is characterized by low levels of the truncated isoforms of two genes, C/EBP-α and -β, which are translated by a process of reinitiation after the ribosome translates the short uORF present in their 5′ UTR. SBDS protein function has been demonstrated to be required for reinitiation of translation on these two transcripts, therefore the loss of SBDS protein leads to the absence of the truncated isoforms of *C/EBP-α* and *-β*.[116,117]

uORF deregulation has been linked also to the predisposition and development of cancer, via different mechanisms.[118] For instance, point mutations can enhance the inhibitory effect on the mORF of oncosuppressor genes. This is the case of *CDKN2A* (cyclin-dependent kinase inhibitor 2A), which is the main gene associated with predisposition to melanoma and encodes cdk4/cdk6 kinase inhibitors. One of the causative mutations is in the non-coding part of the gene (G-34T) and gives rise to an alternative TIS. Translation of the resulting uORF creates a truncated protein that can decrease expression from the mORF and predispose to cancer.[119] Another example is provided by the cyclin-dependent kinase inhibitor p27KIP1 (*CDKN1B*), a cell cycle inhibitor. In a patient affected by multiple endocrine neoplasia syndrome (MEN4), Occhi and colleagues identified a germline mutation in the 5′ UTR of this gene, causing the deletion of its uORF stop codon.[120] The combination of the resulting longer uORF-encoded peptide and shorter intercistronic distance leads to the down-regulation of p27KIP1 expression.[120] More recently, Schulz and colleagues applied a PCR-based deep sequencing to screen uORFs of proto-oncogenes in more than 300 human samples from 7 different types of cancer and found rare loss-of-function uORF mutations, such as *EPHB1* in two cases of breast and colon cancer, and *MAP2K6* in colon adenocarcinoma.[121]

These examples show that uORFs mutations have a major impact on human disease, and display the need to further investigate common or non-recurrent mutations in humans.

4.3 uORFs as genome editing targets

One of the most challenging therapeutic aims of gene therapy for a variety of disorders is achieving a safe and selective increase in the level of endogenous proteins. Given the substantial presence of uORFs in human mRNAs and their general inhibitory activity, disruption of these short open reading frames can in principle be a valid approach to obtain such a precise increase.

In 2016 Liang and colleagues used antisense oligonucleotides (ASOs) to target uORFs in order to inhibit their expression and achieve translational enhancement of the downstream mORF.[122,123] ASOs are chemically synthesized oligonucleotides usually exploited to downregulate the expression of a target protein by causing its transcript degradation or to modulate the splicing.[124] The authors reasoned that designing ASOs complementary to the uORF region of the mRNA of a gene could disrupt its inhibitory activity and lead to the upregulation of the mORF. In particular, they designed ASOs targeting the uORF of *RNASEH1* and achieved increased protein expression without substantially affecting mRNA levels, global translation, or causing ER stress. Moreover, to prove the efficacy of the approach *in vivo*, they demonstrated translational enhancement in mice, by treating the animals with an ASO targeting the Lrpprc uORF. Overall, they observed significant increases in protein expression (30–150%), but they argue that for each target gene or cell system, the ASO chemistry, affinity, and position should be determined in a tailored fashion.[123]

The same approach was used by Sasaki and colleagues, who designed ASOs to create a steric inhibition of the uORF present in the 5′ UTR of an important therapeutic target, *CFTR*. The authors demonstrated increased protein expression in multiple cell lines and also in Cystic Fibrosis patient-derived primary cells, and proposed the use of ASOs as a mutation-agnostic treatment for a disease characterized by an important heterogeneity.[125]

CRISPR-Cas-mediated genome editing can give important advantages to these approaches aimed at down-regulating the inhibitory potential of uORFs. For example, unlike ASOs, CRISPR-Cas can be used to achieve permanent modifications in the genome at the desired loci. In the last years, indeed, this technology has been applied to uORFs with the purpose of modulating translation. uORF disruption by CRISPR-Cas-mediated genome editing has been achieved either by Cas-induced double-strand breaks, to generate deletions that disrupt the uORF, or by using CRISPR-Cas base editors to create specific point mutations in the uORF able to impair its inhibitory activity (Fig. 5). The first of these attempts was carried out by Gao's group in 2018 in plants.[126] The authors chose as targets 4 genes important for the plant development and that present one uORF in their 5′ UTR region: *AtBRI1*, the phytohormone brassinosteroid receptor in Arabidopsis; *AtVTC2*, GDP-L-galactose phosphorylase, which has a role in the production of ascorbic acid in plants; *LsGGP1* and *LsGGP2*, its homolog in lettuce. They designed sgRNAs targeting the initiation codon of the uORF of each gene and obtained mutant lines with the desired modifications. They observed a significant increase in the protein expression from the downstream mORF of all the tested genes, which was paralleled by the expected phenotype. For example, lettuce mutants with disrupted uORFs presented increased ascorbic acid content up to 150%, and increased resistance to

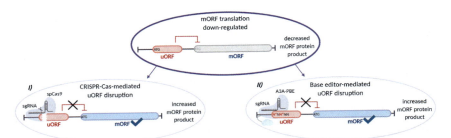

Fig. 5 uORFs genome editing to regulate mORF gene expression. (*I*) SpCas9 creates double-strand breaks (DSB) which generate deletions that disrupt uORF-mediated translation inhibition, leading to increased mORF expression; (*II*) A3A-PBE (APOBEC3A-plant base editor) creates single-point mutations in the uORF, impairing its inhibitory activity on the mORF and increasing protein expression.

oxidative stress due to the higher amount of vitamin C present in the leaves. In this work, the authors demonstrated for the first time that targeting uORFs with CRISPR–Cas genome editing approaches provides a way to up-regulate translation from the mORF. More recently, the Gao group published also a detailed protocol for manipulation of gene translation in plants using CRISPR–Cas and designing sgRNAs targeting the uAUG of target genes or the codons next to it.[127] The authors point out that this approach has the advantage of acting only at the translational level of gene expression, and circumvents the problem of gene silencing encountered when using transgene-mediated overexpression to achieve up-regulation of the desired target. On the other hand, the limitations of this protocol would be that not all the genes are strongly regulated by an upstream open reading frame.

Following these works, Xing and colleagues in a recent paper used the CRISPR–Cas cytosine deaminase base editor A3A-PBE (APOBEC3A-plant base editor)[128] to engineer the uORF present in the 5′ UTR of the transcription factor *FvebZIPs1.1* in strawberry (*Fragaria vesca*), to achieve fine-tuning of the sugar content. They generated seven novel alleles with different phenotypic strengths and combined them to produce transgene-free mutants with diverse sugar contents.[129]

Overall, these examples highlight an increased interest in the possibility to manipulate uORFs and their relevant role in the physiological gene expression regulation in the cell. Moreover, they suggest that more work can be done in this direction, especially to obtain a therapeutic effect on diseases that could benefit from a fine-tuning of protein levels.

5. Internal ribosomal entry sites (IRESs)
5.1 IRES elements and cap-independent translation initiation

Internal ribosomal entry sites (IRESs) are regions in the mRNAs that allow the internal initiation of translation. Indeed, even though the vast majority of cellular mRNAs are translated by the cap-dependent mechanisms, it is now known that, in certain cases, translation can also occur in a cap-independent manner, thanks to specific mRNA sequences able to recruit the 40S ribosomal subunit in the vicinity of an AUG starting codon.[130] The first IRES was identified in the genome of poliovirus (PV) and encephalomyocarditis virus (EMCV), and later in many other viruses.[131,132] During infection, the cap-dependent translation of the cell is impaired, and the viral RNAs use the internal initiation for their gene expression.[133] Viral IRESs are characterized by complex secondary and tertiary structures that allow them to efficiently bind the 40S ribosomal subunit. Moreover, they require little or no binding to cellular eukaryotic initiation factors.[133] For example, in the initiation promoted by type 1 IRESs (such as the ones present in Poliovirus), the 43S ribosome is recruited by an eIF4G/eIF4A complex, whereas in type 4 IRESs (e.g., in Cricket paralysis virus), no initiation factor is needed to start translation.[6]

Soon after their discovery, an IRES was identified in the 5′ UTR of a cellular mRNA, the immunoglobulin heavy chain binding protein (BiP).[134] Since then, many cellular mRNAs have been found to present IRESs in their 5′ UTR, the most well-known being vascular endothelial growth factor (*VEGF*)[135] or *c-Myc*,[136] and the list is growing. A common characteristic of most of those genes is to be required especially during stress response (such as ER stress or hypoxia), and more generally in situations where the cap-dependent translation is compromised.[137,138] Moreover, an important role has been attributed to IRES-mediated translation in the development of cancer.[139–141] Nevertheless, IRESs have been identified also in mRNAs important for normal organism development.[142] Most cellular IRESs are located in the 5′ UTR of the mRNA, although in rare cases IRES structures have been observed downstream of the AUG of the protein.[137] In those latter cases, cap-independent translation via IRES gives rise to truncated forms of the corresponding protein.[143,144] The cellular IRESs generally lack any type of conserved sequence or structures, therefore they are difficult to categorize and also to predict. Like their viral counterparts, cellular IRESs may need the aid of additional factors to promote internal initiation. Such components are termed IRES *trans*-acting factors (ITAFs), and some of them are even shared between cellular and viral mRNAs. ITAFs are believed to help the interaction between the ribosomal subunit and the IRES by stabilizing their complex structure.[145]

In recent years, the importance of cellular IRESs is being gradually recognized also in the phenomenon of polycistronic genes, which were originally discovered in viruses. A growing number of genes are in fact found to contain more than one or overlapping cistrons, and the presence of IRESs can ensure translation also from the downstream ORFs.[146,147] This evidence sheds light on the potential of IRES sequences to allow fine-tuning of translation in response to environmental stimuli, but also in normal cell development.

The ability of IRESs to promote cap-independent translation has made it possible to exploit them in biotechnological applications such as the construction of polycistronic reporters to co-express different proteins of interest. In such vectors, the IRES sequence is inserted between the two cistrons. The first protein is translated by the cap-mediated mechanism, while the IRES promotes the expression of the second one. Before the use of IRESs, one of the most common strategies to achieve co-expression of two genes was the use of separate promoters. However, this approach suffers from the possibility of transcriptional imbalances that could lead for example to a subset of cells expressing only one of the target genes.[148] This disadvantage is overcome in IRES-mediated bicistronic reporters, since only one mRNA is transcribed, avoiding transcription interferences.[149] This feature is of particular interest in both gene transfer applications and also gene therapy protocols.[150,151] The EMCV IRES has been the first used to create bicistronic and polycistronic vectors. However, it is reported that translation of the second cistron directed by this sequence is not always sufficiently high in the commercially available constructs.[152] The efficiency of an IRES-mediated translation relies on its sequence and structure but also on the availability of ITAFs, and so it is strongly influenced by the cell type and its microenvironment.[153] Moreover, it has been noticed that the structure and length of the intercistronic, non-coding sequence between the IRES and the second gene plays a role in the strength of the cap-independent translation.[154] Many efforts have been made to identify better IRESs, screening viral and cellular sequences in order to find optimal expression of both cistrons.[148,153,155–158] In conclusion, the potential of IRESs is of extreme interest, and advancements in their knowledge will help broaden their possible applications.

5.2 IRESs as therapeutic targets

Since their discovery, many attempts have been carried out to target IRESs for therapeutic purposes, mainly because their characteristic complex structures and their interactions with different elements (such as ITAFs or initiation factors) make them appealing therapeutic targets.[159] The efforts made mainly focus on treating viral infections, given the often insufficient options to treat such diseases and the fact that IRESs mostly drive translation of viral mRNAs.[160]

Some examples of these antiviral proof of concepts include the use of antisense oligonucleotides able to mediate the IRES destruction (by triggering a degradation pathway that involves RNAse H[161]) or impair their ability to bind the ribosomal subunit[162]; ribozymes or DNAzymes to directly cleave the IRES,[163,164] and small molecules or peptides.[159,165]

Nonetheless, some approaches have been developed also to target these structures in cellular mRNAs,[166] especially given their role in cancer development. Many studies, indeed, demonstrate that key oncogenes are translated in an IRES-dependent manner, suggesting that tumor cells could use this as a strategy to maintain their expression also in unfavorable conditions, such as the stress induced by cytotoxic therapeutic agents.[135,167,168]

Most examples present in literature to target cellular IRESs consist of small molecules able to inhibit IRES function.[169–172] For example, Vaklavas and colleagues performed a high throughput screening of 135,000 compounds and found 3 candidates able to inhibit IRES-mediated translation in the type I insulin-like growth factor receptor (*IGF-1R*).[167] *IGF-1R* is overexpressed in most human tumors and contributes to malignant phenotypes such as motility and invasiveness. Its 5′ UTR is characterized by an extremely long and complex structure, containing also a uORF which inhibits the activity of the scanning ribosomes. The presence of an IRES ensures the translation by circumventing the 5′ UTR structures.[173]

Pastor and colleagues targeted the IRES-mediated translation in the voltage-gated calcium channel gene *CACNA1A*.[169] A CAG repeat in this gene is responsible for the spinocerebellar ataxia type 6 (SCA6). *CACNA1A* is an example of multicistronic mRNA since it encodes two proteins, α1A and α1ACT, of which the latter is considered to be responsible, in its mutant form, for the pathogenesis of SCA6. α1A is translated via the cap-dependent route, while α1ACT relies on IRES for its expression.[174] The authors performed a screening of FDA-approved molecules and found 10 candidates able to inhibit selectively the *CACNA1A* IRES region.[169] In another work from the same group, the authors identified a miRNA able to selectively impair IRES-mediated translation and tested its efficacy in a mouse model of the disease, where they observed the rescue of the disease phenotype.[175]

6. Concluding remarks

The real potential of translation *cis*-acting elements manipulation to achieve the desired modulation in gene expression has yet to be fully understood. Translational regulation is being increasingly explored as a means of fine-tuning the expression of desired proteins, both in industrial production and with therapeutic aims. In a proof-of-principle work by Ferreira et al., the authors aimed at developing a systematic approach to engineer protein expression by acting on two of these elements, Kozak sequence and uORFs. In particular, they varied the translation initiation site of EGFP in a reporter vector and inserted concurrently dipeptide-encoding uORFs in the 5′ UTR of the same reporter. By doing so, they managed to achieve protein expression levels spanning three orders of magnitude, demonstrating the possibility of fine-tuning translation in mammalian cells.[176] In several cases, a subtle modulation of protein production is more desirable than a knockout. For instance, when a protein is part of a multimeric complex, all the subunits must be tightly regulated in expression to reach optimal levels of the desired protein complex product. In this regard, Eisenhut and colleagues constructed a panel of RNA hairpins (termed Regulation elements, "RgE(s)") that, inserted in the 5′ UTR of target genes, are able to fine-tune translational regulation creating different strengths of inhibition, depending on RgE(s) stability and/or proximity to the 5′ cap.[177] These works suggest a great potential for the manipulation of different *cis*-elements simultaneously, with the aim of fine-tuning gene expression. In terms of potential therapeutic applications, several Mendelian disorders stem from allelic losses, whose compensation should be finely tuned, which in principle could be reached by translational reprogramming. CRISPR-Cas9 genome editing is becoming increasingly recognized as a powerful tool to achieve this manipulation. A deeper knowledge of the biology and function of the elements involved in translation regulation will pave the way to new applications and therapeutic approaches.

Acknowledgments

FG is recipient of a fellowship funded by a grant from the "Ogni giorno per Emma Onlus" Association. All images were created with Biorender.com.

References

1. Doudna JA, Charpentier E. Genome editing. The new frontier of genome engineering with CRISPR-Cas9. *Science*. 2014;346(6213):1258096.
2. Gilbert LA, Horlbeck MA, Adamson B, et al. Genome-scale CRISPR-mediated control of gene repression and activation. *Cell*. 2014;159(3):647–661. https://doi.org/10.1016/j.cell.2014.09.029.
3. Kiani S, Chavez A, Tuttle M, et al. Cas9 gRNA engineering for genome editing, activation and repression. *Nat Methods*. 2015;12(11):1051–1054.
4. Hershey JWB, Sonenberg N, Mathews MB. Principles of translational control. *Cold Spring Harb Perspect Biol*. 2019;11(9):a032607. https://doi.org/10.1101/cshperspect.a032607.
5. Hinnebusch AG. The scanning mechanism of eukaryotic translation initiation. *Annu Rev Biochem*. 2014;83(1):779–812. https://doi.org/10.1146/annurev-biochem-060713-035802.
6. Jackson RJ, Hellen CUT, Pestova TV. The mechanism of eukaryotic translation initiation and principles of its regulation. *Nat Rev Mol Cell Biol*. 2010;11(2):113–127.
7. Kozak M. How do eucaryotic ribosomes select initiation regions in messenger RNA? *Cell*. 1978;15(4):1109–1123.
8. Kozak M. Pushing the limits of the scanning mechanism for initiation of translation. *Gene*. 2002;299(1):1.
9. Das S, Maitra U. Functional significance and mechanism of eIF5-promoted GTP hydrolysis in eukaryotic translation initiation. *Prog Nucleic Acid Res Mol Biol*. 2001;70:207–231.
10. Kozak M. Regulation of translation via mRNA structure in prokaryotes and eukaryotes. *Gene*. 2005;361:13–37.
11. Schuller AP, Green R. Roadblocks and resolutions in eukaryotic translation. *Nat Rev Mol Cell Biol*. 2018;19(8):526–541.
12. Rodnina MV, Wintermeyer W. Protein elongation, co-translational folding and targeting. *J Mol Biol*. 2016;428(10 pt B):2165–2185.
13. Dever TE, Dinman JD, Green R. Translation elongation and recoding in eukaryotes. *Cold Spring Harb Perspect Biol*. 2018;10(8):a032649. https://doi.org/10.1101/cshperspect.a032649.
14. Dever TE, Green R. The elongation, termination, and recycling phases of translation in eukaryotes. *Cold Spring Harb Perspect Biol*. 2012;4(7):a013706.
15. Jackson RJ, Hellen CUT, Pestova TV. Termination and post-termination events in eukaryotic translation. *Adv Protein Chem Struct Biol*. 2012;86:45–93.
16. Kapp LD, Lorsch JR. The molecular mechanics of eukaryotic translation. *Annu Rev Biochem*. 2004;73:657–704.
17. Sonenberg N, Hinnebusch AG. Regulation of translation initiation in eukaryotes: mechanisms and biological targets. *Cell*. 2009;136(4):731–745. https://doi.org/10.1016/j.cell.2009.01.042.
18. Hernández G, Osnaya VG, Pérez-Martínez X. Conservation and variability of the AUG initiation codon context in eukaryotes. *Trends Biochem Sci*. 2019;44(12):1009–1021.
19. Kuersten S, Goodwin EB. The power of the 3′ UTR: translational control and development. *Nat Rev Genet*. 2003;4(8):626–637. https://doi.org/10.1038/nrg1125.
20. Macdonald P. Diversity in translational regulation. *Curr Opin Cell Biol*. 2001;13(3):326–331.
21. Hernández G, Altmann M. Origins and evolution of the mechanisms regulating translation initiation in eukaryotes. *Trends Biochem Sci*. 2010;35(2):63–73.

22. So L, Lee J, Palafox M, et al. The 4E-BP-eIF4E axis promotes rapamycin-sensitive growth and proliferation in lymphocytes. *Sci Signal.* 2016;9(430), ra57.
23. Merrick WC, Pavitt GD. Protein synthesis initiation in eukaryotic cells. *Cold Spring Harb Perspect Biol.* 2018;10(12):a033092. https://doi.org/10.1101/cshperspect.a033092.
24. Korets SB, Czok S, Blank SV, Curtin JP, Schneider RJ. Targeting the mTOR/4E-BP pathway in endometrial cancer. *Clin Cancer Res.* 2011;17(24):7518–7528.
25. Hernández G, Osnaya VG, García A, Velasco MX. On the origin and early evolution of translation in eukaryotes. In: Hernández G, Jagus R, eds. *Evolution of the Protein Synthesis Machinery and Its Regulation.* Springer International Publishing; 2016:81–107.
26. Mignone F, Gissi C, Liuni S, Pesole G. Untranslated regions of mRNAs. *Genome Biol.* 2002;3(3), REVIEWS0004.
27. Kozak M. Possible role of flanking nucleotides in recognition of the AUG initiator codon by eukaryotic ribosomes. *Nucleic Acids Res.* 1981;9(20):5233–5252.
28. Sherman F, Stewart JW, Schweingruber AM. Mutants of yeast initiating translation of Iso-1-cytochrome c within a region spanning 37 nucleotides. *Cell.* 1980;20(1):215–222. https://doi.org/10.1016/0092-8674(80)90249-4.
29. Cordell B, Weiss SR, Varmus HE, Michael BJ. At least 104 nucleotides are transposed from the 5′ terminus of the avian sarcoma virus genome to the 5′ termini of smaller viral mrnas. *Cell.* 1978;15(1):79–91. https://doi.org/10.1016/0092-8674(78)90084-3.
30. Piatak M, Ghosh PK, Bhaskara Reddy V, Lebowitz P, Weissman SM. Complex structures and new surprises in SV40 mRNA11This work was supported by a grant from the American Cancer Society and by grant #CA-16038 from the National Cancer Institute, DHEW. *Extrachromosomal DNA.* 1979;199–215. https://doi.org/10.1016/b978-0-12-198780-0.50019-2. Published online.
31. Preston CM, McGeoch DJ. Identification and mapping of two polypeptides encoded within the herpes simplex virus type 1 thymidine kinase gene sequences. *J Virol.* 1981;38(2):593–605.
32. Kozak M. Comparison of initiation of protein synthesis in procaryotes, eucaryotes, and organelles. *Microbiol Rev.* 1983;47(1):1–45.
33. Kozak M. Compilation and analysis of sequences upstream from the translational start site in eukaryotic mRNAs. *Nucleic Acids Res.* 1984;12(2):857–872.
34. Kozak M. Point mutations define a sequence flanking the AUG initiator codon that modulates translation by eukaryotic ribosomes. *Cell.* 1986;44(2):283–292.
35. Kozak M. An analysis of 5′-noncoding sequences from 699 vertebrate messenger RNAs. *Nucleic Acids Res.* 1987;15(20):8125–8148.
36. Trifonov EN. Translation framing code and frame-monitoring mechanism as suggested by the analysis of mRNA and 16 S rRNA nucleotide sequences. *J Mol Biol.* 1987;194(4):643–652.
37. Kozak M. Influences of mRNA secondary structure on initiation by eukaryotic ribosomes. *Proc Natl Acad Sci USA.* 1986;83(9):2850–2854.
38. Huang HK, Yoon H, Hannig EM, Donahue TF. GTP hydrolysis controls stringent selection of the AUG start codon during translation initiation in Saccharomyces cerevisiae. *Genes Dev.* 1997;11(18):2396–2413.
39. Noderer WL, Flockhart RJ, Bhaduri A, et al. Quantitative analysis of mammalian translation initiation sites by FACS-seq. *Mol Syst Biol.* 2014;10:748.
40. Cuperus JT, Groves B, Kuchina A, et al. Deep learning of the regulatory grammar of yeast 5′ untranslated regions from 500,000 random sequences. *Genome Res.* 2017;27(12):2015–2024.
41. Decoene T, Peters G, De Maeseneire SL, De Mey M. Toward predictable 5′UTRs in Saccharomyces cerevisiae: development of a yUTR calculator. *ACS Synth Biol.* 2018;7(2):622–634.

42. Blanco N, Williams AJ, Tang D, et al. Tailoring translational strength using Kozak sequence variants improves bispecific antibody assembly and reduces product-related impurities in CHO cells. *Biotechnol Bioeng.* 2020;117(7):1946–1960. https://doi.org/10.1002/bit.27347.
43. Xu H, Wang P, You J, et al. Screening of Kozak-motif-located SNPs and analysis of their association with human diseases. *Biochem Biophys Res Commun.* 2010;392(1):89–94.
44. Fisher E, Nitz I, Lindner I, et al. Candidate gene association study of type 2 diabetes in a nested case-control study of the EPIC-Potsdam cohort—role of fat assimilation. *Mol Nutr Food Res.* 2007;51(2):185–191. https://doi.org/10.1002/mnfr.200600162.
45. Thoeringer CK, Binder EB, Salyakina D, et al. Association of a Met88Val diazepam binding inhibitor (DBI) gene polymorphism and anxiety disorders with panic attacks. *J Psychiatr Res.* 2007;41(7):579–584.
46. Baker RI, Eikelboom J, Lofthouse E, et al. Platelet glycoprotein Ibalpha Kozak polymorphism is associated with an increased risk of ischemic stroke. *Blood.* 2001;98(1):36–40.
47. Afshar-Kharghan V, Li CQ, Khoshnevis-Asl M, López JA. Kozak sequence polymorphism of the glycoprotein (GP) Ibα gene is a major determinant of the plasma membrane levels of the platelet GP Ib-IX-V complex. *Blood.* 1999;94(1):186–191. https://doi.org/10.1182/blood.v94.1.186.413k19_186_191.
48. Carter AM, Catto AJ, Bamford JM, Grant PJ. Platelet GP IIIa PlA and GP Ib variable number tandem repeat polymorphisms and markers of platelet activation in acute stroke. *Arterioscler Thromb Vasc Biol.* 1998;18(7):1124–1131.
49. Jacobson EM, Concepcion E, Oashi T, Tomer Y. A Graves' disease-associated Kozak sequence single-nucleotide polymorphism enhances the efficiency of CD40 gene translation: a case for translational pathophysiology. *Endocrinology.* 2005;146(6):2684–2691.
50. Ouahchi K, Arita M, Kayden H, et al. Ataxia with isolated vitamin E deficiency is caused by mutations in the α–tocopherol transfer protein. *Nat Genet.* 1995;9(2):141–145. https://doi.org/10.1038/ng0295-141.
51. Usuki F, Maruyama K. Ataxia caused by mutations in the alpha-tocopherol transfer protein gene. *J Neurol Neurosurg Psychiatry.* 2000;69(2):254–256.
52. Hall JM, Lee MK, Newman B, et al. Linkage of early-onset familial breast cancer to chromosome 17q21. *Science.* 1990;250(4988):1684–1689.
53. Signori E, Bagni C, Papa S, et al. A somatic mutation in the 5'UTR of BRCA1 gene in sporadic breast cancer causes down-modulation of translation efficiency. *Oncogene.* 2001;20(33):4596–4600.
54. Wang X, Liu Z, Li G, et al. Efficient gene silencing by adenine base editor-mediated start codon mutation. *Mol Ther.* 2020;28(2):431–440. https://doi.org/10.1016/j.ymthe.2019.11.022.
55. Chen S, Xie W, Liu Z, et al. CRISPR start-loss: a novel and practical alternative for gene silencing through base-editing-induced start codon mutations. *Mol Ther Nucleic Acids.* 2020;21:1062–1073.
56. Rees HA, Liu DR. Base editing: precision chemistry on the genome and transcriptome of living cells. *Nat Rev Genet.* 2018;19(12):770–788.
57. Komor AC, Kim YB, Packer MS, Zuris JA, Liu DR. Programmable editing of a target base in genomic DNA without double-stranded DNA cleavage. *Nature.* 2016;533(7603):420–424.
58. Gaudelli NM, Komor AC, Rees HA, et al. Programmable base editing of a•T to G•C in genomic DNA without DNA cleavage. *Nature.* 2017;551(7681):464–471.
59. Molla KA, Qi Y, Karmakar S, Baig MJ. Base editing landscape extends to perform transversion mutation. *Trends Genet.* 2020;36(12):899–901.

60. Grünewald J, Zhou R, Lareau CA, et al. A dual-deaminase CRISPR base editor enables concurrent adenine and cytosine editing. *Nat Biotechnol*. 2020;38(7):861–864.
61. Sakata RC, Ishiguro S, Mori H, et al. Base editors for simultaneous introduction of C-to-T and A-to-G mutations. *Nat Biotechnol*. 2020;38(7):865–869. https://doi.org/10.1038/s41587-020-0509-0.
62. Park S, Beal PA. Off-target editing by CRISPR-guided DNA base editors. *Biochemistry*. 2019;58(36):3727–3734.
63. Zhou C, Sun Y, Yan R, et al. Off-target RNA mutation induced by DNA base editing and its elimination by mutagenesis. *Nature*. 2019;571(7764):275–278. https://doi.org/10.1038/s41586-019-1314-0.
64. Grünewald J, Zhou R, Garcia SP, et al. Transcriptome-wide off-target RNA editing induced by CRISPR-guided DNA base editors. *Nature*. 2019;569(7756):433–437.
65. Rees HA, Wilson C, Doman JL, Liu DR. Analysis and minimization of cellular RNA editing by DNA adenine base editors. *Sci Adv*. 2019;5(5). https://doi.org/10.1126/sciadv.aax5717, eaax5717.
66. Jang H-K, Bae S. I-silence, please! An alternative for gene disruption via adenine base editors. *Mol Ther*. 2020;28(2):348–349.
67. Kim B, Sasaki Y, Egawa T. Restriction of nonpermissive RUNX3 protein expression in T lymphocytes by the Kozak sequence. *J Immunol*. 2015;195(4):1517–1523. https://doi.org/10.4049/jimmunol.1501039.
68. Egawa T, Tillman RE, Naoe Y, Taniuchi I, Littman DR. The role of the Runx transcription factors in thymocyte differentiation and in homeostasis of naive T cells. *J Exp Med*. 2007;204(8):1945–1957.
69. Levanon D, Groner Y. Structure and regulated expression of mammalian RUNX genes. *Oncogene*. 2004;23(24):4211–4219.
70. Somers J, Pöyry T, Willis AE. A perspective on mammalian upstream open reading frame function. *Int J Biochem Cell Biol*. 2013;45(8):1690–1700.
71. Silva J, Fernandes R, Romão L. Translational regulation by upstream open reading frames and human diseases. *Adv Exp Med Biol*. 2019;1157:99–116.
72. McGillivray P, Ault R, Pawashe M, Kitchen R, Balasubramanian S, Gerstein M. A comprehensive catalog of predicted functional upstream open reading frames in humans. *Nucleic Acids Res*. 2018;46(7):3326–3338.
73. Johnstone TG, Bazzini AA, Giraldez AJ. Upstream ORFs are prevalent translational repressors in vertebrates. *EMBO J*. 2016;35(7):706–723.
74. Chew G-L, Pauli A, Schier AF. Conservation of uORF repressiveness and sequence features in mouse, human and zebrafish. *Nat Commun*. 2016;7(1):1–10.
75. Lin Y, May GE, Kready H, et al. Impacts of uORF codon identity and position on translation regulation. *Nucleic Acids Res*. 2019;47(17):9358.
76. Kozak M. Structural features in eukaryotic mRNAs that modulate the initiation of translation. *J Biol Chem*. 1991;266(30):19867–19870.
77. Ye Y, Liang Y, Yu Q, et al. Analysis of human upstream open reading frames and impact on gene expression. *Hum Genet*. 2015;134(6):605–612. https://doi.org/10.1007/s00439-015-1544-7.
78. Thomas KR, Capecchi MR. Introduction of homologous DNA sequences into mammalian cells induces mutations in the cognate gene. *Nature*. 1986;324(6092):34–38.
79. Tzani I, Ivanov IP, Andreev DE, et al. Systematic analysis of the PTEN 5′ leader identifies a major AUU initiated proteoform. *Open Biol*. 2016;6(5):150203. https://doi.org/10.1098/rsob.150203.
80. Davuluri RV, Suzuki Y, Sugano S, Zhang MQ. CART classification of human 5′ UTR sequences. *Genome Res*. 2000;10(11):1807–1816. https://doi.org/10.1101/gr.gr-1460r.
81. Chen H-H, Tarn W-Y. uORF-mediated translational control: recently elucidated mechanisms and implications in cancer. *RNA Biol*. 2019;16(10):1327–1338.

82. Calvo SE, Pagliarini DJ, Mootha VK. Upstream open reading frames cause widespread reduction of protein expression and are polymorphic among humans. *Proc Natl Acad Sci USA*. 2009;106(18):7507–7512.
83. Kozak M. Constraints on reinitiation of translation in mammals. *Nucleic Acids Res*. 2001;29(24):5226–5232.
84. Law GL, Raney A, Heusner C, Morris DR. Polyamine regulation of ribosome pausing at the upstream open reading frame of S-adenosylmethionine decarboxylase. *J Biol Chem*. 2001;276(41):38036–38043.
85. Rebbapragada I, Lykke-Andersen J. Execution of nonsense-mediated mRNA decay: what defines a substrate? *Curr Opin Cell Biol*. 2009;21(3):394–402.
86. Mendell JT, Sharifi NA, Meyers JL, Martinez-Murillo F, Dietz HC. Nonsense surveillance regulates expression of diverse classes of mammalian transcripts and mutes genomic noise. *Nat Genet*. 2004;36(10):1073–1078.
87. Young SK, Willy JA, Wu C, Sachs MS, Wek RC. Ribosome reinitiation directs gene-specific translation and regulates the integrated stress response. *J Biol Chem*. 2015;290(47):28257–28271.
88. Pöyry TAA, Kaminski A, Connell EJ, Fraser CS, Jackson RJ. The mechanism of an exceptional case of reinitiation after translation of a long ORF reveals why such events do not generally occur in mammalian mRNA translation. *Genes Dev*. 2007;21(23):3149–3162.
89. Young SK, Wek RC. Upstream open reading frames differentially regulate gene-specific translation in the integrated stress response. *J Biol Chem*. 2016;291(33):16927–16935. https://doi.org/10.1074/jbc.r116.733899.
90. Harding HP, Zhang Y, Zeng H, et al. An integrated stress response regulates amino acid metabolism and resistance to oxidative stress. *Mol Cell*. 2003;11(3):619–633.
91. Pakos-Zebrucka K, Koryga I, Mnich K, Ljujic M, Samali A, Gorman AM. The integrated stress response. *EMBO Rep*. 2016;17(10):1374–1395.
92. Sajjanar B, Deb R, Raina SK, et al. Untranslated regions (UTRs) orchestrate translation reprogramming in cellular stress responses. *J Therm Biol*. 2017;65:69–75.
93. Nerlov C. The C/EBP family of transcription factors: a paradigm for interaction between gene expression and proliferation control. *Trends Cell Biol*. 2007;17(7):318–324. https://doi.org/10.1016/j.tcb.2007.07.004.
94. Wethmar K, Smink JJ, Leutz A. Upstream open reading frames: molecular switches in (patho)physiology. *Bioessays*. 2010;32(10):885–893.
95. Palam LR, Baird TD, Wek RC. Phosphorylation of eIF2 facilitates ribosomal bypass of an inhibitory upstream ORF to enhance CHOP translation. *J Biol Chem*. 2011;286(13):10939–10949. https://doi.org/10.1074/jbc.m110.216093.
96. Lee Y-Y, Cevallos RC, Jan E. An upstream open reading frame regulates translation of GADD34 during cellular stresses that induce eIF2α phosphorylation. *J Biol Chem*. 2009;284(11):6661–6673. https://doi.org/10.1074/jbc.m806735200.
97. Baird TD, Palam LR, Fusakio ME, et al. Selective mRNA translation during eIF2 phosphorylation induces expression of IBTKα. *Mol Biol Cell*. 2014;25(10):1686–1697.
98. Hinnebusch AG. Translational regulation of GCN4 and the general amino acid control of yeast. *Annu Rev Microbiol*. 2005;59(1):407–450. https://doi.org/10.1146/annurev.micro.59.031805.133833.
99. Kozak M. Effects of intercistronic length on the efficiency of reinitiation by eucaryotic ribosomes. *Mol Cell Biol*. 1987;7(10):3438–3445. https://doi.org/10.1128/mcb.7.10.3438.
100. Vattem KM, Wek RC. Reinitiation involving upstream ORFs regulates ATF4 mRNA translation in mammalian cells. *Proc Natl Acad Sci USA*. 2004;101(31):11269–11274.

101. Hronová V, Mohammad MP, Wagner S, et al. Does eIF3 promote reinitiation after translation of short upstream ORFs also in mammalian cells? *RNA Biol.* 2017;14 (12):1660–1667.
102. Morris DR, Geballe AP. Upstream open reading frames as regulators of mRNA translation. *Mol Cell Biol.* 2000;20(23):8635–8642.
103. Sherry ST, Ward MH, Kholodov M, et al. dbSNP: the NCBI database of genetic variation. *Nucleic Acids Res.* 2001;29(1):308–311.
104. Wethmar K, Schulz J, Muro EM, Talyan S, Andrade-Navarro MA, Leutz A. Comprehensive translational control of tyrosine kinase expression by upstream open reading frames. *Oncogene.* 2015;35(13):1736–1742.
105. Karczewski KJ, Francioli LC, Tiao G, et al. The mutational constraint spectrum quantified from variation in 141,456 humans. *Nature.* 2020;581(7809):434–443.
106. Whiffin N, Genome Aggregation Database Production Team, Karczewski KJ, et al. Characterising the loss-of-function impact of 5′ untranslated region variants in 15,708 individuals. *Nat Commun.* 2020;11(1):1–12. https://doi.org/10.1038/s41467-019-10717-9.
107. Barbosa C, Peixeiro I, Romão L. Gene expression regulation by upstream open reading frames and human disease. *PLoS Genet.* 2013;9(8):e1003529.
108. Kondo S, Schutte BC, Richardson RJ, et al. Mutations in IRF6 cause Van der Woude and popliteal pterygium syndromes. *Nat Genet.* 2002;32(2):285.
109. Cazzola M, Skoda RC. Translational pathophysiology: a novel molecular mechanism of human disease. *Blood.* 2000;95(11):3280–3288. https://doi.org/10.1182/blood.v95. 11.3280.011k41_3280_3288.
110. Ghilardi N, Skoda RC. A single-base deletion in the thrombopoietin (TPO) gene causes familial essential thrombocythemia through a mechanism of more efficient translation of TPO mRNA. *Blood.* 1999;94(4):1480–1482.
111. Kikuchi M, Tayama T, Hayakawa H, Takahashi I, Hoshino H, Ohsaka A. Familial thrombocytosis. *Br J Haematol.* 1995;89(4):900–902.
112. Wiestner A, Schlemper RJ, van der Maas AP, Skoda RC. An activating splice donor mutation in the thrombopoietin gene causes hereditary thrombocythaemia. *Nat Genet.* 1998;18(1):49–52.
113. Ghilardi N, Wiestner A, Kikuchi M, Ohsaka A, Skoda RC. Hereditary thrombocythaemia in a Japanese family is caused by a novel point mutation in the thrombopoietin gene. *Br J Haematol.* 1999;107(2):310–316. https://doi.org/10.1046/j.1365-2141.1999.01710.x.
114. Sivagnanasundaram S, Morris AG, Gaitonde EJ, McKenna PJ, Mollon JD, Hunt DM. A cluster of single nucleotide polymorphisms in the 5′-leader of the human dopamine D3 receptor gene (DRD3) and its relationship to schizophrenia. *Neurosci Lett.* 2000;279 (1):13–16.
115. Pasaje CFA, Bae JS, Park B-L, et al. WDR46is a genetic risk factor for aspirin-exacerbated respiratory disease in a korean population. *Allergy Asthma Immunol Res.* 2012;4(4):199. https://doi.org/10.4168/aair.2012.4.4.199.
116. In K, Zaini MA, Müller C, Warren AJ, von Lindern M, Calkhoven CF. Shwachman–Bodian–diamond syndrome (SBDS) protein deficiency impairs translation re-initiation from C/EBPα and C/EBPβ mRNAs. *Nucleic Acids Res.* 2016;44(9):4134.
117. Silva J, Fernandes R, Romão L. Gene expression regulation by upstream open reading frames in rare diseases. *J Rare Dis Res Treat.* 2017;2(4):33–38.
118. Diederichs S, Bartsch L, Berkmann JC, et al. The dark matter of the cancer genome: aberrations in regulatory elements, untranslated regions, splice sites, non-coding RNA and synonymous mutations. *EMBO Mol Med.* 2016;8(5):442–457. https://doi.org/10.15252/emmm.201506055.

119. Liu L, Dilworth D, Gao L, et al. Mutation of the CDKN2A 5' UTR creates an aberrant initiation codon and predisposes to melanoma. *Nat Genet.* 1999;21(1):128–132.
120. Occhi G, Regazzo D, Trivellin G, et al. A novel mutation in the upstream open reading frame of the CDKN1B gene causes a MEN4 phenotype. *PLoS Genet.* 2013;9(3):e1003350.
121. Schulz J, Mah N, Neuenschwander M, et al. Loss-of-function uORF mutations in human malignancies. *Sci Rep.* 2018;8(1):1–10. https://doi.org/10.1038/s41598-018-19201-8.
122. Liang X-H, Shen W, Sun H, Migawa MT, Vickers TA, Crooke ST. Translation efficiency of mRNAs is increased by antisense oligonucleotides targeting upstream open reading frames. *Nat Biotechnol.* 2016;34(8):875–880.
123. Liang X-H, Shen W, Crooke ST. Specific increase of protein levels by enhancing translation using antisense oligonucleotides targeting upstream open frames. In: Li L-C, ed. *RNA Activation.* Singapore: Springer; 2017:129–146.
124. Bennett CF. Therapeutic antisense oligonucleotides are coming of age. *Annu Rev Med.* 2019;70:307–321. https://doi.org/10.1146/annurev-med-041217-010829.
125. Sasaki S, Sun R, Bui H-H, Crosby JR, Monia BP, Guo S. Steric inhibition of 5' UTR regulatory elements results in upregulation of human CFTR. *Mol Ther.* 2019;27(10):1749–1757. https://doi.org/10.1016/j.ymthe.2019.06.016.
126. Zhang H, Si X, Ji X, et al. Genome editing of upstream open reading frames enables translational control in plants. *Nat Biotechnol.* 2018;36(9):894–898.
127. Si X, Zhang H, Wang Y, Chen K, Gao C. Manipulating gene translation in plants by CRISPR–Cas9-mediated genome editing of upstream open reading frames. *Nat Protoc.* 2020;15(2):338–363. https://doi.org/10.1038/s41596-019-0238-3.
128. Zong Y, Song Q, Li C, et al. Efficient C-to-T base editing in plants using a fusion of nCas9 and human APOBEC3A. *Nat Biotechnol.* 2018;36(10):950–953.
129. Xing S, Chen K, Zhu H, et al. Fine-tuning sugar content in strawberry. *Genome Biol.* 2020;21:1–14. https://doi.org/10.1186/s13059-020-02146-5.
130. Fitzgerald KD, Semler BL. Bridging IRES elements in mRNAs to the eukaryotic translation apparatus. *Biochim Biophys Acta.* 2009;1789(9–10):518–528.
131. Jang SK, Davies MV, Kaufman RJ, Wimmer E. Initiation of protein synthesis by internal entry of ribosomes into the 5' nontranslated region of encephalomyocarditis virus RNA in vivo. *J Virol.* 1989;63(4):1651–1660.
132. Jang SK, Kräusslich HG, Nicklin MJ, Duke GM, Palmenberg AC, Wimmer E. A segment of the 5' nontranslated region of encephalomyocarditis virus RNA directs internal entry of ribosomes during in vitro translation. *J Virol.* 1988;62(8):2636–2643.
133. Komar AA, Hatzoglou M. Internal ribosome entry sites in cellular mRNAs: mystery of their existence. *J Biol Chem.* 2005;280(25):23425–23428.
134. Macejak DG, Sarnow P. Internal initiation of translation mediated by the 5' leader of a cellular mRNA. *Nature.* 1991;353(6339):90–94.
135. Huez I, Créancier L, Audigier S, Gensac MC, Prats AC, Prats H. Two independent internal ribosome entry sites are involved in translation initiation of vascular endothelial growth factor mRNA. *Mol Cell Biol.* 1998;18(11):6178–6190.
136. Paulin FE, West MJ, Sullivan NF, Whitney RL, Lyne L, Willis AE. Aberrant translational control of the c-myc gene in multiple myeloma. *Oncogene.* 1996;13(3):505–513.
137. Komar AA, Hatzoglou M. Cellular IRES-mediated translation: the war of ITAFs in pathophysiological states. *Cell Cycle.* 2011;10(2):229–240.
138. Gerlitz G, Jagus R, Elroy-Stein O. Phosphorylation of initiation factor-2 alpha is required for activation of internal translation initiation during cell differentiation. *Eur J Biochem.* 2002;269(11):2810–2819. https://doi.org/10.1046/j.1432-1033.2002.02974.x.

139. Svitkin YV, Herdy B, Costa-Mattioli M, Gingras A-C, Raught B, Sonenberg N. Eukaryotic translation initiation factor 4E availability controls the switch between cap-dependent and internal ribosomal entry site-mediated translation. *Mol Cell Biol.* 2005;25(23):10556–10565.
140. Xi S, Zhao M, Wang S, et al. IRES-mediated protein translation overcomes suppression by the p14ARF tumor suppressor protein. *J Cancer.* 2017;8(6):1082.
141. Marina D, Arnaud L, Noel LP, Felix S, Bernard R, Natacha C. Relevance of translation initiation in diffuse glioma biology and its therapeutic potential. *Cells.* 2019;8(12):1542. https://doi.org/10.3390/cells8121542.
142. Xue S, Tian S, Fujii K, Kladwang W, Das R, Barna M. RNA regulons in Hox 5′UTRs confer ribosome specificity to gene regulation. *Nature.* 2015;517(7532):33.
143. Komar AA, Lesnik T, Cullin C, Merrick WC, Trachsel H, Altmann M. Internal initiation drives the synthesis of Ure2 protein lacking the prion domain and affects [URE3] propagation in yeast cells. *EMBO J.* 2003;22(5):1199–1209.
144. Grover R, Candeias MM, Fåhraeus R, Das S. p53 and little brother p53/47: linking IRES activities with protein functions. *Oncogene.* 2009;28(30):2766–2772.
145. Hellen CUT. IRES-induced conformational changes in the ribosome and the mechanism of translation initiation by internal ribosomal entry. *Biochim Biophys Acta.* 2009;1789(9–10):558–570.
146. Karginov TA, Pastor DPH, Semler BL, Gomez CM. Mammalian polycistronic mRNAs and disease. *Trends Genet.* 2017;33(2):129.
147. Chou AC, Aslanian A, Sun H, Hunter T. An internal ribosome entry site in the coding region of tyrosyl-DNA phosphodiesterase 2 drives alternative translation start. *J Biol Chem.* 2019;294(8):2665.
148. Al-Allaf FA, Abduljaleel Z, Athar M, et al. Modifying inter-cistronic sequence significantly enhances IRES dependent second gene expression in bicistronic vector: construction of optimised cassette for gene therapy of familial hypercholesterolemia. *Non-coding RNA Res.* 2019;4(1):1.
149. de Felipe P, Izquierdo M. Construction and characterization of pentacistronic retrovirus vectors. *J Gen Virol.* 2003;84(pt 5):1281–1285.
150. Ngoi SM, Chien AC, Lee CGL. Exploiting internal ribosome entry sites in gene therapy vector design. *Curr Gene Ther.* 2004;4(1):15–31.
151. Morgan RA, Couture L, Elroy-Stein O, Ragheb J, Moss B, Anderson WF. Retroviral vectors containing putative internal ribosome entry sites: development of a polycistronic gene transfer system and applications to human gene therapy. *Nucleic Acids Res.* 1992;20(6):1293–1299.
152. Sadikoglou E, Daoutsali E, Petridou E, Grigoriou M, Skavdis G. Comparative analysis of internal ribosomal entry sites as molecular tools for bicistronic expression. *J Biotechnol.* 2014;181:31–34.
153. Licursi M, Christian SL, Pongnopparat T, Hirasawa K. In vitro and in vivo comparison of viral and cellular internal ribosome entry sites for bicistronic vector expression. *Gene Ther.* 2011;18(6):631–636.
154. Attal J, Théron MC, Houdebine LM. The optimal use of IRES (internal ribosome entry site) in expression vectors. *Genet Anal.* 1999;15(3–5):161–165. https://doi.org/10.1016/s1050-3862(99)00021-2.
155. Wong E-T, Ngoi S-M, Lee CGL. Improved co-expression of multiple genes in vectors containing internal ribosome entry sites (IRESes) from human genes. *Gene Ther.* 2002;9(5):337–344. https://doi.org/10.1038/sj.gt.3301667.
156. Harries M, Phillipps N, Anderson R, Prentice G, Collins M. Comparison of bicistronic retroviral vectors containing internal ribosome entry sites (IRES) using expression of human interleukin-12 (IL-12) as a readout. *J Gene Med.* 2000;2(4):243–249.

157. Qiao J, Roy V, Girard MH, Caruso M. High translation efficiency is mediated by the encephalomyocarditis virus internal ribosomal entry sites if the natural sequence surrounding the eleventh AUG is retained. *Hum Gene Ther*. 2002;13(7):881–887.
158. Lee JC, Wu TY, Huang CF, Yang FM, Shih SR, Hsu JT. High-efficiency protein expression mediated by enterovirus 71 internal ribosome entry site. *Biotechnol Bioeng*. 2005;90(5):656–662. https://doi.org/10.1002/bit.20440.
159. Komar AA, Hatzoglou M. Exploring internal ribosome entry sites as therapeutic targets. *Front Oncol*. 2015;5:233. https://doi.org/10.3389/fonc.2015.00233.
160. Jubin R, Hepatitis C. IRES: translating translation into a therapeutic target. *Curr Opin Mol Ther*. 2001;3(3):278–287.
161. Martinand-Mari C, Lebleu B, Robbins I. Oligonucleotide-based strategies to inhibit human hepatitis C virus. *Oligonucleotides*. 2003;13(6):539–548.
162. Dibrov SM, Parsons J, Carnevali M, et al. Hepatitis C virus translation inhibitors targeting the internal ribosomal entry site. *J Med Chem*. 2014;57(5):1694–1707.
163. Kumar D, Chaudhury I, Kar P, Das RH. Site-specific cleavage of HCV genomic RNA and its cloned core and NS5B genes by DNAzyme. *J Gastroenterol Hepatol*. 2009;24(5):872–878. https://doi.org/10.1111/j.1440-1746.2008.05717.x.
164. Romero-López C, Berzal-Herranz B, Gómez J, Berzal-Herranz A. An engineered inhibitor RNA that efficiently interferes with hepatitis C virus translation and replication. *Antiviral Res*. 2012;94(2):131–138. https://doi.org/10.1016/j.antiviral.2012.02.015.
165. Lozano G, Trapote A, Ramajo J, et al. Local RNA flexibility perturbation of the IRES element induced by a novel ligand inhibits viral RNA translation. *RNA Biol*. 2015;12(5):555.
166. Holcik M. Targeting translation for treatment of Cancer—a novel role for IRES? *Curr Cancer Drug Targets*. 2004;4(3):299–311. https://doi.org/10.2174/1568009043333005.
167. Vaklavas C, Meng Z, Choi H, Grizzle WE, Zinn KR, Blume SW. Small molecule inhibitors of IRES-mediated translation. *Cancer Biol Ther*. 2015;16(10):1471.
168. Dobson T, Chen J, Krushel LA. Dysregulating IRES-dependent translation contributes to overexpression of oncogenic aurora a kinase. *Mol Cancer Res*. 2013;11(8):887–900. https://doi.org/10.1158/1541-7786.mcr-12-0707.
169. Pastor PDH, Du X, Fazal S, Davies AN, Gomez CM. Targeting the CACNA1A IRES as a treatment for spinocerebellar ataxia type 6. *Cerebellum*. 2018;17(1):72–77.
170. Didiot M-C, Hewett J, Varin T, et al. Identification of cardiac glycoside molecules as inhibitors of c-Myc IRES-mediated translation. *J Biomol Screen*. 2013;18(4):407–419. https://doi.org/10.1177/1087057112466698.
171. Shi Y, Yang Y, Hoang B, et al. Therapeutic potential of targeting IRES-dependent c-myc translation in multiple myeloma cells during ER stress. *Oncogene*. 2016;35(8):1015.
172. Holmes B, Lee J, Landon KA, et al. Mechanistic target of rapamycin (mTOR) inhibition synergizes with reduced internal ribosome entry site (IRES)-mediated translation of cyclin D1 and c-MYC mRNAs to treat glioblastoma. *J Biol Chem*. 2016;291(27):14146.
173. Meng Z, Jackson NL, Choi H, King PH, Emanuel PD, Blume SW. Alterations in RNA-binding activities of IRES-regulatory proteins as a mechanism for physiological variability and pathological dysregulation of IGF-IR translational control in human breast tumor cells. *J Cell Physiol*. 2008;217(1):172–183.
174. Du X, Wang J, Zhu H, et al. A second cistron in the CACNA1A gene encodes a transcription factor that mediates cerebellar development and SCA6. *Cell*. 2013;154(1):118.
175. Miyazaki Y, Du X, Muramatsu S, Gomez CM. An miRNA-mediated therapy for SCA6 blocks IRES-driven translation of the CACNA1A second cistron. *Sci Transl Med*. 2016;8(347):347ra94. https://doi.org/10.1126/scitranslmed.aaf5660.

176. Ferreira JP, Wesley Overton K, Wang CL. Tuning gene expression with synthetic upstream open reading frames. *Proc Natl Acad Sci USA*. 2013;110(28):11284–11289.
177. Eisenhut P, Mebrahtu A, Moradi Barzadd M, et al. Systematic use of synthetic 5'-UTR RNA structures to tune protein translation improves yield and quality of complex proteins in mammalian cell factories. *Nucleic Acids Res*. 2020;48(20):e119. https://doi.org/10.1093/nar/gkaa847. Published online October.

CHAPTER FOURTEEN

Synthetic genomics for curing genetic diseases

Simona Grazioli[a,†] and Gianluca Petris[b,*]

[a]Scuola Superiore Sant'Anna, Pisa, Italy
[b]Medical Research Council Laboratory of Molecular Biology (MRC LMB), Cambridge, United Kingdom
*Corresponding author: e-mail address: gpetris@mrc-lmb.cam.ac.uk

Contents

1. Introduction to synthetic genomics		478
1.1 Reading genomes		478
1.2 DNA writing: From gene to genome synthesis		480
1.3 Synthetic genomics in human and animal cells		484
2. Synthetic genomics techniques		487
2.1 Yeast and bacterial artificial chromosomes		487
2.2 Mammalian (and human) artificial chromosomes		492
2.3 Transfer of natural and artificial chromosomes		496
2.4 Chromosome elimination		497
3. Synthetic genomics applications for the treatment of genetic diseases		500
3.1 Modeling genetic diseases		501
3.2 Synthetic genomics approaches for gene and genome therapy		503
3.3 Synthetic genomics to understand the human genome		507
4. Perspectives		507
Acknowledgments		508
References		508

Abstract

From the beginning of the genome sequencing era, it has become increasingly evident that genetics plays a role in all diseases, of which only a minority are single-gene disorders, the most common target of current gene therapies. However, the majority of people have some kind of health problems resulting from congenital genetic mutations (over 6000 diseases have been associated to genes, https://www.omim.org/statistics/geneMap) and most genetic disorders are rare and only incompletely understood. The vision and techniques applied to the synthesis of genomes may help to address

---

[†] Present address: Medical Research Council Laboratory of Molecular Biology (MRC LMB), Cambridge, United Kingdom.

unmet medical needs from a chromosome and genome-scale perspective. In this chapter, we address the potential therapy of genetic diseases from a different outlook, in which we no longer focus on small gene corrections but on higher-order tools for genome manipulation. These will play a crucial role in the next years, as they prelude to a much deeper understanding of the architecture of the human genome and a more accurate modeling of human diseases, offering new therapeutic opportunities.

Abbreviations

AAV	adeno-associated viral vector
ARS	autonomously replicating sequence
BAC	bacterial artificial chromosome
bp	basepair
CDS	coding sequences
DMD	Duchenne muscular dystrophy
Gb	gigabase
HAC	human artificial chromosome
hiPSC	human induced pluripotent stem cells
iPSC	induced pluripotent stem cells
kb	kilobase
MAC	mammalian artificial chromosome
Mb	megabase
mESC	mouse embryonic stem cells
MMCT	microcell-mediated chromosomal transfer
nt	nucleotide
NGS	next generation sequencing
TAR	transformation-associated recombination
YAC	yeast artificial chromosome

1. Introduction to synthetic genomics
1.1 Reading genomes

Rapid advances in the field of genomics have brought the knowledge of the architecture of life to unprecedented depth. The ability to decipher the sequence of genomes has massively grown, thanks to a rapid development in high-throughput DNA reading technology. In less than 50 years from the sequencing of the first phage genomes (PhiX174, 5.3 kilobases (kb) in 1977[1] and bacteriophage lambda, 49 kb in 1982[2]) by Sanger and colleagues (Fig. 1), more recent endeavors led to the sequencing of the first human genomes in 2001 (3 Gb)[3,4] and the genomes of thousands of people worldwide during the 1000 Genomes Project (2015)[5] and the 100,000 Genomes Project (2018).[6] Sequencing has been applied to reveal the genome of organisms from all kingdoms, including plants, for example the rice genome (*Oryza sativa* L., 389 Mb) was sequenced in 2005[7] and the grapevine genome (*Vitis vinifera*, 487 Mb) in 2007.[8]

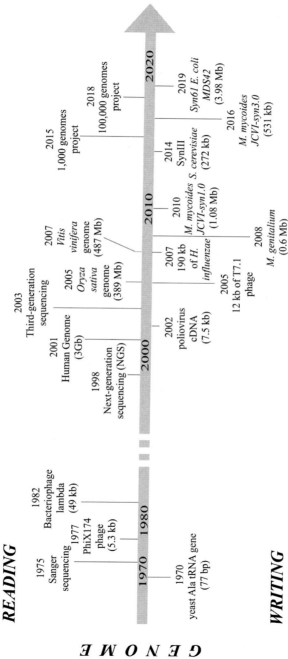

Fig. 1 Selected milestones in genome reading and writing.

Sequencing technologies evolved significantly over the years: the first-generation sequencing methods were developed in the 1970s and included the still-used chain termination-based Sanger sequencing[9] and the less popular Maxam-Gilbert chemical cleavage method.[10] Optimization of such techniques now produces average read lengths up to 800–1200 nt. In more recent times and for genome sequencing purposes, more high-throughput techniques based on either short reads (50–600 bp) or long reads (from 500 bp to 2.3 Mb) have been developed.[11] In the first category, the Next- or Second-Generation Sequencing (NGS) methods include Roche's pyrosequencing, Illumina's bridge-amplification sequencing by synthesis and others such as the sequencing by oligonucleotide ligation and detection (SOLiD) system by Applied Biosystems.[12–15] In the second category, the third-generation sequencing methods include pacific bioscience (PacBio)'s single-molecule real time (SMRT) sequencing, based on light-guiding nanostructures called Zero-Mode Waveguides, and Oxford Nanopore Technologies (ONT), which relies on the characteristic decrease in current generated by the passage of a specific nucleotide across a nanoscopic biological or solid-state pore.[16–18] Third-generation sequencing reads have no theoretical upper length boundary.

These advances in DNA reading technologies resulted in an important time and cost drop: from the 10 years and £2 billion (€2.4 billion; US$3 billion) required for the initial sequencing of a full human genome, it is now possible to get the genome sequence of a patient in about 1 day for less than £700,[6] even though the overall cost can actually be two-three times higher.[19]

The availability of DNA sequences and functional genome information offered the possibility to identify DNA alterations that are correlated with genetic diseases and to develop ideas of how to modify genomes to directly explore genotype-phenotype causality. This is now leading to new treatments which work by correcting aberrant DNA sequences in specific areas of the body, such as the recently developed CRISPR-based strategy to treat LCA10 blindness (https://clinicaltrials.gov/ct2/show/NCT03872479). In parallel, while technologies improved for genome reading, the idea of *writing* entire genomes became more and more plausible and, despite numerous technical, economical and practical challenges, can now be regarded as growing quickly. In the next paragraph, we will describe some remarkable achievements in this field.

1.2 DNA writing: From gene to genome synthesis

Unlike traditional biotechnology, whose starting points are naturally occurring DNA sequences, synthetic genomics is a nascent field that makes use of

DNA synthesis methods to write genes and genomes. The first step towards synthetic genomics was the synthesis, in 1970, of the 77-bp yeast alanine tRNA gene, obtained by aligning very short chemically synthesized deoxyribonucleotide segments through base pairing and by joining them together through the enzymatic activity of polynucleotide kinase and DNA ligase[20] (Fig. 1). The first genome made by chemical synthesis was the 7.5-kb poliovirus cDNA created by Wimmer and colleagues in 2002, which produced infectious viruses.[21] The approach initially joined overlapping complementary DNA oligonucleotides (length ~69 nt) into 0.4- to 0.6-kb fragments, which were further ligated into three larger DNA segments cloned in intermediate plasmid vectors. These were eventually used to assemble the full-length cDNA in a plasmid by means of unique restriction sites. In 2005, about 12 kb of the T7 phage genome (total genome size 40 kb) were heavily re-designed avoiding overlapping genes, selecting essential elements and assigning a singular function to each DNA sequence to obtain the T7.1 artificial phage.[22] The assembly was achieved by use of both PCR-amplified T7 regions and synthetic DNA fragments, combined with standard recombinant DNA techniques, thereby yielding an infective virus. However, the simple use of restriction enzymes and plasmids cannot satisfy the challenges posed by the assembly of a megabase (Mb)-sized bacterial genome (Fig. 2). A promising strategy, instead, relied on progressive assembly *in vitro*, combined with recombination of precursor DNA fragments in an evolutionary divergent host. This ensured that the synthetic DNA would remain largely transcriptionally silent and interfere less with host viability.[23] In this way, the 3.5-Mb genome of the bacterium *Synechocystis PCC6803* was built within the 4.2-Mb genome of *Bacillus subtilis 168*.[24] This was obtained by joining four separate 0.8- to 0.9-Mb DNA mega-chunks, assembled by stepwise integration of several PCR fragments.[24] Stepwise fragment assembly and

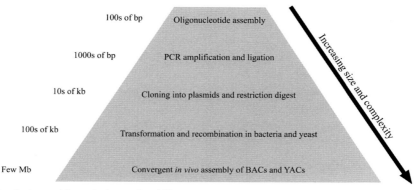

Fig. 2 Assembly techniques for different DNA sizes.

recombination has also been used to build organelle genomes. Mouse mitochondrial (16.3 kb) and rice chloroplast (134.5 kb) genomes were built within the *Bacillus subtilis* genome,[25] followed by their circularization by homologous recombination in a pBR322 vector to create a DNA sequence suitable for transfer to other hosts (e.g. *E. coli* for the mouse mitochondrial genome) and having an architecture similar to the natural genome of these two eukaryotic organelles.[26,27]

An important step in synthetic genomics was the development of lambda red homology-based recombination: electroporated linear DNA from a library of bacterial artificial chromosomes (BACs) was shown to be efficiently integrated by lambda red recombination into the *E. coli* chromosome.[28] Lambda red recombineering was exploited to introduce about one tenth (~190 kb) of the *Haemophilus influenzae* genome from two non-contiguous regions as episomes in *E. coli* host[29]. Despite this genome assembly was in a divergent host, the *H. influenzae* genes were transcribed in *E. coli*. Although this finding offers the possibility of host cell genome replacement, at the same time it showed that recipient bacterial strains were unable to tolerate some exogenous DNA sequences including ribosomal RNA (rRNA), ribosomal proteins and other toxic genes.[29]

In 2008 the whole bacterial *M. genitalium* genome (nearly 0.6 Mb) was completely assembled starting from chemically synthesized oligonucleotides.[30] These were ligated into overlapping 5- to 7-kb DNA fragments, which were joined by *in vitro* recombination to make intermediate 24-kb, 72-kb and finally four 144-kb fragments (each approximately one quarter of the final genome). The large fragments were cloned into Bacterial Artificial Chromosomes (BACs) in *E. coli*, and transferred to yeast for final joining by homologous recombination.[30] Although a clone with the correct sequence was identified, the creation of a bacterial cell controlled by a chemically synthesized genome was demonstrated only later, in 2010, with the assembly of the 1.08-Mbp *Mycoplasma mycoides* JCVI-syn1.0 genome.[31] The *M. mycoides* genome was assembled convergently through three rounds of transformation-assisted recombination in yeast, first creating 109 partially overlapping fragments of 10 kb each, then 11 fragments of 100 kb, and finally the complete 1.1-Mb genome. This genome was different from the wild type *M. mycoides* strain for the presence of watermark sequences (deletions and polymorphisms).[31] The synthetic *M. mycoides* genome was isolated and transplanted into restriction-minus *M. capricolum* recipient cells, forming, after cell divisions, new *M. mycoides* cells controlled solely by the synthetic genome and its derived transcriptome and proteome.

Several studies have now demonstrated that it is possible to heavily modify bacterial genomes. The *E. coli* MDS42 and *B. subtilis* genomes could be reduced by nearly a quarter, while a 0.5-Mb reduction was achieved in the 14-Mb *Saccharomyces pombe* yeast genome.[32–34] These efforts culminated to the synthesis of the rationally designed JCVI-syn3.0 *M. mycoides* genome of only 531 kb, which is to date smaller than any other autonomously replicating cell genome found in nature.[35] JCVI-syn3.0 cells have a genome only half the size of wild-type *M. mycoides*, which includes more than a hundred quasi-essential genes with unknown function which were unexpectedly required for cell viability.[35] Synthetic genomics groups have also partially[36] or completely[37] synthesized recoded *E. coli* genomes. Church and co-workers tested different recoding schemes with 57 codons,[36] while Chin and co-workers created the viable bacterial strain Syn61, in which the TAG stop codon and the TCG and TCA serine codons were removed, leaving only 61 codons in all coding sequences (CDS).[37] The recently published method for programmable chromosome fission and fusion clearly showed that it is now possible to quickly and efficiently manipulate Mb-scale DNA fragments in *E. coli*.[38]

In contrast, attempts of large-scale synthetic genomics in eukaryotic cells are still limited. The most impressive examples of rational large-scale genome manipulation have been performed in yeast (mainly *Saccharomyces cerevisiae*).[39,40] This is because yeast is particularly suitable for genome manipulation thanks to its high homologous recombination rate and the size of its genome (the *S. cerevisiae* genome is 14 Mb, with 12 Mb non-redundant DNA)[41] is a reasonable progressive target after the assembly of synthetic bacterial genomes. A first synthesis of a functional yeast chromosome was reported only 7 years ago,[39] with the assembly of the synIII chromosome (0.272 Mb). SynIII is about 44 kb smaller than the wild type *S. cerevisiae* chromosome III and was engineered to include TAG/TAA stop-codon replacements, deletion of subtelomeric regions, introns, tRNAs, transposons and silent mating loci, and insertion of recombination sites to enable further genome reduction.[39] The synIII chromosome was obtained by PCR of overlapping 60- to 79-mer oligonucleotides to make 750-bp building blocks, which were assembled in plasmids, before yeast recombination into 2- to 4-kb minichunks. Since then, several other *S. cerevisiae* synthetic chromosomes have been created,[40,42] but the convergent *Sc2.0* synthetic yeast strain, containing all 16 engineered chromosomes and having a genome reduced by 8% from the initial 14 Mb, is yet to be completed. Other efforts in yeast succeeded in building a single-chromosome strain, in which the entire 12-Mb haploid genome of the *S. cerevisiae* BY4742 strain was combined into a single giant chromosome by 15 cycles of end-to-end fusions and centromere and telomere eliminations.[43] This monochromosomal yeast strain was still able to complete sexual life cycles, although with slower growth rates and provides a useful model to study the advantages of having more than a single chromosome in eukaryotic organisms.[43]

When considering the opportunity to create genomes or large synthetic DNA fragments, it is essential to have in mind the costs, which for DNA synthesis are not yet as affordable as in the case of DNA reading.

The price of DNA synthesis is still about US$0.10-0.20 per bp, which is more than five orders of magnitude higher than for DNA sequencing. As a consequence, the estimated cost of all synthetic DNA required for Sc2.0 is, excluding all the expenses required for genome assembly, about US$1.25 million[40]; it is therefore clear that the total synthesis of a Gb-sized animal or human genome would be prohibitively expensive at current costs.

1.3 Synthetic genomics in human and animal cells

The development of DNA sequencing, synthesis and assembly technologies has brought the possibility of genome assembly and extensive genome editing (e.g. repair of damaged genomes, genome recoding, genome-wide elimination of mobile genetic elements) of more complex organisms, such as humans, considerably closer: it is clear that similar achievements would disclose an unprecedented deep level of understanding of the human genome as well as a whole new range of therapeutic possibilities. At the same time, it is undeniable that several difficulties need to be overcome to be able to build a human genome or even a human chromosome. The same applies also when considering the use of synthetic genomics approaches to correct complex chromosomal rearrangements, restore large chromosomal deletions or modify several independent genes causing genetic diseases. As mentioned before, chemical synthesis of DNA is currently too expensive and, in many cases, probably unnecessary for the correction of genomic defects, so efficient technologies for copying or isolating healthy donor DNA sequences, to be assembled and delivered to correct damaged genes and chromosomes, are required (the available techniques to isolate, modify and deliver large fragments of the human genome or even entire chromosomes are described below in Section 2).

It should be intuitively evident that the DNA assembly and delivery technologies used for bacteria and yeast genome synthesis, or for the early effort of human genome sequencing, are not sufficient for the more ambitious aim of creating entire or important portions of human and mammalian genomes without adaptations: in the first place, these are approximately three orders of magnitude bigger than *E. coli* (2.5–3 Gb compared to 5 Mb) and more than a hundred times bigger than yeast. Also, mammalian cells are diploid and are characterized by a highly repetitive genome and by a considerable portion of non-coding elements (many of which incompletely investigated) not present in microbial organisms. Gene expression in human cells is no longer organized in operons, but, instead, every gene is driven by its own promoter and precisely regulated at the transcriptional, post-transcriptional, translational and post-translational level.[44] In recent years, the important role of complex epigenetic signatures has also become evident and needs to be accounted for in the reconstruction of complex genomes.[45,46]

Despite these challenges, technologies for the manipulation of complex genomes have rapidly been developed over the last 30 years.[47,48] The first steps toward synthetic genomics in higher eukaryotic cells can be traced back to early gene targeting experiments in the late 1980s, which relied on the low spontaneous homologous recombination rate in mice to modify endogenous genes.[49] Today, a wide toolbox of efficient and programmable gene editing technologies is available: these have proven capable of modifying the human genome in a number of ways, from introducing point mutations and small indels, to making large deletions, insertions and chromosomal translocations, and even making epigenetic modifications.[47,48,50]

Even though some of the most significant progress in genome editing was achieved since the development of CRISPR-derived technologies in 2012,[51,52] many of the most remarkable examples of complex genomes manipulation are predated. Such examples are the multi-Mb engineering of the mouse genome for therapeutic aims. Having considerable synteny and homology to human genomes, *Mus musculus* has been exploited multiple times in the past to produce human or humanized antibodies[53–57] and T cell receptors.[58] In some cases, the final mouse humanization included additional human genes encoding for other important immune proteins such as human class I major histocompatibility complex (*HLA-A*0201*) and β2-microglobulin (*B2M*) genes.[58] These humanized animals were obtained either by delivering Mb-size fragments of human chromosomes, using microcell-mediated chromosomal transfer,[54] or using YAC and BAC vectors carrying fragments of the human genome, which were integrated either randomly or by homologous recombination and recombinase insertion in repetitive and convergent parallel steps.[53,55–58]

In 2010, humanized mice with a diverse human T-cell antigen receptor (TCR) repertoire were obtained by assembling in yeast two couples of overlapping YACs, containing the variable regions of the TCRα (820 kb and 900 kb YACs) and TCRβ (620 kb and 240 kb) human gene loci. The final YAC vectors, 1.4 Mb (TCRα) and 0.7 Mb (TCRβ) in size, were independently delivered to mouse embryonic stem cells (mESCs) by yeast fusion and randomly integrated into the genome, yielding, after appropriate crossing, functional humanized T-cell receptors.[58] Similar YAC-based approaches had been described earlier, for example in 1994 for the integration in mESCs of two YACs (170–220 kb) capable of producing human immunoglobulins in the derived mice.[55] Then, in 1997, four YACs (345 kb, 290 kb, 280 kb and 240 kb) were recombined in yeast into a 1.02-Mb YAC containing three human heavy chain antibody constant regions and about two thirds of the variable regions, and other three YACs (180 kb, 380 kb, 480 kb) were recombined in two rounds in an 800-kb YAC containing κ light chain constant region and nearly half of variable domains; both YACs were delivered to mESCs by yeast fusion and random genomic integration.[53] In more recent years, similar results were achieved, but the Mb-size human immunoglobulin loci DNA assembly was performed directly by consecutive steps in the mouse genome, rather than in yeast. Also, BACs were used as transfer vectors: in 2014 Lee and colleagues engineered the mouse genome by consecutive cycles of electroporation delivery and Cre-mediated BAC integrations (five BACS for each IgH, IgL and IgK) into a landing-pad, followed by marker recycle using piggyBac transposase and final inactivation of mouse variable regions by Cre-mediated genomic inversion.[56] Finally, the entire human immunoglobulin variable-gene repertoire (2.7 Mb) was cloned into the mouse genome upstream of intact functional mouse constant regions. In the same year Macdonald and colleagues, in probably the largest and most remarkable functional in situ replacement of a mammalian genome, humanized 6 Mb of the mouse immunoglobulin loci. This has been achieved for both the heavy and light chain genes in mESCs with a strategy based on BAC carrier vectors electroporation, spontaneous homologous recombination and Cre/loxP-driven elimination of selectable markers.[57,59] In parallel cells, the mouse IgH and IgK loci were first deleted by insertion of LoxP sites and Cre-mediated elimination of 2.5–3 Mb of mouse genome; then the human variable region immunoglobulin loci were inserted by several repetitive steps, delivering each up to 210 kb of human DNA sequences. At the end of the process, and following the removal of residual selectable markers by Flp-recombination, mESCs were engineered to contain nearly 1 and 0.5 Mb of human DNA for IgH and IgK loci, respectively. Despite the relatively low efficiency of most recombination steps obtained with pre-CRISPR technologies, and after appropriate crossing, this example showed the final elimination of about 6 Mb of the mouse genome and its replacement with 3 Mb of assembled human DNA.[57] These large-scale genome editing efforts have produced a remarkable impact on the treatment of many diseases through the development of therapeutic antibodies.[60,61]

If these examples of multi-Mb animal genome manipulations have been obtained in the pre-CRISPR era, it is quite plausible that very extensive genome manipulation can be achieved by applying the most recent genome editing technologies[47,48,50] and the increased understanding of chromosome biology.[62–65] An extensive description of several of these recent genome editing techniques is given in other chapters of this book[66]; for this reason, the next section will focus on the DNA engineering methods suitable for the more ambitious and challenging synthetic genomics purposes, which may, in the future, advance the gene and cell therapy field to potentially correct any kind of patient's damaged genome. Some of these technologies and knowledge have been developed during the Human Genome Project and have already provided important tools to cut and reassemble genomic segments of different origins, such as yeast and bacterial genomes.

2. Synthetic genomics techniques

In this section we will not enter into the details of all the synthetic biology cloning and assembly methods reviewed elsewhere,[67] instead, we focus on the main vectors and approaches suitable to assemble and deliver large DNA fragments from of at least 50–100 kb up to an entire mammalian chromosome.

2.1 Yeast and bacterial artificial chromosomes

Yeast and bacterial artificial chromosomes (YACs and BACs) have been commonly used to carry long genomic fragments (0.1–2 Mb), which cannot be accommodated in plasmids or cosmids.[68] YAC and BAC vectors have advantages and disadvantages that can determine which of them is a more suitable platform for carrying, engineering or delivering a large DNA fragment. A description of YAC and BAC features and applications is provided in the next two sections.

2.1.1 Yeast artificial chromosomes

Developed in the late 1980s,[69] YACs are *Saccharomyces cerevisiae*-derived vectors able to carry Mb-sized DNA sequences, which were used at first for the cloning of genomic DNA during the Human Genome Project.[68] The essential functional units of a YAC are yeast origins of replication (Autonomously Replicating Sequences or ARS), yeast centromeres[70] and, depending on their circular or linear form, yeast telomeric sequences.[71]

YACs have a high insert capacity (more than 2 Mb has been achieved)[71–73] and they have been used to clone and study full-length human genes in their natural genomic context.[74] Although YACs can be amenable for the cloning of some sequences that may be unstable in bacterial cells (e.g. AT-rich sequences, repetitive regions such as telomeres and centromeres), their main weakness is the actual instability and the high frequency of chimerism affecting the carried exogenous DNA. This instability is very likely due to the high level of homologous recombination of yeast cells, which will affect the YAC causing deletions and other rearrangements. Moreover, DNA cloned in YACs must contain at least one autonomously replicating sequence (ARS)-like sequence functioning like an origin of replication site in yeast.[71] ARS-like sequences are common in animal genomes (every 20 kb in mammalian DNA), but in chromosomal regions with multiple repetitive elements, such as the centromere and telomere, the ARS frequency may be reduced, requiring that the ARS must be present in the YAC backbone.[75] However, to avoid selection of a self-circularized empty YAC, the ARS-containing YAC backbones require the addition of a negative selectable marker whose expression must be eliminated by the genomic DNA insertion: this can be a URA3 gene under a split ADH1 promoter which tolerates up to 130-bp insertions (such as TAR homology hooks, see below) but gets inactivated upon successful cloning of the larger region of interest.[75]

As mentioned above, the main advantage of YACs is when manipulating large DNA fragments that are too big for BACs or bacteriophage P1-derived Artificial Chromosomes (PACs). Yeast can be transformed with Mb-sized DNA fragments carrying whole mammalian genes. At first most YAC libraries were obtained by limited restriction enzyme digestion of genomic DNA and ligation to a sticky-end compatible YAC backbone *in vitro* before yeast transformation. This semi-random cloning strategy can produce YAC libraries with an average size above 1 Mb, but they also show a high frequency of unwanted recombination.[71] To enhance YAC stability, recombination-deficient yeast strains, in which recombination genes are dysfunctional, can be used.[76,77] This is a strategy that proved to be especially effective with RAD52 mutants or, even better, in inducible RAD52 strains.[76]

Ensuring the YAC stability while maintaining the remarkably high rate of yeast homologous recombination is essential for a programmable yeast cloning approach called Transformation-Associated Recombination (TAR), developed by Larionov, Kouprina and colleagues.[78] TAR is an efficient strategy for the isolation of large genomic tracts, full-length genes and loci up to 250–300 kb.[79,80] TAR cloning had been exploited in the sequencing of genomes, in evolutionary studies, in the study of chromosomal mutations and rearrangements. TAR exploits the high homologous recombination rate of *Saccharomyces cerevisiae* to integrate specific genomic regions between two unique homology hooks (down to 60 bp in length) at the 5′- and 3′-termini of a linearized YAC. The genomic DNA is co-transformed with the YAC backbone into yeast spheroplasts and cloned *in vivo*.[78,81]

A significant improvement in the recovery of the desired insert was obtained later, by coupling Cas9 cleavage of complex genomes and TAR cloning; this programmable approach has been described to improve the specificity of TAR by about 10 fold.[82]

TAR cloning has already been used for the assembly of the *M. genitalis* and *M. mycoides* genomes,[30,31] the assembly of several genetic pathways in yeast,[83] and the synthesis of yeast chromosomes.[39] The most recent total synthesis of an *E. coli* genome, Syn61 (4 Mb), also exploited yeast recombination to assemble multiple 10-kb DNA fragments to create 100-kb YACs as essential building blocks to create the entire bacterial genome.[37]

Indeed, beyond the payload size, the main advantage of choosing yeast and YACs is the extremely flexible and multiplex DNA engineering capabilities of this organism.[39,43,84] Despite being mainly popular among synthetic biologists, DNA engineering in yeast is simple and competitive with *in vitro* cloning platforms (e.g. Golden Gate[85] or Gibson cloning[86,87]) both in term of the size and the number of fragments (up to at least 25)[84] that can be assembled in a single step.

An important issue when working with YACs and yeast assembly is the laborious isolation of YAC DNA from the yeast genome,[88] whose chromosomes (0.2–1.5 Mb) have a size comparable to most YAC vectors, especially when linear. As YACs are also maintained at one copy per cell, YAC vector purification has poor yields and represents a critical step. Often, YAC isolation by pulsed-field gel electrophoresis (PFGE) is needed to obtain sufficient purity, increasing the risk that shearing forces may break the YAC in the process. This is particularly relevant when increasing the YAC size. Engineered yeast strains have been created to avoid the presence of endogenous chromosomes overlapping with YACs.[89]

While alternative delivery methods (e.g. cell-cell fusion)[53,55,58,90] into higher eukaryotic cells have been explored, bypassing the critical purification step, other strategies have focused on *retrofitting* YACs to BACs for large-scale production and easier purification.[91]

2.1.2 Bacterial artificial chromosomes

The stability and purification limits of YACs in yeast had favored the use of alternative vectors over YACs during the Human Genome Project.[68] These were phage P1-derived Artificial Chromosomes (PACs) and *E. coli*-derived BACs, which had a smaller insert capacity (up to approximately 350 kb) compared to YACs,[92,93] but appear to be much more stable. This upper size limit might be simply the consequence of transformation limits, rather than the maximal payload actually achievable. Indeed, it is possible to rearrange the *E. coli* genome into different Mb-sized BACs.[38,94]

BACs are particularly useful for the handling of large DNA fragments as (i) they are generally stable with essentially no chimerism, (ii) they can be more efficiently isolated intact because they are circular and supercoiled and can be easily discriminated from the *E. coli* genome, (iii) they can be purified in higher quantity and purity thanks to the faster growth and smaller genome of *E. coli* compared to yeast, and (iv) they can be delivered into different cells by electroporation with 10–100 times higher efficiency than yeast (at least up to 300 kb) or conjugation. Moreover, many more laboratories are used to working with *E. coli* rather than yeast.

BAC propagation is controlled by an appropriate origin of replication (e.g. OriS, OriV) and the F plasmid partitioning elements (parA and parB), maintaining a low copy number (one or two copies per cell) and reducing the potential for unwanted recombination and insert instability.

The DNA payload was historically cloned into BACs by partial restriction enzyme digestion, followed by ligation of genomic DNA.[92,95] However, programmable, not semi-random, cloning of large DNA sequences into BACs can be cumbersome. Past approaches have tried to clone big gene clusters by a series of annealing and hybridization steps from overlapping PCR amplifications[96]; also, selected regions of bacterial genomes up to 48 kb have been successfully cloned into plasmids by homologous recombination and a resistance split-fusion strategy.[97]

The advent of lambda phage (Redαβγ) and Rac phage-derived (RecE/RecT) recombination provided a method to engineer large BACs (200 kb), but did not provide a directly programmable cloning method for very large exogenous DNA fragments directly into BACs,[98–100] probably due to the low transformation efficiency of large linear DNA fragments. However, exogenous fragments can be assembled from smaller pieces into large BACs (310 kb) *in vitro* by Gibson assembly before transformation into *E. coli*.[86] In recent years, Cas9-assisted targeting of chromosome segments (CATCH) has been developed to provide a powerful, programmable tool for the isolation of specific long genomic sections.[101] The system was employed on bacterial chromosomes to isolate defined DNA regions up to 100 kb of bacterial DNA, which were then cloned into BACs with appropriate 30-bp homology arms by Gibson assembly and electroporated into recipient cells.[101] The enzymatic digestion of the bacterial cell wall and the Cas9-guided DNA cleavage are performed in low-melting agarose plugs, which prevent excessive shearing due to DNA manipulation. The analysis of these big vectors can be routinely performed by pulse field gel electrophoresis (PFGE) and Nanopore/short-read sequencing.[38,102] Thus, programmable nuclease-based approaches can provide a very precise tool to isolate and clone DNA sequences of interest into BACs.

Larger BACs can be obtained *in vivo* by assembling different BACs in a single chromosome; *in vivo* Cas9 cleavage, lambda-red-mediated recombination (REXER)[103] and conjugation have been exploited to join different BACs and assemble Mb-sized bacterial genomes.[37,38] These examples suggest how to increase the length of a DNA insert cloned into a BAC to Mb-scales, even if this is currently not achievable by direct cloning in a single step in *E. coli*.

For genome editing purposes, BACs up to 300 kb can be isolated and purified from bacteria and can be electroporated or transfected with low efficiency into mammalian cells, serving also as template for homologous recombination.[57,59] Alternative BAC delivery/integration methods have also been explored including the 150-kb BAC delivery by transposition into human embryonic stem cells[104] and conjugation-mediated horizontal gene transfer.[105]

BACs and YACs have also been used to assemble and carry large DNA payloads within viral genomes like herpesviruses (e.g. herpes simplex virus type-I and human cytomegalovirus) and baculovirus, which in turn can be exploited as efficient vectors to deliver large (50–100 kb) DNA payloads to animal and human cells.[106–109] However, these vectors still carry viral genes and, excluding some cancer-related exceptions, are not as advanced and clinically validated as other gene therapy platforms like AAV and retroviruses.[110,111]

2.2 Mammalian (and human) artificial chromosomes

Mammalian and human artificial chromosomes (MACs and HACs) have interesting properties as tools for gene therapy applications and have unique characteristics if compared to the standard gene therapy vectors. While these latter, such as AAV and retrovirus, are usually used for gene and cell therapy, either by integration (semi-randomly or targeted) into the human genome or maintained as episomes in non-dividing cells, MACs and HACs are designed to stably maintain large portions of exogenous DNA outside of the endogenous chromosomes in actively replicating cells. The DNA in HACs is replicated and segregated autonomously, in parallel to the normal genome replication, without altering the host chromosomes. From this perspective, HACs should allow stable addition of genetic material without introducing safety issues concerning DNA double-strand breaks (DSBs) into the target cell genome.

How much DNA can be carried in a single artificial chromosome is unknown, but it is very likely that a single HAC may stably carry at least the same amount of DNA as in the largest human chromosomes (e.g. chromosome 1, 250 Mb). Indeed, stability has been reported to be lower for small HACs (below 1 Mb) than for larger artificial chromosomes.[112] In addition, if HACs are not carrying essential genes, their progressive loss could be easily tolerated by the cell.

MACs and HACs have been used to carry the largest human genes, such as *DMD* (2.4 Mb),[113] and to address major goals of synthetic biology, including the implementation of synthetic gene regulation circuits into mammalian cells and the development of biosensor cell lines.[62] While the integration of the genetic elements for these functions into mammalian chromosomes may lead to genotoxicity due to the perturbation of existing regulatory elements, chromatin state and gene expression, their localization into HACs can minimize such potential problems. Also, while the expression of integrated transgenes can be influenced by the surrounding chromatin state, this effect is usually mitigated in artificial chromosomes. Therefore, MACs and HACs have been used as orthogonal non-integrating and autonomously replicating episomal vectors to carry large payloads (e.g. 10- to 60-Mb MACs have been employed).[114,115] This has been advantageous when trying to study gene expression reproducibly, mimicking near endogenous gene contexts. Indeed, even the largest virus-derived vectors (e.g. herpesvirus) are capable of delivering only up to 150-kb full-length genes.[106] These genes are often cDNA-derived or missing enhancers and other promoter regulatory regions, therefore not recapitulating a wild-type genetic environment and regulation. Moreover, these very large viral-derived vectors can be lost during cell replication, can lead to immune response from the host organism, cannot be precisely controlled in copy number, and can lead to undesired integrations as mentioned above for smaller vectors. On the contrary, MACs and HACs harbor centromeric regions and mammalian replication origins, thereby duplicating synchronously with the cell cycle and being maintained in single copy. However, the biggest limitation of HACs for gene therapy approaches is their problematic delivery, mostly constraining their application to ex vivo genome engineering.

The essential functional units of MACs and HACs are: centromeres, telomeres, origins of replication and the genes of interest payload. Centromeric repeats have been considered indispensable for the correct segregation of MACs and HACs in mitosis.[116] In natural chromosomes, the spindle microtubules connect to the centromere through the kinetochore, whose structure is formed by the centromere proteins (CENPs).[117] The telomeres are bound by a protein cap which avoids the recruitment of double-strand break repair mechanisms and fusion between distinct chromosomes. The telomeres and centromere ensure HAC maintenance and stability,[62,118] which has been demonstrated *in vitro* in human cells, and both *in vitro* and *in vivo* in mouse (so, both in cell lines and transchromosomal animals).[119–121]

Another advantage of HACs is that they can be identified and isolated by flow cytometry[116]; there are established techniques to transfer them into specific cells (e.g. chicken DT40, hamster CHO, mouse A9, human HT1080, hiPSC),[119–122] such as microcell-mediated chromosome transfer,[123] as explained below in Section 2.3.

HACs can be obtained in two possible ways: either a bottom-up or a top-down approach.

2.2.1 Bottom-up artificial chromosomes

The bottom-up (*centromere seeding*) approach was initially developed in human HT1080 fibrosarcoma cells. It is based on centromere formation after transfection of BACs or YACs containing synthetic alphoid repeats (centromeric alpha-satellite DNA), from 30 to 200 kb in size, which are amplified by the cell to form a new centromere and thus a new mitotically stable artificial chromosome.[124–126] Bottom-up HACs are usually circular, but they could potentially be modified to be linear by telomere insertion.[127,128] The initial constraint of bottom-up HAC formation exclusively in HT1080 cells was overcome by controlling epigenetic modifications (H3K9me3) in other cell lines, as shown by Masumoto and colleagues by tethering histone acetyltransferases to the input alphoid DNA, allowing kinetochore formation.[129]

In the most sophisticated version of a bottom-up HAC, the circular alphoid^{tetO} HAC, the centromere is made conditional by embedding tetracycline operator (tetO) sequences inside the alphoid repeats. In this way, HAC-specific centromere decondensation and replication-mediated HAC loss can be precisely controlled through the expression of tet-repressor protein (tetR) fusions with chromatin remodeling factors.[130,131] This system is ideal for gene function studies, allowing phenotype comparison with and without stable gene expression in the same cell.

Alphoid^{tetO} HACs have been successfully used for whole-gene delivery into different cell lines by inserting these into the HAC backbone via Cre-mediated recombination in CHO cells.[132] Therapeutic gene expression levels were shown to be maintained stably without significant gene silencing effects.[133] Importantly, these artificial chromosomes retain the same epigenetic structure even after several passages and transfer between different cell types, as confirmed by immuno-FISH for the assembly of CENP-A and chromatin immunoprecipitation (ChIP), detecting the same level of histone modifications (H3K4me3, H3K4me2 and H3K9me3) and CENP-A protein on alphoid^{tetO} DNA in the original donor and final acceptor cells. No rearrangements or unwanted integrations were detected. These findings show the potential translational relevance of the newest generation of HACs.[133]

Recently, *de novo* HACs have been obtained bypassing the two usual requirements for centromere formation, i.e. the need for centromeric alpha satellite repeats, containing the 17-nt "CENP-B boxes," and the high expression of CENP-B protein.[64] Centromere seeding relies on the presence of the CENP-A variant of histone H3 in centromeric nucleosomes, which is recognized by the protein CENP-B. HJURP is a centromere chromatin assembly protein which is responsible for CENP-A propagation with a well-investigated epigenetic mechanism.[134] This interaction, together with CENP-C, is responsible for the formation of a functional centromere. In a new way to develop HACs, the need for alpha satellite repeats, CENP-B boxes and CENP-B expression was overcome by artificially localizing HJURP fused to Lac-I to a transfected BAC containing LacO sequences.[64] HJUPR localization on a 200-kb BAC of non-alphoid DNA enriches for CENP-A nucleosomes and establishes HAC formation, via multimerization and in some cases capture of additional genomic DNA sequences up to a size of 5–10 Mb. Thus, these *de novo* HACs, stably inherited across generations, can be obtained in a CENP-B-independent manner using non-repetitive sequences indicating the possibility of develop fully synthetic centromeres and chromosomes.[135]

2.2.2 Top-down artificial chromosomes

In the top-down approach to create artificial chromosomes, the starting point is represented by a natural human chromosome, which is modified by telomere-associated chromosome truncation not far from the centromeric region.[136–138] In this procedure artificial telomeres, usually about 1 kb of TTAGGG repeats, are inserted by HDR in the desired chromosomal location. The inserted artificial telomeres are then extended by cellular telomerase to full-length (2–15 kb) functional telomeres.[139,140] This procedure was historically performed in homologous recombination-proficient cells such as DT40, previously engineered to contain a human chromosome. The linear HACs (minichromosomes) conserve functional centromeres and mitotic stability in human and mouse cells.[141] To date, minichromosomes have been made from human chromosomes 14, 21, X and Y.[136,138,142–144] However, in some cases the minichromosome copy number was less stable.[136] It is possible that the stability of artificial chromosomes in cell culture and *in vivo* is simply dependent on the payload influence on the fitness of each trans-chromosomal cell type,[145] as also seen in natural aneuploidies.[146] In the most advanced version of minichromosomes, the human minichromosome 21 (named 21HAC, 5 Mb), all residual genes including the ones located in pericentromeric regions were removed.[118] For the accurate control of cellular differentiation and death, this minichromosome has been further engineered to contain loxP sites, a reversible immortalization cassette, encoding the catalytic subunit of human telomerase (hTERT), the cell cycle regulator Bmi1, the negative selectable marker Timidine Kinase, an inducible caspase (iCasp9) and the MyoD cDNA fused with the estrogen receptor (ER) in order to activate its expression upon tamoxifen administration (MYOD-ERT2).[113] This, and simpler version of the minichromosome, were engineered to carry a 2.4-Mb DNA fragment from human chromosome X containing the *Duchenne muscular dystrophy*

(*DMD*) gene (DYS-HAC).[113,147,148] The DYS-HAC was proposed as a potential tool for cell therapies, producing muscle progenitor cells and mesoangioblasts, which, after transplantation, resulted in long-lasting morphological and functional amelioration of DMD model mice. Of note, a small minichromosome for the *DMD* gene has also been developed having human chromosome X centromere.[136]

Overall, both top-down and bottom-up HACs represent advanced cloning platforms. They can contain large gene-sized payloads without the need to alter endogenous cellular chromosomes and thus, in this sense, preventing insertional mutagenesis and genomic instability.

However, the main challenge in artificial chromosome manipulation is the difficulty of transferring them easily from one cell type to another: up to now, there is no efficient method to shuttle chromosomes between cells without a considerable risk of damaging them and in some cases also the genome of the receiving cells.

2.3 Transfer of natural and artificial chromosomes

Because of their size, chromosomes are not usually delivered by traditional lipid-based transfection methods. This is true for HACs as well as for natural chromosomes. This means that, most likely, the gene therapy use of artificial chromosomes has to be limited to ex vivo applications followed by a cell therapy.

The most used technique, microcell-mediated chromosomal transfer (MMCT) was established as an effective way to transfer intact chromosomes from donor cells to specific acceptor cells (typically chicken DT40, hamster CHO, mouse A9 and human fibrosarcoma-derived HT1080 cells, mESC, miPSCs).[149–153] MMCT is based on the production of small membrane-coated "microcells" around one or a few chromosomes. It involves a preliminary arrest of the cells in metaphase, in which the chromosomes are most condensed, by means of cell cycle inhibitors (e.g. colcemid). In the most recent version, the actin cytoskeleton is depolymerized by polymerization-inhibiting drugs (e.g. cytochalasin B, latrunculin B).[154] The microcells are isolated via high-speed centrifugation in either liquid medium (only for adherent cells),[154,155] or in Percoll/Ficoll density gradients,[151,156,157] exploiting the shearing force with the surrounding liquid or gel to form microcells. The final transfer is accomplished by microcell-cell fusion. The low efficiency of MMCT requires large input quantities for a successful experiment; also, MMCT has been optimized with reasonable efficiency only for a limited number of cell types, whereas transfer to other lines has not been documented extensively. Moreover, because microcells can be efficiently produced mainly from rodent cells (e.g. CHO, mouse A9), which are capable of tolerating the required 48 h of colcemide treatment, several groups first fused the cell of interest with a rodent cell line and then generated microcells for MMCT from the hybrid cells.[63,158]

The success of MMCT mostly depends on efficient fusion between the microcells and the recipient cells. This was historically achieved by a short incubation in poly-ethylene glycol (PEG) and use of phytohemagglutinin,[150] but virus-induced cell fusion using inactivated Sendai virus,[159] wild-type or engineered measles virus HN-F protein or murine leukemia virus (MLV) envelope protein has also been explored.[149,160,161]

As an alternative to MMCT, direct artificial chromosome isolation and transfection by cationic lipids has been reported,[115,116] but it presents the risk of damaging the chromosomes by shearing during the transfection and delivery step into the nucleus of the acceptor cell. Despite the fact that in MMCT chromosomes are always protected inside cell-derived vesicles, MMCT also presents a risk of chromosomal damage due to the formation of the micro-nucleated cells that can result in chromosomal rearrangements.[162]

2.4 Chromosome elimination

The ability to transfer chromosomes from cell to cell is an opportunity to replace endogenous chromosomes when they are damaged, as occurs in many genetic diseases. The replacement would be necessary in order to avoid aneuploidy problems and requires the development of chromosome elimination techniques. These techniques could also be used to correct cells affected by supernumerary chromosomes as occurs in several, and sometimes quite common, genetic diseases such as the Down syndrome (Table 1).

Different techniques are available to induce chromosome loss. Recently, multiplex CRISPR nuclease targeting of chromosomes has been used to drive chromosome loss by fragmentation.[163] Despite being powerful and efficient, however, this strategy may risk generating several DNA fragments that might integrate into the cell genome.

In some cases, it might be possible that the simple positive selection for a transferred chromosome might drive the loss of one of the respective endogenous pair, albeit at low rate, as reported for chromosome X in mESC.[153] Alternatively, introducing a negative selectable marker into a

Table 1 Examples of syndromes involving large genome alterations.

Disease	OMIM	Genetic defect	Treatment	Symptoms	Frequency (live births)	Lethal within 2 years
Fragile X	300624	FMR1 mutation/deletion (Xq27.3)	Palliative	Cognitive deficit, anxiety, attention deficit hyperactivity disorder (ADHD), light-/sound-sensitivity, aggressivity, malformations (sometimes)	1:5000	No
Smith–Magenis syndrome	182290	17p11.2 deletion	Palliative	Tracheobronchial abnormalities, anxiety, ADHD, brachycephaly, brachydactyly	1:15000	No
Potocki–Lupski syndrome	610883	17p11.2 duplication	Palliative	Pharynx abnormalities, ADHD, autism, dysarthria	1:25000	No
1p36 deletion syndrome	607872	1p36 deletion	Palliative	Cognitive deficit, delayed growth, hypotonia, seizures, speech disability, malformations, hearing/vision impairment, facial features	1:5000	No
DiGeorge syndrome	188400	22q11.2 deletion	Palliative	Cyanosis, high incidence of infections, facial features, cleft palate, delayed growth, breathing problems, poor muscle tone, cognitive deficit	1:4000	No
Wolf–Hirschhorn syndrome	194190	4p16.3 deletion/translocation	Palliative	Delayed development, cognitive deficit, hypotonia, seizures, skeletal abnormalities, heart defects, hearing loss, urinary tract malformations	1:50000	No
Angelman syndrome	105830	15q11.2-q13 deletion	Palliative	Excitability, hyperactivity, attention deficit, less sleep need, strabismus, scoliosis, behavioral features	1:12000	No
Prader–Willi syndrome	176270	15q11.2-q13 deletion	Palliative	Delayed sexual development, diabetes, respiratory problems, heart failure, hypotonia, growth delay, cognitive delay, excessive appetite	1:10000	No

Syndrome	OMIM/ORPHA	Chromosomal abnormality	Treatment	Symptoms	Prevalence	Prenatal screening
Williams syndrome	194050	7q11.2 deletion (up to 1.5 Mb)	Palliative	Facial features, ADHD, phobias, little growth, sunken chest, cognitive deficit, kidney abnormalities, farsightedness, speech delay, low muscle tone	1:10000	No
Charcot–Marie–Tooth	606482	19p13.2 duplication	Palliative	Muscle weakness (feet), flat or arched feet, curled toes, lack of sensation in arms and feet, tremor, scoliosis, dysphagia (rare)	1:2500	No
WAGR syndrome/11p deletion	194072	11p13 deletion	Palliative	Wilms' kidney tumor, aniridia, genitourinary problems, mental retardation (WAGR), hyperphagia (sometimes)	1:500000	No
Patau syndrome	ORPHA:3378	Trisomy 13	None	Ceft lip and palate, microphthalmia or anophthalmia, microcephaly, deafness, capillary hemangiomas	1:12000	Yes
Mosaic trisomy 15 (MVA1)	257300	Trisomy 15	Surgery, orthopedic	Growth delays, cognitive deficit, craniofacial malformation, cardiac defects, finger/toe malformations, skeletal malformations, genital abnormalities	Not available	No
Edward's syndrome	ORPHA:3380	Trisomy 18	None	Cleft palate, heart defects, developmental delays, chest deformity, microcephaly, weak cry, deformed feet, lung, kidney and stomach defects	1:8000	Yes
Down syndrome	190685	Trisomy 21	Palliative	Facial features, short neck, white spots in the iris, small stature, small hands and feet, poor muscle tone and loose joints	1:6000	No

target chromosome is sufficient to trigger its selective loss with an efficiency of about 10^{-4}, as shown in chromosome 21 triploid human iPSCs.[164]

A recombinase-based strategy to induce chromosome loss has also been investigated in human and mouse cells.[165–168] This approach requires the incorporation of consecutive loxP sites in opposite orientations into a chromosome of interest. Upon chromosome duplication in mitosis and induction of Cre recombinase expression, a recombination event results in a dicentric and an acentric chromosome that are subsequently lost. The efficiency of this strategy varied a lot in the different reports, however it can be very high under the right settings, similar to the CRISPR-nuclease approach mentioned above.

Since an essential determinant for chromosomal stability is the presence of a functional centromeric region, which is controlled at the epigenetic level, alterations of centromeric chromatin state can lead to selective chromosomal loss. For example, a model bottom-up HAC, designed with several tet-operator (tetO) sequences embedded in the centromeric alphoid repeats (alphoid^{tetO} HAC),[158] can be eliminated from the cell upon expression of a tet-repressor protein (tetR) carrying either a transcriptional activator or a repressor.[130,131] Extending this principle, altering centromere condensation and epigenetics with programmable CRISPR-transcription regulators and epigenetic modifiers[48] might be an effective way to induce chromosome loss, even though gRNA selection might be quite complex due to the presence of similar sequences in many centromeres and the incomplete sequencing of all centromeric regions.[169]

3. Synthetic genomics applications for the treatment of genetic diseases

The ability to assemble and replace large fragments of the human genome, or even of entire human chromosomes, could shorten the list of incurable genetic diseases in direct and indirect ways. On one hand, Mb-scale genome surgery may offer the opportunity to study and model genetic diseases in an isogenic setting and in humanized animal models, on the other hand, it may allow the removal of the origin of diseases that affect multiple DNA regions, an extended section of DNA, or whole chromosomes by correcting the damaged genome. Some examples of diseases which are considered nowadays incurable, such as big chromosomal inversions, deletions and rearrangements are provided in Table 1.

3.1 Modeling genetic diseases

The ability to *rewrite* the genome is very useful to study and model genetic disease. From a practical point of view, it can enable the creation of model cells and animals without the need of having patient-derived cells, which can have limited availability for rare diseases. It can also enable the isolation of specific genetic features and their application to different genetic contexts to assess them independently. Indeed, although sequencing technologies provide powerful tools to correlate genotype and phenotype, they do not always establish a clear causality link and are limited to the correlation of already discovered mutations.[170] Gene editing enables more detailed study of the links between genotype and phenotype, and makes it possible to model the effect of mutations beyond those already reported. Synthetic genomics promises to go further and more systematically in this direction, aiming at the opportunity not only to recreate a particular mutation, but also to rewrite long DNA sequences up to the size of the entire genome. In this way, it will be possible to dissect and uncouple the impact of many genetic variations present in a population, that modulate, or in some case prevent, the development of genetic diseases even in the presence of a pathological mutation. For polygenic disorders (e.g. hypertension, coronary heart disease, diabetes and cancer) this deeper analysis may potentially lead to new targets for the future development of genetic or pharmacological therapies.

One of the big challenges in modeling polygenic genetic disorders is that their genetic determinants are not simply localized in one or a few specific genes or loci: they can be scattered genome-wide, involving many genes or non-coding sequences in one or more chromosomes. They would therefore benefit from synthetic genomics studies, both by engineering large genome fragments containing several variants and by multiplexed gene editing.[171]

Even when considering diseases caused by small genes, the best *in vivo* modeling in preclinical settings (e.g. in mice, rats, or non-human primates) may require large genetic engineering, as it is often preferable, if not required, to have the complete human gene sequences comprising distal regulatory regions to recapitulate patients' phenotypes and potential outcomes of gene editing therapies. This can be true also for model human cells and organoids, where appropriate engineering of surrounding small nucleotide polymorphisms and genetic variants represents an important modulator of gene expression and disease features.[172] Such models are particularly important to recapitulate rare genetic diseases and genetic defects that are lethal during development or soon after birth, for which patient samples may be less available.[173,174]

Synthetic genomics approaches may help to develop or improve the modeling of several currently incurable diseases involving large genetic aberrations, such as fragile X syndrome, Huntington's disease, Smith-Magenis syndrome, Potocki-Lupski syndrome, 1p36 deletion syndrome, DiGeorge's syndrome, Wolf-Hirschhorn syndrome, Angelman syndrome, Prader-Willi syndrome, Williams syndrome, Charchot-Marie-Tooth, WAGR syndrome (also known as 11p deletion syndrome), and the most common trisomies: Patau's syndrome (chromosome 13), mosaic trisomy 15, Edward's syndrome (chromosome 18) and Down syndrome (chromosome 21) (Table 1).

Some examples of disease modeling involving large-scale genome editing that for their size prelude applications of synthetic genomics are described below. For instance, a 450-kb YAC, carrying the entire *FMR1* gene (with different CGG expansions) and upstream and downstream regulatory sequences, was injected into mouse oocytes to generate mice modeling the fragile X syndrome.[175] With the same rationale, transgenic mice were also obtained to model Huntington's disease with YACs expressing different versions of the Huntingtin protein from human promoter and regulatory sequences.[176,177] In the case of the Charcot-Marie-Tooth syndrome, an overall frequent (1/2500) hereditary peripheral neuropathy, subtype 1A is the most common form and is associated with a partial duplication in a 1.5-Mb region of chromosome 17, containing the *PMP22* gene. Model mice have been created by pronuclear injection of a 0.56-Mb YAC including the *PMP22* gene.[178] These mice are now valuable assets to evaluate gene editing treatments in a fully humanized gene context.[179] YACs have also been used to create a mouse strain carrying the full size human *DMD* gene; a very large 2.7-Mb YAC was delivered by fusion into mESC knock-out for mouse *DMD*.[73] This massive genome engineering rescued mice locomotion compared to parental knock-out strains and further engineering may be exploited to test gene editing approaches targeting patient mutations.

Similar to YACs, also BAC vectors, despite their usually smaller size, have been used for mouse transgenesis when it was required to maintain human regions surrounding selected genes.[180] For instance, a BAC (190 kb) transgenic mouse model for Parkinson's disease has been recently established.[181] It contains the full-size (A53T mutated) α-synuclein (*SNCA*) gene and surrounding sequences, including the human *SNCA* promoter, with a dinucleotide polymorphism in the 5' Rep1 region and two risk-associated SNPs. Compared to previous animal strains, this new highly refined model better recapitulates the prodromal stage of Parkinson disease.[181] Similarly, a 173-kb BAC transgenic mouse containing the complete human C9orf72 gene, with either a normal allele or disease-associated expansions (100–1000 G_4C_2 repeats), displayed the pathologic features seen in patients. These features include widespread RNA foci, repeat-associated non-ATG (RAN) translated dipeptides, altered nucleolin distribution and a loss of the repressive mark H3K9me3 in astrocytes and neurons.[182,183] Despite these patient similarities, behavioral abnormalities and neurodegeneration were not reported, supporting that these molecular changes precede disease manifestation and offering the opportunity to study the onset of a disease. Finally, *SHANK3* locus duplications characteristic of some neuropsychiatric disorders in patients have been modeled by BAC integration in a transgenic mouse, showing around a 50% increase in *SHANK3* expression, recapitulating maniac-like behaviors and contributing to better understand mechanisms and potential therapies that might benefit individuals with *SHANK3* overexpression.[184]

Beyond these examples of BAC and YAC use, the humanization of animal models has been extended to the creation of transchromosomic animals. Such chromosomes can be HACs,[63] which for instance have been used for modeling diseases like Duchenne muscular dystrophy,[147] or even full-sized human chromosomes as in the case of modeling Down's syndrome in the transchromosomic mouse strain carrying human chromosome 21.[185] Further and more extensive humanization of the mouse genome and of other models are expected in the future.[186] They will likely also include applications for safer and more compatible xenogenic organs with multiple and complete inactivation of potentially harmful animal endogenous retroviruses[171,187] and they will be facilitated by the development of more accurate and efficient synthetic genomics tools.

3.2 Synthetic genomics approaches for gene and genome therapy

The advances in the field of synthetic genomics may offer the opportunity to approach genetic diseases, including the most complex and currently incurable ones, from a different perspective and at a different scale. On one hand, gene therapy and gene editing are traditionally relying on single-base or size-limited interventions[188]; the clinically most advanced gene therapy approaches are based on gene addition to compensate for, not to correct, a defective single gene with the delivery of an exogenous cDNA.[189] However, many genes produce different transcripts and splicing variants, often containing different exons and coding sequences which may have different functions that cannot always be recapitulated in a single cDNA.[190,191] Such cDNA can either be delivered to cells as an episome, for example when using adeno-associated viral vector (AAV), or integrated into host patient's chromosomes. This integration of exogenous DNA into the genome is often obtained semi-randomly, using retro/lentiviral vectors, or in safe harbor loci, taking advantage of homology-directed recombination. More recently, genetic surgery technologies have been developed for the correction or compensation of endogenous deleterious mutations using programmable nucleases or DNA editors[47] and positive results have been reported from in programmable-nuclease clinical trials.[192] However, as you can read in the previous chapter of this book,[66] the ambition of most genetic interventions is usually focused on the editing of a single gene, a single nucleotide or small mutations, in particular those more common among patients or located in important regulatory regions.[192,193] Small- or cDNA-sized interventions are well suited for the effective payload capacity of AAV (up to 4.5–4.7 kb, even less than 3.5 kb if homology arms are necessary) or retro/lentiviral vectors (8 kb).[111,194,195] However, of the 38,214 genes (coding and non-coding) reported in the NCBI database for the entire human genome, in 11,655 (30.50%) of these genes the mRNA is longer than 3.5 kb, in 8039 (21.04%) longer than 4.5 kb, in 2275 (5.95%) longer than 8 kb (in agreement with previous reports indicating median mRNA size).[196] Even considering only the size of the CDS several important therapeutic genes (and several programmable nucleases) are beyond the capacity of the standard gene therapy vectors.[47,197,198]

There are exceptions to these small size interventions, such as the attempts to develop a cure for 62% of *DMD* mutations causing Duchenne muscular dystrophy by deleting more than 300 kb from the *DMD* genomic sequence to produce partially functional Becker muscular dystrophy-like dystrophin.[199] Nevertheless, even when stem cell genomes are manipulated, usually the attempt is mainly focused on the repair of size-limited defects in a single gene or, when feasible, on the deletion of a damaged sequence. *Per se* these are sufficient and probably the best strategies for many genetic diseases, but they leave out of research focus a number of patients and disorders where the genomic damage, that needs to be repaired or rebuilt, is too large or complex (for some examples see Table 1). In such circumstances, a classical alternative might be an allogenic transplant, where possible and therapeutically plausible, such as for hematopoietic defects or immunodeficiency associated with a particular disorder, and when compatible donors are available.

On the other hand, synthetic genomics aims at developing the technologies to write or massively edit an important portion or even an entire eukaryotic genome. Thus, for the ambitions of synthetic genomics, there should not be a mutation or genetic defect that is too big to be fixed or a DNA fragment that should be considered too big to be delivered. However, it is evident that potential therapeutic applications of synthetic genomics will be strongly affected by delivery issues and limited to *in vitro* manipulation and re-engraftment of the engineered cells (either stem cells, progenitor cells or differentiated highly proliferative cells such as keratinocytes, B and T-cells).[200-202] Embracing a synthetic genomic perspective would suggest to establish technologies pointing to either correct the complex genetic defect in patient cells for autologous transplantations or engineer heterologous cells to match the histocompatibility profile of the patient. Notably, several genome editing efforts in these directions have been published,[203] including different knock-outs of MHC class I and II in hiPSCs and hematopoietic stem cells (HSC) and overexpression of immunomodulatory proteins such as CD47.[204-206]

It must be considered that many of these diseases with unmet medical needs lead to systemic or multi-organ developmental defects requiring probably embryonic or *intra utero* delivery, which are not exempt from important ethical concerns, as described in this volume by Andy Greenfield in chapter 1.[207] However, even focusing only on the incidence of genetic defects to cytogenetically visible autosomal deletions, a huge number (1 every 7000 new-borns) is affected.[174] *In utero* genome manipulation has thus the potential to prenatally treat genetic diseases that result in significant morbidity and mortality before or shortly after birth.[173] Alternatively, when possible, the next-generation genome therapeutics efforts might be focused on healing the most affected and major organs.

One important and unique feature of synthetic genomics approaches applied to gene therapy is that they offer a new alternative way to stably modify the genome of dividing cells, given by the use of HACs; conventional gene therapy strategies mainly focus on adding genes in ectopic chromosomal positions or repairing the natural endogenous gene. However, gene therapy using HAC technologies promises new opportunities for therapy: full-length human genes can be added to the genome without the requirement of altering existing chromosomes and potentially leading to DSBs and karyotype aberrations.[63] HACs have several interesting features as vectors for gene therapy: (i) they are long-term stably maintained in host cells without integrating into the host genome,[133] (ii) they can contain entire genes and gene clusters including distal regulatory regions to provide natural-like levels of tissue and temporal gene expression, but they may even contain several copies of the same gene to stably express supraphysiological protein levels (e.g. Factor VIII)[208] and (iii) the presence of HACs has not been reported to be a risk for cellular transformation or immune responses.[62] HACs have been created for several full length genes (0.2–2.3 Mb) such as CFTR, DMD and ß-globin, maintaining transgene expression and stable transmission for generations also in the absence of selection.[148,209,210]

Nevertheless, the main issue with HACs is that they are too bulky to be efficiently delivered inside target cells and, unlike plasmids, BACs and YACs, they cannot be amplified in bacteria or yeast. For these reasons, most examples of HAC use are associated with stem cells or other highly proliferating cells. Whereas some types of highly stable and proliferating adult stem cells can be employed as delivery cells (Lgr5+ intestinal, nephron and hepatic stem cells or blood mesangioblasts), induced pluripotent stem cells (iPSCs) or embryonic stem cells are surely ideal carriers because of their high proliferation capacity and translational capabilities.[211,212] For these applications, artificial minichromosomes have been built and engineered in rodent cells and in avian DT40 cells and transferred via MMCT to the final iPSC recipient cell, showing effective genetic correction and translational potential in mouse models for the treatment of muscular dystrophy.[62,122,148] In addition to HACs, also the transplant of natural chromosomes might be used in the future for the therapy of genetic disease. An interesting recently published example shows the correction of chronic granulomatous disease (CGD) defect in mouse iPSC knock-out for *CYBB* gene on chromosome X.[152] The *CYBB* gene is often affected by large deletions in (CGD) patients, thus the authors proposed to replace the pathological chromosome with a wild-type chromosome X from donor mouse cells. To do so, they first knocked-out the *HPRT* gene on the damaged X chromosome, then used MMCT to transfer the new X chromosome (having functional *HPRT* gene) and selected for the loss of the damaged chromosome using HAT media. When these largely modified iPSC were differentiated, functional granulocytes were obtained.[152]

Beyond these still limited examples, it is plausible to expect that in the future, with technological improvements in the ability to deliver and engineer DNA fragments at Mb- and chromosomal-scale in human cells, eliminating at the same time damaged endogenous chromosomes, synthetic genomic techniques may provide new therapeutic solutions to unmet medical needs.

3.3 Synthetic genomics to understand the human genome

Synthetic genomics offers incredible opportunities to deepen our knowledge of the human genome and to engineer it on unprecedented scale. Chromosome shuttling, replacement and modification techniques could be adopted in the modeling and correction of complex genetic aberrations. At the same time, the knowledge gained from synthetic genomics approaches will be valuable for the understanding of the architecture and function of specific genomic regions. For instance, for several genetic diseases there is an incomplete understanding of the mutations causing them and also genome-wide association studies did not fully unravel the causality between several genetic variants and specific phenotypic traits. Synthetic genomics can provide the methods to dissect, delete and rebuild genome portions, while potentially uncovering the function and role of the *loci* of interest. Genome dissection and reassembly might also help to understand a huge number of biological aspect, from the susceptibility to viruses to the role of endogenous retroviruses, transposons and the rest of the genomic *dark matter* (i.e. those seemingly meaningless regions which have only been poorly investigated up to now), thereby paving the way for the attempt to find the minimal set of genes to make a functional higher eukaryotic genome. With this perspective, programmed minimal synthetic cells can be envisaged for highly specialized functions. Also, as synthetic genomics will allow the shuffling of large genetic regions, it will also elucidate the relationship between close and distant genetic elements, enabling the study of the importance of genome architecture on an unprecedented scale.[43]

It is therefore clear that three main applications of synthetic genomics, i.e. restructuring, recoding and resizing genomes will provide great contributions also for the development of therapies. Indeed, through a more complete understanding of the human genome, it seems likely that genetic diseases will be more efficiently and precisely treated, also with conventional pharmacological, gene therapy and gene editing approaches.

4. Perspectives

The main idea implied by synthetic genomics is that we should develop the possibility to build, repair and modify entire genomes, moving genetic engineering from a gene to genome scale. In this sense a genome therapy could be developed to understand and correct polygenic disorders and chromosomal abnormalities, and to envision futuristic applications, such as the creation of designer cells resistant to viruses or created with a minimal required genome locked to perform a desired biological task. The advent of gene and genome therapies is not free from profound ethical considerations and regulatory boundaries. However, these ethical concerns are continuously considered and addressed by the scientific community, as it was also immediately realized after the first conceptualization of DNA manipulation and genetic surgery in the late 1960s.[213,214]

Acknowledgments

G.P. gratefully acknowledges funding from the European Union's Horizon 2020 research and innovation programme under the Marie Skłodowska-Curie grant agreement RE-GENESis no. 897663. We thank Sarah Montgomery, Jakob Birnbaum and Daniel de la Torre for fruitful discussions.

References

1. Sanger F, Air GM, Barrell BG, et al. Nucleotide sequence of bacteriophage phi X174 DNA. *Nature*. 1977;265:687–695.
2. Sanger F, Coulson AR, Hong GF, Hill C, Petersen GB. Nucleotide sequence of bacteriophage lambda DNA. *J Mol Biol*. 1982;162(4):729–773.
3. Craig Venter J, Adams MD, Myers EW, et al. The sequence of the human genome. *Science*. 2001;291(5507):1304–1351. https://doi.org/10.1126/science.1058040.
4. International Human Genome Sequencing Consortium. Initial sequencing and analysis of the human genome. *Nature*. 2001;409:860–921. https://doi.org/10.1038/35087627.
5. Auton A, Abecasis GR, Altshuler DM, et al. A global reference for human genetic variation. *Nature*. 2015;526(7571):68–74. https://doi.org/10.1038/nature15393.
6. Turnbull C, Scott RH, Thomas E, et al. The 100 000 genomes project: bringing whole genome sequencing to the NHS. *Br Med J*. 2018;361:k1687. https://doi.org/10.1136/bmj.k1687.
7. Matsumoto T, Wu J, Kanamori H, et al. The map-based sequence of the rice genome. *Nature*. 2005;436(7052):793–800. https://doi.org/10.1038/nature03895.
8. Jaillon O, Aury JM, Noel B, et al. The grapevine genome sequence suggests ancestral hexaploidization in major angiosperm phyla. *Nature*. 2007;449(7161):463–467. https://doi.org/10.1038/nature06148.
9. Sanger F, Coulson AR. A rapid method for determining sequences in DNA by primed synthesis with DNA polymerase. *J Mol Biol*. 1975;94(3):441–448. https://doi.org/10.1016/0022-2836(75)90213-2.
10. Maxam AM, Gilbert W. A new method for sequencing DNA. *Proc Natl Acad Sci USA*. 1977;74(2):560–564.
11. Amarasinghe SL, Su S, Dong X, Zappia L, Ritchie ME, Gouil Q. Opportunities and challenges in long-read sequencing data analysis. *Genome Biol*. 2020;21(1):1–16. https://doi.org/10.1186/s13059-020-1935-5.

12. Ronaghi M, Karamohamed S, Pettersson B, Uhlén M, Nyrén P. Real-time DNA sequencing using detection of pyrophosphate release. *Anal Biochem.* 1996;242(1):84–89. https://doi.org/10.1006/abio.1996.0432.
13. Quail MA, Smith M, Coupland P, et al. A tale of three next generation sequencing platforms: comparison of Ion Torrent, Pacific Biosciences and Illumina MiSeq sequencers. *BMC Genomics.* 2012;13(341):13.
14. Voelkerding KV, Dames SA, Durtschi JD. Next-generation sequencing: from basic research to diagnostics. *Clin Chem.* 2009;55(4):641–658. https://doi.org/10.1373/clinchem.2008.112789.
15. Heather JM, Chain B. The sequence of sequencers: the history of sequencing DNA. *Genomics.* 2016;107(1):1–8. https://doi.org/10.1016/j.ygeno.2015.11.003.
16. Li J, Stein D, McMullan C, Branton D, Aziz MJ, Golovchenko JA. Ion-beam sculpting at nanometre length scales. *Nature.* 2001;412(6843):166–169. https://doi.org/10.1038/35084037.
17. Dekker C. Solid-state nanopores. *Nat Nanotechnol.* 2007;2(4):209–215. https://doi.org/10.1038/nnano.2007.27.
18. Levene MJ, Korlach J, Turner SW, Foquet M, Craighead HG, Webb WW. Zero-mode waveguides for single-molecule analysis at high concentrations. *Science.* 2003;299(5607):682–686. https://doi.org/10.1126/science.1079700.
19. Schwarze K, Buchanan J, Fermont JM, et al. The complete costs of genome sequencing: a microcosting study in cancer and rare diseases from a single center in the United Kingdom. *Genet Med.* 2020;22(1):85–94. https://doi.org/10.1038/s41436-019-0618-7.
20. Agarwal KL, Buchi H, Caruthers MH, et al. Total synthesis of the gene for an alanine transfer ribonucleic acid from yeast. *Nature.* 1970;227(5253):27–34. https://doi.org/10.1038/227027a0.
21. Cello J, Cello J, Paul AV, Wimmer E. Chemical synthesis of poliovirus cDNA: generation of infectious virus in the absence of natural template. *Science.* 2002;297:1016–1018. https://doi.org/10.1126/science.1072266.
22. Chan LY, Kosuri S, Endy D. Refactoring bacteriophage T7. *Mol Syst Biol.* 2005;1:2005.0018. https://doi.org/10.1038/msb4100025.
23. Annaluru N, Ramalingam S, Chandrasegaran S. Rewriting the blueprint of life by synthetic genomics and genome engineering. *Genome Biol.* 2015;16(1):125. Published 2015 Jun 16. https://doi.org/10.1186/s13059-015-0689-y.
24. Itaya M, Tsuge K, Koizumi M, Fujita K. Combining two genomes in one cell: stable cloning of the Synechocystis PCC6803 genome in the Bacillus subtilis 168 genome. *PNAS.* 2005;102(44):15971–15976.
25. Itaya M, Fujita K, Kuroki A, Tsuge K. Bottom-up genome assembly using the Bacillus subtilis genome vector. *Nat Methods.* 2008;5(1):2001–2003. https://doi.org/10.1038/NMETH1143.
26. Fu Y, Tigano M, Sfeir A. Safeguarding mitochondrial genomes in higher eukaryotes. *Nat Struct Mol Biol.* 2020;27(8):687–695. https://doi.org/10.1038/s41594-020-0474-9.
27. Tong W, Kim TS, Park YJ. Rice chloroplast genome variation architecture and phylogenetic dissection in diverse oryza species assessed by whole-genome resequencing. *Rice (N Y).* 2016;9(1):57. https://doi.org/10.1186/s12284-016-0129-y.
28. Datsenko KA, Wanner BL. One-step inactivation of chromosomal genes in Escherichia coli K-12 using PCR products. *Proc Natl Acad Sci USA.* 2000;97(12):6640–6645. https://doi.org/10.1073/pnas.120163297.
29. Holt RA, Warren R, Flibotte S, Missirlis PI, Smailus DE. Rebuilding microbial genomes. *BioEssays.* 2007;29:580–590. https://doi.org/10.1002/bies.20585.
30. Gibson DG, Benders GA, Andrews-pfannkoch C, et al. Complete chemical synthesis, assembly, and cloning of a Mycoplasma genitalium genome. *Science.* 2008;319 (February):1215–1221.

31. Gibson DG, Glass JI, Lartigue C, et al. Creation of a bacterial cell controlled by a chemically synthesized genome. *Science*. 2010;329(5987):52–56. https://doi.org/10.1126/science.1190719.
32. Giga-hama Y, Tohda H, Takegawa K, Kumagai H. Schizosaccharomyces pombe minimum genome factory. *Biotechnol Appl Biochem*. 2007;46:147–155. https://doi.org/10.1042/BA20060106.
33. Mizoguchi H, Mori H, Fujio T. Escherichia coli minimum genome factory. *Biotechnol Appl Biochem*. 2007;46:157–167. https://doi.org/10.1042/BA20060107.
34. Ara K, Ozaki K, Nakamura K, Yamane K, Sekiguchi J, Ogasawara N. Bacillus minimum genome factory: effective utilization of microbial genome information. *Biotechnol Appl Biochem*. 2007;46:169–178. https://doi.org/10.1042/BA20060111.
35. Hutchison CA, Chuang RY, Noskov VN, et al. Design and synthesis of a minimal bacterial genome. *Science*. 2016;351(6280). https://doi.org/10.1126/science.aad6253, aad6253.
36. Ostrov N, Landon M, Guell M, et al. Design, synthesis, and testing toward a 57-codon genome. *Science*. 2016;353(6301):1–5.
37. Fredens J, Wang K, De Torre D, et al. Total synthesis of Escherichia coli with a recoded genome. *Nature*. 2019;569:514–518. https://doi.org/10.1038/s41586-019-1192-5.
38. Wang K, de la Torre D, Robertson WE, Chin JW. Programmed chromosome fission and fusion enable precise large-scale genome rearrangement and assembly. *Science*. 2019;365(6456):922–926. https://doi.org/10.1126/science.aay0737.
39. Annaluru N, Muller H, Mitchell LA, et al. Total synthesis of a functional designer eukaryotic chromosome. *Science*. 2014;344(6179):55–58. https://doi.org/10.1126/science.1249252.
40. Richardson SM, Mitchell LA, Stracquadanio G, et al. Design of a synthetic yeast genome. *Science*. 2017;355(6329):1040–1044. https://doi.org/10.1126/science.aaf4557.
41. Pretorius IS, Boeke JD. Yeast 2.0—connecting the dots in the construction of the world's first functional synthetic eukaryotic genome. *FEMS Yeast Res*. 2018;18(4):foy032. https://doi.org/10.1093/femsyr/foy032.
42. Mitchell LA, Wang A, Stracquadanio G, et al. Synthesis, debugging, and effects of synthetic chromosome consolidation: synVI and beyond. *Science*. 2017;355(6329):eaaf4831. https://doi.org/10.1126/science.aaf4831.
43. Shao Y, Lu N, Wu Z, et al. Creating a functional single-chromosome yeast. *Nature*. 2018;560(7718):331–335. https://doi.org/10.1038/s41586-018-0382-x.
44. Jackson M, Marks L, May GHW, Wilson JB. The genetic basis of disease. *Essays Biochem*. 2018;62(5):643–723. https://doi.org/10.1042/EBC20170053.
45. Bohacek J, Mansuy IM. Epigenetic inheritance of disease and disease risk. *Neuropsychopharmacology*. 2013;38(1):220–236. https://doi.org/10.1038/npp.2012.110.
46. Park M, Keung AJ, Khalil AS. The epigenome: the next substrate for engineering. *Genome Biol*. 2016;17(1):183. https://doi.org/10.1186/s13059-016-1046-5.
47. Anzalone AV, Koblan LW, Liu DR. Genome editing with CRISPR–Cas nucleases, base editors, transposases and prime editors. *Nat Biotechnol*. 2020;38(7):824–844. https://doi.org/10.1038/s41587-020-0561-9.
48. Adli M. The CRISPR tool kit for genome editing and beyond. *Nat Commun*. 2018;9(1):1911. https://doi.org/10.1038/s41467-018-04252-2.
49. Capecchi MR. Altering the genome by homologous recombination. *Science*. 1989;244(4910):1288–1292. https://doi.org/10.1126/science.2660260.
50. Choi PS, Meyerson M. Targeted genomic rearrangements using CRISPR/Cas technology. *Nat Commun*. 2014;5:1–6. https://doi.org/10.1038/ncomms4728.
51. Jinek M, Chylinski K, Fonfara I, Hauer M, Doudna JA, Charpentier E. A programmable dual-RNA-guided DNA endonuclease in adaptive bacterial immunity. *Science*. 2012;337(August):816–822.

52. Gasiunas G, Barrangou R, Horvath P, Siksnys V. Cas9-crRNA ribonucleoprotein complex mediates specific DNA cleavage for adaptive immunity in bacteria. *Proc Natl Acad Sci USA*. 2012;109(39):2579–2586. https://doi.org/10.1073/pnas.1208507109.
53. Mendez MJ, Green LL, Corvalan JRF, et al. Functional transplant of megabase human immunoglobulin loci recapitulates human antibody response in mice. *Nature*. 1997;15 (February):146–156.
54. Tomizuka K, Shinohara T, Yoshida H, et al. Double trans-chromosomic mice: maintenance of two individual human chromosome fragments containing Ig heavy and κ loci and expression of fully human antibodies. *Proc Natl Acad Sci USA*. 2000;97 (2):722–727. https://doi.org/10.1073/pnas.97.2.722.
55. Green LL, Hardy MC, Maynard-Currie CE, et al. Antigen–specific human monoclonal antibodies from mice engineered with human Ig heavy and light chain YACs. *Nat Genet*. 1994;7(1):13–21. https://doi.org/10.1038/ng0594-13.
56. Lee E, Liang Q, Ali H, et al. Complete humanization of the mouse immunoglobulin loci enables efficient therapeutic antibody discovery. *Nat Biotechnol*. 2014;32 (4):356–363. https://doi.org/10.1038/nbt.2825.
57. Macdonald LE, Karow M, Stevens S, Auerbach W, Poueymirou WT. Precise and in situ genetic humanization of 6 Mb of mouse immunoglobulin genes. *Proc Natl Acad Sci USA*. 2014;111(14):5147–5152. https://doi.org/10.1073/pnas.1323896111.
58. Li LP, Lampert JC, Chen X, et al. Transgenic mice with a diverse human T cell antigen receptor repertoire. *Nat Med*. 2010;16(9):1029–1034. https://doi.org/10.1038/nm.2197.
59. Valenzuela DM, Murphy AJ, Frendewey D, et al. High-throughput engineering of the mouse genome coupled with high-resolution expression analysis. *Nat Biotechnol*. 2003;21(6):652–660.
60. Nelson AL, Dhimolea E, Reichert JM. Development trends for human monoclonal antibody therapeutics. *Nat Rev Drug Discov*. 2010;9(10):767–774. https://doi.org/10.1038/nrd3229.
61. Kaplon H, Muralidharan M, Schneider Z, Reichert JM. Antibodies to watch in 2020. *MAbs*. 2020;12(1):1–24. https://doi.org/10.1080/19420862.2019.1703531.
62. Kouprina N, Tomilin AN, Masumoto H, Earnshaw WC, Larionov V. Human artificial chromosome-based gene delivery vectors for biomedicine and biotechnology. *Expert Opin Drug Deliv*. 2014;11(4):517–535. https://doi.org/10.1517/17425247.2014.882314.
63. Oshimura M, Uno N, Kazuki Y, Katoh M, Inoue T. A pathway from chromosome transfer to engineering resulting in human and mouse artificial chromosomes for a variety of applications to bio-medical challenges. *Chromosom Res*. 2015;23(1):111–133. https://doi.org/10.1007/s10577-014-9459-z.
64. Logsdon GA, Gambogi CW, Liskovykh MA, et al. Human artificial chromosomes that bypass centromeric DNA. *Cell*. 2019;178(3):624–639. https://doi.org/10.1016/j.cell.2019.06.006.
65. Barra V, Fachinetti D. The dark side of centromeres: types, causes and consequences of structural abnormalities implicating centromeric DNA. *Nat Commun*. 2018;9(1):4340. Published 2018 Oct 18. https://doi.org/10.1038/s41467-018-06545-y.
66. Petris G. Preface. In: Petris G, ed. *Progress in Molecular Biology and Translational Science*. vol. 182. Academic Press; 2021.
67. Casini A, Storch M, Baldwin GS, Ellis T. Bricks and blueprints: methods and standards for DNA assembly. *Nat Rev Mol Cell Biol*. 2015;16(9):568–576. https://doi.org/10.1038/nrm4014.
68. Lander ES, Linton LM, Birren B, et al. Initial sequencing and analysis of the human genome. *Nature*. 2001;409(6822):860–921. https://doi.org/10.1038/35057062.

69. Burke DT, Carle GF, Olson MV. Cloning of large segments of exogenous DNA into yeast by means of artificial chromosome vectors. *Science*. 1987;236:806–812.
70. Murray AW, Szostak JW. Construction of artificial chromosomes in yeast. *Nature*. 1983;305(5931):189–193. https://doi.org/10.1038/305189a0.
71. Alasdair M. *YAC Protocols*. 2nd ed. Humana Press Inc.; 2006. https://doi.org/10.1385/1597451584.
72. Ramsay M. Yeast artificial chromosome cloning. *Mol Biotechnol*. 1994;1(2):181–201. https://doi.org/10.1007/BF02921558.
73. T'Hoen PAC, De Meijer EJ, Boer JM, et al. Generation and characterization of transgenic mice with the full-length human DMD gene. *J Biol Chem*. 2008;283(9):5899–5907. https://doi.org/10.1074/jbc.M709410200.
74. Kouprina N, Lee NCO, Kononenko AV, Samoshkin A, Larionov V. From selective full-length genes isolation by TAR cloning in yeast to their expression from HAC vectors in human cells. In: Narayanan K, ed. *Bacterial Artificial Chromosomes*. 2nd ed. New York, NY: Humana Press; 2014. https://doi.org/10.1007/978-1-4939-1652-8_1.
75. Noskov VN, Kouprina N, Leem SH, Ouspenski I, Barrett C, Larionov V. A general cloning system to selectively isolate any eukaryotic or prokaryotic genomic region in yeast. *BMC Genomics*. 2003;4(1):16. https://doi.org/10.1186/1471-2164-4-16.
76. Kouprina N, Nikolaishvili N, Graves J, Koriabine M, Resnick MA, Larionov V. Integrity of human YACs during propagation in recombination-deficient yeast strains. *Genomics*. 1999;56(3):262–273. https://doi.org/10.1006/geno.1998.5727.
77. Le Y, Dobson MJ. Stabilization of yeast artificial chromosome clones in a rad54-3 recombination-deficient host strain. *Nucleic Acids Res*. 1997;25(6):1248–1253. https://doi.org/10.1093/nar/25.6.1248.
78. Larionov V, Kouprina N, Graves J, Chen XN, Korenberg JR, Resnick MA. Specific cloning of human DNA as yeast artificial chromosomes by transformation-associated recombination. *Proc Natl Acad Sci USA*. 1996;93(1):491–496. https://doi.org/10.1073/pnas.93.1.491.
79. Kouprina N, Larionov V. Transformation-associated recombination (TAR) cloning for genomics studies and synthetic biology. *Chromosoma*. 2016;125(4):621–632. https://doi.org/10.1007/s00412-016-0588-3.Transformation-associated.
80. Kouprina N, Noskov VN, Larionov V. Selective isolation of large chromosomal regions by transformation-associated recombination cloning for structural and functional analysis of mammalian genomes. *Methods Mol Biol*. 2006;349:85–101. https://doi.org/10.1385/1-59745-158-4:85.
81. Larionov V, Kouprina N, Solomon G, Barrett JC, Resnick MA. Direct isolation of human brca2 gene by transformation-associated recombination in yeast. *Proc Natl Acad Sci USA*. 1997;94(14):7384–7387. https://doi.org/10.1073/pnas.94.14.7384.
82. Lee NCO, Larionov V, Kouprina N. Highly efficient CRISPR/Cas9-mediated TAR cloning of genes and chromosomal loci from complex genomes in yeast. *Nucleic Acids Res*. 2015;43(8). https://doi.org/10.1093/nar/gkv112, e55.
83. Mitchell LA, Chuang J, Agmon N, et al. Versatile genetic assembly system (VEGAS) to assemble pathways for expression in S. cerevisiae. *Nucleic Acids Res*. 2015;43(13):6620–6630. https://doi.org/10.1093/nar/gkv466.
84. Gibson DG, Benders GA, Axelrod KC, et al. One-step assembly in yeast of 25 overlapping DNA fragments to form a complete synthetic Mycoplasma genitalium genome. *Proc Natl Acad Sci USA*. 2008;105(51):20404–20409. https://doi.org/10.1073/pnas.0811011106.
85. Pryor JM, Potapov V, Pokhrel N, Lohman GJS. Rapid 40 kb genome construction from 52 parts. *bioRxiv*. 2020. https://doi.org/10.1101/2020.12.22.424019.

86. Gibson DG, Young L, Chuang RY, Venter JC, Hutchison CA, Smith HO. Enzymatic assembly of DNA molecules up to several hundred kilobases. *Nat Methods.* 2009;6(5):343–345. https://doi.org/10.1038/nmeth.1318.
87. Gibson DG. Enzymatic assembly of overlapping DNA fragments. *Methods Enzymol.* 2011;498:349–361. https://doi.org/10.1016/B978-0-12-385120-8.00015-2.
88. Noskov VN, Chuang RY, Gibson DG, Leem SH, Larionov V, Kouprina N. Isolation of circular yeast artificial chromosomes for synthetic biology and functional genomics studies. *Nat Protoc.* 2011;6(1):89–96. https://doi.org/10.1038/nprot.2010.174.
89. Hamer L, Johnston M, Green ED. Isolation of yeast artificial chromosomes free of endogenous yeast chromosomes: construction of alternate hosts with defined karyotypic alterations. *Proc Natl Acad Sci USA.* 1995;92(25):11706–11710. https://doi.org/10.1073/pnas.92.25.11706.
90. Brown DM, Chan YA, Desai PJ, et al. Efficient size-independent chromosome delivery from yeast to cultured cell lines. *Nucleic Acids Res.* 2017;45(7):e50. https://doi.org/10.1093/nar/gkw1252.
91. Kouprina N, Annab L, Graves J, et al. Functional copies of a human gene can be directly isolated by transformation-associated recombination cloning with a small 3′ end target sequence. *Proc Natl Acad Sci USA.* 1998;95(8):4469–4474. https://doi.org/10.1073/pnas.95.8.4469.
92. Shizuya H, Birren B, Kim UJ, et al. Cloning and stable maintenance of 300-kilobase-pair fragments of human DNA in Escherichia coli using an F-factor-based vector. *Proc Natl Acad Sci USA.* 1992;89(18):8794–8797. https://doi.org/10.1073/pnas.89.18.8794.
93. Monaco AP, Larin Z. YACs, BACs, PACs and MACs: artificial chromosomes as research tools. *Trends Biotechnol.* 1994;12(7):280–286. https://doi.org/10.1016/0167-7799(94)90140-6.
94. Mukai T, Yoneji T, Yamada K, Fujita H, Nara S, Su'etsugu M. Overcoming the challenges of megabase-sized plasmid construction in Escherichia coli. *ACS Synth Biol.* 2020;9(6):1315–1327. https://doi.org/10.1021/acssynbio.0c00008.
95. Shizuya H, Kouros-Mehr H. The development and applications of the bacterial artificial chromosome cloning system. *Keio J Med.* 2001;50(1):26–30.
96. Wang RY, Shi ZY, Chen JC, Chen GQ. Cloning large gene clusters from E. coli using in vitro single-strand overlapping annealing. *ACS Synth Biol.* 2012;1(7):291–295. https://doi.org/10.1021/sb300025d.
97. Zhu Y, Yang Y, Den P, Huang Y, Ni M, Fang H. Direct cloning and transplanting of large DNA fragments from Escherichia coli chromosome. *Sci China Life Sci.* 2016;59(10):1034–1041. https://doi.org/10.1007/s11427-016-5100-z.
98. Rivero-Müller A, Lajić S, Huhtaniemi I. Assisted large fragment insertion by Red/ET-recombination (ALFIRE) - An alternative and enhanced method for large fragment recombineering. *Nucleic Acids Res.* 2007;35(10): e78. https://doi.org/10.1093/nar/gkm250.
99. Zhang Y, Muyrers JPP, Stewart AF. DNA cloning by homologous recombination in Escherichia coli. *Nat Biotechnol.* 2000;18(December):1314–1317. https://doi.org/10.1038/82449.
100. Narayanan K, Williamson R, Zhang Y, Stewart A, Ioannou P. Efficient and precise engineering of a 200 kb β-globin human/bacterial artificial chromosome in E. coli DH10B using an inducible homologous recombination system. *Gene Ther.* 1999;6:442–447.
101. Jiang W, Zhao X, Gabrieli T, Lou C, Ebenstein Y, Zhu TF. Cas9-assisted targeting of chromosome segments CATCH enables one-step targeted cloning of large gene clusters. *Nat Commun.* 2015;6:1–8. https://doi.org/10.1038/ncomms9101.

102. Gabrieli T, Sharim H, Fridman D, Arbib N, Michaeli Y, Ebenstein Y. Selective nanopore sequencing of human BRCA1 by Cas9-assisted targeting of chromosome segments (CATCH). *Nucleic Acids Res.* 2018;46(14): e87. https://doi.org/10.1093/nar/gky411.
103. Wang K, Fredens J, Brunner SF, Kim SH, Chia T, Chin JW. Defining synonymous codon compression schemes by genome recoding. *Nature.* 2016;539(7627):59–64. https://doi.org/10.1038/nature20124.
104. Rostovskaya M, Fu J, Obst M, et al. Transposon-mediated BAC transgenesis in human ES cells. *Nucleic Acids Res.* 2012;40(19):e150. https://doi.org/10.1093/nar/gks643.
105. Narayanan K, Warburton PE. DNA modification and functional delivery into human cells using Escherichia coli DH10B. *Nucleic Acids Res.* 2003;31(9):1–7. https://doi.org/10.1093/nar/gng051.
106. Hibbitt OC, Wade-martins R. Delivery of large genomic DNA inserts > 100 kb Using HSV-1 amplicons. *Curr Gene Ther.* 2006;6:325–336.
107. Close WL, Bhandari A, Hojeij M, Pellett PE. Generation of a novel human cytomegalovirus bacterial artificial chromosome tailored for transduction of exogenous sequences. *Virus Res.* 2017;242(June):66–78. https://doi.org/10.1016/j.virusres.2017.09.007.
108. Mansouri M, Bellon-Echeverria I, Rizk A, et al. Highly efficient baculovirus-mediated multigene delivery in primary cells. *Nat Commun.* 2016;7(May):1–13. https://doi.org/10.1038/ncomms11529.
109. Mansouri M, Berger P. Baculovirus for gene delivery to mammalian cells: past, present and future. *Plasmid.* 2018;98(May):1–7. https://doi.org/10.1016/j.plasmid.2018.05.002.
110. Kwang TW, Zeng X, Wang S. Manufacturing of AcMNPV baculovirus vectors to enable gene therapy trials. *Mol Ther—Methods Clin Dev.* 2016;3:15050. https://doi.org/10.1038/mtm.2015.50.
111. Anguela XM, High KA. Entering the modern era of gene therapy. *Annu Rev Med.* 2019;70:273–288. https://doi.org/10.1146/annurev-med-012017-043332.
112. Kouprina N, Liskovykh M, Petrov N, Larionov V. Human artificial chromosome (HAC) for measuring chromosome instability (CIN) and identification of genes required for proper chromosome transmission. *Exp Cell Res.* 2020;387(2):111805. https://doi.org/10.1016/j.yexcr.2019.111805.
113. Benedetti S, Uno N, Hoshiya H, et al. Reversible immortalisation enables genetic correction of human muscle progenitors and engineering of next-generation human artificial chromosomes for Duchenne muscular dystrophy. *EMBO Mol Med.* 2018;10(2):254–275. https://doi.org/10.15252/emmm.201607284.
114. Lindenbaum M, Perkins E, Csonka E, et al. A mammalian artificial chromosome engineering system (ACE System) applicable to biopharmaceutical protein production, transgenesis and gene-based cell therapy. *Nucleic Acids Res.* 2004;32(21): e172. https://doi.org/10.1093/nar/gnh169.
115. de Jong G, Telenius A, Vanderbyl S, Meitz A, Drayer J. Efficient in-vitro transfer of a 60-Mb mammalian artificial chromosome into murine and hamster cells using cationic lipids and dendrimers. *Chromosom Res.* 2001;9(6):475–485. https://doi.org/10.1023/a:1011680529073.
116. DeJong G, Telenius AH, Telenius H, Perez CF, Drayer JI, Hadlaczky G. Mammalian artificial chromosome pilot production facility: large-scale isolation of functional satellite DNA-based artificial chromosomes. *Cytometry.* 1999;35(2):129–133. https://doi.org/10.1007/s10577-014-9459-z.
117. Bergmann JH, Martins NMC, Larionov V, Masumoto H, Earnshaw WC. HACking the centromere chromatin code: insights from human artificial chromosomes. *Chromosom Res.* 2012;20(5):505–519. https://doi.org/10.1007/s10577-012-9293-0.

118. Kazuki Y, Hoshiya H, Takiguchi M, et al. Refined human artificial chromosome vectors for gene therapy and animal transgenesis. *Gene Ther.* 2010;18(4):384–393. https://doi.org/10.1038/gt.2010.147.
119. Hasegawa Y, Ishikura T, Hasegawa T, et al. Generating a transgenic mouse line stably expressing human MHC surface antigen from a HAC carrying multiple genomic BACs. *Chromosoma.* 2015;124(1):107–118. https://doi.org/10.1007/s00412-014-0488-3.
120. Liskovykh M, Ponomartsev S, Popova E, et al. Stable maintenance of de novo assembled human artificial chromosomes in embryonic stem cells and their differentiated progeny in mice. *Cell Cycle.* 2015;14(8):1268–1273. https://doi.org/10.1080/15384101.2015.1014151.
121. Suzuki N, Nishii K, Okazaki T, Ikeno M. Human artificial chromosomes constructed using the bottom-up strategy are stably maintained in mitosis and efficiently transmissible to progeny mice. *J Biol Chem.* 2006;281(36):26615–26623. https://doi.org/10.1074/jbc.M603053200.
122. Kazuki Y, Hiratsuka M, Takiguchi M, et al. Complete genetic correction of iPS cells from duchenne muscular dystrophy. *Mol Ther.* 2010;18(2):386–393. https://doi.org/10.1038/mt.2009.274.
123. Iida Y, Kim JH, Kazuki Y, et al. Human artificial chromosome with a conditional centromere for gene delivery and gene expression. *DNA Res.* 2010;17(5):293–301. https://doi.org/10.1093/dnares/dsq020.
124. Csonka E, Cserpan I, Fodor K, et al. Novel generation of human satellite DNA-based artificial chromosomes in mammalian cells. *J Cell Sci.* 2000;113(18):3207–3216.
125. Keresõ J, Praznovszky T, Cserpán I, et al. De novo chromosome formations by large-scale amplification of the centromeric region of mouse chromosomes. *Chromosom Res.* 1996;4(3):226–239. https://doi.org/10.1007/BF02254964.
126. Grimes BR, Schindelhauer D, McGill NI, Ross A, Ebersole TA, Cooke HJ. Stable gene expression from a mammalian artificial chromosome. *EMBO Rep.* 2001;2(10):910–914. https://doi.org/10.1093/embo-reports/kve187.
127. Ebersole TA, Ross A, Clark E, et al. Mammalian artificial chromosome formation from circular alphoid input DNA does not require telomere repeats. *Hum Mol Genet.* 2000;9(11):1623–1631. https://doi.org/10.1093/hmg/9.11.1623.
128. Ikeno M, Grimes B, Okazaki T, et al. Construction of YAC-based mammalian artificial chromosomes. *Nat Biotechnol.* 1998;16(5):431–439. https://doi.org/10.1038/nbt0598-431.
129. Ohzeki JI, Bergmann JH, Kouprina N, et al. Breaking the HAC barrier: histone H3K9 acetyl/methyl balance regulates CENP-A assembly. *EMBO J.* 2012;31(10):2391–2402. https://doi.org/10.1038/emboj.2012.82.
130. Nakano M, Cardinale S, Noskov VN, et al. Inactivation of a human kinetochore by specific targeting of chromatin modifiers. *Dev Cell.* 2008;14(4):507–522. https://doi.org/10.1016/j.devcel.2008.02.001.
131. Ebersole T, Okamoto Y, Noskov VN, et al. Rapid generation of long synthetic tandem repeats and its application for analysis in human artificial chromosome formation. *Nucleic Acids Res.* 2005;33(15):1–8. https://doi.org/10.1093/nar/gni129.
132. Kim JH, Kononenko A, Erliandri I, et al. Human artificial chromosome (HAC) vector with a conditional centromere for correction of genetic deficiencies in human cells. *Proc Natl Acad Sci USA.* 2011;108(50):20048–20053. https://doi.org/10.1073/pnas.1114483108.
133. Ponomartsev SV, Sinenko SA, Skvortsova EV, et al. Human AlphoidtetO artificial chromosome as a gene therapy vector for the developing hemophilia A model in mice. *Cells.* 2020;9(4):879. https://doi.org/10.3390/cells9040879.

134. Fukagawa T. Critical histone post-translational modifications for centromere function and propagation. *Cell Cycle*. 2017;16(13):1259–1265. https://doi.org/10.1080/15384101.2017.1325044.
135. Willis AB, Foltz DR. Cell biology: hacking alpha satellites out of the HAC. *Curr Biol*. 2019;29(22):R1194–R1196. https://doi.org/10.1016/j.cub.2019.09.059.
136. Mills W, Critcher R, Lee C, Farr CJ. Generation of an ~2.4 Mb human X centromere-based minichromosome by targeted telomere-associated chromosome fragmentation in DT40. *Hum Mol Genet*. 1999;8(5):751–761. https://doi.org/10.1093/hmg/8.5.751.
137. Kuroiwa Y, Shinohara T, Notsu T, et al. Efficient modification of a human chromosome by telomere-directed truncation in high homologous recombination-proficient chicken DT40 cells. *Nucleic Acids Res*. 1998;26(14):3447–3448. https://doi.org/10.1093/nar/26.14.3447.
138. Kuroiwa Y, Tomizuka K, Shinohara T, et al. Manipulation of human minichromosomes to carry greater than megabase-sized chromosome inserts. *Nat Biotechnol*. 2000;18(10):1086–1090. https://doi.org/10.1038/80287.
139. Kahl VFS, Allen JAM, Nelson CB, et al. Telomere length measurement by molecular combing. *Front Cell Dev Biol*. 2020;8(June):1–14. https://doi.org/10.3389/fcell.2020.00493.
140. Itzhaki JE, Barnet MA, MacCarthy AB, Buckle VJ, Brown WRA, Porter ACG. Targeted breakage of a human chromosome mediated by cloned human telomeric DNA. *Nat Genet*. 1992;2(4):283–287.
141. Kuroiwa Y, Yoshida H, Ohshima T, et al. The use of chromosome-based vectors for animal transgenesis. *Gene Ther*. 2002;9(11):708–712. https://doi.org/10.1038/sj.gt.3301754.
142. Katoh M, Ayabe F, Norikane S, et al. Construction of a novel human artificial chromosome vector for gene delivery. *Biochem Biophys Res Commun*. 2004;321(2):280–290. https://doi.org/10.1016/j.bbrc.2004.06.145.
143. Kakeda M, Nagata K, Osawa K, et al. A new chromosome 14-based human artificial chromosome (HAC) vector system for efficient transgene expression in human primary cells. *Biochem Biophys Res Commun*. 2011;415(3):439–444. https://doi.org/10.1016/j.bbrc.2011.10.088.
144. Brown KE, Barnett MA, Burgtorf C, Shaw P, Buckle VJ, Brown WRA. Dissecting the centromere of the human Y chromosome with cloned telomeric DNA. *Hum Mol Genet*. 1994;3(8):1227–1237. https://doi.org/10.1093/hmg/3.8.1227.
145. Shinohara T, Tomizuka K, Takehara S, et al. Stability of transferred human chromosome fragments in cultured cells and in mice. *Chromosom Res*. 2000;8(8):713–725. https://doi.org/10.1023/A:1026741321193.
146. Wilson MG, Towner JW, Forsman I. Decreasing mosaicism in Down's syndrome. *Clin Genet*. 1980;17(5):335–340. https://doi.org/10.1111/j.1399-0004.1980.tb00159.x.
147. Hoshiya H, Kazuki Y, Abe S, et al. A highly stable and nonintegrated human artificial chromosome (HAC) containing the 2.4 Mb entire human dystrophin gene. *Mol Ther*. 2009;17(2):309–317.
148. Tedesco FS, Hoshiya H, D'Antona G, et al. Stem cell-mediated transfer of a human artificial chromosome ameliorates muscular dystrophy. *Sci Transl Med*. 2011;3(96). https://doi.org/10.1126/scitranslmed.3002342, 96ra78.
149. Hiratsuka M, Ueda K, Uno N, et al. Retargeting of microcell fusion towards recipient cell-oriented transfer of human artificial chromosome. *BMC Biotechnol*. 2015;15(1):1–8. https://doi.org/10.1186/s12896-015-0142-z.
150. McNeill CA, Brown RL. Genetic manipulation by means of microcell-mediated transfer of normal human chromosomes into recipient mouse cells. *Proc Natl Acad Sci USA*. 1980;77(9 II):5394–5398. https://doi.org/10.1073/pnas.77.9.5394.

151. Dicken ES, Epner EM, Fiering S, Fournier REK, Groudine M. Efficient modification of human chromosomal alleles using recombination-proficient chicken/human microcell hybrids. *Nat Genet.* 1996;12(2):174–182. https://doi.org/10.1038/ng0296-174.
152. Castelli A, Susani L, Menale C, et al. Chromosome transplantation: correction of the chronic granulomatous disease defect in mouse induced pluripotent stem cells. *Stem Cells.* 2019;37(7):876–887. https://doi.org/10.1002/stem.3006.
153. Paulis M, Castelli A, Susani L, et al. Chromosome transplantation as a novel approach for correcting complex genomic disorders. *Oncotarget.* 2015;6(34):35218–35230. https://doi.org/10.18632/oncotarget.6143.
154. Liskovykh M, Lee NC, Larionov V, Kouprina N. Moving toward a higher efficiency of microcell-mediated chromosome transfer. *Mol Ther—Methods Clin Dev.* 2016; 3(April):16043. https://doi.org/10.1038/mtm.2016.43.
155. Betz BL, Weissman BE. Hybrid capture of putative tumor suppressor genes. *Methods Mol Biol.* 2003;222:365–373. https://doi.org/10.1385/1-59259-328-3:365.
156. Doherty AMO, Fisher EMC. Microcell-mediated chromosome transfer (MMCT): small cells with huge potential. *Mamm Genome.* 2003;14(9):583–592. https://doi.org/10.1007/s00335-003-4002-0.
157. Stubblefield E, Pershouse M. Direct formation of microcells from mitotic cells for use in chromosome transfer. *Somat Cell Mol Genet.* 1992;18(6):485–491. https://doi.org/10.1007/BF01232645.
158. Kouprina N, Earnshaw WC, Masumoto H, Larionov V. A new generation of human artificial chromosomes for functional genomics and gene therapy. *Cell Mol Life Sci.* 2013;70(7):1135–1148. https://doi.org/10.1007/s00018-012-1113-3.
159. Fournier REK, Ruddle FH. Microcell mediated transfer of murine chromosomes into mouse, Chinese hamster, and human somatic cells. *Proc Natl Acad Sci USA.* 1977;74 (1):319–323. https://doi.org/10.1073/pnas.74.1.319.
160. Katoh M, Kazuki Y, Kazuki K, et al. Exploitation of the interaction of measles virus fusogenic envelope proteins with the surface receptor CD46 on human cells for microcell-mediated chromosome transfer. *BMC Biotechnol.* 2010;10:1–11. https://doi.org/10.1186/1472-6750-10-37.
161. Suzuki T, Kazuki Y, Oshimura M, Hara T. Highly efficient transfer of chromosomes to a broad range of target cells using Chinese hamster ovary cells expressing murine leukemia virus-derived envelope proteins. *PLoS One.* 2016;11(6):1–11. https://doi.org/10.1371/journal.pone.0157187.
162. Kneissig M, Keuper K, De Pagter MS, et al. Micronuclei-based model system reveals functional consequences of chromothripsis in human cells. *elife.* 2019;8. https://doi.org/10.7554/eLife.50292, e50292.
163. Zuo E, Huo X, Yao X, et al. CRISPR/Cas9-mediated targeted chromosome elimination. *Genome Biol.* 2017;18(1):1–18. https://doi.org/10.1186/s13059-017-1354-4.
164. Li LB, Chang KH, Wang PR, Hirata RK, Papayannopoulou T, Russell DW. Trisomy correction in down syndrome induced pluripotent stem cells. *Cell Stem Cell.* 2012;11 (5):615–619. https://doi.org/10.1016/j.stem.2012.08.004.
165. Lewandowski M, Martin GR. Cre-mediated chromosome loss in mice. *Nat Genet.* 1997;17(October):223–225.
166. Sato H, Kato H, Yamaza H, et al. Engineering of systematic elimination of a targeted chromosome in human cells. *Biomed Res Int.* 2017;2017:6037159. https://doi.org/10.1155/2017/6037159.
167. Matsumura H, Tada M, Otsuji T, et al. Targeted chromosome elimination from ES-somatic hybrid cells. *Nat Methods.* 2007;4(1):23–25. https://doi.org/10.1038/nmeth973.

168. Grégoire D, Kmita M. Recombination between inverted loxP sites is cytotoxic for proliferating cells and provides a simple tool for conditional cell ablation. *Proc Natl Acad Sci USA*. 2008;105(38):14492–14496. https://doi.org/10.1073/pnas.0807484105.
169. Jain M, Olsen HE, Turner DJ, et al. Linear assembly of a human centromere on the y chromosome. *Nat Biotechnol*. 2018;36(4):321–323. https://doi.org/10.1038/nbt.4109.
170. Momozawa Y, Mizukami K. Unique roles of rare variants in the genetics of complex diseases in humans. *J Hum Genet*. 2021;66(1):11–23. https://doi.org/10.1038/s10038-020-00845-2.
171. Yang L, Güell M, Niu D, et al. Genome-wide inactivation of porcine endogenous retroviruses (PERVs). *Science*. 2015;350(6264):1101–1104. https://doi.org/10.1126/science.aad1191.
172. Dekkers JF, Berkers G, Kruisselbrink E, et al. Characterizing responses to CFTR-modulating drugs using rectal organoids derived from subjects with cystic fibrosis. *Sci Transl Med*. 2016;8(344). https://doi.org/10.1126/scitranslmed.aad8278, 344ra84.
173. Rossidis AC, Stratigis JD, Chadwick AC, et al. In utero CRISPR-mediated therapeutic editing of metabolic genes. *Nat Med*. 2018;24(10):1513–1518. https://doi.org/10.1038/s41591-018-0184-6.
174. Brewer C, Holloway S, Zawalnyski P, Schinzel A, Fitzpatrick D. A chromosomal deletion map of human malformations. *Am J Hum Genet*. 1998;63(4):1153–1159. https://doi.org/10.1086/302041.
175. Peier AM, Nelson DL. Instability of a premutation-sized CGG repeat in FMR1 YAC transgenic mice. *Genomics*. 2002;80(4):423–432. https://doi.org/10.1006/geno.2002.6849.
176. Hodgson JG, Agopyan N, Gutekunst CA, et al. A YAC mouse model for Huntington's disease with full-length mutant huntingtin, cytoplasmic toxicity, and selective striatal neurodegeneration. *Neuron*. 1999;23(1):181–192. https://doi.org/10.1016/S0896-6273(00)80764-3.
177. Van Raamsdonk JM, Warby SC, Hayden MR. Selective degeneration in YAC mouse models of Huntington disease. *Brain Res Bull*. 2007;72(2-3 SPEC. ISS):124–131. https://doi.org/10.1016/j.brainresbull.2006.10.018.
178. Huxley C, Passage E, Manson A, et al. Construction of a mouse model of Charcot-Marie-Tooth disease type 1A by pronuclear injection of human YAC DNA. *Hum Mol Genet*. 1996;5(5):563–569. https://doi.org/10.1093/hmg/5.5.563.
179. Lee JS, Lee JY, Song DW, et al. Targeted PMP22 TATA-box editing by CRISPR/Cas9 reduces demyelinating neuropathy of Charcot-Marie-Tooth disease type 1A in mice. *Nucleic Acids Res*. 2020;48(1):130–140. https://doi.org/10.1093/nar/gkz1070.
180. Beil J, Fairbairn L, Pelczar P, Buch T. Is BAC transgenesis obsolete? State of the art in the era of designer nucleases. *J Biomed Biotechnol*. 2012;2012:308414. https://doi.org/10.1155/2012/308414.
181. Taguchi T, Ikuno M, Hondo M, et al. α-synuclein BAC transgenic mice exhibit RBD-like behaviour and hyposmia: a prodromal Parkinson's disease model. *Brain*. 2020;143(1):249–265. https://doi.org/10.1093/brain/awz380.
182. Jury N, Abarzua S, Diaz I, et al. Widespread loss of the silencing epigenetic mark H3K9me3 in astrocytes and neurons along with hippocampal-dependent cognitive impairment in C9orf72 BAC transgenic mice. *Clin Epigenetics*. 2020;12(1):1–15. https://doi.org/10.1186/s13148-020-0816-9.
183. O'Rourke JG, Bogdanik L, Muhammad AKMG, et al. C9orf72 BAC transgenic mice display typical pathologic features of ALS/FTD. *Neuron*. 2015;88(5):892–901. https://doi.org/10.1016/j.neuron.2015.10.027.
184. Han K, Holder JL, Schaaf CP, et al. SHANK3 overexpression causes manic-like behaviour with unique pharmacogenetic properties. *Nature*. 2013;503(7474):72–77. https://doi.org/10.1038/nature12630.

185. O'Dohorerty A, Ruf S, Mulligan C, et al. An aneuploid mouse strain carrying human chromosome 21 with Down syndrome phenotypes. *Science*. 2005;309(September): 2033–2037.
186. Zhu F, Nair RR, Fisher EMC, Cunningham TJ. Humanising the mouse genome piece by piece. *Nat Commun*. 2019;10(1):1–13. https://doi.org/10.1038/s41467-019-09716-7.
187. Scobie L, Denner J, Schuurman HJ. Inactivation of porcine endogenous retrovirus in pigs using CRISPR-Cas9, editorial commentary. *Xenotransplantation*. 2017;24(6): 1303–1307. https://doi.org/10.1111/xen.12363.
188. Maule G, Casini A, Montagna C, et al. Allele specific repair of splicing mutations in cystic fibrosis through AsCas12a genome editing. *Nat Commun*. 2019;10(1):3556. https://doi.org/10.1038/s41467-019-11454-9.
189. High KA. Turning genes into medicines—what have we learned from gene therapy drug development in the past decade? *Nat Commun*. 2020;11(1):2–5. https://doi.org/10.1038/s41467-020-19507-0.
190. Xiong HY, Alipanahi B, Lee LJ, et al. The human splicing code reveals new insights into the genetic determinants of disease. *Science*. 2015;347(6218):1254806. https://doi.org/10.1126/science.1254806.
191. Buratti E, Baralle FE. Influence of RNA secondary structure on the pre-mRNA splicing process. *Mol Cell Biol*. 2004;24(24):10505–10514. https://doi.org/10.1128/mcb.24.24.10505-10514.2004.
192. Frangoul H, Altshuler D, Cappellini MD, et al. CRISPR-Cas9 gene editing for sickle cell disease and β-thalassemia. *N Engl J Med*. 2020;384(3):252–260. https://doi.org/10.1056/nejmoa2031054.
193. Hirakawa MP, Krishnakumar R, Timlin JA, Carney JP, Butler KS. Gene editing and CRISPR in the clinic: current and future perspectives. *Biosci Rep*. 2020;40(4). https://doi.org/10.1042/BSR20200127, BSR20200127.
194. Wu Z, Yang H, Colosi P. Effect of genome size on AAV vector packaging. *Mol Ther*. 2010;18(1):80–86. https://doi.org/10.1038/mt.2009.255.
195. Kumar M, Keller B, Makalou N, Sutton RE. Systematic determination of the packaging limit of lentiviral vectors. *Hum Gene Ther*. 2001;12(15):1893–1905. https://doi.org/10.1089/104303401753153947.
196. Piovesan A, Antonaros F, Vitale L, Strippoli P, Pelleri MC, Caracausi M. Human protein-coding genes and gene feature statistics in 2019. *BMC Res Notes*. 2019;12 (1):1–5. https://doi.org/10.1186/s13104-019-4343-8.
197. Novikova IV, Hennelly SP, Sanbonmatsu KY. Sizing up long non-coding RNAs. *BioArchitecture*. 2012;2(6):189–199.
198. Chamberlain K, Riyad JM, Weber T. Expressing transgenes that exceed the packaging capacity of adeno-associated virus capsids. *Hum Gene Ther Methods*. 2016;27(1):1–12. https://doi.org/10.1089/hgtb.2015.140.
199. Ousterout DG, Kabadi AM, Thakore PI, Majoros WH, Reddy TE, Gersbach CA. Multiplex CRISPR/Cas9-based genome editing for correction of dystrophin mutations that cause duchenne muscular dystrophy. *Nat Commun*. 2015;6:1–13. https://doi.org/10.1038/ncomms7244.
200. Sánchez A, Schimmang T, García-Sancho J. Cell and tissue therapy in regenerative medicine. *Adv Exp Med Biol*. 2012;741:89–102. https://doi.org/10.1007/978-1-4614-2098-9_7.
201. Katsuda T, Matsuzaki J, Yamaguchi T, et al. Generation of human hepatic progenitor cells with regenerative and metabolic capacities from primary hepatocytes. *elife*. 2019;8:1–31. https://doi.org/10.7554/eLife.47313.
202. Fraietta JA, Nobles CL, Sammons MA, et al. Disruption of TET2 promotes the therapeutic efficacy of CD19-targeted T cells. *Nature*. 2018;558(7709):307–312. https://doi.org/10.1038/s41586-018-0178-z.

203. Kelton W, Waindok AC, Pesch T, et al. Reprogramming MHC specificity by CRISPR-Cas9-Assisted cassette exchange. *Sci Rep*. 2017;7(November 2016):1–12. https://doi.org/10.1038/srep45775.
204. Jang Y, Choi J, Park N, et al. Development of immunocompatible pluripotent stem cells via CRISPR-based human leukocyte antigen engineering. *Exp Mol Med*. 2019;51(1):1–11. https://doi.org/10.1038/s12276-018-0190-2.
205. Torikai H, Mi T, Gragert L, et al. Genetic editing of HLA expression in hematopoietic stem cells to broaden their human application. *Sci Rep*. 2016;6(February):1–11. https://doi.org/10.1038/srep21757.
206. Deuse T, Hu X, Gravina A, et al. Hypoimmunogenic derivatives of induced pluripotent stem cells evade immune rejection in fully immunocompetent allogeneic recipients. *Nat Biotechnol*. 2019;37(3):252–258. https://doi.org/10.1038/s41587-019-0016-3.
207. Greenfield A. Making sense of heritable human genome editing: Scientific and ethical considerations. In: Petris G, ed. *Curing Genetic Diseases through Genome Reprogramming. Progress in Molecular Biology and Translational Science*. Academic Press; 2021. https://doi.org/10.1016/bs.pmbts.2020.12.008.
208. Kurosaki H, Hiratsuka M, Imaoka N, et al. Integration-free and stable expression of FVIII using a human artificial chromosome. *J Hum Genet*. 2011;56(10):727–733. https://doi.org/10.1038/jhg.2011.88.
209. Basu J, Compitello G, Stromberg G, Willard HF, Van Bokkelen G. Efficient assembly of de novo human artificial chromosomes from large genomic loci. *BMC Biotechnol*. 2005;5:1–11. https://doi.org/10.1186/1472-6750-5-21.
210. Rocchi L, Braz C, Cattani S, et al. Escherichia coli-cloned CFTR loci relevant for human artificial chromosome therapy. *Hum Gene Ther*. 2010;21(9):1077–1092. https://doi.org/10.1089/hum.2009.225.
211. Deinsberger J, Reisinger D, Weber B. Global trends in clinical trials involving pluripotent stem cells: a systematic multi-database analysis. *NPJ Regen Med*. 2020;5(1):1–13. https://doi.org/10.1038/s41536-020-00100-4.
212. Haake K, Ackermann M, Lachmann N. Concise review: towards the clinical translation of induced pluripotent stem cell-derived blood cells—ready for take-off. *Stem Cells Transl Med*. 2019;8(4):332–339. https://doi.org/10.1002/sctm.18-0134.
213. Nirenberg MW. Will society be prepared? *Science*. 1967;157(3789):633. https://doi.org/10.1126/science.157.3789.633.
214. Berg P. Asilomar 1975: DNA modification secured. *Nature*. 2008;455(7211):290–291. https://doi.org/10.1038/455290a.

Index

Note: Page numbers followed by "*f*" indicate figures and "*t*" indicate tables.

A

AAV. *See* Adeno-associated viruses (AAVs)
AAV.dCas9-VP64 system, 54–55
AAV vector-based gene augmentation therapies, 58
ABCA4 coding sequence, 41
ABE. *See* Adenine base editors (ABE)
ABE-dCas9 fusion, 43
ABE-mediated DNA editing, 43
Academies of Sciences International Commission, 4–5
AD. *See* Alzheimer's disease (AD)
ADAR. *See* Adenosine deaminase acting on RNA (ADAR)
ADAR2-E488Q mutation, 47
Adenine base editors (ABE), 43–44, 140–141, 199–200, 453–454
Adeno-associated viruses (AAVs), 129, 336
  episomes, durability of, 245–246
  toxicity, 246–247
Adeno-associated virus 8 (AAV8) vectors, 342–343
Adeno-associated virus (AAV) vectors, 34–35, 194–195, 375–376*t*, 503–504
  Cas proteins, cellular responses to, 279–280
  clinical trials, 259, 260–262*t*
  homologous recombination (HR), 266
  limitation of, 373
  OTC deficiency, 273–274
  pediatric population, 265–266
  zonated metabolic pathway, 263–265
Adenosine deaminase acting on RNA (ADAR), 45–46
Adenosine deaminase deficiency (ADA-SCID)
  biochemistry, 119
  symptoms, 118–119
Adenovirus (AdV), 209–210, 374, 375–376*t*, 377
Albumin locus, 315–316

Alemtuzumab, 125–126
Allele-specific gene silencing approach, 35–37
Allogeneic HSC transplantation, 156, 312
  BM transplants, 125–126
  common complications, 127–128
  PID complications, 128
  treatment, PIDS, 126–127
Allotransplants, 126
Alpha globin locus, 312
Alstrom syndrome (ALMS), 359
Alzheimer's disease (AD), 428–429
Amelioration, 165
Amyloid precursor protein (APP), 428
Anterograde transport, 355
Antisense oligonucleotides (ASOs), 203–204, 356–357, 425, 460–462
  exon skipping, 232, 234–235
  gene upregulation, 378–379
  in vivo delivery, 377–379
  microRNA-17 inhibition, 378
  in monogenic disorders treatment, 378–379
  sepofarsen, 382
  as splice modulators, 378–379
Apolipoprotein B MRNA editing enzyme catalytic subunit 1 (APOBEC1), 42
APP. *See* Amyloid precursor protein (APP)
ART. *See* Assisted reproductive technology (ART)
Artemis deficiency, 127
Artemis-deficient SCID (SCID-A)
  biochemistry, 120
  chronic granulomatous disease (CGD), 120–121
  symptoms, 120
Artemis protein, 119
ASD. *See* Autism spectrum disorder (ASD)
ASOs. *See* Antisense oligonucleotides (ASOs)

Index

Assisted reproductive technology (ART)
  clinic, 23–24
  genetically modified (GM) humans in, 10–12
  HHGE as, 8–13
  sector, 18–19
Ataxia with vitamin E deficiency (AVED)
  α-tocopherol transfer protein (α-TTP), mutations in, 452
  Kozak sequence, 452
AUG codon, 445–448
Autism spectrum disorder (ASD), 419–421
Autologous HSCs, transplantation of, 156–157
Autonomously replicating sequences (ARS), 488
Autosomal dominant polycystic kidney disease (ADPKD), 365
Autosomal recessive polycystic kidney disease (ARPKD), 365

B

Bacterial artificial chromosomes (BAC), 482, 486–487, 490–492, 506
Bardet-Bield syndrome (BBS), 359
Basal membrane zone (BMZ), 82–85, 95–98
Base editing strategies, 175
Base editors (BEs), 42–45, 132–133, 161–162, 199–201
  adenine, 43–44
  advantages and drawbacks of, 200–201
  cytosine, 42–43
  prime editors, 44–45
B cell disorders, 115
BCL11A BS, 172–173
BCL11A erythroid-specific enhancer, 174–177
Becker muscular dystrophy (BMD), 230–231
BEs. See Base editors (BEs)
β-globin gene mutations, to treat β-hemoglobinopathies
  β-thalassemia mutations, correction of, 166–168
  SCD mutation, correction of, 162–166, 163f
β-globin locus, 155f
β-hemoglobinopathies, 155, 157, 169
  causing mutations, 157
  on genome editing approaches, 176–177
β-like globin genes, 155f
β-thalassemia mutations, correction of, 166–168
β-thalassemias, 155–156
Bispecific antibodies (bsAbs), 450
Blistering skin disease epidermolysis bullosa, 82–86, 83f
  dystrophic EB (RDEB), 85–86
  epidermolysis bullosa simplex (EBS), 84–85
  junctional epidermolysis bullosa (JBS), 84–85
BMZ. See Basal membrane zone (BMZ)
Bone marrow transplant (BMT), 125
BoxB-λN-ADAR, 48
BRCA1 gene, 452–453
Busulfan, 125–126

C

Cancer, uORF deregulation, 460
Candida albicans, 117–118
Cap-binding protein, 442–443
CAR T-cells. See Chimeric antigen receptor (CAR T-cells)
Cas9, 407–408, 429
Cas12a, 407
Cas9 endonuclease expression in vivo, 62
Cas9-induced DSBs, 38–39
Cas9 nickase (Cas9n), 199–200
Cas9 nuclease-mediated HDR, 166
Cas nucleases, 192–195
Cas9-sgRNA ribonucleoproteins, 63–64
Cationic liposomes, 344
CBE. See Cytosine base editor (CBE)
Cell autonomous liver diseases, 271–274, 272f
Cell lines, 303–304
Cell replacement, 414–415
Central nervous system (CNS), 338–339, 409–413
Centromere proteins (CENPs), 493
CEP290 gene, 381
CF. See Cystic fibrosis (CF)
CFTR gene. See Cystic fibrosis transmembrane conductance regulator (CFTR) gene

Index

CHAMPIONS trial, 317–318
Charcot-Marie-Tooth syndrome, 502
CHD8 gene, 421
Chemokine (C-C motif) receptor 5 (CCR5), 312–313
Chimeric antigen receptor (CAR T-cells), 3–4
Chromatin immunoprecipitation sequencing (ChIP-Seq), 60–61
Chronic granulomatous disease (CGD), 115, 124–127, 506
  autosomal recessive form (AR-CGD), 120
  biochemistry, 121
  symptoms, 121
Cilium, primary, 355, 356f
CIRCLE-Seq, 60–61
Cis-regulatory regions, 157–158
Classical/canonical nonhomologous end-joining (C-NHEJ), 133
Clustered regularly interspaced short palindromic repeat (CRISPR)-associated Cas9 nucleases, 157, 160–161, 168, 171–173, 176, 194–195, 258–259, 267, 268f, 275–276, 302, 307–311, 333
  approaches, 5–6
  genome editing, 3–4
  POU5F1 gene, 6–7
Clustered regularly interspaced short palindromic repeats (CRISPR), 35–37, 91–98, 135–138, 192, 301–302, 407–408, 429
  delivery of, 209–211
  developments, 98–100
  reagents to retina, 63–65, 66–69t
  system, 101
CNS. See Central nervous system (CNS)
COL7A1 gene, 85–87
Common Terminology Criteria for Adverse Events (CTCAE), 317–318
Complementary DNA (cDNA), 342
Computer algorithms, 60–61
Cone photoreceptors, 33–34
Coronary artery disease (CAD), Kozak sequence, 451
Covid-19 pandemic, 18

CRISPR. See Clustered regularly interspaced short palindromic repeats (CRISPR)
CRISPR activation (CRISPRa), 52–57
CRISPR-Cas13, 47–48
CRISPR/Cas systems, 60–70, 357, 440–441
  base editor systems, 42
  challenges, 369–371
  efficacy and specificity, 369
  epigenetic editing for
    renal ciliopathies, 386–387
    retinal ciliopathies, 384–385
  ex vivo, 370, 370f
  gene editing for
    renal ciliopathies, 382–383
    retinal ciliopathies, 379–382
  in vivo, 370–379, 370f
    ASOs, 377–379
    kidney, 376–377
    retina, 374–376
  single base pair editing systems, 40–41
  and therapeutic applications, 365–379, 367f
CRISPR genome engineering for retinal diseases, 31–40, 36f, 58–70
  DNA base editing, 40–45
  base editor systems, 42–45
  epigenetic editing, 52–58
    challenges, 57–58
    dominant diseases, targeting pathogenic genes in, 54–55
    targeting disease pathways, 55–56
  RNA editing, 45–52
    ADAR physiology, 46
    endogenous ADAR, 50
    exogenous ADAR, 47–50
    of inherited retinal disease, 51
    tools, 47
CRISPRi interference (CRISPRi), 52–57
CRISPR-mediated adenine base-editors, 43–44
CRISPR RNA (crRNA), 38, 365–366
Cystic fibrosis (CF), 187–192
  base conversion for, 201
  gene addition/editing for, 191–192
  gene therapy for, 191
  mutations, 188t
  therapies in pipeline, 190–191

Cystic fibrosis transmembrane conductance regulator (*CFTR*) gene, 188–190, 189f
  application of nucleases, 196–197
  gene editing, 192–202
  modulation, 203–206
Cytomegalovirus (CMV), 124
Cytosine base editors (CBE), 42–43, 140, 199–201, 453–454

D

DARPP-32, 424
dCas9 protein, 53
Delivery vehicles, 197–198, 209–212
Deoxygenation, 156
Diazepam-binding inhibitor (*DBI*) gene, 451
Digenome-Seq, 60–61
DISC1 model, 422
DISCOVER-Seq, 60–61
Disease modeling, 414–415
Distal appendages, 355
DMD. *See* Duchenne muscular dystrophy (DMD)
DMD gene, 226–229
DNA base editing, 40–45
  base editor systems, 42–45
DNA cross-link repair 1C (*DCLRE1C*) gene, 119
DNA methylation, 383–385
DNA methyltransferases (DNMTs), 368
Dominant diseases, targeting pathogenic genes in, 54–55
Dominant-negative alleles, 91–92
Dominant-negative mutations, 54–55
Double-strand breaks (DSBs), 87, 132–133, 157, 267–270, 269f, 333, 334f, 357–358, 405, 440–441, 492
  EJ-based repair of, 93–94
  generation, 91–92
  inductions, 101–103
  repair outcomes, 98–100
Down-regulation of transcription factors, in HbF expression, 175–176
Down syndrome, 497, 502
Doxycycline (or tetracycline)-inducible Tet-ON expression system, 62
DSB-free genome editing tools, 159
DSB-free strategy, 168
DSBs. *See* Double-strand breaks (DSBs)
Duchenne muscular dystrophy (DMD), 190–191, 226–228, 504–505
  current and future therapies
    challenges for, 244–247
    durability of adeno-associated virus episomes, 245–246
    immunity issues, 244–245
    safety, 246–247
  DMD gene, 228–229
  dystrophin
    functional domains, 229–230
    miniaturized, 233–234
    structure and function, 227f, 229–230
  dystrophinopathies, 230–232
  first-generation gene therapies for, 232
    exon skipping with antisense oligonucleotides (AONs), 234–235
    gene replacement, 233–234
    vectorized exon-skipping with modified U7 snRNA, 235–237, 236f
  second-generation approaches for, 237–239
    homology-directed gene repair, 242–243
    multiplex gene editing for exon deletion, 239–241
    single-cut gene editing for exon skipping and reframing, 241–242
Dystrophic EB (RDEB), 85–86
Dystrophinopathies, 230–232
Dystrophins
  functional domains, 229–230
  miniaturized, 233–234
  structure and function, 227f, 229–230

E

EB. *See* Epidermolysis bullosa (EB)
EBS. *See* Epidermolysis bullosa simplex (EBS)
E. coli, 482–483, 490
EcTadA. *See* *Escherichia coli* tRNA adenosine deaminase (EcTadA)
EJ. *See* End-joining (EJ)
EJ-based repair, 91–92
EJC. *See* Exon junction complexes (EJC)
Eliglustat, 300

ELSI. *See* Ethical, legal and social implications (ELSI)
Embryonic stem cells (ESCs), 408–409
EMPOWERS trial, 317–318
Encephalomyocarditis virus (EMCV), 463
End-joining (EJ), 87–91
Endogenous ADAR, 50
   Leveraging Endogenous ADAR for Programmable Editing of RNA (LEAPER), 50
   RESTORE and GluR2 recruitment, 50
Endogenous fetal γ-globin genes, de-repression of, 169–177
   β-hemoglobinopathies, on genome editing approaches, 176–177
   γ-globin transcriptional repressors
   BCL11A, erythroid-specific enhancer, 174–175
     down-regulation of transcription factors, in *HbF* expression, 175–176
   mimicking HPFH mutations, in *HBG* promoters, 169–174
   γ-globin activator BSs, 173–174
   γ-globin repressor BSs, 171–173
Endothelial nitric oxide synthase (eNOS), 304
End-stage kidney disease (ESKD), 365
Enzyme activity assessments, lysosomal disorders (LDs), 293–296
Enzyme replacement therapy (ERT), 124, 126–127, 297–300
Epidermolysis bullosa (EB)
   blistering skin disease, 82–86
   gene editing applications in, 101–103
   gene therapeutic applications for, 86–87
Epidermolysis bullosa simplex (EBS), 84–85
Epidermolytic ichthyosis (EI), 84–85
Epigenetic editing, 52–58
   challenges, 57–58
   CRISPR/Cas systems
     renal ciliopathies, 386–387
     retinal ciliopathies, 384–385
   dominant diseases, targeting pathogenic genes in, 54–55
   targeting disease pathways, 55–56
Episomes, 265–266
Escherichia coli tRNA adenosine deaminase (EcTadA), 43

Ethical, legal and social implications (ELSI), 8
Ethics
   of genome editing, 18–19
   of HHGE, 13–19
Eukaryotes, 38–39
   protein synthesis in, 442–444
   translational regulation, 444–445
     internal ribosomal entry sites (IRESs), 463–466
     Kozak sequence, 445–455
     upstream open reading frames (uORFs), 455–462
Exogenous ADAR, 47–50
   BoxB-λN-ADAR, 48
   CRISPR-Cas13, 47–48
   GluR2-ADAR, 49
   MS2-MCP-ADAR, 49
   SNAP-ADAR, 48–49
   synthetic RNA targeting system, 49–50
Exon deletion
   multiplex gene editing for, 239–241
   single-cut strategy, 241
Exon junction complexes (EJC), 205–206
Exon skipping
   antisense oligonucleotides (AONs), 232, 234–235
   single-cut gene editing for, 241–242
   vectorized, with modified U7 snRNA, 235–237, 236f
Extrachromosomal elements (ECEs), 232
Ex vivo
   gene editing, 206–209
   gene therapy, 208–209

F

Fabry disease
   cell lines, 304
   CRISPR-Cas9 system, 311
   diagnosis, 293–296
   enzyme replacement therapy, 297–299
   newborn screening studies, 291–292
   small molecules, 300–301
Familial Alzheimer disease (fAD), 428–429
Fanconi anemia, 115–116, 125–126
FDA-approved gene therapy, 58
Fetal hemoglobin, 157
First-generation base editor (CBE1), 42

FIX TENDZ Study, 278–279
Flavobacterium okeanokoites, 159
Fludarabine, 125–126
Fluorescence-activated cell sorting (FACS), 448–449
FokI cleavage domain, 406–407
FokI-dependent DNA cleavage, 159
Footprint-less correction, of pathogenic alleles, 95–100
Francisella novicida Cas9 (FnCas9), 366
Fumarylacetoacetate hydrolase (FAH), 276–277, 371

G

α-Galactosidase A (GLA), 304
γ-globin activator BSs, 173–174
γ-globin repressor BSs, 171–173
γ-globin transcriptional repressors
  BCL11A, erythroid-specific enhancer, 174–175
  down-regulation of transcription factors, in HbF expression, 175–176
GATA motif, 174–175
Gaucher disease
  CCR5 locus, 313
  enzyme replacement therapy, 297–300
  macrophage model, 304–305
  microglia-like cell model, 305
  phagocytes, 292
  pseudogenes, 296
  small molecules, 300
  targeted mutation analysis, 296
  Zebrafish model, 306–307
Gene addition therapy, retroviral and lentiviral-based corrective therapies, 131–132
Gene delivery systems, 63
Gene depletion, 91–92
Gene editing
  applications in, 101–103
  for CF, 191–192
  CFTR, 192–202, 211
  CRISPR/Cas systems
    renal ciliopathies, 382–383
    retinal ciliopathies, 379–382
  delivery of CRISPR, 209–211
  development for treatment of genodermatoses, 87–100
  ex vivo or in vivo, 206–209
  gene therapeutic applications for, 86–87
  programmable nuclease-based, 211–212
  reading frame restoration of pathogenic alleles, 92–94
  strategies for epidermolysis bullosa, 88f, 89–90t (*see also* Blistering skin disease epidermolysis bullosa)
  viral-vector mediated, 210
Gene-editing tools, next generation of
  adenine base editor (ABE) mechanism, 140–141
  BE translational research, 141–145
  Cas9 base editors (BEs), 139–140
  CRISPR/Cas9, 135–136
  cytosine base editor (CBE) mechanism, 140
  limitations and off-target effects, 138–139
  PID, CRISPR for, 136–138
  prime editors (PEs), 142–145
  transcription activator-like effector nucleases (TALENs), 134–135
  zinc-finger nucleases (ZFN), 134–135
Gene replacement therapy, 232
  Duchenne muscular dystrophy (DMD), 233–234
GeneRide strategy, 277–278, 280
Gene supplementation strategies, 36f
Gene therapy, 85–87, 259, 263, 279, 318
  for cystic fibrosis, 191, 211–212
  ex vivo, 208–209
  gene addition, 207
  in vivo, 196–197
  for retinal disease, 36f
  target, 34–35
Genetically modified (GM) humans in ART, 10–12
Genodermatoses, gene editing development for treatment of, 87–100
Genome editing, 3–4, 332–337
  adeno-associated viruses (AAVs), 336
  in animal models, 335, 337f
  CRISPR-based, 335, 336f
  endogenous cellular DNA repair mechanisms, 334f
  ethics of, 18–19
  ex vivo, 335, 337f
  germline, 8–10

Index

and human dignity, 16–18
human research embryos, 6–7
in vivo, 335, 337*f*, 341–344
for mucolipidoses (ML), 338–341
  clinical trials, 344–347
  in vitro, 338–341
  in vivo, 341–344
for mucopolysaccharidoses (MPS), 338–341
  clinical trials, 344–347, 346*t*
  in vitro, 338–341
  in vivo, 341–344
of neurodegenerative disorders, 423–429
for neurodevelopmental disorders, 416–423
protein-based, 333, 334*f*
stem cells, 408–415
technologies, 405–408
Genome editing targets
  Kozak sequence as, 453–455, 453*f*
  uORFs as, 460–462, 462*f*
Genome editing tools, 158*f*
  base and prime editing, 161–162
  nuclease-mediated genome editing, 157–161
    CRISPR-Cas9 nuclease, 160–161
    ZFNs and TALENs, 159–160
Genome engineered cells, selective expansion of, 276–277
Genome engineering, 38, 405
Genotoxicity, 61–62
Germline genome editing, 8–10
GFP. *See* Green fluorescent protein (GFP)
Gibson cloning, 489
Glucosylceramide, 300
GluR2-ADAR, 49
GluR2 recruitment, 50
Glycosaminoglycans (GAGs), degradation of, 328
GMP production. *See* Good manufacturing practice (GMP) production
GNPTAB gene, 331
Golden gate cloning, 489
Golden retriever DMD model (GRDM), 236–237
Good manufacturing practice (GMP) production, 98–100, 211–212

Governance
  of genome editing, 22–24
  of HHGE, 7–8
Graft-*vs.*-host disease (GvHD), 125–128
Graves' disease (GD), Kozak sequence
  CD40 deregulation, 452
  modulation by SNPs, 452
Green fluorescent protein (GFP), 95–98, 448–449
Guide RNA (gRNA), 160

H

Haploinsufficiency, 35–37, 54–55
HD. *See* Huntington's disease (HD)
HDR. *See* Homology-directed repair (HDR)
HDR-mediated gene editing, 40–41
Heavy chains (HCs), 450
Hematopoietic stem cells, 338
Hematopoietic stem cell transplantation (HSCT), 123–124, 127–128, 156, 297, 312
Hemoglobin (Hb), 155
Hemophilia, 271–272, 272*f*
Hepatic artery blood flow, 263–265
Hepatic lobule structure and function, 263–265, 264*f*
Hepatocytes, 259, 263–266, 280
Hereditary persistence of fetal hemoglobin (HPFH), 169
Hereditary tyrosinemia type I (HT1), 276–277
Heritable human genome editing (HHGE), 4–8
  acceptability and governance of, 7–8
  assisted reproductive technology, 8–13
  ethics of, 13–19
  impacts on societal norms, 19–21
High-fidelity Cas variants, 61–62
High InDel frequency, 171–172
hiPSCs. *See* Human induced pluripotent stem cells (hiPSCs)
Histone lysine acetylation (HKAc), 383–384
Histone methylation (HKme), 383–385
Histone modification, 383–384
Homologous recombination (HR), 95–98, 266–270, 269*f*
Homology-dependent repair, 95–98

Homology-directed repair (HDR), 5–7, 87–91, 157–159, 195, 267–271, 269f, 273–274, 274f, 301, 333, 334f, 343, 366–368, 405, 440–441
  for DMD mutation correction, 242–243
  mechanisms, 406
Homology-independent targeted integration (HITI), 243, 270, 274, 275f
HPFH. See Hereditary persistence of fetal hemoglobin (HPFH)
HPFH mutations, in *HBG* promoters, 169–174
  γ-globin activator BSs, 173–174
  γ-globin repressor BSs, 171–173
HR. See Homologous recombination (HR)
HSCT. See Hematopoietic stem cell transplantation (HSCT)
HSCT-associated immunological complications, 156
HTT. See Huntingtin gene (HTT)
Human artificial chromosomes (HACs), 505–506
  bottom-up artificial chromosomes, 493–495
  top-down artificial chromosomes, 495–496
Human dignity, genome editing and, 16–18
Human Fertilization & Embryology Act, 22
Human Fertilization & Embryology Authority (HFEA), 6–7, 18–19
Human Gene Mutation Database, 205–206
Human induced pluripotent stem cells (hiPSCs), 163–164, 414, 417f
  application of patient derived, 419, 422, 424, 428–429
  DISC1, 422
  HD, 424–425
  microcephalic patient-derived, 419
  mutations in, 422–423
  neurodegenerative processes with, 423
  PD, 427–428
  SMA-derived, 426
  synergy of, 419
Human research embryos, 6–7
Huntingtin gene (HTT), 424–425
Huntington's disease (HD), 424–425, 502

I

ICSI. See Intracytoplasmic sperm injection (ICSI)
IF. See Intermediate filament (IF)
Illumina's bridge-amplification sequencing, 480
InDels. See Insertions/deletions (InDels)
Induced pluripotent stem cells (iPSCs), 196–197, 303, 314, 338, 339f, 409, 506
Informed consent of participants, 2–3
Inherited retinal disease (IRD), 31–33
  mutations, 41
  RNA editing of, 51
Insertions/deletions (InDels), 405–406
Integrated stress response (ISR) pathway, 456–457
Interferon regulatory factor 6 (IRF6), 459
Interleukin 2 receptor subunit gamma (*IL2RG*) gene, 117–118
Intermediate filament (IF), 84–85
Internal ribosomal entry sites (IRESs), 441f
  and cap-independent translation initiation, 463–464
  as therapeutic targets, 465–466
Intracytoplasmic sperm injection (ICSI), 6–7
Intraflagellar transport (IFT), 355
In vitro fertilization (IVF), 6–9, 12–13, 15–16
In vitro modeling
  of neurodegenerative diseases, 424–429
  of neurodevelopmental disorders, 418–423
In vitro mutation-specific genome editing, 307–311
In vitro neural stem cell systems, 412–414
In vivo genome editing, 335, 337f, 341–344
  CRISPR/Cas-based therapy, 371–379
    ASOs, 377–379
    hydrodynamic delivery, 371
    kidney, 376–377
    retina, 374–376
  gene therapy, 196–197
iPSCs. See Induced pluripotent stem cells (iPSCs)
IRES trans-acting factors (ITAFs), 463–464
IVF. See In vitro fertilization (IVF)

Index 529

J

JEB. *See* Junctional epidermolysis bullosa (JEB)
Joubert syndrome (JBTS), 359
Junctional epidermolysis bullosa (JEB), 84–85

K

Keratin 5 (*KRT5*) gene, 84–85, 91–92
Keratin 14 (*KRT14*) gene, 84–85
Kidney, CRISPR/Cas delivery to, 376–377
KLF1 gene disruption, 175
KLF1 haploinsufficiency, 175
Kozak sequence, 441*f*, 445–455, 449*f*
  and disease, 450–453
  as genome editing target, 453–455, 453*f*
  translational influences, 445–450
Krabbe disease
  clinical manifestations, 292–293
  hematopoietic stem cell transplantation, 297
  neural stem cells (NSCs), 313–314
KRAB repressor, 53
KRT10 gene, 91–92

L

Lateral ganglionic eminence (LGE), 411–412
LCA. *See* Leber congenital amaurosis (LCA)
LCR. *See* Locus control region (LCR)
LDs. *See* Lysosomal disorders (LDs)
Leaky scanning, 442–443, 457
Leber congenital amaurosis (LCA), 33–34, 356–357, 364
Lentivirus (LV), 43–44, 129–131, 374, 375–376*t*, 377
Leveraging Endogenous ADAR for Programmable Editing of RNA (LEAPER), 50
LGE. *See* Lateral ganglionic eminence (LGE)
Light chains (LCs), 450
Lipid nanoparticle technology, 281–282
Liposomes, 314–315
Liver
  blood supply sources, 263
  clotting factors, 277
  functions, 259

Liver-targeted genome editing
  disease specific challenges, 270–274
  liver biology and implications, 263–266
  strategies
    genome engineered cells, selective expansion of, 276–277
    locus-specific gene disruption, 275–276
    therapeutic transgenes, targeted insertion of, 277–279
  translational considerations
    Cas proteins, cellular responses to, 279–280
    logistic and commercial constraints, 281
    unwanted mutations, 280
LncRNAs. *See* Long-noncoding RNA (LncRNAs)
Locus control region (LCR), 155
Locus-specific gene disruption, 275–276
Long-noncoding RNA (LncRNAs), 203–204
LV-based gene therapy, 156–157
LV-derived therapeutic β-globin, 156–157
Lysosomal disorders (LDs)
  clinical manifestations, 292–293, 294–295*t*
  definition, 291
  diagnosis, 293–297
  genome editing
    animal models, 306–307
    applications, 319
    cell models, 303–305, 306*f*
    clinical trials, 317–318
    editing studies, 307, 308–309*t*
    ex vivo system, 311–314
    in vitro mutation-specific, 307–311
    in vivo system, 314–317
    tools, 301–302
  molecular genetics, 293–297
  natural history, 292–293
  physiopathology, 292–293
  prevalence, 291–292
  treatments, 294–295*t*
    enzyme replacement therapy, 297–300
    gene therapy, 301
    hematopoietic stem cell transplantation, 297
    small molecules, 300–301

Lysosomal storage disorders (LSDs), iPSCs models for, 338, 339f
Lysosomes, 292

M

Macrophages, 304–305
Macula, 33–34
Madin-Darby canine kidney (MDCK) cells, 386
Main ORF (mORF), 455–456
Mammalian artificial chromosomes (MACs), 492–496
Mammalian retina, 32–33f
Mannose-6-phosphate, 331
Maxam-Gilbert chemical cleavage method, 480
MCPH. See Microcephaly primary hereditary (MCPH)
Meckel syndrome (MKS), 359
Melphalan, 125–126
Metabolic zonation, 263–265
Metachromatic leukodystrophy
  clinical manifestations, 292–293
  hematopoietic stem cell transplantation, 297
Methionyl-tRNA$_i$, 442–443
Methylated cytosines, 43
5-Methylcytosine (5mC), 383–384
MH. See Microhomology (MH)
mHTT. See Mutant huntingtin (mHTT)
Microcell-mediated chromosome transfer (MMCT), 496–497
Microcephaly, 418–419
Microcephaly primary hereditary (MCPH), 418–419
Microdystrophins, 230, 230f, 233–234
Microhomology (MH), 87–91
Microhomology-mediated end-joining (MMEJ) repair, 270
MicroRNA (MiRNAs), 204–205
Migalastat, 300–301
miRNA recognition elements (MREs), 205
MiRNAs. See MicroRNA (MiRNAs)
Mitochondrial DNA(mtDNA), 11–12
Mitochondrial replacement techniques (MRT), 12–13
  acceptability of, 11–12
  safety and efficacy of, 18–19
  therapeutic fallacy, 12–13
ML. See Mucolipidoses (ML)
Moloney murine leukemia virus, 129–131
Monogenic disease, prevent transmission of, 14–16
MPS. See Mucopolysaccharidoses (MPS)
MRT. See Mitochondrial replacement techniques (MRT)
MS2 bacteriophage coat protein (MCP), 49
MS2-MCP-ADAR, 49
mtDNA. See Mitochondrial DNA(mtDNA)
Mucolipidoses (ML)
  clinical and molecular characteristics, 331–332
  genome editing for
    clinical trials, 344–347
    in vitro, 338–341
    in vivo, 341–344
  ML II, 331–332, 332t
  ML III α/β, 332, 332t
  ML III gamma, 332, 332t
  neuraminidase, alterations in, 331
Mucolipidosis IV, Zebrafish model of, 306–307
Mucopolysaccharidose I (MPS I), 310, 316–317
Mucopolysaccharidose II (MPS II)
  clinical trials, 317–318
  Zebrafish model, 306–307
Mucopolysaccharidoses (MPS)
  clinical and molecular characteristics, 328–331, 329t
  genome editing for
    clinical trials, 344–347, 346t
    in vitro, 338–341
    in vivo, 341–344
  glycosaminoglycans (GAGs), degradation of, 328
  incidence of, 330–331
  residual enzyme activity, 330, 330f
  symptoms and onset, 330, 330f
Multiple endocrine neoplasia syndrome (MEN4), uORF mutations, 460
Multiplex gene editing, for exon deletion, 239–241
Mutant huntingtin (mHTT), 425

MYBPC3 gene, 6–7
Mycoplasma mycoides, 482–483
Myeloablative conditioning (MAC), 125–126

N
N-acetylglucosamine-1-phosphotransferase, 331
Nanopore/short-read sequencing, 491
ncRNAs. *See* Noncoding RNAs (ncRNAs)
Neisseria meningitidis Cas9 (NmCas9), 366, 373
Neocortical (NCX) NES, 413
Nephronophthisis (NPHP), 356–357, 365
Neural retina-specific leucine zipper protein (NRL), 380
Neural stem cells (NSCs), 313–314, 338–339, 408–415
  in vitro, 412–414
  neurodegenerative processes with human, 423
  neurodevelopmental programs with, 416–417, 417*f*
Neurodegenerative diseases, 415, 423, 429
  in vitro modeling of, 424–429
Neurodevelopmental disorders
  genome editing for, 416–423
  in vitro modeling of, 418–423
Neuronal ceroid lipofuscinosis (NCL1)
  microglia-like cell model, 305
  ovine model, 307
Neurons, 421
  ASD, 420
  cells, 409–411, 416
  death of dopaminergic (DA), 427–428
  familial Alzheimer disease (fAD) iPSC-derived, 428–429
  forebrain, 422–423
  GABAergic medium-spiny, 411–413
  generated in vitro, 424
  HD hiPSC-derived, 424–425
  late-born, 416
  loss of dopaminergic, 415
  mature, 411, 413–414, 418, 430
  orientation of, 421–422
  patient-derived fibroblasts and, 425
Next- or second-generation sequencing (NGS), 480

NHEJ. *See* Non-homologous end-joining (NHEJ)
NHEJ DNA repair pathway, 40
Niemann Pick type C diseases
  cell models, 305
  hematopoietic stem cell transplantation, 297
  macrophage model, 304–305
  phagocytes, 292
  small molecules, 300
  Zebrafish model, 306–307
Non-cell autonomous liver diseases, 271–272, 272*f*
Noncoding RNAs (ncRNAs), 203
Non-homologous end-joining (NHEJ), 5–6, 98–100, 120, 157–158, 195, 197, 267–271, 269*f*, 301, 333, 334*f*, 366, 405–406, 440–441
Nonsense mediated mRNA, 205–206
Nonviral vectors, 211
Novel therapies, 2–3
NSCs. *See* Neural stem cells (NSCs)
Nuclease-dead Cas proteins (dCas), 368
Nuclease-independent gene editing, 198–199
Nuclease-induced DSBs, 169
Nuclease-induced InDels, 172–173
Nuclease-mediated genome editing, 157–161
  CRISPR-Cas9 nuclease, 160–161
  ZFNs and TALENs, 159–160
Nucleosome remodeling and deacetylase (NuRD), 176

O
Off-target editing, 52
Off-target effects, 196
Off-target mutations, 60–61
Opsin mislocalization, 379–380
Ornithine transcarbamylase (OTC) deficiency, 272–274, 272*f*
Orofacial-digital syndrome (OFD), 359
Outer Retinal ganglion cells (oRGCs), 409–411

P
Pachyonychia congenita (PC), 84–85
PAM. *See* Protospacer adjacent motif (PAM)

Parkinson's disease (PD), 427–428
Pathogenic alleles
  footprint-less correction of, 95–100
  genome editing-mediated reading frame restoration of, 92–94
Patient-derived cells, 303
PC. See Pachyonychia congenita (PC)
PD. See Parkinson's disease (PD)
P1-derived artificial chromosomes (PACs), 488–490
Pediatric population, liver-targeted genome editing, 279–281
People with cystic fibrosis (PwCF), 187–190, 211–212
  benefit, 206
  case of, 207–208
  CFTR modulators, 191–192
  effect in lung function, 211
  ex vivo gene-corrected autologous cells, 208–209
  fraction of, 197–198
  phase II trials for, 190–191
Peptide nucleic acids (PNA), 198–199
Petri-dish patient, 206–211
PGT. See Preimplantation genetic testing (PGT)
Phagocytes, 292
Pharmacological chaperones (PCs), 298–299f, 300–301
Phosphorodiamidate morpholino oligomer (PMO), 234–235
PKD1 gene, 386
Pluripotent stem cells (PSCs), 408–415
  human neurodevelopment with, 409–412, 410f
  neurodegenerative processes with human, 423
  neurodevelopmental programs with, 416–417
PNA. See Peptide nucleic acids (PNA)
Pneumocystis jirovecii, 117–118
PNT. See Prenatal testing (PNT)
Poliovirus (PV), internal ribosomal entry sites (IRESs), 463
Poly-A binding protein (PABP), 442–443
Poly(rC)-binding protein 1 (PCBP1), 190
Polycystic kidney and hepatic disease 1 (PKHD1), 365

Polycystic kidney disease 1 (PKD1), 365
Polycystic kidney disease 2 (PKD2), 365
Polyethylene glycol-conjugated ADA (PEG-ADA), 124
Polygenic genetic disorders, 501
Porcine endogenous retrovirus (PERV), 383
Portal venous blood, 263
Post-transcriptional mRNA modifications, 45–46
POU5F1 gene, 6–7
Precise integration into target chromosome (PITCh), 270
Preimplantation genetic testing (PGT), 8–11
  cases of, 21
  HHGE, 12–14, 19–20
Premature termination codons (PTCs), 43–44, 91–92, 190–191
Prenatal screening, 116
Prenatal testing (PNT), 8–9
Presenilin (PSEN), 428
Primary ciliopathies, 356–357, 359–365
  associated genes, 360–363t
  phenotypes, 360–363t
  phenotypical manifestations, 359
  renal ciliopathies, 364–365
    epigenetic editing for, 386–387
    gene editing for, 382–383
  retinal ciliopathies, 364
    epigenetic editing for, 384–385
    gene editing for, 379–382
Primary immunodeficiencies (PID), targeted genome editing
  categorization, 115
  classifications of
    adenosine deaminase deficiency (ADA-SCID), 118–119
    artemis-deficient SCID (SCID-A), 119–120
    autoimmune lymphoproliferative syndrome (ALPS), 123
    common variable immunodeficiency (CVID), 123
    Epstein-Barr virus (EBV), 123
    hematopoietic stem and progenitor cell (HSPC) transplant, 123
    leukocyte adhesion deficiency (LAD), 123

major histocompatibility complex (MHC) Class II Deficiency, 123
Wiskott-Aldrich syndrome (WAS), 122–123
X-linked immunodeficiency magnesium defect, 123
X-linked immunodeficiency with magnesium defect, Epstein-Barr virus infection, and neoplasia (XMEN), 123
X-linked lymphoproliferative disease (XLP), 123
X-linked severe combined immunodeficiency (X-SCID), 117–118
complications, 114
diagnosis, 116
gene-editing tools, next generation of, 132–145
one-size-fits-all method, 114–115
phenotypes, 115–116
prevalence of, 114
significance, 116–117
therapies and clinical trials
  allogeneic HSC transplant, 125–128
  gene addition therapy, 129–132
  non-transplant therapies, 124–125
  treatment, 114–115
Prime editing guide RNA (pegRNA), 44–45, 98–100, 202
Prime editors (PEs), 44–45, 98–100, 132–133, 142–145, 202
Primer binding site (PBS), 98–100
Programmable nuclease-based gene editing, 192–199, 211–212
Programmed cell death protein 1 (PD-1), 453–454
Proprotein convertase subtilisin/kexin 9 (PCSK9) gene, 275–276
Protein synthesis, in eukaryotes, 442–444
Protospacer adjacent motif (PAM), 38, 93–94, 194–195, 200–201, 407
PSCs. See Pluripotent stem cells (PSCs)
PSEN. See Presenilin (PSEN)
Pseudoexons, 231
Pseudogenes, 296
PTCs. See Premature termination codons (PTCs)

Pulsed-field gel electrophoresis (PFGE), 489–491
PwCF. See People with cystic fibrosis (PwCF)

R

rAAV6. See Recombinant adeno-associated virus serotype 6 (rAAV6)
Radial glia cells (RGC), 409–411
RBC. See Red blood cell (RBC)
RDEB. See Recessive form of dystrophic EB (RDEB)
Reactive oxygen species (ROS), 121
Recessive form of dystrophic EB (RDEB), 85–87, 92–93
  CRISPR/Cas9 studies in, 98–101
  patient population, 93–94
Recombinant adeno-associated virus serotype 6 (rAAV6), 164–165
Red blood cell (RBC), 155–156
Red fluorescent protein (RFP), 448–449
Reduced-intensity conditioning (RIC), 125–126
Renal ciliopathies, 364–365
  autosomal dominant polycystic kidney disease (ADPKD), 365
  autosomal recessive polycystic kidney disease (ARPKD), 365
  epigenetic editing for, 386–387
  gene editing for, 382–383
  nephronophthisis (NPHP), 365
REPAIR system (RNA Editing for Programmable A to I Replacement), 48
Repeat-variable diresidues (RVDs), 333
RESCUE system, 48
RESTORE, 50
Retina, CRISPR/Cas delivery to, 374–376
Retinal ciliopathies, 364
  epigenetic editing for, 384–385
  gene editing for, 379–382
  Leber congenital amaurosis (LCA), 364
  retinitis pigmentosa (RP), 364
Retinal degeneration, 379
Retinal disease, CRISPR genome editing of, 58–65
Retinal gene augmentation therapies, 62
Retinal organoids, 379–380

Retinal pigment epithelium (RPE), 376
Retinitis pigmentosa (RP), 364
Reverse transcriptase (RT), 202
RGC. *See* Radial glia cells (RGC)
RGENs. *See* RNA-guided endonucleases (RGENs)
Rhodopsin kinase (RK) promoter, 376
Ribonucleoprotein (RNP), 444–445
RNA-binding proteins (RBPs), 444–445
RNA editing, 45–52
  ADAR physiology, 46
  endogenous ADAR, 50
  exogenous ADAR, 47–50
  of inherited retinal disease, 51
  tools, 47
RNA-guided endonucleases (RGENs), 405
RNA interference (RNAi), 425
Roche's pyrosequencing, 480
RT. *See* Reverse transcriptase (RT)
RUNX3 protein, 454–455

S

Safe harbors loci, 311–314
Safety and efficacy
  assessments of, 18
  of MRT, 18–19
Sandhoff disease
  in vivo genome editing, 316
  Zebrafish model, 306–307
Sanger sequencing, 480
Scanning mechanism, 445–446
SCC. *See* Squamous cell carcinoma (SCC)
SCD. *See* Sickle cell disease (SCD)
SCD mutation, correction of, 162–166, 163*f*
Schizophrenia (SCZ), 421–423
Scientific assessment, 18–19
SCNT. *See* Somatic cell nuclear transfer (SCNT)
scRNA-seq. *See* Single-cell RNA-sequencing (scRNA-seq)
SCZ. *See* Schizophrenia (SCZ)
Secondary immunodeficiencies (SIDs), 115–116
Self-Limiting Cas9 circuit for Enhanced Safety and specificity (SLiCES) system, 63
Senior-Loken syndrome (SLSN), 359

Sequencing by oligonucleotide ligation and detection (SOLiD) system, 480
Severe combined immunodeficiency (SCID), 115
sgRNA. *See* Single guide RNA (sgRNA)
SHH. *See* Sonic Hedgehog (SHH)
Short-rib thoracic dysplasias (SRTDs), 359
Shwachman-Bodian-Diamond syndrome gene *(SBDS)*, 459–460
Shwachman-Diamond syndrome, uORF mutations, 459–460
Sialidosis, 331
Sickle cell disease (SCD), 125–126, 156, 164–166
Single-cell RNA-sequencing (scRNA-seq), 207, 413
Single-cut gene editing, for exon skipping and reframing, 241–242
Single guide RNA (sgRNA), 194–195, 197–198, 366, 373
Single homology arm donor mediated intron-targeting Integration (SATI), 270
Single-molecule real time (SMRT) sequencing, 480
Site-directed RNA editing, 45–46
Site-specific epigenetic editing, 368
SMA. *See* Spinal muscular atrophy (SMA)
Small molecules, 300–301
SNAP-ADAR, 48–49
Somatic cell nuclear transfer (SCNT), 11–12
Sonic Hedgehog (SHH), 411–412
SOX6, 175–176
SpCas9. *See* *Streptococcus pyogenes* (SpCas9)
Spinal muscular atrophy (SMA), 191, 426–427
Spinocerebellar ataxia type 6 (SCA6), 465
Squamous cell carcinoma (SCC), 85–86
Staphylococcus aureus Cas9 (SaCas9), 279–280, 366, 373
Steroid-inducible *GRE5* promoter, 62
Streptococcus canis Cas9 (ScCas9), 366
Streptococcus pyogenes (SpCas9), 63, 160, 194–195
Streptococcus thermophilus Cas9 (St1Cas9), 366
Stromal mesenchymal cells, 338
Substrate reduction therapy (SRT), 298–299*f*, 300

Index

SUNRISE trial, 277–278
Super-exon codes, 197–198
Synthetic genomics, genetic diseases
  bacterial artificial chromosomes (BAC), 482, 486–487, 490–492, 506
  chromosome elimination, 497–500
  DNA writing, gene to genome synthesis, 480–484
  genetic disease treatment, applications for gene and genome therapy, 503–507
  human genome, 507
  modeling genetic diseases, 501–503
  in human and animal cells, 484–487
  human artificial chromosomes (HACs), 505–506
    bottom-up artificial chromosomes, 493–495
    top-down artificial chromosomes, 495–496
  mammalian artificial chromosomes (MACs), 492–496
  natural and artificial chromosomes, transfer of, 496–497
  perspectives, 507–508
  reading genomes, 478–480
  yeast artificial chromosomes (YAC), 486–490, 506
Synthetic RNA sequences, 258–259
Synthetic RNA targeting system, 49–50

T

TALENs. *See* Transcription activator-like effector nucleases (TALENs)
Targeted mutation analysis, 296
Targeted next-generation sequencing pipelines, 296
Targeting disease pathways, 55–56
Target site blockers (TSBs), 205
Tay-Sachs disease
  in vivo genome editing, 316
  prime editing, 311
  targeted mutation analysis, 296
T-cell antigen receptor (TCR), 486–487
T cell biomarkers, 116
T cell disorders, 115
T cell receptor excision circles (TRECs), 116
T cells, 3–4

Tetracycline operator (tetO) sequences, 494, 500
TFs. *See* Transcription factors (TFs)
Therapeutic fallacy, 12–13
Therapeutic transgenes, targeted insertion of, 277–279
Thrombopoietin gene (*TPO*), 459
Transcription activator-like effector nucleases (TALENs), 38, 91–92, 134–135, 157, 159–160, 164, 167–168, 192–197, 266–267, 268*f*, 301–302, 333, 334*f*, 357, 405–407
Transcription activator-like effector proteins (TALEs), 333, 357
Transcriptional activation, of *CFTR*, 203
Transcription factors (TFs), 406
Transformation-associated recombination (TAR), 489
Translation
  eukaryotes, 444–445
    internal ribosomal entry sites (IRESs), 463–466
    Kozak sequence, 445–455
    upstream open reading frames (uORFs), 455–462
  reinitiation, 457–458
Translational readthrough inducing drugs (TRIDs), 190–191
Translation initiation sequence (TIS), 445–446
Treosulphan, 125–126
TRIDs. *See* Translational readthrough inducing drugs (TRIDs)
Tri-pronuclear (3PN) embryos, 6–7
TSBs. *See* Target site blockers (TSBs)
Type 2 AAV (AAV2), 374
Type II IF protein, 84–85
Type I insulin-like growth factor receptor (IGF-1R), 465

U

UABCs. *See* Upper airway basal cells (UABCs)
UDG. *See* Uracil-DNA glycosylase (UDG)
Undesired genome editing, 60–61
Universal Declaration on the Human Genome and Human Rights of UNESCO, 16

Untranslated regions (UTRs), 444–445
Upper airway basal cells (UABCs), 208–209
Upstream open reading frames (uORFs), 441f
  as genome editing targets, 460–462, 462f
  impact on translation, 455–458
  mutations, diseases caused by, 458–460
Uracil, 42
Uracil-DNA glycosylase (UDG), 42
Urea cycle defects, 280
US Food and Drug Administration (FDA), 426–427
USH2A coding sequence, 41

V

Van der Woude syndrome, uORF mutations, 459
Vascular endothelial growth factor (VEGF), 463–464
Vector-mediated gene delivery, 41
Ventricular RGCs (vRGCs), 409–411
Viral vectors, 210, 314–315
Virus like particles (VLPs), 210–211, 374

W

Watson-Crick base pairing, 266–267
Wiskott-Aldrich syndrome (WAS), 115, 125
  biochemistry, 122–123
  symptoms, 122
World Health Organization (WHO), 7–8

X

Xanthomonas spp., 406–407
Xenopus, 408–409
Xenotransplatation, 383
X-linked agammaglobulinemia (XLA), 115
X-linked severe combined immunodeficiency (X-SCID)
  biochemistry, 118
  symptoms, 117–118

Y

Yeast artificial chromosomes (YAC), 486–490, 506

Z

Zinc-finger nucleases (ZFNs), 38, 95–98, 134–135, 157, 159–160, 192–197, 266–267, 268f, 277–278, 301–302, 317, 333, 334f, 341–342, 357, 405–406
Zinc finger proteins (ZFPs), 266–267
Zinc finger protein transcription factors (ZFP-TFs), 425
Zolgensma, 3–4

Printed in the United States
by Baker & Taylor Publisher Services